SUPERSTRINGS, SUPERGRAVITY AND UNIFIED THEORIES

THE ICTP SERIES IN THEORETICAL PHYSICS — VOLUME 2

SUPERSTRINGS, SUPERGRAVITY AND UNIFIED THEORIES

Proceedings of the Summer Workshop in High Energy Physics and Cosmology

Trieste, Italy
10 June — 19 July 1985

editors

G FURLAN
R JENGO
J C PATI
D W SCIAMA
Q SHAFI

World Scientific

Published by

World Scientific Publishing Co Pte Ltd.
P. O. Box 128, Farrer Road, Singapore 9128

Library of Congress Cataloging-in-Publication data is available.

SUPERSTRINGS, SUPERGRAVITY AND UNIFIED THEORIES

ISBN 9971-50-035-3
9971-50-036-1 (pbk)

Printed in Singapore by Kim Hup Lee Printing Co. Pte. Ltd.

PREFACE

The International Center for Theoretical Physics has been organizing a workshop in High Energy Physics and Cosmology every summer starting in 1981. The primary aim of these workshops has been to bring together active physicists from all over the world — in particular those from the developing countries — and to provide a platform where the frontier developments in the field are presented by leading workers and pioneers in the field over a period of four to five weeks.

The 1985 workshop was particularly active and we believe stimulating. There were over one hundred participants from thirty-three different countries. The lecturers came from the west as well as from the east. The highlight of the workshop in 1985, as may have been expected, was the exposition of the superstring theories. There were in addition intensive discussions on problems pertaining to grand unified theories, quantum chromodynamics, supersymmetry, supergravity, preonic theories and cosmology.

In 1985, as in the previous years, we were fortunate that each of the areas mentioned above was presented by several leading physicists in the field. We gratefully acknowledge the participation of all the lecturers. We thank in particular all those who have taken the trouble of making their lecture notes available for publication in the proceedings. We hope that these proceedings prove to be useful for all physicists working in the field, in particular to the many who could not participate in the workshop. Unfortunately, the lecture notes of C. Rebbi and J. Schwarz, could not be available for publication in the proceedings.

The workshop has benefitted immensely from the constant inspiration provided by Professor Abdus Salam. The constant help of the ICTP staff, in particular of Dr. K. Delafi, Louissa Sossi and Janet Varnier, was essential for the smooth running of the workshop.

G. Furlan
R. Jengo
J. C. Pati
D. W. Sciama
Q. Shafi

CONTENTS

Preface v

Introduction to Kaluza-Klein Theories 1
J. Strathdee

Basic Features of String Theories 22
J.-L. Gervais

Free Strings 57
L. Brink

Lectures on Superstrings 95
M. B. Green

Heterotic String Theory 158
D. J. Gross

Kaluza-Klein and Superstring Theories 182
P. H. Frampton

Path Integral and Anomalies 230
K. Fujikawa

Vertex Operators and Algebras 255
P. Goddard

Superstring Phenomenology 292
P. Candelas, G. T. Horowitz, A. Strominger & E. Witten

Low Energy Superstring Theory 308
B. A. Ovrut

String Theory and Conformal Invariance: A Review of Selected Topics 331
S. R. Wadia

Lowering the Critical Dimension for Heterotic Strings 350
E. Sezgin

Conformally Invariant Quantum Field Theories in 2 Dimensions 364
W. Nahm

Superstrings and Preons 377
J. C. Pati

Particle Physics and Cosmology 428
D. W. Sciama

Lectures on Particle Physics and Cosmology 444
E. W. Kolb

Particle Physics and the Standard Cosmology 465
S. Sarkar

Dark Matter and Galaxy Formation 494
R. Valdarnini

Supersymmetry, Monojets and Dark Matter 509
G. L. Kane

The Observation and Phenomenology of Glueballs 548
S. J. Lindenbaum

SUPERSTRINGS, SUPERGRAVITY AND UNIFIED THEORIES

INTRODUCTION TO KALUZA-KLEIN THEORIES

J. Strathdee
International Centre for Theoretical Physics, Trieste, Italy.

ABSTRACT

General features of Kaluza-Klein theories are discussed with emphasis on the problem of determining the ground state geometry and its symmetry. A technique is described for setting up harmonic expansions in cases where the internal space is a coset space.

1. New dimensions

Over the last dozen years our appreciation of the Kaluza-Klein idea [1] has evolved significantly. Before 1975 one thought of it more as a mathematical artifice than as a realistic picture of spacetime. The extra dimensions were not quite "real". This outlook changed when the notion of <u>spontaneous compactification</u> was proposed [2]. The idea is that the ground state geometry is determined dynamically - somewhat like the Higgs field vacuum configuration in gauge theories - by solving some generalized version of the 4-dimensional Einstein equation. In the light of one's experience with the spontaneous breakdown of internal symmetries through the Nambu-Goldstone mechanism, it was easy to believe that the physical vacuum could assume a non-trivial geometry (and topology). Space-time might indeed be some kind of cylinder with four flat dimensions and any number of highly curved ones.

At first the renewed interest in Kaluza-Klein mechanism centred on supergravity theories in more than four dimensions. In particular the $N = 1$ supergravity in $D = 11$ dimensions was studied intensively. It was hoped that the gauge symmetries of low energy physics could be explained in terms of the symmetries of the internal, compact manifold [3]. This approach was not successful. To explain the chiral type of symmetry it appears to be necessary that the underlying higher dimensional theory should include Yang-Mills fields as well as gravity [4),5]. In order to deal with chirality, therefore, one had to sacrifice the idea that low energy gauge symmetries could be understood through the Kaluza-Klein mechanism.

Even in four dimensions it has proved difficult to quantize gravity-containing theories. In more than four dimensions the problem is exacerbated. It is therefore not reasonable to regard any Kaluza-Klein type of model as a fundamental theory. These models must be viewed as "low energy" effective theories descending from something more fundamental. It now seems that a superstring theory [6] might provide the underlying quantum dynamics and that its field theory approximation is $N = 1$, $D = 10$ supergravity coupled to a Yang-Mills supermultiplet of either $SO(32)$ or $E_8 \times E_8$ plus small corrections. It is hoped that the superstring theory can be consistently quantized and that the 10-dimensional supergravity together with quantum corrections can be extracted from it. Spontaneous compactification might then follow the Kaluza-Klein pattern, yielding a vacuum geometry with the structure of 4-dimensional Minkowski spacetime multiplied by a 6-dimensional compact manifold, perhaps one of the Calabi—Yau spaces [7]. These topics will be discussed by others. My purpose now is to discuss general features of Kaluza-Klein theories with some emphasis on the problem of ground state symmetries.

The geometry of 4+K-dimensional spacetime must factorize, at least locally, into $M^4 \times B^K$, where M^4 is the flat, or very nearly flat, spacetime of conventional physics and B^K is compact and very small (i.e. small enough to be invisible at present). Low energy states are associated with the propagation of particles on M^4. Any propagation on B^K will require energies $\sim \hbar c/a_K$ where a_K measures the size of B^K. Such states would be unexcited by probes of energy $\ll \hbar c/a_K$ and the internal space would therefore be invisible. Above this threshold, spacetime would be seen to have 4+K dimensions. Below threshold, the evidence is indirect and relies entirely on symmetry. The scale of B^K may not be detectable but its symmetries would be those of low energy ($\ll \hbar c/a_K$) particle physics.

The internal symmetries of the low-energy sector may be either global or local (first or second kind gauge symmetries). In the original model of Kaluza and Klein, the local U(1) of electromagnetism was related to the rotational symmetry of a 5-dimensional spacetime "cylinder", $M^4 \times S^1$. The vector potential, A_m, was to correspond to the metric components, g_{m5}. Another example of this is global supersymmetry which results when the internal space is parametrized by anticommuting rather than ordinary c-number co-ordinates [8].

What of the scale, a_K? At present we have only that $a_K \lesssim 10^{-17}$ cm ($\hbar c/a_K \gtrsim 10^3$ GeV), but it could of course be much smaller. In the Kaluza-Klein model, where the electromagnetic coupling is related to Newton's constant, it is necessary to take $a_K \sim 10^{-33}$ cm. This would force all the interal excitations to have masses on the Planck scale ($\gtrsim 10^{19}$ GeV).

For a dynamical explanation of the geometry, $M^4 \times B^K$, it is natural to invoke the principles of general relativity in 4+K dimensions. The metric tensor and connection are then treated as dynamical variables whose ground state expectation values define the geometry of $M^4 \times B^K$. How to determine this ground state geometry? There are several aspects to this problem. Firstly, since general relativity is not renormalizable one cannot use the methods of perturbative quantum field theory. At best one can ignore quantum effects and try to solve the problem classically. But even the classical problem is not well defined. Apart from metric and connection, what are the dynamical variables? What is the dimension, K? There is a wide range of candidate models to choose from. If the internal space has non-vanishing curvature then the dynamical set must include "matter" fields to contribute on the right-hand side of Einstein's equations. We have no very clear guide as to what kind of matter should be incorporated.

If supersymmetry is imposed as well as general covariance, then the choices are quite limited. Indeed, there seems to be a unique 11-dimensional supergravity: all fields, fermionic and bosonic, belong, together with the metric, to a single representation [9]. This theory admits only one parameter in the Lagrangian, Newton's constant. Only somewhat less restrictive are the 10-dimensional supergravities, which can be coupled to local gauge fields [10].

Conversely, it can be argued that supergravity theories are most naturally formulated in many-dimensional spacetimes and thereby give substance to the Kaluza-Klein picture. In other words, if supersymmetry is a fundamental symmetry of the world, then spacetime should have more than four dimensions. If this is true, then the Kaluza-Klein mechanism must play an essential role in our understanding of low energy physics.

Once the model is chosen the main problem is to determine the ground state geometry. Since we cannot quantize the theory, this amounts to solving the classical field equations. One's general expectation is that the more symmetrical solutions will be the more stable, and it is therefore reasonable to search among the most symmetrical cases, i.e. those solutions which admit the largest invariance groups. Any particular solution can be tested for stability against small perturbations (but this will not guarantee its stability against tunneling phenomena, spontaneous dissolution) and such classical stability is essential if a sensible particle physics is to be mounted on it. A second requirement is that the solution should describe a manifold of the form $M^4 \times B^K$ (locally), where M^4, with Lorentzian signature, is effectively flat and B^K, with Euclidean signature, is compact. The curvature of $M^4 \sim 1/a_4^2$ defines a scale, a_4, which must be large compared to the scale of B^K, $a_4 \gg a_K$. This is necessary if the theory is to have a low energy sector in which B^K remains invisible. The emergence of a classical solution with these features is sometimes referred to as "spontaneous compactification".

In testing for classical stability it is necessary to analyze the spectrum of excitation modes in order to ensure that no ghosts or tachyons are possible. (A more direct test would be to demonstrtae the existence of a conserved, non-negative, fluctuation energy. But the information gained by analyzing the excitation spectrum is useful also for interpreting the theory.) In dealing with the excitation modes it is important to know the ground state symmetry since the modes can be classified in irreducible representations of it. Now the 4+K-dimensional theory is covariant with respect to general co-ordinate transformations, tangent space rotations, and perhaps various local or global gauge transformations. Many of these symmetries will not be present in the ground state: they are spontaneously broken. In particular, from among the general co-ordinate transformations and tangent space rotations, only a finite dimensional (global) group will leave the ground state metric invariant.

Having found a stable classical approximation to the ground state the next problem is to extract an effective theory of the low energy phenomena on M^4. Now, it is always possible to reduce the 4+K-dimensional theory to 4-dimensional form by expanding all the field variables in a complete set of functions on B^K and then integrating the Lagrangian over this space. The coefficients in the expansions are fields on M^4, an infinite number of them, and their equations of motion are obtainable from the integrated Lagrangian. The mass-like terms in this Lagrangian will generally be of order $1/a_K^2$, i.e. large. Hence it is important to find the so-called zero-modes, the combinations of fields on M^4 which do not carry these large masses. They will be the only relevant fields for the low energy sector.

Once the zero-mode structure is understood, it is in principle relatively simple to obtain the effective Lagrangian for the low energy sector. One replaces the complete expansions on B^K by what may be called the <u>zero-mode ansatz</u>: truncated expansions in which only the contributions of zero-mode fields are retained. Integration of the 4+K-dimensional Lagrangian over B^K now yields a 4-dimensional Lagrangian for the long range fields and their interactions. (In practice it may not be possible to obtain this in closed form if the interactions are non-polynomial).

Is the classical approximation to the ground state really relevant? Unfortunately, we are not at present able to improve on it. Since these theories are not renormalizable, the use of perturbative approximations is questionable. It may turn out, as some authors hope, that one of the supergravities is ultra-violet finite, but we do not yet have such a theory. In any case, even if the problem of perturbative quantization is solved, there will remain the question of non-perturbative effects. These might include such phenomena as composite gauge fields and fermionic condensates, which could radically change the vacuum structure. For example, it is widely believed that the notorious cosmological constant, which appears in spontaneously compactified supergravities, may be an artefact of the classical approximation.

Finally, one might ask whether such theories can be fundamental. Is it not perhaps an idealization to treat the metric tensor of general relatively, with its dimensional coupling constant, as a truly local field? If gravity itself is really an effective theory, valid only at energies $\ll 10^{19}$ GeV, what are we to make of the excitation modes in Kaluza-Klein theories? In the context of a consistently quantized superstring theory, such questions could be disposed of.

In the following sections we present an analysis of the excitation spectrum of the original Kaluza-Klein model, and then go on to discuss some of the formalism needed for generalization to higher dimensional spacetimes. This is mainly concerned with symmetry questions.

2. Spectrum of 4+1 Kaluza-Klein theory

The simplest Kaluza-Klein model is the original one: pure gravity in a spacetime of five dimensions. The vacuum geometry is flat and one can impose the topology $M^4 \times S^1$. That is, the new co-ordinate is to be thought of as periodic. Because of the simplicity of this theory one can easily obtain its excitation spectrum. Write the metric tensor in the form

$$g_{MN} = \eta_{MN} + h_{MN} \quad , \tag{2.1}$$

where $M,N = 0,1,2,3,4$ and η_{MN} is the flat (vacuum) metric

$$\eta_{MN} = \mathrm{diag}(-1,+1,+1,+1,+1) \quad . \tag{2.2}$$

Treating the components h_{MN} as small quantities, one obtains the second order terms in the Einstein Lagrangian,

$$\mathcal{L}_2 = -\frac{1}{4}\left(h_{MN,L}\, h_{MN,L} - 2h_{MN,L}\, h_{ML,N} + 2\, h_{MN,N}\, h_{LL,M} - h_{MM,L}\, h_{NN,L}\right) + \frac{1}{2} T_{MN}\, h_{MN} \quad , \tag{2.3}$$

where T_{MN} is an external source. The connections vanish so that simple partial derivatives are indicated, $h_{MN,L} = \partial_L h_{MN}$. The equations of motion derived from (2.3) take the form

$$h_{MN,LL} - h_{ML,NL} - h_{NL,ML} + h_{LL,MN} + \eta_{MN}(h_{KL,KL} - h_{KK,LL}) = - T_{MN} \quad . \qquad (2.4)$$

These equations are compatible only if T is conserved,

$$T_{MN,N} = 0 \quad . \qquad (2.5)$$

This condition results from the 5-dimensional general covariance. The usual way to solve (2.4), subject to the compatibility requirement (2.5), is to impose a co-ordinate condition. For example, in the gauge

$$h_{MN,N} = 0 \qquad (2.6)$$

they are solved by

$$h_{MN} = - \frac{1}{\partial^2} \left[T_{MN} - \frac{1}{3}\left(\eta_{MN} - \frac{\partial_M \partial_N}{\partial^2}\right) T_{LL}\right] \quad , \qquad (2.7)$$

where ∂^2 represents the 5-dimensional d'Alembertian.

On substituting the solution (2.7) into the Lagrangian (2.3), discarding total derivatives, one finds that $\mathcal{L}_2$ reduces to

$$\mathcal{L}_2(T) = \frac{1}{4} T_{MN} h_{MN} = - \frac{1}{4} \left[T_{MN} \frac{1}{\partial^2} T_{MN} - \frac{1}{3} T_{MM} \frac{1}{\partial^2} T_{NN}\right] \quad . \qquad (2.8)$$

This expression represents the effective interaction between conserved (i.e. physical) sources due to the exchange of Kaluza-Klein particles.

Now consider the pole, $\partial^2 = 0$, in (2.8). The residue simplifies on using the conservation of T_{MN}. Firstly, if $\partial_4 \neq 0$ we can use a frame in which

$$\partial_0 = \partial_4 \qquad \text{and} \qquad \partial_i = 0, \qquad i = 1,2,3 \quad . \qquad (2.9)$$

In this frame we have

$$T_{0M} = T_{4M} \quad , \quad M = 0,1,2,3,4 \qquad (2.10)$$

so that $T_{MM} = T_{ii}$ and $T_{MN}^2 = T_{ij}^2$. It follows that, in the neighbourhood of the pole, the effective interaction (2.8) reduces to

$$\mathcal{L}_2(T) \simeq \frac{1}{4} T^t_{ij} \frac{1}{\partial_0^2 - \partial_4^2} T^t_{ij} \quad . \tag{2.11}$$

It thus appears that the physically significant part of the source is the traceless 3-dimensional tensor

$$T^t_{ij} = T_{ij} - \frac{1}{3} \delta_{ij} T_{kk} \quad . \tag{2.12}$$

This result indicates that the massive states ($\partial_4 \neq 0$) carry spin 2. Their masses are given by the eigenvalues of $-\partial_4^2$. If the S^1 of the Kaluza-Klein vacuum is taken to be a circle of radius, a, then the eigenvalues of $-\partial_4^2$ equal $(n/a)^2$, $n = 1,2,\ldots$ It should be noted that the radius cannot be calculated in this model, it is simply an integration constant.

On the other hand, if $\partial_4 = 0$ we have the massless states and they can be analyzed conveniently by choosing a frame where

$$\partial_0 = \partial_3 \quad \text{and} \quad \partial_1 = \partial_2 = \partial_4 = 0 \quad . \tag{2.13}$$

In this frame the effective interaction reduces, in the neighbourhood of the pole, to the form

$$\frac{1}{4} \sum_{\lambda=-2}^{2} T_{-\lambda} \frac{1}{\partial_0^2 - \partial_3^2} T_\lambda \quad , \tag{2.14}$$

where λ refers to the O(2)

$$\begin{aligned} T_2 &= \frac{1}{2}(T_{11} - T_{22}) + i\, T_{12} \\ T_1 &= T_{41} + i\, T_{42} \\ T_0 &= \frac{1}{\sqrt{6}} (T_{11} + T_{22} - 2\, T_{44}) \\ T_{-1} &= T_{41} - i\, T_{42} \\ T_{-2} &= \frac{1}{2}(T_{11} - T_{22}) - i\, T_{12} \quad . \end{aligned} \tag{2.15}$$

The terms in (2.14) correspond to the exchange of graviton, photon and Brans-Dicke scalar.

There are no tachyons or ghosts in the spectrum. This indicates that the situation is at least classically stable.

For this model the zero modes are particularly simple. They are the terms in the harmonic expansion on S^1 which are independent of the new co-ordinate $x^4 \equiv y$. The zero mode ansatz can be written in the form

$$g_{MN}\, dx^M dx^N = g_{mn}(x)\, dx^m dx^n + \phi(x)^2\, (dy - \kappa\, A_m(x)\, dx^m)^2 \quad , \tag{2.16}$$

where $m,n = 0,1,2,3$ and the fields g_{mn}, ϕ, A_m depend only on x^m. The parameter, κ , which appears in this ansatz is proportional to the Planck length.

The 1-form $dy - \kappa\, A_m(x)\, dx^m$ is invariant under the special class of 5-dimensional co-ordinate transformations

$$x^m \to x^m \quad , \qquad y \to y + \Lambda(x) \tag{2.17}$$

provided the 4-vector $A_m(x)$ transforms according to the rule

$$A_m \to A_m + \frac{1}{\kappa}\partial_m \Lambda \quad , \tag{2.18}$$

i.e., like a vector potential. By considering the coupling of A_m to fields which depend on y one can see that the corresponding charge operator is given by

$$Q = i\,\kappa\,\partial_y \tag{2.19}$$

with eigenvalues $n\kappa/a$, $n = 0, \pm1, \pm2, \ldots$ Thus, the unit of electric charge in this model is κ/a, where κ is the Planck length and a is the radius of S^1 .

3. Ground state symmetry

In order to discuss general relatively in 4+K dimensions we must first establish some notation. Since we shall want to be able to deal with fermions as well as bosons, it is necessary to use a generalized vierbein formalism. We therefore introduce the 4+K-dimensional multibein, $E_M{}^A(z)$, and connection, $B_{M[AB]}(z)$. The 4+K co-ordinates are denoted by z^M, and mid-alphabet letters, M, N are to be read as world indices. Early alphabet letters, A, B,... denote frame labels. These distinctions refer to the transformation properties of E and B, of which there are two kinds. Firstly, under general co-ordinate transformations, $z \to z'$,

$$E_M{}^A(z) \to E_M'{}^A(z') = \frac{\partial z^N}{\partial z'^M} E_N{}^A(z) \ ,$$

$$B_{M[AB]}(z) \to B'_{M[AB]}(z') = \frac{\partial z^N}{\partial z'^M} B_{N[AB]}(z) \ , \tag{3.1}$$

i.e. both E and B are covariant 4+K vectors. Secondly, under frame rotations (the tangent space group),

$$E_M{}^A(z) \longrightarrow E_M'{}^A(z) = E_M{}^B(z)\, a_B{}^A(z)$$

$$B_{M[AB]}(z) \longrightarrow B'_{M[AB]}(z) = (a^{-1})_A{}^C (a^{-1})_B{}^D B_{M[CD]}(z) + (a^{-1})_A{}^C \partial_M a_{CB} \ . \tag{3.2}$$

where $a_A{}^B(z)$ is the 4+K-dimensional pseudo-orthogonal matrix,

$$a_A{}^C a_B{}^D \eta_{CD} = \eta_{AB} = \mathrm{diag}(-1,1,1,\ldots,1) \ . \tag{3.3}$$

Out of E and B and their derivatives are made the various geometrical quantities:

(1) metric $$g_{MN} = E_M{}^A E_N{}^B \eta_{AB} \tag{3.4}$$

(2) torsion $$T_{MN}{}^A = \partial_M E_N{}^A - \partial_N E_M{}^A - E_N{}^C B_{MC}{}^A + E_M{}^C B_{NC}{}^A \tag{3.5}$$

(3) curvature $$R_{MN[AB]} = \partial_M B_{N[AB]} - \partial_N B_{M[AB]} + B_{M[AC]} B_{N[CB]} - B_{N[AC]} B_{M[CB]} \tag{3.6}$$

all of which transform as tensors with respect to the two groups.

To generate equations of motion for E and B it is usual in Kaluza-Klein theories to take the Lagrangian density

$$\mathcal{L} = |\det E|\, E_A{}^M E_B{}^N R_{MN[AB]} \ , \tag{3.7}$$

where the contravariant multibein, $E_A{}^M$, is defined as the matrix inverse of $E_M{}^A$. (To save space we use the convention that repeated frame labels denote the invariant contraction: $X_A Y_A \equiv \eta^{AB} X_A Y_B$, etc.). In the Lagrangian (3.7) it is possible to vary E and B independently or, alternatively, to eliminate B by imposing the constraint, $T_{MN}{}^A = 0$. When matter fields are present, these alternatives are not equivalent since the connection, B, will appear also in the covariant derivatives of matter fields, and so could give rise to non-vanishing torsion.

Typical of matter fields would be the fermion, $\psi(z)$, belonging to a spinor representation of the tangent space group, and the gauge field $A_A(z)$ which would be a 4+K vector. Both are scalars with respect to the group of general co-ordinate transformations. The gauge field 1-form, $A = E^A A_A = dz^M E_M{}^A A_A$, transforms as a Yang-Mills potential with respect to the gauge group, i.e.

$$A(z) \rightarrow A'(z') = k(z)^{-1} A(z)\, k(z) + k(z)^{-1}\, d\, k(z) \,. \tag{3.8}$$

Now suppose that a Lagrangian for the complete system has been constructed by adjoining to (3.17) the appropriate matter field terms. Suppose further that a candidate ground state solution to the equations of motion has been found. Denote this solution by $\langle E_M{}^A(z)\rangle$, $\langle A_A(z)\rangle$, etc. Of particular interest is the invariance group of the solution, i.e. the subgroup of the direct product of general co-ordinate, tangent space, and gauge groups for which

$$\langle E_M'{}^A(z)\rangle = \langle E_M{}^A(z)\rangle \,, \qquad \langle A_A'(z)\rangle = \langle A_A(z)\rangle \tag{3.9}$$

etc. In other words, the elements $\partial z/\partial z'$, $a(z)$ and $k(z)$ must be constrained in a way such that

$$\frac{\partial z^N}{\partial z'^M} \langle E_N{}^B(z)\rangle\, a_B{}^A(z) = \langle E_M{}^A(z')\rangle$$

$$k(z)^{-1} \langle A(z)\rangle\, k(z) + k(z)^{-1}\, d\, k(z) = \langle A(z')\rangle \tag{3.10}$$

etc. The most well-known example of this is the Poincaré group which emerges as the symmetry of the flat vacuum solution in 4-dimensional general relativity:- linear co-ordinate transformations are associated with tangent space Lorentz transformations in such a way as to leave invariant the vierbein, $\langle E_m{}^a(x)\rangle = \delta_m^a$. At present we are interested in solutions for which the geometry factorizes,

$$\langle E_M{}^A(z)\rangle = \begin{pmatrix} e_m{}^a(x) & 0 \\ 0 & e_\mu{}^\alpha(y) \end{pmatrix} , \qquad (3.11)$$

where $e_m{}^a(x)$ denotes the vierbein of some 4-dimensional spacetime M^4, and $e_\mu{}^\alpha(y)$ is the K-bein of a compact space, B^K. (We shall use lower case Latin indices for M^4 and Greek indices for B^K. The co-ordinates are correspondingly distinguished, $z^M = (z^m, y^\mu)$.) Since M^4 will have to be quite symmetric (i.e. be Poincaré or de Sitter invariant) in order to represent the vacuum, it cannot be expected to support an invariant 4-vector. We shall therefore assume that the ground state value of the 4+K-vector, $A_A(z)$, is confined to B^K, i.e.

$$\langle A_A(z)\rangle = \begin{pmatrix} 0 \\ a_\alpha(y) \end{pmatrix} . \qquad (3.12)$$

The group which leaves (3.11) and (3.12) invariant in the sense of equations (3.10) will be important for classifying the fluctuations about this solution. In particular, the fields associated with zero modes which are relevant to the low energy sector, will belong to irreducible representations of this group. The idea would be to expand all fields $\phi(x,y)$ in complete sets of functions, $Y_n(y)$,

$$\phi(x,y) = \sum \phi_n(x)\, Y_n(y) , \qquad (3.13)$$

where the Y_n belong to irreducible representations of the symmetry of B^K. The coefficient fields $\phi_n(x)$ will then transform accordingly. Integration of the Lagrangian density over B^K will yield a 4-dimensional Lagrangian for the $\phi_n(x)$ whose structure reflects the symmetry of B^K. In practice this step can be rather tedious to take and we shall therefore specialize further by assuming that B^K is a coset space, i.e. a space of "maximal" symmetry. For these spaces the Y_n are, in principle, known and many of their properties are easily exploited.

4. Mode analysis on coset spaces

Since one purpose of the Kaluza-Klein picture is to explain internal symmetries in terms of the symmetry of an internal space, we shall confine our considerations to the most symmetrical category - the coset spaces. If H is a subgroup of the compact continuous group, G, then the space of left cosets, denoted G/H, is invariant under the action of G. We shall suppose that the internal space, B^K, obtained by solving the 4+K-dimensional equations is such a quotient space, and consider the technical question of how to express its symmetry and extract consequences. Harmonic expansions are familiar in problems where the symmetry is U(1) or SU(2), and we shall be dealing with straightforward generalizations of these.

For coset spaces there is a simple procedure for constructing the K-bein e^α. Before describing it, however, we must introduce some notations. Let the group G be generated by a set of charges, $Q_{\hat\alpha}$, which span the Lie algebra,

$$[Q_{\hat\alpha}, Q_{\hat\beta}] = c_{\hat\alpha\hat\beta}{}^{\hat\gamma} Q_{\hat\gamma} . \tag{4.1}$$

From among the $Q_{\hat\alpha}$ let there be singled out a sub-set, $Q_{\bar\alpha}$, which spans the Lie algebra of H. The remainder, Q_α, will span the K-dimensional tangent space of G/H. We shall assume that G/H is reductive, i.e.

$$[Q_\alpha, Q_{\bar\beta}] = c_{\alpha\bar\beta}{}^{\gamma} Q_\gamma \tag{4.2}$$

but not necessarily "symmetric",

$$[Q_\alpha, Q_\beta] = c_{\alpha\beta}{}^{\bar\gamma} Q_{\bar\gamma} + c_{\alpha\beta}{}^{\gamma} Q_\gamma . \tag{4.3}$$

(For a symmetric space we would have $c_{\alpha\beta}{}^{\gamma} = 0$.)

Suppose now that G has been separated into left cosets, gH. From each coset select one element, L_y, to represent it. This can be done according to any convenient prescription. For example, one might choose $L_y = \exp(y^\alpha Q_\alpha)$. The connection between one prescription and another is effected by a right translation, $L'_y = L_y k_y$, where $k_y \in H$. The action of G on G/H is now defined: multiplication from the left by any $g \in G$ will carry L_y into another element belonging to a coset whose representative element is $L_{y'}$. In other words we can write

$$g\, L_y = L_{y'}\, h \quad , \quad h \in H \quad . \tag{4.4}$$

This equation can be solved unambiguously for y' and h as functions of y and g. It defines the "left translation" of G/H by g.

To find the K-bein construct the 1-form $L_y^{-1}\, d\, L_y$. This 1-form necessarily belongs to the Lie algebra of G and so can be expanded in the basis $Q_{\hat\alpha}$,

$$L_y^{-1}\, d\, L_y = e^{\alpha}(y)\, Q_{\alpha} + e^{\bar\alpha}(y)\, Q_{\bar\alpha} \tag{4.5}$$

which defines the K-bein, $e^{\alpha}(y)$. To see how e^{α} and $e^{\bar\alpha}$ transform under left translations one can use (4.4). Thus,

$$\begin{aligned} L_{y'}^{-1}\, d\, L_{y'} &= h\, L_y^{-1}\, g^{-1}\, d\,(g\, L_y\, h^{-1}) \\ &= h(L_y^{-1}\, d\, L_y)\, h^{-1} + h\, d\, h^{-1} \\ &\quad + h\, L_y^{-1}\,(g^{-1}\, d\, g)\, L^y\, h^{-1} \quad . \end{aligned} \tag{4.6}$$

To interpret this equation it is helpful to expand $g^{-1}\, d\, g$ and $h\, d\, h^{-1}$, which belong to the Lie algebras of G and H, respectively,

$$\begin{aligned} g^{-1}\, d\, g &= (g^{-1}\, d\, g)^{\hat\alpha}\, Q_{\hat\alpha} \\ h\, d\, h^{-1} &= (h\, d\, h^{-1})^{\bar\alpha}\, Q_{\bar\alpha} \quad . \end{aligned} \tag{4.7}$$

Introduce the matrices $D_{\hat\alpha}{}^{\hat\beta}$ of the adjoint representation of G,

$$g^{-1}\, Q_{\hat\alpha}\, g = D_{\hat\alpha}{}^{\hat\beta}(g)\, Q_{\hat\beta} \quad . \tag{4.8}$$

With the notation of (4.7) and (4.8) we can separate (4.6) into two pieces,

$$e^{\alpha}(y') = e^{\beta}(y)\, D_{\beta}{}^{\alpha}(h^{-1}) + (g^{-1}\, d\, g)^{\hat\beta}\, D_{\hat\beta}{}^{\alpha}(L_y\, h^{-1}) \tag{4.9}$$

$$e^{\bar{\alpha}}(y') = e^{\bar{\beta}}(y)\, D_{\bar{\beta}}{}^{\bar{\alpha}}(h^{-1}) + (h\, d\, h^{-1})^{\bar{\alpha}}$$

$$+ (g^{-1}\, d\, g)^{\hat{\beta}}\, D_{\hat{\beta}}{}^{\bar{\alpha}}\,(L_y\, h^{-1})\,. \qquad (4.10)$$

Both of these equations are significant. Firstly, if $dg = 0$ one sees from (4.9) that $e^{\alpha}(y)$ indeed satisfies the requirement (3.10) for an invariant K-bein. In K-K theory g can depend on the co-ordinates of M^4, in which case $dg \neq 0$. To understand the second term in (4.9) it is really necessary to take into account the fields associated with fluctuations of the ground state. Among these will be found a Yang-Mills vector $A^{\hat{\alpha}}$ which undergoes the usual inhomogeneous transformations of a gauge potential. It turns out that the second term in (4.9) is precisely cancelled by the inhomogeneous term in the transformation of A^{α}. The equation (4.10) indicates that $e^{\bar{\alpha}}(y)$ is a connection form on G/H. (The last term in (4.10) is compensated by the transformation of $A^{\bar{\alpha}}$ as in (4.9).)

The role of $e^{\bar{\alpha}}$ as a connection 1-form is clarified by rearranging (4.5) to read

$$d\, L_y^{-1} + e^{\bar{\alpha}}(y)\, Q_{\bar{\alpha}}\, L_y^{-1} = -\, e^{\alpha}\, Q_{\alpha}\, L_y^{-1}\,. \qquad (4.5')$$

The left-hand side can be interpreted as the covariant differential of L_y^{-1}. Indeed one can define the covariant derivative ∇_{α} by

$$d + e^{\bar{\alpha}}(y)\, Q_{\bar{\alpha}} = a e^{\alpha}\, \nabla_{\alpha}\,, \qquad (4.11)$$

where a is a parameter with the dimensions of length which represents the scale of G/H. With this definition of covariant derivative one can easily derive the identity

$$\nabla_{\alpha}\, \nabla_{\beta}\, L_y^{-1} = \frac{1}{a^2}\, Q_{\beta}\, Q_{\alpha}\, L_y^{-1} \qquad (4.12)$$

which expresses two important properties. Firstly, with the connection $e^{\bar{\alpha}}$ the Laplacian on G/H is expressed in terms of the quadratic Casimir invariants of G and H,

$$\nabla_\alpha \nabla_\alpha L_y^{-1} = \frac{1}{a^2} Q_\alpha Q_\alpha L_y^{-1}$$

$$= \frac{1}{a^2} (C_2(G) - C_2(H)) L_y^{-1} \quad . \tag{4.13}$$

Secondly, the curvature and torsion are given by

$$[\nabla_\alpha, \nabla_\beta] L_y^{-1} = - \frac{1}{a^2} [Q_\alpha, Q_\beta] L_y^{-1}$$

$$= - \frac{1}{a^2} (c_{\alpha\beta}{}^{\bar\gamma} Q_{\bar\gamma} + c_{\alpha\beta}{}^{\gamma} Q_\gamma) L_y^{-1}$$

$$= - \frac{1}{a^2} c_{\alpha\beta}{}^{\bar\gamma} Q_{\bar\gamma} L_y^{-1} + \frac{1}{a} c_{\alpha\beta}{}^{\gamma} \nabla_\gamma L_y^{-1} \quad . \tag{4.14}$$

Hence the curvature and torsion 2-forms are

$$\mathcal{R} = -\frac{1}{2} e^\alpha \wedge e^\beta \, c_{\alpha\beta}{}^{\bar\gamma} Q_{\bar\gamma}$$

$$T^\gamma = \frac{a}{2} e^\alpha \wedge e^\beta \, c_{\alpha\beta}{}^{\gamma} \quad . \tag{4.15}$$

In particular, if G/H is symmetric then the connection $e^{\bar\alpha}$ is torsionless. On the other hand, if G/H is not symmetric then the torsionless connection (which takes its values in the Lie algebra of O(K) will contain a part lying outside the image of H in O(K), viz.

$$\mathcal{B}_{\alpha\beta} = e^{\bar\gamma} c_{\alpha\bar\gamma\beta} + e^\gamma \frac{1}{2} c_{\alpha\gamma\beta} \quad . \tag{4.16}$$

In Kaluza-Klein systems it is normally the torsionless connection which is generated by the field equations. (For a fuller discussion of these matters, see Ref. 11.

In many cases of interest the 4+K-dimensional gravitational field is coupled to a gauge field. The gauge fields, associated with a group, K, may acquire non-vanishing but G-invariant values on G/H. This can happen whenever it is possible to embed H in K. Let $q_{\bar\gamma}$ denote the image of $Q_{\bar\gamma}$ in the Lie algebra of K. Then the gauge potential 1-form

$$A = \frac{1}{f} e^{\bar{\gamma}} q_{\bar{\gamma}} \tag{4.17}$$

and its associated field strength 2-form

$$F = \frac{1}{2f} e^{\alpha} \wedge e^{\beta} c_{\alpha\beta}{}^{\bar{\gamma}} q_{\bar{\gamma}} \tag{4.18}$$

can be shown to satisfy the Yang-Mills equations[12)]. Moreover, the 2-form F is clearly invariant under left translations of G/H associated with gauge transformations corresponding to the embedding of H in K. This generalized ground state symmetry, which involves associating both a gauge transformation and a tangent space rotation with a left translation of G/H, implies the use of a generalized covariant derivative,

$$\nabla = d + e^{\bar{\gamma}}(y)(Q_{\bar{\gamma}} + q_{\bar{\gamma}}) \ . \tag{4.19}$$

For mode expansions on a quotient space G/H, it is natural to use the matrices of the unitary representations of G. These are well known to provide a complete set for representing functions on G and, as we shall see, it is quite simple to restrict them for functions on G/H. To begin, let $\phi(g)$ be a square integrable function on G. Then it can be represented by the expansion

$$\phi(g) = \sum_{n} \sum_{p,q} \sqrt{d_n}\, D^{n}_{pq}(g)\, \phi^{n}_{qp} \ , \tag{4.20}$$

where D^{n}_{pq} is a unitary matrix of dimension d_n and the sum includes all matrix elements of the unitary irreducible representations, $g \to D^{n}(g)$. The coefficients, ϕ^{n}_{pq} are projected out by integrating over the group,

$$\phi^{n}_{pq} = \frac{\sqrt{d_n}}{V_G} \int_G d\mu\, D^{n}_{pq}(g^{-1})\, \phi(g) \ , \tag{4.21}$$

where $d\mu$ is the invariant measure normalized to volume V_G.

For functions on the quotient space G/H the expansion (4.20) is subject to some restrictions. In practice one has to deal with multiplets of functions, $\phi_i(g)$, which satisfy a constraint of the form

$$\phi_i(h\,g) = \mathbb{D}_{ij}(h)\cdot\phi_j(g) \quad , \tag{4.22}$$

where $h \in H$ and $\mathbb{D}(h)$ is some particular representation of H. This means that the $\phi_i(g)$, though not strictly constant over the points in a coset, are related by a linear rule, i.e.

$$\phi_i(g_1) = \mathbb{D}_{ij}(g_1 g_2^{-1})\ \phi_j(g_2) \tag{4.23}$$

for g_1 and g_2 in the same coset. The appropriate restriction of the expansion (4.20) is now clear: it must include only those terms for which

$$D^n(h\,g) = \mathbb{D}(h)\ D^n(g) \quad . \tag{4.24}$$

In other words, on restriction to the subgroup H, the particular representation $\mathbb{D}(h)$ must be included in $D^n(h)$. (If the irreducible representation, $\mathbb{D}(h)$, is contained more than once in a particular D^n then an appropriate labelling must be introduced to distinguish them.) If the cosets are parametrized by L_y then we arrive at the expansion formula[11]

$$\phi_i(g) = \sum_n \sum_{\zeta,q} \sqrt{\frac{d_n}{d_{\mathbb{D}}}}\ D^n_{i\zeta,q}(L_y^{-1})\ \phi^n_{q\zeta} \quad . \tag{4.25}$$

The notation $D^n_{i\zeta,q}$ means that from the matrix D^n we take only those rows that correspond to the subspace which carries $\mathbb{D}(h)$,

$$D^n_{i\zeta,q}(h\,L_y^{-1}) = \mathbb{D}_{ij}(h)\ D^n_{j\zeta,q}(L_y^{-1}) \quad .$$

The supplementary label, ζ , is needed whenever this subspace occurs more than once. (The dimension of $\mathbb{D}$ is denoted $d_{\mathbb{D}}$.) The expansion (4.25) can be inverted to give

$$\phi^n_{q\zeta} = \frac{1}{V_K} \sqrt{\frac{d_n}{d_{\mathbb{D}}}} \int_{G/H} d\mu\ D^n_{q,i\zeta}(L_y)\ \phi_i(y) \quad , \tag{4.26}$$

where V_K denotes the volume of G/H and the invariant measure is $d\mu = d^K y\ |\det e_\mu{}^\alpha|$.

To apply these formulae in the 4+K-dimensional theory it is necessary to find the H-content of the various fields in the system. This amounts to fixing the embedding of H in the tangent space group, SO(1,3+K). Now, under the assumption that the geometry factorizes into $M^4 \times B^K$, the relevant part of the tangent space group is SO(1,3) $\times$ SO(K), and H must be embedded in SO(K). In fact the embedding is determined by the response of the K-bein to left translations,

$$e^{\alpha}(y') = e^{\beta}(y)\, D_{\beta}{}^{\alpha}(h) \quad ,$$

where $D_{\beta}{}^{\alpha}$ is an element of SO(K), and $h \in H$. Infinitesimally we can write

$$\begin{aligned} D_{\alpha\beta}(h) &= \delta_{\alpha\beta} + \delta h^{\bar{\gamma}}\, c_{\alpha\bar{\gamma}\beta} \\ &= \delta_{\alpha\beta} + \omega_{\alpha\beta} \quad , \end{aligned}$$

where $\omega_{\alpha\beta} = -\omega_{\beta\alpha}$, and $c_{\alpha\bar{\gamma}\beta}$ are structure constants of G. Viewed as an infinitesimal SO(K) transformation this implies

$$\frac{1}{2}\,\omega_{\alpha\beta}\, \Sigma^{\alpha\beta} = \delta h^{\bar{\gamma}}\, Q_{\gamma}$$

with SO(K) generators $\Sigma^{\alpha\beta}$, i.e.

$$Q_{\bar{\gamma}} = \frac{1}{2}\, c_{\alpha\bar{\gamma}\beta}\, \Sigma^{\alpha\beta} \quad . \tag{4.27}$$

This implies, for example, that the K-vector of SO(K) has the H-content of G/H. But, quite generally, it governs the decomposition of every tangent space multiplet into irreducible representations of H. Each irreducible piece can then be harmonically analyzed according to the method described above. Thus (4.25) is interpreted as,

$$\phi_i(x,y) = \sum_{n} \sum_{\zeta,q} \sqrt{\frac{d_n}{d_D}}\; D^{n}_{i\zeta,q}(L_y^{-1})\; \phi^{n}_{q\zeta}(x) \quad . \tag{4.28}$$

It is important to realize that the sum over representations in (4.25) is restricted to those which include the particular $D(h)$ appropriate to the multiplet $\phi_i(y)$. For example, in the familiar example of the 2-sphere SU(2)/U(1), one uses the SU(2) matrix elements $D^{j}_{\lambda m}$, $j = |\lambda|\ |\lambda|+1,\ldots; m = -j,\ldots,j$ with fixed λ, to expand the function ϕ_{λ} . [It is perhaps more common to begin with the spherical harmonics $Y_{jm} \sim D^{j}_{Om}$, appropriate to the scalar ϕ_O, and then construct from them the so-called vector harmonics by applying a vector operator. From our point of view this is equivalent to constructing the functions $D^{j}_{\lambda m}$, but the latter have much simpler orthogonality properties and are therefore easier to use.]

The orthogonality condition satisfied by the functions $D^n(L_y^{-1})$ takes the form

$$\frac{1}{V_K} \int_{G/H} d\mu \, D^{n}_{q,i\zeta}(L_y) \, D^{n'}_{i\zeta',q'}(L_y^{-1}) = $$

$$= \frac{d_{\mathbb{D}}}{d_n} \delta_{nn'} \delta_{qq'} \delta_{\zeta\zeta'} \tag{4.29}$$

Invariant integrals involving three or more factors can in principle be evaluated as easily, but some information about the Clebsch-Gordan coefficients of G would be needed.

A useful property of the $D^n(L_y^{-1})$ is their simple response to covariant differentation. With the connection $e^{\bar{\alpha}}(y)$ it follows from (4.5') and (4.11) that

$$\nabla_\alpha L_y^{-1} = -\frac{1}{a} Q_\alpha L_y^{-1} \tag{4.30}$$

and therefore

$$\nabla_\alpha D^{n}_{i\zeta,q}(L_y^{-1}) = -\frac{1}{a} D^{n}_{i\zeta,q}(Q_\alpha L_y^{-1}) \ . \tag{4.31}$$

This formula can be used to reduce all differential operations on G/H to algebraic manipulations with the matrices $D^n(Q_\alpha)$. For example, the spectrum of values of the operator $\nabla_\alpha \nabla_\alpha$ acting on $\phi_i(y)$ is given by

$$\nabla_\alpha \nabla_\alpha D^{n}_{i\zeta,q}(L_y^{-1}) = \frac{1}{a^2} D^{n}_{i\zeta,q}(Q_\alpha Q_\alpha L_y^{-1})$$

$$= \frac{1}{a^2} (C_G(D^n) - C_H(\mathbb{D})) \, D^{n}_{i\zeta,q} (L_y^{-1}) \ , \tag{4.32}$$

where $C_G(D^n)$ and $C_H(\mathbb{D})$ denote the values taken by the quadratic Casimir operators of G and R in the representations D^n and $\mathbb{D}$, respectively.

References

1) Th. Kaluza, Sitzungsber, Preuss. Akad. Wiss. Berlin, Math. Phys. Kl, 966 (1921); O. Klein, Z. Phys. 37, 895 (1926).

2) J. Scherk and J.H. Schwarz, Phys. Lett. 57B, 463 (1975); E. Cremmer and J. Scherk, Nucl. Phys. B103, 393 (1967); B108, 409 (1976); J.F. Luciani, Nucl. Phys. 135B, 111 (1978).

3) E. Witten, Nucl. Phys. B186, 412 (1981).

4) Z. Horvath, L. Palla, E. Cremmer and J. Scherk, Nucl. Phys. B127, 57 (1977).

5) E. Witten, Fermion Quantum Numbers in Kaluza-Klein Theories, Proc. 1983 Shelter Island II Conf. (MIT Press, 1984).

6) See the reports by M. Green, J. Schwarz and D. Gross in this volume.

7) P. Candelas, G. Horowitz, A. Strominger and E. Witten, Nucl. Phys. B258, 46 (1985).

8) Abdus Salam and J. Strathdee, Nucl. Phys. B79, 477 (1974).

9) E. Cremmer and B. Julia, Nucl. Phys. B159, 141 (1979).

10) E. Bergshoeff, M. de Roo, B. De Wit and P. van Nieuwenhuizen, Nucl. Phys. B195, 97 (1982); G.F. Chapline and N.S. Manton, Phys. Letts. 120B, 105 (1983).

11) Abdus Salam and J. Strathdee, Ann. Phys. 141, 316 (1982).

12) S. Randjbar-Daemi and R. Percacci, Phys. Lett. 117B, 41 (1982).

BASIC FEATURES OF STRING THEORIES

Jean-Loup Gervais
Laboratoire de Physique Théorique de l'Ecolė Normale Supérieure
24, rue Lhomond, 75231 Paris cedex 05, France

At the present time it is hardly necessary to emphasize the fundamental importance of string models since superstring theories are the most promising candidates for a completely unified theory of all interactions. Moreover string concepts have played an important role in the recent developments of theoretical physics and mathematics by suggesting many new important ideas, such as in particular supersymmetry[1)], and have led to very interesting progress in the related critical models in two dimensions.

It is hopeless to try and cover the whole subject of string theories. These lectures will mostly concentrate on the basic features of free strings and tree string scattering amplitudes. The bosonic string will be used in order to illustrate, without too much technical complications, the common features of all string models.

The old covariant operator formalism is presented in section (I) as a warming up. It is based on clever manipulations of trivial free fields in the two dimensional space of the string parameters. Our discussion will allow to see the fundamental role of the conformal algebra in two dimensions explicitly, and to discuss in a simple fashion the ghost-killing mechanism and the emergence of the critical space time dimension. In section (II), we shall show how the above conformal invariance appears as a remnant of the reparametrization invariance of the world sheet swept out by the string. We shall review the functional approach to quantization of the string modes, from the viewpoint of gauge theory in two dimensions. The relationship between the path integrals based on the new and the old reparametrization invariant actions will be discussed. The connection of the latter with the string scattering amplitudes in the light cone gauge will be summarized. Finally we shall review, in section (III), the possibility of building new string models in the Lorentz covariant formalism from interacting field theories in two dimensions. This brings in the question of conformally invariant field theories in general, and of the related critical systems.

There already exists a large number of review articles[2)3)4)5)6)], and two reprint volumes are about to appear[7)]. The viewpoint chosen here is somewhat different.

1. The operator formalism.

Before discussing the more sophisticated viewpoints which were later developed, it is useful to review the operator formalism where the relevant features are derived from simple harmonic oscillators in a pedestrian way.

a) Free bosonic modes.

The relevant quantity is a field $X^\mu(\sigma,\tau)$ describing the space-time position of the point of the string, which is characterized by the parameter σ, at time τ. μ is thus a Lorentz index ($\mu = 0, 1, \ldots, \mathcal{D}-1$) the space-time dimension $\mathcal{D}$ is kept as a free parameter. Since the string has a finite extension, σ varies over a finite range. For closed strings, which we shall discuss first, X^μ is a periodic function of σ. We can take, by convention, the period to be equal to 2π and introduce the simple action

$$S = \frac{1}{4\pi}\int_0^{2\pi} d\sigma \int d\tau \left((\dot{X}^\mu)^2 - (X'^\mu)^2 \right); \quad \dot{X} = \frac{\partial X}{\partial \tau}; \quad X' = \frac{\partial X}{\partial \sigma} \tag{1.1}$$

The factor in front of S can be modified by a simple rescaling of X^μ. Imposing the periodicity condition

$$X^\mu(\sigma + 2\pi, \tau) = X^\mu(\sigma,\tau) \tag{1.2}$$

gives a simple set of harmonic oscillators with $\omega_m = |m|$, m integer $\neq 0$, and we can write the standard free field decomposition in a box

$$X^\mu(\sigma,\tau) = q^\mu + p^\mu \tau + \sum_{m\neq 0} \frac{1}{\sqrt{2|m|}} \left[\alpha_m^\mu e^{i(m\sigma - |m|\tau)} + \alpha_m^{\mu +} e^{-i(m\sigma - |m|\tau)} \right] \tag{1.3}$$

$$[\alpha_m^\mu, \alpha_m^\nu] = [q^\mu, q^\nu] = [p^\mu, p^\nu] = 0 \tag{1.4}$$

$$[\alpha_m^\mu, \alpha_m^{\nu +}] = \delta_{m,m}\, \eta^{\mu\nu} \tag{1.5}$$

$$[q^\mu, p^\nu] = i\,\eta^{\mu\nu} \tag{1.6}$$

q^μ and p^μ are the center of mass position and total momentum, respectively. The space-time flat metric, $\eta^{\mu\nu}$, is equal to +1 for the diagonal space like components. On (1.5) one sees that, as usual, the time components α_m^0, $\alpha_m^{0\dagger}$ generate a Fock space with a non positive definite metric. We shall come back to this later on. For $m > 0$, $\alpha_m^\mu (\alpha_{-m}^\mu)$ annihilate right movers (left movers). It is convenient to separate these modes and rewrite, for $m > 0$,

$$a_m^\mu \equiv i\sqrt{m}\,\alpha_{-m}^\mu \quad ; \quad a_{-m}^\mu \equiv -i\sqrt{m}\,\alpha_{-m}^{\mu\dagger}$$

$$\bar{a}_m^\mu \equiv i\sqrt{m}\,\alpha_m^\mu \quad ; \quad \bar{a}_{-m}^\mu \equiv -i\sqrt{m}\,\alpha_m^{\mu\dagger} \tag{1.7}$$

in such a way that

$$X^\mu = q^\mu + p^\mu \tau + \frac{i}{\sqrt{2}} \sum_{m \neq 0} \left(a_m^\mu e^{-imu} + \bar{a}_m^\mu e^{-imv} \right) \frac{1}{m}$$

$$u \equiv \tau + \sigma \; ; \quad v \equiv \tau - \sigma \tag{1.8}$$

$$[a_m^\mu, a_m^\nu] = [\bar{a}_m^\mu, \bar{a}_m^\nu] = m \eta^{\mu\nu} \delta_{m,-m}$$

$$[a_m^\mu, \bar{a}_m^\nu] = 0 \quad ; \quad a_m^{\mu\dagger} = a_{-m}^\mu \quad ; \quad \bar{a}_m^{\mu\dagger} = \bar{a}_{-m}^\mu \tag{1.9}$$

The point of (1.7) (1.8) is to introduce the following simple expansions

$$P^\mu(z) \equiv \frac{1}{\sqrt{2}} (\dot{X}^\mu + X'^\mu) = \sum_m a_m^\mu z^{-m}$$

$$\bar{P}^\mu(z) \equiv \frac{1}{\sqrt{2}} (\dot{X}^\mu - X'^\mu) = \sum_m \bar{a}_m^\mu \bar{z}^{-m}$$

$$z = e^{iu} ; \; \bar{z} = e^{iv} ; \quad a_0^\mu = \bar{a}_0^\mu \equiv \frac{p^\mu}{\sqrt{2}} \tag{1.10}$$

The energy-momentum tensor is such that

$$T_0^0 \pm T_0^1 = \frac{1}{4\pi} :(\dot{X} \pm X')^2 \tag{1.11}$$

According to (1.10) its Fourier modes are such that

$$\frac{1}{2}(T_0^0 + T_0^1) = \frac{1}{2\pi} \sum_m L_m z^{-m}$$

$$\frac{1}{2}(T_0^0 - T_0^1) = \frac{1}{2\pi} \sum_m \bar{L}_m \bar{z}^{-m} \tag{1.12}$$

$$L_m = \frac{1}{2}\sum_n : a_n^\mu a_{m-n}^\mu : \quad ; \quad \bar{L}_m = \frac{1}{2}\sum_n : a_n^\mu a_{m-n}^\mu : \tag{1.13}$$

From (1.8), (1.9), one deduces that

$$[L_m, X^\mu] = z^{m+1}\frac{\partial}{\partial z}X^\mu$$

$$[\bar{L}_m, X^\mu] = \bar{z}^{m+1}\frac{\partial}{\partial \bar{z}}X^\mu \tag{1.14}$$

and L_m, $\bar{L}_m$ respectively generate the infinitesimal transformations

$$z \to z(1+\varepsilon z^m) \quad ; \quad \bar{z} \to \bar{z}$$

$$z \to z \quad ; \quad \bar{z} \to \bar{z}(1+\varepsilon \bar{z}^m) \tag{1.15}$$

They are all the transformations which preserve the angles with the metric $g_{\sigma\sigma} = -g_{\tau\tau} = 1$, $g_{\sigma\tau} = 0$, and are compatible with the periodicity condition (1.2). This last fact is an immediate consequence of the fact that the infinitesimal change being only a function of

$$z = e^{i(\sigma+\tau)} \quad ; \quad \bar{z} = e^{i(\tau-\sigma)} \tag{1.16}$$

is a periodic function of σ with period 2π. The subset of the operators $L_{\pm1}, L_0$ ($\bar{L}_{\pm1}, \bar{L}_0$) are the infinitesimal generators of the Möbius transformations of $z(\bar{z})$: $z \to (az+b)/(cz+d)$, $(\bar{z} \to (a\bar{z}+b)/(c\bar{z}+d))$ In particular, it is clear from (1.15) and (1.16) that $L_0 + \bar{L}_0$ and $L_0 - \bar{L}_0$ respectively generate τ and σ translations. The algebra of the L_m or $\bar{L}_m$ operators (the Virasoro algebra) closes apart from a c-number term (central charge) which plays a crucial role and we now discuss its calculation in a way which will easily generalize to different cases. This is best done by going to the Euclidean 2-D space such that one replaces τ by $-i\nu$, ν real. Equations (1.16) become

$$z = e^{\nu+i\sigma} \quad ; \quad \bar{z} = e^{\nu-i\sigma} \tag{1.17}$$

z, $\bar{z}$ are now the two complex conjugate variables and the transformations (1.15) simply become analytic transformations. We shall only consider the L_m algebra since the discussion of $\bar{L}_m$ is exactly the same. Clearly, equation (1.13) leads to

$$L_n = \frac{1}{4\pi i} \oint \frac{dz}{z} z^n : P^2(z) : \tag{1.18}$$

where the integral is in the complex z plane around an arbitrary contour surrounding the origin. We start the evaluation of $[L_n, L_m]$ by writing

$$L_n L_m = \frac{1}{(4\pi i)^2} \oint \frac{dx}{x} \oint \frac{dy}{y} x^n y^m : P^2(x) :: P^2(y) : \tag{1.19}$$

$$L_m L_n = \frac{1}{(4\pi i)^2} \oint \frac{dx}{x} \oint \frac{dy}{y} x^n y^m : P^2(y) :: P^2(x) : \tag{1.20}$$

Apply Wick's theorem to each expression. The contraction formula

$$\langle 0 | P^\mu(z_1) P^\nu(z_2) | 0 \rangle = \eta^{\mu\nu} \frac{z_1 z_2}{(z_1 - z_2)^2} \tag{1.21}$$

makes sense only if $|z_1| > |z_2|$ (otherwise the series diverges). Hence we have to choose $|x| > |y|$ in (1.19) and $|y| > |x|$ in (1.20). In the difference between (1.19) and (1.20) we thus pick up the poles of the Wick ordered expansion at x=y. This leads to[5)]

$$[L_n, L_m] = (n-m) L_{n+m} + \frac{D}{12}(n^3 - n)\, \delta_{n,-m} \tag{1.22}$$

The first term, which is linear in the L_n is the standard Lie algebra contribution. The second inhomogeneous term is the central charge which is proportional to D .

We now turn to open strings. It is convenient to choose $0 \leq \sigma \leq \pi$. The boundary conditions are

$$X'^\mu = 0 \quad , \text{ at } \sigma = 0, \pi \tag{1.23}$$

The action

$$S = \frac{1}{2\pi} \int_0^\pi d\sigma \int d\tau \left((\dot{X}^\mu)^2 - (X'^\mu)^2 \right) \tag{1.24}$$

only describes one set of modes since right and left movers mix at the end points. One now has

$$X^\mu = q^\mu + p^\mu \tau + i \sum_{m \neq 0} \frac{a_m^\mu}{m} e^{-im\tau} \cos(m\sigma) \tag{1.25}$$

$$\frac{1}{2}(T_0^0 \pm T_0^1) = \frac{1}{4\pi} :\left(\sum_m a_m^\mu e^{-im(\tau \pm \sigma)}\right)^2: \tag{1.26}$$
$$a_0^\mu \equiv p^\mu$$

Obviously there is a single set of Virasoro generators. It satisfies the algebra (1.22).

b) The dual tree amplitude between tachyons.

We shall mostly consider open strings which are simpler. For an arbitrary momentum vector k^μ, one introduces the basic vertex operator

$$V_k(z) = :e^{ik^\mu X^\mu}: \equiv e^{ik\cdot q} z^{k\cdot p} e^{-\sum_{m<0} \frac{k\cdot a_m}{m} z^{-m}} e^{-\sum_{m>0} \frac{k\cdot a_m}{m} z^{-m}} \tag{1.27}$$

Consider the ground state $|0\rangle$ such that

$$a_m^\mu |0\rangle = 0 \quad m>0 \quad ; \quad p^\mu |0\rangle = 0 \tag{1.28}$$

One easily obtains

$$\langle 0| V_{k_N}(z_N) V_{k_{N-1}}(z_{N-1}) \cdots V_{k_1}(z_1)|0\rangle = \prod_{\ell > m} (z_\ell - z_m)^{k_\ell \cdot k_m} \tag{1.29}$$

if the k_ℓ are conserved

$$\sum_{\ell=1,N} k_\mu^\ell = 0 \quad . \tag{1.30}$$

Equation (1.29), however, is derived by a summing series which converges only if $|z_N| > |z_{N-1}| > \cdots > |z_1|$. As we shall immediately see, the Möbius transformations of all z 's play a crucial role. To investigate this feature we introduce the cross ratios $(m > \ell)$

$$u_{m\ell} \equiv \frac{(z_m - z_{\ell-1})(z_{m-1} - z_\ell)}{(z_m - z_\ell)(z_{m-1} - z_{\ell-1})} \tag{1.31}$$

One obtains, assuming $(k^\ell)^2 = k^2$ independent of ℓ ,

$$\prod_{m>\ell}(z_m - z_\ell)^{k_\ell\cdot k_m} = \prod_{m>\ell} u_{m\ell}^{-\alpha_{m\ell}-1}\prod_{\ell}(z_{\ell+1}-z_\ell)^{(k^2-\alpha_0-1)}(z_{\ell+2}-z_\ell)^{(\frac{k^2}{2}-\alpha_0-1)} \tag{1.32}$$

where

$$\alpha_{m\ell} = \alpha_0 - \frac{1}{2}\Big(\sum_{\ell}^{m-1} k_\ell\Big)^2 . \tag{1.33}$$

α_0 and k^2 are arbitrary so far, but if we choose

$$\alpha_0 = 1 \tag{1.34}$$
$$k^2 = 2 \tag{1.35}$$

we arrive at a Möbius invariant integrand

$$\prod_{\ell} dz_\ell \prod_{m>\ell}(z_m - z_\ell)^{k_\ell\cdot k_m} = \prod_{\ell}\frac{dz_\ell}{z_{\ell+2}-z_\ell}\prod_{m>\ell} u_{m\ell}^{-\alpha_{m\ell}-1} \tag{1.36}$$

The multi-Veneziano formula is basically the corresponding integral. Some more care is needdd however. First, the vertex (1.27) describes a source k^μ located at the point $z = e^{i(\sigma+\tau)}$. As will become clearer in the next section, open string scattering amplitudes are described by external sources located at the end points $\sigma = 0$ or π. In the Euclidean, $z_\ell = e^{\nu_\ell + i\sigma_\ell}$, we choose $\sigma_\ell = 0$ and all z_ℓ are on the real axis. Remembering the above condition of validity of (1.29), one integrates for $z_N > z_{N-1} > \dots > z_1$. This is still not all, however, since (1.36) is Möbius invariant and we must divide out the integration over all Möbius transformations which preserve the real axis. This is done by giving arbitrary fixed values to three z variables (the choice is irrelevant). The multi-Veneziano formula reads, finally,

$$\mathcal{S}_N = \int d\mu(z) \langle 0|\, V_{k_N}(z_N)\, V_{k_{N-1}}(z_{N-1}) \cdots V_{k_1}(z_1)|0\rangle$$

$$d\mu(z) = \prod_{\ell \neq a,b,c}(dz_\ell)\,|z_a - z_b||z_b - z_c||z_c - z_a| \prod_{r}\theta(z_{r+1}-z_r) \tag{1.37}$$

By Möbius transformation we can map the real line onto the unit circle.

This shows that (1.37) is invariant under cyclic permutations of the z's. The complete, crossing symmetric S-matrix is obtained by summing over non cyclic permutations. The pole structure of (1.37) arises from the coincidence of z variables. From (1.31)(1.36) one sees that all poles are located at

$$\alpha(s) \equiv 1 + \frac{1}{2} s = n \quad ; \quad n \text{ integer} > 0 \tag{1.38}$$

where $\sqrt{s}$ is the center of the mass energy. The integer n is the highest spin of the intermediate particles at this level and one has a family of Regge trajectories spaced by integers, the leading trajectory being given by $J = \alpha(s)$. Conditions (1.34)(1.35) are such that the external particle with mass $-k^2 = -2$ is the lightest particle of the spectrum. Unfortunately this is a tachyon. From (1.38) one sees that the Regge trajectories have a slope $\alpha' = 1/2$. This is due to our particular choice of factors in front of the actions (1.1) and (1.24).

The pole structure is exhibited by changing variables from z_ℓ to N-3 independent $u_{m\ell}$'s. There are many ways to do this which lead to different dual pole configurations. In particular, let us choose $z_N \to \infty$, $z_{N-1} = 1$, $z_1 = 0$. In these limits

$$\begin{aligned} V_{k_1}(z)|0\rangle &\sim |k_1, 0\rangle \\ \langle 0| V_{k_N}(z_N) &\sim z_N^{-2} \langle k_N, 0| \end{aligned} \tag{1.39}$$

where, in general, we introduce the states $|k, 0\rangle$ such that

$$a_m^\mu |k, 0\rangle = 0 \, , \quad m > 0 \quad ; \quad p^\mu |k, 0\rangle = k^\mu |k, 0\rangle \tag{1.40}$$

Equation (1.37) becomes

$$\mathcal{S}_N = \int dz_2 \cdots dz_{N-2} \langle k_N, 0| V_{k_{N-1}}(1) V_{k_{N-2}}(z_{N-2}) \cdots V_{k_2}(z_2) | k_1, 0\rangle \tag{1.41}$$

For L_0, formula (1.13) gives, simply

$$L_0 = \frac{p^2}{2} + \sum_{m>0} a_{-m}^\mu a_m^\mu \tag{1.42}$$

and it is easy to check that (remember condition (1.35))

$$V_k(z) = z^{L_0} V_k(1) z^{-L_0} z^{-1} \tag{1.43}$$

Next, one changes variables by letting

$$z_{N-2} = x_{N-1}\ ;\ z_{N-3} = x_{N-1}x_{N-2}\ ;\ \cdots;\ z_2 = x_{N-1}\cdots x_3 \tag{1.44}$$

which is such that

$$0 \leq x_\ell \leq 1 \quad ; \quad \prod_\ell \frac{dz_\ell}{z_\ell} = \prod_\ell \frac{dx_\ell}{x_\ell} \tag{1.45}$$

One can now integrate out the x variables obtaining

$$\mathcal{I}_N = \langle k_N, 0|\, V_{k_{N-1}}(1) \frac{1}{L_0 - 1} V_{k_{N-2}}(1) \frac{1}{L_0-1} \cdots\cdots \frac{1}{L_0 - 1} V_{k_2}(1)\, |k_1, 0\rangle \tag{1.46}$$

In this formula the so-called multiperipheral pole structure is explicit. The associated diagram is drawn on figure (1):

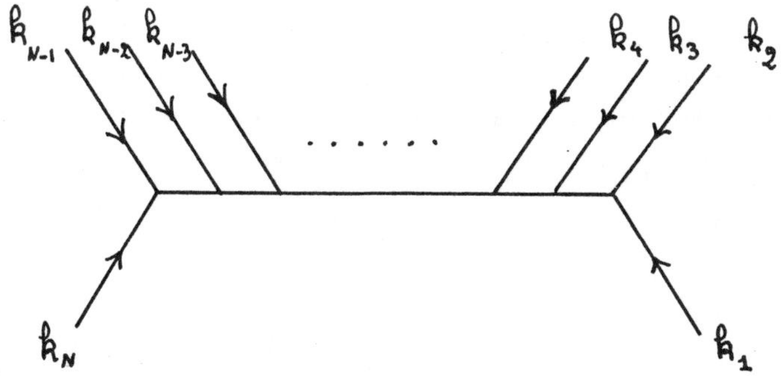

Fig. (1)

Next, a fundamental property of $V_k(z)$ for general k is its conformal covariance. One can check that

$$[L_m, V_k(z)] = z^m\left[z\frac{\partial}{\partial z} + (m+1)\frac{k^2}{2}\right] V_k(z) \tag{1.47}$$

For $k^2 = 2$ (condition (1.35)) we can write

$$(L_m - L_0 - m)\, V_k(1) = V_k(1)\,(L_m - L_0) \tag{1.48}$$

Moreover, it is simple to verify that

$$(L_m - L_0 + 1)\frac{1}{L_0 - 1} = \frac{1}{L_0 + m - 1}(L_m - L_0 - m + 1) \tag{1.49}$$

and, therefore

$$(L_m - L_0 + 1)\frac{1}{L_0 - 1} V_k(1) = \frac{1}{L_0 + m - 1} V_k(1)(L_m - L_0 + 1) \tag{1.50}$$

Consider inserting $(L_m - L_0 + 1)$ in the left of a propagator $1/(L_0 - 1)$ in (1.46). One can move it to right step by step using (1.50) until it reaches the state $|k_1, 0\rangle$ on the left hand side. There

$$\begin{aligned} (L_0 - 1)|k_1, 0\rangle &= (\tfrac{1}{2}k_1^2 - 1)|k_1, 0\rangle = 0 \\ L_m|k_1, 0\rangle &= 0 \quad , \quad m > 0 \end{aligned} \tag{1.51}$$

One thus sees that for $m > 0$, $L_m - L_0 + 1$ gives zero when it is inserted in the left of any propagator $1/(L_0 - 1)$ in (1.46). The poles of $\mathcal{S}_N$ arise from the so-called on-shell intermediate states which are eigenvectors of L_0 with eigenvalue 1. From the above argument it follows that they have non vanishing residue only if they are in addition such that $(L_m - L_0 + 1)$ applied to them gives zero. Hence the poles only appear from the so-called on-shell physical states which satisfy

$$L_m|phys\rangle = 0 \qquad (L_0 - 1)|phys\rangle = 0 \tag{1.52}$$

As we already noticed, the Fock space of the oscillators a_m^μ is not positive definite. One hopes that (1.52) does define a positive definite physical subspace. Let us check this at the lowest levels. L_0 is of the form

$$L_0 = \frac{p^2}{2} + N \tag{1.53}$$

where N has positive integer or zero eigenvalues. From (1.52) these are related to the mass μ^2 through the relation

$$(\frac{\mu^2}{2} + 1)|phys\rangle = N|phys\rangle \tag{1.54}$$

The ground state $|0, k\rangle$ is such that

$$N|0, k\rangle = 0$$

and (1.54) gives $\mu^2 = -2$. It is the lightest particle which is a tachyon. For N=1, we have the state

$$|k,\varepsilon\rangle \equiv \varepsilon_\mu a^\mu_{-1} |k,0\rangle \tag{1.55}$$

Its mass is zero according to (1.54). The rest of conditions (1.52) is trivially satisfied except for n =1 which gives

$$k^\mu \varepsilon_\mu = 0 \tag{1.56}$$

This state is a massless spin one particle and, as usual, condition (1.56) ensures the positivity of the norm

$$\langle k,\varepsilon|k,\varepsilon\rangle = \varepsilon^2 \tag{1.57}$$

The next level is more intriguing. For N=2, the general eigenstate of N is of the form

$$|k,\alpha,\beta\rangle \equiv \left(\tfrac{1}{2}\alpha_{\mu\nu}\, a^\mu_{-1} a^\nu_{-1} + \beta_\lambda a^\lambda_{-2}\right)|k,0\rangle \tag{1.58}$$

Condition (1.54) gives $k^2 = -2$ and k being time like, we go to its rest frame $k=(\sqrt{2},\vec{0})$. $|k,\alpha,\beta\rangle$ is physical if α and β satisfy

$$\sqrt{2}\,\beta_\mu = \alpha_{\mu 0} \quad ; \quad 5\,\alpha_{00} = \alpha_{ii} \tag{1.59}$$

This leaves us with a spin 2 state with components

$$\alpha^{(2)}_{ij} = \alpha_{ij} - \frac{\delta_{ij}\,\alpha_{ll}}{D-1} \quad ; \quad \alpha^{(2)}_{00} = 0 \tag{1.60}$$

and a spin zero state with $\alpha^{(0)}_{ij} = 0$, $\alpha^{(0)}_{00} \neq 0$. After some computation, one finds

$$\langle k,\alpha,\beta|k,\alpha,\beta\rangle = \frac{1}{2}\left(\alpha^{(2)}_{ij}\right)^2 + \left(\alpha^{(0)}_{00}\right)^2 \frac{26-D}{2(D-1)} \tag{1.61}$$

which is positive definite only if $D \leq 26$. The free string formalism and the tree amplitude (1.46) are acceptable only for $D \leq 26$. At D = 26 the scalar state has zero norm and one can show that it decouples from the S -matrix. As a matter of fact the bosonic string theory we have just described is completely consistent only at D = 26. In the present covariant formalism, this appears only at the level of loops where unitarity breaks down except at this critical dimension. This can be seen by looking at the so-called one loop Pomeron diagram which has an unphysical cut except at $\mathfrak{D} = 26$.

The particular role of $\mathfrak{D} = 26$ is already apparent at the level of free strings as the above discussion shows. In general, for $\mathfrak{D} \neq 26$, conditions (1.52) remove one string degree of freedom while at $\mathfrak{D} = 26$, only $26-2 = 24$ degrees of freedom remain. A general argument which indicates that only $\mathfrak{D} = 26$ is consistent will now be given following Kato and Ogawa[8]. This is based on the BRS transformations associated with the Virasoro algebra. Consider in general a Lie algebra G with generators T^a and structure constants C^{ab}_c

$$[T^a, T^b] = C^{ab}_c T^c \tag{1.62}$$

The BRS transformation is constructed from a ghost field η_a which is such that, under BRS,

$$\delta\eta_a = -\frac{\omega}{2} C^{bc}_a \eta_b \eta_c \tag{1.63}$$

η_a is anticommuting, ω is the infinitesimal anticommuting parameter. For a quantity ψ which transforms under G as $\delta_G\psi = \varepsilon_a T^a \psi$ the BRS transformation reads

$$\delta\psi = \omega\eta_a T^a \psi \tag{1.64}$$

It is just the group action with $\omega\eta_a$ as parameter. It is easy to check that (1.63) and (1.64) are such that the BRS transformation is nilpotent ($\delta^2 = 0$). This only follows from the Lie algebra (1.62) together with the Jacobi identity. If we introduce an additional ghost field ρ such that

$$\{\rho^a, \rho^b\} = 0 \qquad \{\rho^a, \eta_b\} = \delta_{ab}. \tag{1.65}$$

we can write the generator of (1.63), (1.64) as

$$Q = \sum_a T^a \eta_a - \frac{1}{2} \sum_{abc} C^{bc}_a \rho^a \eta_b \eta_c \tag{1.66}$$

Going back to the Virasoro algebra we notice two fundamental features: the Lie algebra (1.22) has a central term and there are an infinite number of generators. Let us proceed, anyhow. We now have two ghost fields η_n , ρ_m , n , m integers such that

$$\{\eta_n , \rho_m\} = \delta_{n,-m} \tag{1.67}$$

Following ref.8), we generalize (1.66) as

$$Q = \sum_m L_m \eta_{-m} + \sum_{n,m} m : \rho_n \eta_m \eta_{-n-m} : - \alpha_0 \eta_0 \tag{1.68}$$

The last term is the ambiguity arising from normal ordering. In order to compute Q^2 it is useful to rewrite

$$Q = \frac{1}{2\pi i} \oint dz/z \left[\frac{1}{2} : P^2(z) : \eta(z) + : \rho(z) \eta(z) z \eta'(z) : \right] \tag{1.69}$$

where

$$\eta(z) = \sum_n \eta_n z^{-n} \quad ; \quad \rho(z) = \sum_n \rho_n z^{-n} \tag{1.70}$$

The calculation proceeds very much like the above derivation of $[L_n, L_m]$. One needs the Wick contraction for ρ and η . Specify the vacuum to be such that

$$\rho_n |0\rangle = 0 \quad ; \quad \eta_n |0\rangle = 0 \quad ; \quad n > 0 \tag{1.71}$$

define the normal ordering of zero modes as

$$: \rho_0 \eta_0 : = - : \eta_0 \rho_0 : = \rho_0 \eta_0 - \eta_0 \rho_0 \tag{1.72}$$

Then

$$\rho(x) \eta(y) = : \rho(x) \eta(y) : + \frac{1}{2} \frac{x+y}{x-y}$$
$$\eta(x) \rho(y) = : \eta(x) \rho(y) : + \frac{1}{2} \frac{x+y}{x-y} \tag{1.73}$$

In computing $\mathcal{Q}^2$ from (1.69) by Wick's theorem one must again order the contours and one picks up only the pole at x=y. This gives

$$\mathcal{Q}^2 = \frac{\mathcal{D}-26}{48\pi i} \oint dz\, (z\eta)''' z\rho + \frac{1-\alpha_0}{2\pi i} \oint dz\, \eta\eta' \tag{1.74}$$

This vanishes only if $\mathcal{D} = 26$; $\alpha_0 = 1$. It follows from (1.71) that the on shell physical states(1.52) satisfy

$$\mathcal{Q}\,|phys\rangle = 0 \tag{1.75}$$

and $\mathcal{Q}$ is really the relevant operator for the consistent string quantization. In the next section we discuss string quantization from the gauge theory view point. In this type of approach BRS invariance is known to play a crucial role. Hence it is natural that the present string theory only makes sense at $\mathcal{D} = 26$, $\alpha_0 = 1$.

2. Functional methods - Gauge approach.

From (1.27) and (1.37), one can rewrite the dual amplitude as a functional integral over surfaces; indeed, standard manipulations show that

$$\langle 0|V_{k_N}(z_N)\cdots V_{k_1}(z_1)|0\rangle \propto \int \mathcal{D}X\, e^{\frac{i}{\pi}\int d\sigma d\tau \left[\frac{1}{2}(\dot{X}^2 - X'^2) + J.X\right]} \tag{2.1}$$

$$J^\mu = \pi \sum_\ell k_\ell^\mu\, \delta(\sigma)\, \delta(\tau-\tau_\ell) \tag{2.2}$$

$$z_\ell = e^{i\tau_\ell} \tag{2.3}$$

The right member of (2.1) involves the self energies of the external sources which are infinite. Disregarding them is equivalent to the normal ordering prescription (1.27). The most important point about (2.1) is that it involves all $\mathcal{D}$ components of X^μ while we know that at $\mathcal{D} = 26$ only 24 degrees of freedom are physical. This decoupling comes in naturally if we functionally integrate over geometric surfaces in $\mathcal{D}$ dimensions and not over $\mathcal{D}$ functions of σ, τ . Indeed a geometrical surface being reparametrization invariant only depends on $\mathcal{D}$ -2 functions. Reparametrization invariance is used as a gauge principle. One starts from an action invariant under reparametrization which is the two dimensional version of general coordinate invariance (2-$\mathcal{D}$ general relativity). Denote by g_{ab} ($a,b=\sigma,\tau$) the 2x2 metric tensor of the surface. The associated 2-D Einstein action $\sqrt{g}\,R$ is a total derivative and can be forgotten. Under reparametrization $\sigma \to \tilde{\sigma}(\sigma,\tau)$, $\tau \to \tilde{\tau}(\sigma,\tau)$; the X^μ transform as

$$\tilde{X}(\tilde{\sigma},\tilde{\tau}) = X(\sigma,\tau) \tag{2.4}$$

They behave as 2-D spin zero matter fields and the associated invariant action reads

$$S_{BDH} = \frac{1}{2\pi}\int d\sigma d\tau \sqrt{g}\; g^{ab}\,\partial_a X^\mu \partial_b X^\mu \tag{2.5}$$

$$g = -\det(g_{ab}) \qquad g^{ab} = (g^{-1})_{ab} \tag{2.6}$$

This action proposed by Brink, Di Vecchia and Howe[9] is also invariant under Weyl transformation

$$X \to X \quad ; \quad g_{ab} \to \Lambda(\sigma,\tau)\, g_{ab} \tag{2.7}$$

where Λ is an arbitrary function. The point of (2.5), as compared to the Nambu Goto action introduced earlier, is that one can regularize it without breaking reparametrization invariance. One can, for instance, introduce Pauli-Villars regulators Y_ℓ^μ, $\ell=1,..,N$, which are 2-D matter fields as X^μ but with masses $M_\ell \neq 0$. One adds to (2.5) the action

$$S_R = \frac{1}{2\pi}\int d\sigma d\tau \sqrt{g} \sum_\ell \left[g^{ab}\partial_a Y_\ell^\mu \partial_b Y_\ell^\mu - M_\ell^2\, Y_\ell^\mu Y_\ell^\mu \right] \tag{2.8}$$

with the understanding that M_ℓ will tend to infinity at the end. The Y_ℓ^μ are alternatively bosons and fermions so as to cancel the infinities. By adjusting the masses M_ℓ we can remove the divergences. S_R is still reparametrization invariant but the mass term breaks the Weyl invariance

(2.7) and an anomalous term appears in general. In the functional integral

$$\int \mathcal{D}g\, \mathcal{D}X\, \delta(F_1)\delta(F_2)\, \Delta_{FP}\, e^{iS_{BDH}} \tag{2.9}$$

one only introduces gauge fixing terms for the reparametrization invariance. The choice of F_1 and F_2 specifies the parametrization of the surface. The basic point here is that all two-dimensional manifolds are conformally flat and one can choose orthonormal coordinates on the surface. However, one considers g_{ab} and X^μ as independent variables. Hence we have two independent objects which behave as metric tensors, i.e.

g_{ab} and

$$G_{ab} = \partial_a X^\mu \partial_b X^\mu \tag{2.10}$$

One can follow two paths from this point:

a) Following Polyakov[10], we can choose the gauge where

$$F_1 = g_{\tau\tau} + g_{\sigma\sigma} \quad ; \quad F_2 = g_{\sigma\tau}$$

$$g_{ab} = e^{\varphi} \begin{pmatrix} 1 & 0 \\ 0 & -1 \end{pmatrix} \tag{2.11}$$

With this choice, it is easy to integrate X^μ out, since the action (2.5) is quadratic in X^μ. One can regularize using (2.8). A counter term is needed:

$$S_C = \frac{\mu}{\pi} \int d\sigma\, d\tau\, \sqrt{g} \tag{2.12}$$

where μ behaves as $\sum_e \varepsilon_e M_e^2 \ln(M_e)$; $\varepsilon_e = \pm 1$; for $M_e \to \infty$. One further determines the Faddeev Popov determinant. The final result reads[10], for the partition function,

$$\int \mathcal{D}X\, \mathcal{D}g\, e^{i(S_{BDH} + S_C)}\, \delta(g_{\tau\tau} + g_{\sigma\sigma})\, \delta(g_{\sigma\tau})\, \Delta_{FP} = \int \mathcal{D}\varphi\, e^{iS_L} \tag{2.13}$$

$$S_L = \frac{26 - D}{48\pi} \int d\sigma\, d\tau \left[\frac{1}{2}(\dot{\varphi}^2 - \varphi'^2) - e^{\varphi} \right] \tag{2.14}$$

S_L is the so-called Liouville action. We put it in a nice looking form by choosing the coefficient of e^{φ} equal to 1 inside the bracket. Originally it was a divergent quantity but we can give it any value by shifting φ by a constant. Obviously, at $\mathcal{D} = 26$, S_L disappears and it seems that the anomaly becomes irrelevant, thus supporting the direct operator treatment described in sect. 1. This discussion is still rather formal however. One integrates over X^{μ} with fixed φ thus ignoring the quantum effects on this latter field. Moreover possible difficulties in choosing the gauge (2.10) are ignored. We shall have more to say on this in section 3.

b) Following A. Neveu and myself[11)], one can choose the gauge fixing functions F_1 and F_2 to be only functions of X^{μ}. Then it is possible to integrate over g_{ab} point by point since the action (2.5) does not involve any of its derivatives. We shall proceed very formally so that $\mathcal{D} \neq 26$ will not be needed explicitly. Since no derivative of g appears in the action, the functional integral over g can be rewritten as

$$\prod_{\sigma,\tau}\int d[g_{ab}]\; e^{\frac{i}{\pi}\left[d\sigma d\tau\sqrt{g}\,g^{cd}G_{cd}(\sigma,\tau)+\mu\sqrt{g}\,d\sigma d\tau\right]}$$

$$\equiv \prod_{\sigma,\tau} F\left(G_{\ell m}(\sigma,\tau)\right) \qquad (2.15)$$

Consider one particular factor of this formal infinite product. It is given by the ordinary integral

$$F(X,Y,Z)=\int dx\,dy\,dz\;\exp\left\{\frac{i}{\sqrt{xy-z^2}}\left[Xy+Yx-2zZ\right]\right\} \qquad (2.16)$$

where we have let

$$X=\frac{d\sigma d\tau}{2\pi}G_{11}\;;\quad Y=\frac{d\sigma d\tau}{2\pi}G_{22}\;;\quad Z=\frac{d\sigma d\tau}{2\pi}G_{12}=\frac{d\sigma d\tau}{2\pi}G_{21}$$

$$x=g_{11}\;;\quad y=g_{22}\;;\quad z=g_{12}=g_{21} \qquad (2.17)$$

We next separate the integral over the determinant of g by inserting into (2.16) the expression

$$1=\int d\rho\;\delta(\rho-xy+z^2) \qquad (2.18)$$

which can also be cast into the form

$$1 = \int d\rho\, dk\; e^{ik(\rho - xy + z^2)} \tag{2.19}$$

Integrating over x, y, z, for fixed ρ and k, one obtains

$$F(X,Y,Z) = \int d\rho \int \frac{dk}{k^{3/2}}\; e^{i\left(k\rho + \frac{D}{k\rho}\right)} \tag{2.20}$$

where

$$D = Z^2 - XY. \tag{2.21}$$

As is usual for path integral with real time, the remaining integrations are only semi-convergent. The integral over k will be replaced by the convergent integral

$$\int_0^\infty \frac{dk}{k^{3/2}}\; e^{-\left(k\rho + \frac{\lambda}{k\rho}\right)} \qquad \lambda = -D \tag{2.22}$$

which can be obtained by deforming the contour of integration. From the formula[12]

$$\int_0^\infty \frac{dx}{\sqrt{x}}\; e^{-\left(x + \frac{\lambda}{x}\right)} = \sqrt{\pi}\; e^{-2\sqrt{\lambda}} \tag{2.23}$$

we deduce

$$\int_0^\infty \frac{dk}{k^{3/2}}\; e^{-\left(k\rho - \frac{D}{k\rho}\right)} = i\sqrt{\frac{\pi\rho}{D}}\; e^{2i\sqrt{D}} \tag{2.24}$$

Recalling (2.17), we rewrite this as

$$F(G) \propto \frac{e^{\frac{i}{\pi}\sqrt{G}\, d\sigma\, d\tau}}{\sqrt{G}\; d\sigma\, d\tau}$$

$$G = -\det(G_{ab}) \tag{2.25}$$

and equation (2.9) becomes

$$\int \mathcal{D}X \frac{\delta(F_1)\delta(F_2)}{\prod\limits_{\sigma,\tau} d\sigma d\tau \sqrt{G}} \Delta_{FP}\, e^{\frac{i}{\pi}\int d\sigma d\tau \sqrt{G}} \tag{2.26}$$

We have found back the original Nambu-Goto action. The measure is however non trivial. It precisely coincides with the one proposed by Sakita and myself[13] from the study of the light-cone gauge in functional integral. Indeed the additional factor $\prod\limits_{\sigma,\tau} (d\sigma d\tau \sqrt{G})^{-1}$ is precisely needed to integrate out two components in this gauge in order to remain with only physical degrees of freedom. We now summarize this point briefly. Define the light-cone notation as

$$X^{\pm} = \frac{1}{\sqrt{2}}(X^0 \mp X^1) \;;\; \underset{\sim}{X} = X^{\mu} \quad \mu = 2,\ldots,\mathcal{D}-1 \tag{2.27}$$

Choose the gauge fixing conditions

$$F_1 \equiv X^+ - f(\sigma,\tau) \;;\; F_2 = (\dot{X} - X')^2 \tag{2.28}$$

f is a function which we shall suitably determine below. Consider the generating functional

$$\mathcal{Z}(J) \equiv \int \mathcal{D}X \frac{\delta(X^+ - f)\delta((\dot{X} - X')^2)}{\prod d\sigma d\tau \sqrt{G}}\, e^{\frac{i}{\pi}\int d\sigma d\tau [\sqrt{G} + J\cdot X]} \tag{2.29}$$

The integration over X^+ is of course trivial. In order to also integrate out the X^- component, we choose f in such a way that after substituting $X^+ = f$ into the exponent, all dependence in X^- disappears from the action including the source term. This is possible since, due to the F_2 gauge condition, the Nambu-Goto action is now linearized:

$$\frac{1}{\pi}\int d\sigma d\tau \sqrt{G} = \frac{1}{2\pi}\int d\sigma d\tau [(\dot{X})^2 - (X')^2] \tag{2.30}$$

Introducing the free field Green function $N(\vec{\xi},\vec{\xi}')$ such that

$$\left(\frac{\partial^2}{\partial\sigma^2}-\frac{\partial^2}{\partial\tau^2}\right)N(\vec{\xi},\vec{\xi}')=-i\delta_2(\vec{\xi}-\vec{\xi}') \tag{2.31}$$

$$\partial_{n_\xi}N(\vec{\xi},\vec{\xi}')\Big|_{\text{boundaries}}=\rho(\vec{\xi}') \quad ; \quad \vec{\xi}\equiv(\sigma,\tau) \tag{2.32}$$

we choose

$$f(\vec{\xi})=-i\int d_2\vec{\xi}'N(\vec{\xi},\vec{\xi}')J^+(\vec{\xi}') \tag{2.33}$$

and the integration over X^- reduces to

$$\int\mathcal{D}X^-\frac{\delta((\mathring{X}-X')^2)\Delta_{FP}}{\prod_{\sigma,\tau}d\sigma d\tau\sqrt{G}}=1 \tag{2.34}$$

This is proven in ref.13). One is left with

$$Z(J)=\exp\left\{-\frac{1}{\pi}\int d_2\vec{\xi}d_2\vec{\xi}'J^-(\xi)J^+(\xi')N(\vec{\xi},\vec{\xi}')\right\}$$
$$\int\mathcal{D}\underset{\sim}{X}\exp\left\{\frac{i}{\pi}\int d_2\xi\left[(\mathring{\underset{\sim}{X}}^2-\underset{\sim}{X}'^2)\frac{1}{2}+\underset{\sim}{J}\cdot\underset{\sim}{X}\right]\right\} \tag{2.35}$$

and the functional integral is only over the $\mathcal{D}$ -2 physical degrees of freedom. The gauge choice $X^+=f$ may seem peculiar since it explicitly depends upon the external sources J^+. Next we briefly show that, on the contrary, this dependence is precisely needed to describe string scattering in the frame where X^+ is taken as time. With this motivation we come back for a moment to the free string formalism of sect. 1 but where one explicitly eliminates the $X^\pm$ components following ref. 14). In the covariant operator formalism one works in a Fock space larger than the physical one which is defined by $(L_0-1)|phys\rangle=L_n|phys\rangle=0$. In the GGRT formalism[14)] one imposes as operator identities

$$L_0=1 \qquad L_n=0 \qquad n\neq 0 \tag{2.36}$$

From (1.26) one sees that this corresponds to

$$\frac{1}{2}:(\dot{X}\pm X')^2: -1 = 0 \tag{2.37}$$

Moreover one eliminates the a_m^+ modes by letting

$$X^+ = p^+\tau \tag{2.38}$$

These last two conditions determine X^- as

$$X^- = q^- + p^-\tau + i\sum_{m\neq 0}\frac{1}{m}\tilde{L}_m\, e^{-im\tau}\cos m\sigma \tag{2.39}$$

$$\tilde{L}_m = \frac{1}{2}\sum_n :\underset{\sim}{a}_n\, \underset{\sim}{a}_{m-n}: \tag{2.40}$$

The only quantum degrees of freedom are the transverse modes $\underset{\sim}{X}$ and q^-, p^+. p^- is determined as

$$p^- = \frac{1}{p^+}(\underset{\sim}{L}_0 - 1) \tag{2.41}$$

Going back to the interacting string theory we remark, following Mandelstam[15)] that (2.38) does not allow to adjust the surface representing different strings since the definition of τ depends on p^+. Hence in the above free string formalism we perform a rescaling of σ and τ by p^+ in such a way that equation (2.38) becomes simply

$$X^+ = \tau \tag{2.42}$$

and the new σ variable now goes from 0 to $\pi p^+ = \alpha$. With this parameters string interactions become very simple. Strings break or combine as shown, typically, on figure (2)

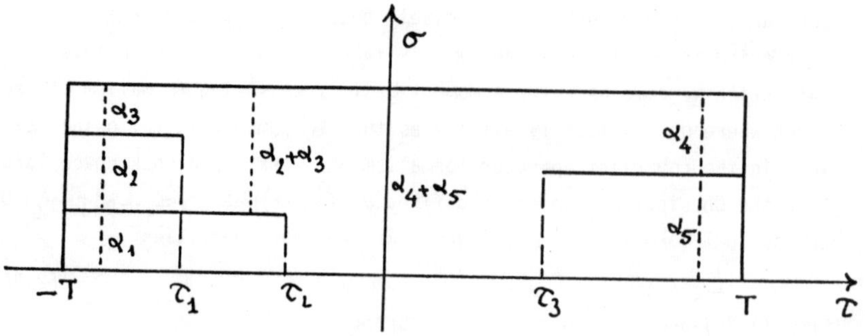

Fig. (2)

Since $\alpha = \pi p^+$ is a conserved quantity, the domains of variation of the various σ automatically adjust at each interaction in such a way that the total range of σ variation remains the same. The scattering process associated with fig. (2) is obtained from (2.35) by putting as sources momenta distributions at -T and +T which represent the initial and final string states. For each incoming (time -T) or outgoing (time T) strings we can use the free field formalism if T is so large that all interactions take place between -T and +T. Equation (2.30) shows that the corresponding p^+ density is

$$p^+ = \frac{1}{\pi}\frac{\partial X^+}{\partial \tau} = \frac{1}{\pi} \tag{2.43}$$

We set, accordingly $J^+ = 1$ at $\tau = \pm T$ in (2.35). At this point it becomes more convenient to work in the Euclidean formulation where (2.31) becomes

$$\Delta N(\vec{\xi},\vec{\xi}') = -\delta_2(\vec{\xi}-\vec{\xi}') \tag{2.44}$$

so that for $\vec{\xi} \sim \vec{\xi}'$

$$N(\vec{\xi},\vec{\xi}') \sim \frac{1}{4\pi} \ln[(\vec{\xi}-\vec{\xi}')^2] \tag{2.45}$$

With this information, it is easy to show that, according to (2.33),

$$\partial_\tau f = i \qquad \text{at} \quad \tau = \pm T \tag{2.46}$$

Moreover, (2.31) shows that f has a vanishing Laplacian inside the domain $-T \leq \tau \leq T$ since J^+ is only at the boundary. This combined with (2.32) shows that $f \equiv i\tau$ everywhere and in the domain of fig (2), the F_1 gauge condition is precisely equivalent to the light-cone condition (2.42) after Wick's rotation. By conformal transformation we can map fig (2) to different σ,τ domains where X^+ will not be equal to $i\tau$ in general. This leads back to the general form (2.28).

Going back to fig (2) we remark that the factor in front of (2.35) can be written as

$$\exp\left[-\frac{1}{\pi}\int d\xi\, J^-(\xi) f(\xi)\right] = \exp\{-iT\mathcal{E}\}$$

$$\mathcal{E} = \sum_{\substack{\ell \\ \text{initial}}} p_\ell^- = \sum_{\substack{\ell \\ \text{final}}} p_\ell^- \tag{2.47}$$

This equation follows immediately from the fact that $f(\xi)$ is a constant on the boundary $\tau = \pm T$ so that only the total J^- appear. Finally, the functional integral (2.35) computed in the domain of fig (2) represents the transition probability between the initial and the final state with strings joining at times τ_1, τ_2 and later breaking at τ_3. In the light-cone formalism, p^- plays the role of the energy. Hence, the factor (2.47) is precisely such that one gets a finite limit for $T \to \infty$. This is the standard way of deriving the S-matrix from the finite time evolution operator. One has to further integrate over the interaction times τ_1, τ_2, τ_3. In the limit $T \to \infty$, time translation invariance is recovered and only interaction time differences are relevant. This leaves us with 2 = 5-3 variables.

Finally one can map fig (2) onto fig (3) by a transformation which is conformal inside the domain

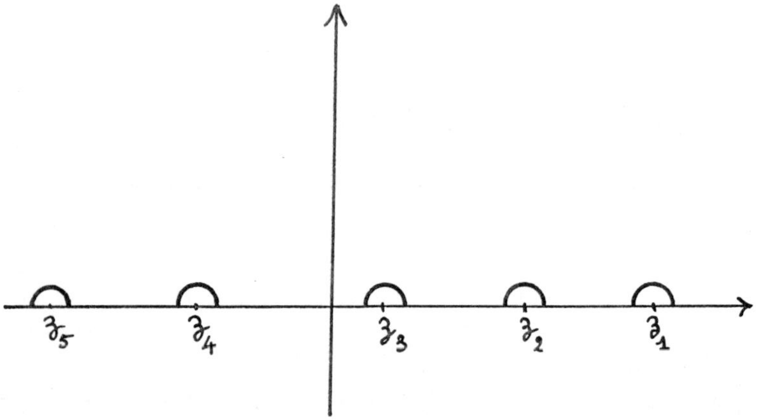

Fig. (3)

The new domain is the upperhalf plane minus the inside of the half circles which shrink to points $z_1, \cdots, z_5$ as $T \to \infty$. We therefore find back the procedure discussed in sect. 1 since the momentum distributions are located on these small circles. Integrating over the time differences becomes equivalent to the integrating over the N-3 Koba-Nielsen variables of sect. 1.

3. Beyond free two-dimensional field theories.

As we just saw, conformal transformations play a crucial role. Using complex variables z and z^* a general conformal transformation reads

$$z' = F(z) \tag{3.1}$$

where F is a function of a single variable. In general a quantity $\mathcal{O}(z,z^*)$ is called conformally covariant[16)] if it transforms according to

$$\mathcal{O}'(z',z'^*) = \left(\frac{dF}{dz}\right)^{-\delta} \left(\frac{dF^*}{dz^*}\right)^{-\bar{\delta}} \mathcal{O}(z,z^*) \tag{3.2}$$

where δ and $\bar{\delta}$ are parameters depending on the quantity considered which are called conformal weights. This notion was rediscovered recently[17)] and the covariant fields were called primary. If we separate the real and imaginary parts according to

$$z = x_1 + i x_2 \tag{3.3}$$

the differential transforms as

$$dx'_\ell = \mu\, R_{\ell m}(\theta)\, dx_m \tag{3.4}$$

where $R_{\ell m}$ is the rotation matrix with angle θ and where μ is the dilatation factor. These two quantities are given by

$$\frac{\partial x'_1}{\partial x_1} = \frac{\partial x'_2}{\partial x_2} = \mu \cos\theta \qquad \frac{\partial x'_1}{\partial x_2} = -\frac{\partial x'_2}{\partial x_1} = \mu \sin\theta \tag{3.5}$$

Formula (3.2) becomes

$$\mathcal{O}'(x'_1, x'_2) = \mu^{-d} e^{-iJ\theta}\, \mathcal{O}(x_1, x_2)$$
$$d = \delta + \bar{\delta} \quad ; \quad J = \delta - \bar{\delta} \tag{3.6}$$

Since μ and θ are the local dilatation and rotation parameters, d is the dimension and J is the spin of the quantity considered.

In a conformally invariant field theory the improved energy momentum tensor is traceless symmetric and conserved. For two-dimensional field theories this has strong consequences. If we have in real σ, τ space

$$\dot{T}^0_0 + \acute{T}^1_0 = \dot{T}^0_1 + \acute{T}^1_1 = 0$$
$$T^0_0 + T^1_1 = 0 \quad ; \quad T^1_0 = T^0_1 \tag{3.7}$$

we conclude that

$$\left(\frac{\partial}{\partial\tau} \mp \frac{\partial}{\partial\sigma}\right)\left(T_0^0 \pm T_0^1\right) = 0 \tag{3.8}$$

Imposing periodicity in σ with period 2π leads to

$$2\pi\left(T_0^0 + T_0^1\right) = \sum_m L_m z^{-m} \quad ; \quad z = e^{i(\tau+\sigma)}$$

$$2\pi\left(T_0^0 - T_0^1\right) = \sum_m \bar{L}_m \bar{z}^{-m} \quad ; \quad \bar{z} = e^{i(\tau-\sigma)} \tag{3.9}$$

In the same way as in the free case of section 1, the L_m and $\bar{L}_m$ are the Virasoro generators. In quantum theory, a covariant operator transforming according to (3.2) satisfies

$$[L_m, \theta] = z^m\left[z\frac{\partial}{\partial z} + (m+1)\delta\right]\theta$$

$$[\bar{L}_m, \theta] = \bar{z}^m\left[\bar{z}\frac{\partial}{\partial \bar{z}} + (m+1)\bar{\delta}\right]\theta \tag{3.10}$$

In conformally invariant quantum field theories there exist several such covariant operators θ_α with weights δ_α and $\bar{\delta}_\alpha$ which depend upon the model considered. At the classical level the above structure is realized by Poisson brackets. The weights are in general modified by quantum effects. An example is the operator $V_k(z)$ of section 1 (formula (1.27)) which has $\delta = \bar{\delta} = 0$ at the classical level and $\delta = k^2/2$, $\bar{\delta} = 0$ in the quantum theory.

One must consider all covariant operators such that, together with all their derivatives, they form a family $\mathcal{A}^{(M)}$ which is closed by short distance expansion

$$\mathcal{A}^{(M)}(z)\, \mathcal{A}^{(N)}(z') \underset{z\to z'}{\sim} \sum_L C_L^{MN}(z-z')\, \mathcal{A}^{(L)}\left(\frac{z+z'}{2}\right) \tag{3.11}$$

In this expansion, the singularity structure of the c-number coefficients C_L^{MN} at short distance is determined by the difference between the dimension of the operator $\mathcal{A}^{(L)}$ and the sum of the dimensions of $\mathcal{A}^{(M)}$ and $\mathcal{A}^{(N)}$.

As is well known[16)17)] the derivative of a covariant operator is not covariant in general. This is why the expansion (3.11) does not include covariant operators only. An important exception is the case of an operator of vanishing weight. Its derivative with respect to z (or z*) is a covariant operator with weights $\delta=1$, $\bar{\delta}=0$ (or $\delta=0$, $\bar{\delta}=1$). Conversely, let us assume there exists an operator I(z) with weights $\delta=1$, $\bar{\delta}=0$. It is obvious that

$$\left[L_n , \oint dz\, I(z) \right] = 0 \tag{3.12}$$

At this point it is useful to recall our derivation of the Virasoro algebra for the free case of section 1. Basically we used the short distance expansion of the product $\mathcal{U}(z)\mathcal{U}(z')$. It is clear that the same computation can be repeated, once the short distance expansion is given, quite generally. Deforming the contour will always peak up the pole term which therefore gives the commutator of the operators considered. Classically $\mathcal{U}$ and $\bar{\mathcal{U}}$ have weights $\delta=2$, $\bar{\delta}=0$ and $\delta=0$, $\bar{\delta}=2$ respectively. Let us now show that if this is also true at the quantum level the mode coefficients of $\mathcal{U}$ and $\bar{\mathcal{U}}$ satisfy the Virasoro algebra. Consider $\mathcal{U}$ for instance. Its short distance expansion reads (A and B are constant C-numbers)

$$\mathcal{U}(z)\,\mathcal{U}(z') \underset{z\to z'}{\sim} A\left[\frac{zz'}{(z-z')^2}\right]^2 + B\,\frac{zz'}{(z-z')^2}\,\mathcal{U}\!\left(\frac{z+z'}{2}\right) + \cdots \tag{3.13}$$

Following the same path as in the free case, one starts from

$$L_n L_m = \frac{1}{(2\pi i)^2}\oint \frac{dx}{x}\,\frac{dy}{y}\, x^n y^m\, \mathcal{U}(x)\,\mathcal{U}(y)$$
$$L_m L_n = \frac{1}{(2\pi i)^2}\oint \frac{dx}{x}\,\frac{dy}{y}\, x^n y^m\, \mathcal{U}(y)\,\mathcal{U}(x) \tag{3.14}$$

From the positivity of the energy it follows that the product $\mathcal{U}(x)\,\mathcal{U}(y)$ is analytic in y/x for $|x| > |y|$ and again we have to choose $|x| > |y|$ and $|y| > |x|$ in the first and in the last equations (3.14) respectively. The commutator is again given by the pole at x = y which is solely specified from (3.13). One obtains

$$[L_n, L_m] = \frac{B}{2}(n-m)\, L_{n+m} + \frac{A}{12}\, \delta_{n,-m}(n^3-n) \qquad (3.15)$$

If B ≠ 2, we may redefine L_m so as to obtain the usual form. The crucial point is that (3.13) holds, namely $\mathcal{T}$ has its naive dimension and there are no logarithmic corrections to the short distance expansion.

Conformally invariant field theories in two dimensions have two different physical applications. First we can consider a statistical system without boundary by taking the variables x_1 and x_2 of formula (3.3) as coordinates. This may experimentally describe a thin film. With this choice of coordinates, d and J are the physical dimension and spin. $L_0 + \bar{L}_0$ is the generator of rescaling of x_1 and x_2 and a statistical system at a critical point becomes scale invariant. It has been argued by Polyakov that critical systems are in fact invariant under the full conformal group and hence are associated with conformally invariant field theories. It is easy to see that L_{-1} and $\bar{L}_{-1}$ generate x_1 and x_2 translations. The ground state of the system is thus annihilated by these operators. It must also be scale invariant. This is sufficient to determine the two point function of any covariant operator. One finds, assuming for instance that $\delta = \bar{\delta}$,

$$\langle 0|\, \mathcal{O}(z, z^*)\, \mathcal{O}(z', z'^*)\, |0\rangle \propto (\,|z-z'|^2)^{-2\delta} \qquad (3.16)$$

and it follows that, in general, d and J are critical exponents. In the scaling limit of discrete statistical systems, only operators of non negative dimensions survive. In the continuous theory all physical operators must have non negative dimension. Hence the short distance expansion (3.11) must close only with this type of operators.

If an operator of conformal weight 1 exists, formula (3.12) shows that its integral is conformally invariant. It can be added to the action with an arbitrary coefficient without destroying the conformal invariance. As a result there exists a family of critical models and one has a critical line instead of a critical plot in the space of the couplings. The appearance of these so-called marginal operators is frequent in two dimensional models.

The other physical application is in dual theories. There also conformal invariance is crucial since it allows to remove the ghost states due to the Lorentz metric. We have seen in section 1 an example of the ghost-killing mechanism in the covariant approach where the key point was to choose $k^2=2$ in such a way that V_k has $\delta = 1$. One can summarize the above discussion by saying that if $k^2=2$, $\oint \frac{dz}{z} V_k(z)$ becomes a marginal operator which commutes with $L_m - L_0$ up to boundary terms. In the free Bose case of section 1 this means that we have a tachyon. Suppose that we have other conformally covariant fields beside X^μ. Then in the dual amplitude (1.35) we can replace $V_k(z)$ by

$$\mathcal{V}_k^\alpha(z) = V_k(z)\, \mathcal{O}_\alpha(z) \tag{3.17}$$

where $\mathcal{O}_\alpha(z)$ is one of the covariant operators of the additional field theory. The complete Virasoro generators are, now

$$J_m = \mathcal{L}_m + L_m \quad ; \quad \mathcal{L}_m = \frac{1}{2}\sum_n :a_n^\mu a_{m-n}^\mu: \tag{3.18}$$

where L_m represent the generators of the additional fields. The ghost elimination only requires that $\mathcal{V}$ has dimension 1 with respect to J_m, namely

$$\frac{k^2}{2} = 1 - \delta_\alpha \tag{3.19}$$

δ_α is the weight of $\mathcal{O}_\alpha$. For each δ_α, $\mathcal{O}_\alpha$ we will have a whole family of Regge trajectories, formula (3.19) describing the emission of its lightest members. If we want to avoid tachyon we will like to have

$$\delta_\alpha \geqslant 1 \tag{3.20}$$

and this condition selects the conformally invariant field theories for which all covariant fields have weights larger than one. From the view point of critical systems, this condition is unusual since it means that the corresponding two point function has a Fourier transform which has at most a logarithmic singularity at zero momentum. We shall come back to this later.

In section 1 we argued, by looking at the BRS transformation, that the model based on free bosonic fields is consistent only at $\mathcal{D} = 26$.

Obviously equation (1.68) is now replaced by

$$Q = \sum_m J_m \eta_{-m} + \sum_{mn} m : \rho_m \eta_n \eta_{-m-n} : - \eta_0 \alpha_0 \tag{3.21}$$

By a calculation exactly similar to section 1, one now finds

$$C + \mathcal{D} = 26 \tag{3.22}$$

$$\alpha_0 = 1 \tag{3.23}$$

where C is the central charge of the additional field theory. If $C > 1$ this allows us to go beyond $\mathcal{D} = 26$. A word of caution is needed here. We assume that the additional fields do not themselves bring in new ghosts to be removed. For instance, adding Neveu-Schwarz fields does precisely this and one must then use a larger algebra (the superconformal one) to remove the ghosts. We stick to bosonic string in order to avoid technical complications as much as possible.

At this point is is useful to recall some general properties of representations of the Virasoro algebra

$$[L_n, L_m] = (n-m) L_{n+m} + \frac{C}{12}(n^3-n)\,\delta_{n,-m} \tag{3.24}$$

From the group theory viewpoint, it can be regarded as being in a Weyl-Cartan basis, L_0 being the only operator of the commuting subalgebra, and L_n with $n \neq 0$ being step operators. Hence a highest weight vector will be such that

$$L_0 |\varepsilon, 0\rangle = \varepsilon |\varepsilon, 0\rangle \;;\quad L_n |\varepsilon, 0\rangle = 0, \; n > 0 \tag{3.25}$$

An irreducible representation is characterized by the values of ε and C. The corresponding vector space which is called a Verma module, is spanned by all vectors of the form

$$|\varepsilon, \{m\}\rangle = \prod_{k>0} (L_{-k})^{m_k} |\varepsilon, 0\rangle \tag{3.26}$$

where m_k are arbitrary positive integers. They are eigenstates of L_0

$$L_0 |\varepsilon, \{m\}\rangle = \Big(\varepsilon + \sum_{k>0} k\, m_k\Big) |\varepsilon, \{m\}\rangle \tag{3.27}$$

All eigenvectors with the same eigenvalues are said to belong to the same level N. Kac[19] has considered the matrix of all inner products in a given module which is entirely determined from the Virasoro algebra together with the hermiticity condition

$$L_{-m} = L_m^+ \tag{3.28}$$

Hence it is purely algebraic and only depends upon the values of $\mathcal{E}$ and C. It obviously factorizes into products of finite matrices at each level. Kac has obtained a closed formula for each finite determinant. Define the quantity

$$\mathcal{E}(p,q) \equiv \frac{1}{48}\Big[(13-C)(p^2+q^2) - 24\, pq - 2(1-C) + (p^2-q^2)\sqrt{(C-1)(C-25)}\,\Big] \tag{3.29}$$

where p and q are arbitrary integers. If we consider a highest weight representation with $\mathcal{E} = \mathcal{E}(p,q)$ for some given p,q both larger than zero, the Kac determinant vanishes at the level N= pq. This vanishing shows that the metric of the Verma module need not be positive definite. The unitarity of the representation is thus in question. It is easy to show that the negative values of the highest weights are all excluded. For positive values we note that, for $C > 1$, $\mathcal{E}(p,q)$ is always negative for $p > 1$, $q > 1$, i.e. when it corresponds to a zero of a Kac determinant. Hence, in this region, the Kac determinants never change signs for positive $\mathcal{E}$ and one can show by explicit construction[20] that there exists a unitary representation for all $\mathcal{E} > 0$, $C > 1$. For $C < 1$, on the contrary, there are Kac zeroes for positive $\mathcal{E}$ and the positivity of the metric is not assured. It has been shown that unitary representations only exist for[21]

$$C = 1 - 6/(r+1)r \quad , \quad r \geqslant 3$$

$$\mathcal{E} = \mathcal{E}(p,q) \qquad 1 \leqslant p \leqslant q < r. \tag{3.30}$$

where r,p,q are integers. Hence the allowed values of $\mathcal{E}$ precisely coincide with zeroes of Kac determinants.

Going back to conformally invariant field theories we recall that, given a covariant operator $\mathcal{O}$ with weight δ, it is easy to see that the state

$$\lim_{z \to 0} O_\alpha(z)|0\rangle \tag{3.31}$$

is a highest weight vector with $\varepsilon = \delta_\alpha$. The spectrum of highest weights coincides with the set of conformal weights which, in general, involves more than one value. The representation of the Virasoro algebra is thus reducible since the Hilbert space is the sum of the corresponding Verma modules. The covariant operators are intertweening operators between the different irreducible representations. For arbitrary ε and C one can construct an infinite family of covariant operators with weights given by formula (3.29) for all p and q positive or negative[22] as a natural byproduct of the exact quantum solution of quantum Liouville theory. These operators are not all physical since formula (3.29) is not always positive. This shows nevertheless that the set of critical dimensions must coincide with Kac formula in general. From the above discussion, it is clear that one has to distinguish three regions for the possible values of C.

a) The region $C < 1$.

As we already pointed out, the Kac zeroes occur for positive highest weights in this case. A systematic discussion has been given[17] which uses this fact, and shows the existence of special values of C

$$C = 1 - 6\,\frac{(r+s)^2}{rs} \tag{3.32}$$

where r and s are integers of opposite signs. The virtue of this formula is that then

$$(C-1)(C-25) = 36\,\frac{(r^2-s^2)^2}{r^2 s^2} \tag{3.33}$$

is the ratio of squares of integers. In view of formula (3.30), unitarity is satisfied only for $r+s = -1$. This subseries remarkably reproduces a whole set of standard critical models[21]. In particular for r=3 and 5, one recovers the Ising (or 2-state Potts) model and the 3-state Potts model. A simple calculation shows that if we introduce

$$\sqrt{Q} = 2\cos\left[\frac{\pi}{12}(C-1)\left(1+\sqrt{\frac{C-25}{C-1}}\right)\right] \tag{3.34}$$

we obtain the correct number of spin components, i.e. Q=2,3 for the Ising and the three-state Potts model respectively. The Q-state Potts model can be defined for continuous values of Q if we transform it into the random cluster model. For $C < 1$ there exist various equivalent critical models. In particular the Q-state critical Potts model is equivalent to a Coulomb gas model. For $C < 1$ we have $Q < 4$ and one is in a Coulomb phase. The point Q=4 corresponds to a point of transition of Kosterlitz-Thouless. Above Q=4 one enters into the plasma phase and the transition becomes first order. We shall come back to this below.

b) The region $C > 25$.

This region has some similarities with the region $C < 1$ since in both cases the square root of formula (3.29) is real. A different approach is needed, however, since now the Kac determinants do not vanish for positive ε. The region $C > 1$ is naturally covered by the quantum Liouville field theory since its central charge is given by[23)]

$$C = 1 + 3/\hbar \qquad (3.35)$$

where $\hbar$ is the Planck constant. The region $C > 25$ corresponds to $\hbar < 1/8$, i.e. to the weak coupling regime of Liouville theory which is connected to the semi-classical limit $\hbar \sim 0$. In the exact quantum solution[23)], special values of C were again found

$$C = 1 + 6(N+1)^2/N \qquad (3.36)$$

They can be put under the form of equation (3.32) continued to r=N and s=1. The spectrum of weights is again given by formula (3.29) with

$$\varepsilon = \varepsilon(1, 2m - N) \;, \quad 0 \leq m \leq \nu \qquad N = 2\nu + 1$$

$$\varepsilon = \varepsilon(1, 2m - 1 - N) \;, \quad 1 \leq m \leq \nu \qquad N = 2\nu \qquad (3.37)$$

It is easily checked that all these values are larger than 1, and condition (3.20) is satisfied. The associated string model has no tachyon. However formula (3.22) shows that $\mathcal{D} < 1$ (!) so that one has not gained much from this viewpoint. The existence of conformally invariant field theories for these values of C does however suggest that there are new critical models. These are models of a new type since all the known critical models have $C < 1$, and a spectrum of conformal weights between 0 and 1. The unusual feature of the new models is, as we already pointed out, that the two-point functions are at most logarithmically divergent. The experimental feature of the transition is thus rather different from the standard ones.

c) The region $1 < C < 25$.

In this case formula (3.29) gives complex values except when $p=\pm q$. The choice $p=q$ is unacceptable except for $p=q=1$, since it leads to negative $\mathcal{E}$. For three special values of C

$$C = 7 \; , \; 13 \; , \; 19 \tag{3.38}$$

local fields have been constructed[24] such that the spectrum of weights is given by formula (3.29) for

$$p = -q = 1,2,\ldots \tag{3.39}$$

Condition (3.20) is again satisfied and the associated string theory has no tachyon. Condition (3.22) leads to

$$\mathcal{D} = 19 \; , \; 13 \; , \; 7 \tag{3.40}$$

and there exist new string models for these values of $\mathcal{D}$. From the viewpoint of statistical models, one therefore predicts isolated points of second order phase transition for the above values of C. This may be a bit surprising since for $C > 1$ one is outside of the Coulomb phase. One can directly see, however, that these points must enjoy special properties. Formula (3.34), when continued for $1 < C < 25$, leads to Q complex in general. For the values (3.38) one obtains

$$\sqrt{Q} = 2\cos\left[\frac{\pi}{2} + i\pi\frac{\sqrt{3}}{2}\right]$$

$$\sqrt{Q} = 2\cos\left[\pi + i\pi\right]$$

$$\sqrt{Q} = 2\cos\left[\frac{3\pi}{2} + i\pi\frac{\sqrt{3}}{2}\right] \tag{3.41}$$

and one can verify that the three special values are the only ones for which Q is real even though the argument of the cosine is complex.

Finally we note that, by combining formulae (3.22) and (3.35), one obtains

$$\frac{1}{\hbar} = \frac{25 - \mathcal{D}}{3} \tag{3.42}$$

while formula (2.14) which follows Polyakov's computation[10], leads, instead, to

$$\frac{1}{\hbar_p} = \frac{26 - \mathcal{D}}{3} \tag{3.43}$$

As we pointed out in section 2, this last expression does not take into account the quantum fluctuation of the Liouville field. This explains the discrepancy.

The conclusion of this paragraph is that the study of conformally invariant field theories from the double viewpoint of string theories and critical systems has unravelled an interesting structure. The string theories discussed here are not based on free field theories in two dimensions and, hence, the dual amplitudes are difficult to determine. The common feature of all the new conformally invariant field theories discussed here is the appearance of operators of dimension 1. For the associated string theories, it corresponds to the existence of a massless string state. This fact should play a key role in the complete understanding of these models. On the other hand, we must say that, at the present time, the relevance of these new models to particle physics is not yet clear. The supersymmetric version of the present discussion has been worked out in all details[25].

REFERENCES

1 J.-L. Gervais, B. Sakita, Nucl.Phys. B34 (1971) 832.

2 V. Alessandrini, D. Amati, M. Le Bellac, D. Olive, Phys.Rep. 1C (1971) 269.

3 J.H. Schwarz, Phys.Rep. 8C (1973) 269.

4 S. Mandelstam, Phys.Rep. 13C (1974) 259.

5 J. Scherk, Rev.Mod.Phys. 47 (1975) 123.

6 J. Schwarz, Phys.Rep. 89 (1982) 223.

7 The First 15 Years of Superstring Theories, ed. by J. Schwarz, World Scientific.

8 M. Kato and K. Ogawa, Nucl.Phys. B212 (1983) 443.

9 L. Brink, P. di Vecchia, P. Howe, Phys.Lett. 65B (1976) 471.

10 A.M. Polyakov, Phys.Lett. 103B (1981) 207. For a review, see ref. 14).

11 Proceeding of the International Conference on High Energy Physics, Brighton, 1983, review talk by J.-L. Gervais.

12 I. Gradshteyn, I. Ryzhik, Tables of Integrals, Ac. Press.

13 J.-L. Gervais, B. Sakita, Phys.Rev.Lett. 30 (1973) 716.

14 P. Goddard, J. Goldstone, C. Rebbi, C. Thorn, Nucl.Phys. B56 (1973) 109.

15 S. Mandelstam, Nucl.Phys. B64 (1973) 205.

16 J.-L. Gervais, B. Sakita, Nucl.Phys. B34 (1973) 205.

17 A. Belavin, A. Polyakov, A. Zamolodchikov, Nucl.Phys. B241 (1980) 333.

18 D. Friedan, Les Houches, lecture notes 1982.

19 V. Kac, Proceedings of the International Congress of Mathematicians, Helsinsky, 1978; Lecture Notes in Physics, vol.94, p. 441, Springer Verlag.

20 J.-L. Gervais, A. Neveu, Com.Math.Phys. 100 (1985) 15.

21 D. Friedan, Z. Qiu, S. Shenker, in Vertex Operator in Mathematics and Physics, ed.J. Lepowsky et al., Springer; Phys.Rev.Lett. 52 (1984) 1575.

22 J.-L. Gervais, A. Neveu, Nucl.Phys. B257, FS14 (1985) 59.

23 J.-L. Gervais, A. Neveu, Nucl.Phys. B224 (1983) 329; B238 (1984) 125.

24 J.-L. Gervais, A. Neveu, Phys.Lett. 151B (1985) 271.

25 J.-F. Arvis, Nucl.Phys. B212 (1983) 151; B218 (1983) 303;
O. Babelon, Nucl.Phys. B258 (1985) 680.

FREE STRINGS

Lars Brink
Institute of Theoretical Physics
Chalmers University of Technology
S-412 96 Göteborg
SWEDEN

1. Introduction

In the description of hadronic processes in the 1960's, properties of the S-matrix were studied. A key ingredient was the occurrence of poles describing physical, point-like states. These studies led via Regge theory to the Veneziano (dual) model[1)], which describes an infinity of bosonic states lying on linear Regge trajectories (spin vs mass^2). Shortly after the discovery of this model, Nambu[2)], Nielsen[3)] and Susskind[4)] independently made a most significant observation, namely that the model, in fact, describes the scattering of one-dimensional objects, strings.

This is a rather appealing picture of a meson; a string connecting a quark and an antiquark. Somehow it was believed by many that the missing parts in the Veneziano model were the quarks at the ends.

The model was quite successful as a phenomenological model. It was, however, soon realized by Virasoro[5)] that only by choosing a specific non-physical (from a hadronic point of view) intercept for the leading Regge trajectory, the model possesses a huge symmetry, corresponding to an infinite algebra, the Virasoro algebra. Such a symmetry was needed to avoid negative-norm intermediate states in the amplitudes, and was eventually proven to be enough[6)]. To understand this symmetry Nambu[7)] (and later Hara[8)] and Goto[9)]) constructed an action principle for a string, invariant under reparametrizations of the world-surface traced out by the string. He could then show that this symmetry lead to the Virasoro algebra. The Nambu action was eventually completely quantized by Goldstone, Goddard, Rebbi and Thorn[10)], who both quantized it covariantly reproducing the description of physical states in the Veneziano model and non-covariantly in the light-cone gauge, where all states explicitly have positive norm.

In the meantime Ramond[11)] had generalized the Virasoro algebra to also include anticommuting generators, in fact laying the ground for supersymmetry. This lead to a dual model with fermions. Soon after, Neveu and Schwarz[12)] found another bosonic model realizing such a symmetry. Eventually it was shown that the Ramond fermions fitted into the Neveu-Schwarz model building up one model with both bosons and fermions[13)].

The "spinning string" corresponding to the Ramond-Neveu-Schwarz model had to await the advent of supergravity before its complete action was constructed[14)], but the basic structure became known long before that[15)].

The string formalisms showed clearly that the quantum theories are consistent only if the dimension of space-time takes specific (critical) values, d=26 for the bosonic string, and d=10 for the spinning string. This was an indication that string models should not be used for hadron physics, although we might still today hope to find a model with d=4. Early on Neveu and Scherk[16)] observed that the scattering amplitudes for the massless spin-1 particles are the ones of a Yang-Mills particle, and later on Scherk and Schwarz[17)] showed that the massless spin-2 particle interacts appropriately to be defined as a graviton. They then made the bold suggestion that perhaps string models should be regarded as unified models including gravity!

The Ramond-Neveu-Schwarz model considered in this new light is still not a consistent model. It contains tachyons and a detailed study of some of the couplings would reveal inconsistencies. Gliozzi, Olive and Scherk[18)], however, pointed out that the tachyons could all be eliminated by making a certain projection on the spectrum. The corresponding model, the "superstrings" have subsequently been constructed by Green and Schwarz, to some extent in collaboration with me[19)]. A most beautiful picture has emerged. The models most probably are consistent quantum theories including gravity. The interactions among these strings are quite simple and unique and have a clear geometric understanding. By demanding no anomalies just a few models are consistent[20)]. In this context the "heterotic string"[21)] was constructed combining features from the bosonic string and the superstrings. Also the phenomenological prospects for strings are quite promising[22]. A dream now would be that only one of the models is completely consistent and in such a case Nature better use it!

In this lecture I will concentrate on the description of free strings, showing how to construct quantum theories with tachyonfree spectra. For a description of interactions, see adjoining lectures in this volume by Green, Schwarz and Gross.

2. Bosonic strings

In the description of point-like particles, it is appropriate to start with a free spinless particle. It will follow a one-parameter trajectory $x^\mu(\tau)$. The classical action describing this particle must be independent of how the trajectory is parametrized and is taken to be proportional to the arc-length the particle travels. (I use space-like metric.)

$$S = m\int_{s_i}^{s_f} ds = m\int_{\tau_i}^{\tau_f} d\tau\sqrt{-\dot{x}^2(\tau)} \quad , \qquad (2.1)$$

where m is identified with the mass of the particle and $\dot{x}^\mu = \frac{dx^\mu}{d\tau}$. The fact that the action is reparametrization invariant has a ring of gravity to it. By regarding $x^\mu(\tau)$ as a set of scalar fields in one dimension we can couple them to a metric $g\equiv g_{\tau\tau}$ and write an action just the way scalar fields are coupled to gravity in 4 dimensions.

$$S = -\frac{1}{2}\int d\tau\,\sqrt{g}(g^{-1}\dot{x}^2-m^2) \quad . \qquad (2.2)$$

(Note that the mass term is like a cosmological constant.)

Eliminating the metric field g and inserting its solution back into (2.2) gives (2.1), so the actions (2.1) and (2.2) are equivalent, at least classically. In fact, (2.2) is more general since it allows us to take the limit m=0.

To construct an action for a freely falling string we follow the same lines. We describe the simplest string by just its d-dimensional Minkowski coordinate $x^\mu(\sigma,\tau)$. The parameters σ and τ span the world-sheet traced out by the string when it propagates; σ is space-like and τ time-like. To write an action we can generalize either (2.1) or

(2.2). We choose the second way and introduce a metric $g_{\alpha\beta}$ and its inverse $g^{\alpha\beta}$ on the world-surface. A reparametrization invariant action is then[23)]

$$S = -\frac{1}{2} T \int_0^\pi d\sigma \int_{\tau_i}^{\tau_f} d\tau \sqrt{-g}\, g^{\alpha\beta} \partial_\alpha x^\mu \partial_\beta x_\mu \quad , \tag{2.3}$$

where $g = \det g_{\alpha\beta}$ and T is a proportionality factor (ensuring that x^μ be a length) which will turn out to be the string tension.

The symmetries of Eq. (2.3) are global Poincaré invariance:

$$\delta x^\mu = \ell^\mu{}_\nu x^\nu + a^\mu \quad ; \quad \delta g_{\alpha\beta} = 0 \tag{2.4}$$

and local reparametrization invariance:

$$\delta x^\mu = \xi^\alpha \partial_\alpha x^\mu$$

$$\delta g_{\alpha\beta} = \xi^\gamma \partial_\gamma g_{\alpha\beta} + \partial_\alpha \xi^\gamma g_{\gamma\beta} + \partial_\beta \xi^\gamma g_{\alpha\gamma} \tag{2.5}$$

together with local Weyl invariance:

$$\delta g_{\alpha\beta} = \Lambda g_{\alpha\beta} \quad ; \quad \delta x^\mu = 0 \quad . \tag{2.6}$$

Classically, one can eliminate $g_{\alpha\beta}$ from Eq. (2.3) by solving its field equations algebraically and substituting this solution back in the action. The resulting expression is the Nambu-Hara-Goto action[7)], namely the area of the world-sheet, (definitely a reparametrization invariant expression). Quantum mechanically the elimination of g involves performing a path integral. In general an extra "Liouville mode" is left over (the Weyl invariance is broken) except in the special case of d=26[24)]. Whether or not it is possible to make sense of the bosonic string theory for d<26 is still not completely settled. For the rest of these lectures we will only consider the string theories in their critical dimensions.

Note that we are not allowed to introduce a cosmological constant this time, if we insist on the Weyl invariance (2.6).

In order to quantize the system properly we will perform a hamiltonian treatment of the system. We first compute the canonically conjugate momenta.

$$p^{\alpha\beta} = \frac{\delta L}{\delta \dot{g}_{\alpha\beta}} = 0 \tag{2.7}$$

$$p_\mu = \frac{\delta L}{\delta \dot{x}^\mu} = -T\sqrt{-g}\; g^{\alpha 0}\partial_\alpha x^\mu \quad . \tag{2.8}$$

Eq. (2.7) defines a set of primary constraints

$$\psi^{\alpha\beta} = p^{\alpha\beta} = 0 \quad . \tag{2.9}$$

Computing the hamiltonian density we find

$$H = -\frac{1}{2g^{00}}\{(-g)^{-1/2}\frac{1}{T}p^2 + 2g^{01}x'\cdot p + T(-g)^{-1/2}x'^2\} \quad , \tag{2.10}$$

where $x' = \frac{\partial x}{\partial \sigma}$.

Following the analysis of Dirac[25)] for constrained systems we define the total hamiltonian as

$$H = \int_0^\pi d\sigma(H + A_{\alpha\beta}p^{\alpha\beta}) \quad , \tag{2.11}$$

where $A_{\alpha\beta}$ are arbitrary coefficients. We introduce the Poisson brackets

$$\{g_{\alpha\beta}(\sigma),\; p^{\gamma\delta}(\sigma')\}_\tau = \delta^{\gamma\delta}_{\alpha\beta}\,\delta(\sigma-\sigma') \tag{2.12}$$

$$\{x^\nu(\sigma),\; p_\mu(\sigma')\}_\tau = \delta^\nu_\mu \delta(\sigma-\sigma') \quad . \tag{2.13}$$

If we now check the time-dependence of the constraints by checking the Poisson bracket of H with $p^{\alpha\beta}$, we find two secondary constraints

$$\phi_1 = \frac{1}{T^2}p^2 + x'^2 = 0 \tag{2.14}$$

$$\phi_2 = \frac{1}{T}x'\cdot p = 0 \quad . \tag{2.15}$$

The total hamiltonian is now the sum of all constraints. The next step is then to check the algebra of constraints.

$$\{\psi^{\alpha\beta}, \phi_i\} = 0 \tag{2.16}$$

$$\{\phi_1(\sigma), \phi(\sigma')\} = \frac{1}{T}(\phi_2(\sigma) + \phi_2(\sigma'))\partial_{\sigma'}\delta(\sigma-\sigma') \tag{2.17}$$

$$\{\phi_1(\sigma), \phi_2(\sigma')\} = -\frac{1}{T}(\phi_1(\sigma) + \phi_1(\sigma'))\partial_{\sigma'}\delta(\sigma-\sigma') \tag{2.18}$$

$$\{\phi_2(\sigma), \phi_2(\sigma')\} = \frac{1}{T}(\phi_2(\sigma) + \phi_2(\sigma'))\partial_{\sigma}\delta(\sigma-\sigma') \quad . \tag{2.19}$$

The algebra closes and hence all constraints are first class. It is now straightforward to quantize by letting $\{A,B\} \to -\frac{i}{\hbar}[A,B]$.

At this point there are two alternative approaches to studying the quantum theory. The first one is covariant quantization, where we choose an orthonormal gauge by

$$g_{\alpha\beta} = \eta_{\alpha\beta} \quad . \tag{2.20}$$

This only fixes the gauge such that it allows us to eliminate the $\psi^{\alpha\beta}$-constraints. The remaining ones are kept. The resulting hamiltonian is: (We now choose $T = \frac{1}{\pi}$ for simplicity)

$$H = \frac{1}{2\pi}\int_0^\pi d\sigma[\pi^2 p^2 + x'^2] \tag{2.21}$$

and

$$p^\mu(\sigma) = \frac{1}{\pi}\dot{x}^\mu(\sigma) \quad . \tag{2.22}$$

The equations of motion are:

$$\ddot{x}^\mu - x''^\mu = 0 \quad . \tag{2.23}$$

In deriving them we have assumed boundary conditions such that no surface terms arise. There are then two choices:

(i) open string, $x'^{\mu}(\sigma=0)=x'^{\mu}(\sigma=\pi)=0$
(ii) closed strings, x^{μ} periodic in σ.

The covariant quantization entails using an indefinite-metric Hilbert space obtained from the commutator

$$[x^{\mu}(\sigma,\tau),\ p_{\nu}(\sigma',\tau)] = i\ \delta^{\mu}_{\nu}\ \delta(\sigma-\sigma') \quad . \tag{2.24}$$

A solution to the equations of motion is in the open string case

$$x^{\mu}(\sigma,\tau) = x^{\mu} + p^{\mu}\tau + i \sum_{n\neq 0} \frac{\alpha_n{}^{\mu}}{n} \cos n\sigma\ e^{-in\tau} \quad . \tag{2.25}$$

The canonical commutator (2.24) amounts to

$$[\alpha_m{}^{\mu},\ \alpha_n{}^{\nu}] = m\ \delta_{m+n,0}\ \eta^{\mu\nu} \quad , \tag{2.26}$$

i.e. we have an infinite set of harmonic oscillators, where the time-components generate negative-norm states. The remaining constraints now have to be imposed on the states of the Hilbert space. Let us write the constraints as

$$\phi_{\pm} = \phi_1 \pm 2\phi_2 \quad . \tag{2.27}$$

Introducing (2.25) we find

$$\phi_{\pm} = \phi(\tau\pm\sigma) \tag{2.28}$$

$$\phi(\sigma) = (\sum_{n=-\infty}^{\infty} \alpha_n{}^{\mu} e^{-in\sigma})^2 \tag{2.29}$$

with ($\alpha_0{}^{\mu}\equiv p^{\mu}$).

Imposing the constraints at $\tau=0$ $\phi(\sigma)$, $-\pi\leq\sigma\leq\pi$, is the full constraint function. Consider its Fourier modes

$$L_n = \int_{-\pi}^{\pi} d\sigma\ e^{in\sigma}\phi(\sigma) = \frac{1}{2} \sum_{m=-\infty}^{\infty} \alpha_{n-m}\cdot\alpha_m \tag{2.30}$$

$$L_0 = \frac{1}{2}\alpha_0{}^2 + \sum_{n=1}^{\infty} \alpha_{-n}\cdot\alpha_n \quad . \qquad (2.31)$$

In the quantum case they become operators and one has to normal order to obtain well-defined expressions. Then these operators satisfy the Virasoro algebra (with a central change found by Weis[26])

$$[L_n, L_m] = (n-m)L_{n+m} + \frac{d}{12}(n^3-n)\delta_{n+m,0} \quad . \qquad (2.32)$$

The Hilbert space of states is defined by

$$\alpha_n{}^{\mu}\ |0\rangle = 0 \quad , \qquad n>0 \quad .$$

The constraints are imposed as in the Gupta-Bleuler quantization of QED. We demand

$$\langle phys \mid L_n \mid phys'\rangle = 0 \quad , \qquad n\neq 0 \qquad (2.33)$$

i.e. $L_n\ |phys\rangle = 0 \quad , \qquad n>0 \quad .$

The zero'th component has to satisfy

$$(L_0-1)\ |phys\rangle = 0 \quad . \qquad (2.34)$$

We will discuss this condition at length later.

These conditions on physical states are identical to the conditions that physical states satisfy in the Veneziano model as found by Virasoro[5].

The alternative method used is to quantize in the light-cone gauge[10]. In this case the gauge is specified completely. Introduce light-cone coordinates

$$x^{\pm} = \frac{1}{\sqrt{2}}(x^0 \pm x^{d-1}) \ , \quad x^i \quad , \quad i=1,\ldots,d-2 \quad . \qquad (2.35)$$

We impose the gauge choices (2.20) together with the choices

$$x^+(\sigma,\tau) = x^+ + p^+\tau \tag{2.36}$$

$$\int_0^\sigma d\sigma' p^+(\sigma',\tau) = \frac{1}{\pi} p^+\sigma \quad . \tag{2.37}$$

The latter gauge choices show that in this gauge each point along the string carry the same "light-cone time" x^+ and the same p^+.

The constraints (2.14) and (2.15) can now be completely solved.

$$\dot{x}^- = \frac{1}{2p^+}(\dot{x}^{i^2} + x'^{i^2}) \tag{2.38a}$$

$$x'^- = \frac{1}{p^+} \dot{x}^i x'^i \quad , \tag{2.38b}$$

which can be integrated to give $x^-(\sigma,\tau)$ in terms of the transverse coordinates $x^i(\sigma,\tau)$, p^+ and a single integration constant x^-. The equations of motion are

$$\ddot{x}^i - x''^i = 0 \quad . \tag{2.39}$$

The dynamics of the string in the light-cone gauge can be completely described by the action

$$S = -\frac{1}{2\pi}\int_0^\pi d\sigma \int d\tau \; \eta^{\alpha\beta}\partial_\alpha x^i \partial_\beta x^i \quad . \tag{2.40}$$

The action although not Lorentz covariant must be Poincaré-invariant. The transformations are <u>non-linearly</u> realized. Before giving these transformations, let us check the dynamical content of the action. The canonically conjugate momentum density is

$$p^i(\sigma,\tau) = \frac{1}{\pi}\dot{x}^i(\sigma,\tau) \quad . \tag{2.41}$$

As in the previous case we have two choices of boundary condition

(i) $x'^i(\sigma=0,\pi) = 0$, open strings, solution:

$$x^i(\sigma,\tau) = x^i + p^i\tau + i \sum_{n\neq 0} \frac{1}{n} \alpha_n{}^i \cos n\sigma \, e^{in\tau} \quad . \tag{2.42}$$

(ii) x^i periodic, closed strings, solution:

$$x^i(\sigma,\tau) = x^i + p^i\tau + \frac{i}{2} \sum_{n\neq 0} \frac{1}{n}(\alpha_n{}^i e^{-2in(\tau-\sigma)} + \tilde{\alpha}_n{}^i e^{-2in(\tau+\sigma)}) \quad . \tag{2.43}$$

It is straightforward to quantize the theory

$$[x^i(\sigma,\tau), p^j(\sigma',\tau)] = i\, \delta^{ij}\delta(\sigma-\sigma') \quad . \tag{2.44}$$

The classically free parameters now become operators satisfying

$$[\alpha_n{}^i, \alpha_m{}^j] = n\, \delta_{n+m,0}\, \delta^{ij} \tag{2.45}$$

$$[\tilde{\alpha}_n{}^i, \tilde{\alpha}_m{}^j] = n\, \delta_{n+m,0}\, \delta^{ij} \quad . \tag{2.46}$$

Again we find an infinity of harmonic oscillators. The Hilbert space of states is constructed by introducing a vacuum $|0\rangle$ and demanding

$$\alpha_n{}^i\ |0\rangle = 0 \quad , \qquad n>0 \tag{2.47a}$$

$$\tilde{\alpha}_n{}^i\ |0\rangle = 0 \quad , \qquad n>0 \quad . \tag{2.47b}$$

It is obvious that the Hilbert space only consists of positive-norm states.

In order to construct the Poincaré generators, one starts with the covariant (classical) ones. Then one adds a gauge transformation to stay in the gauge and introduces finally the solution (2.38). The resulting representation is then[10)]

$$p^+ = p^+ \tag{2.48a}$$

$$p^i = \int_0^\pi p^i(\sigma)d\sigma \tag{2.48b}$$

$$p^- = \frac{1}{\pi}\int_0^\pi d\sigma\, \dot{x}^- = \frac{1}{2\pi p^+}\int_0^\pi d\sigma[\pi^2 p^{i^2} + x'^{i^2}] \tag{2.48c}$$

$$j^{ij} = \int_0^\pi d\sigma(x^i p^j - x^j p^i) \tag{2.49a}$$

$$j^{+i} = \int_0^\pi d\sigma(x^+ p^i - x^i p^+) \tag{2.49b}$$

$$j^{+-} = x^+ p^- - x^- p^+ \tag{2.49c}$$

$$j^{-i} = \int_0^\pi d\sigma(x^-(\sigma)p^i(\sigma) - x^i(\sigma)p^-(\sigma)) \quad , \tag{2.49d}$$

where $x^-(\sigma)$ is solved from (2.38b) and $p^-(\sigma)$ is the integrand of p^- (2.48c).

In the quantum case we symmetrize the generators in order to keep hermiticity. This affects the checking of the commutator $[j^{i-}, j^{j-}]=0$, since j^{i-} is cubic. A detailed computation shows that the commutator is satisfied only if d=26[10] (the critical dimension).

Inserting the solution (2.42) into p^- (2.48c) we obtain

$$p^- = \frac{p^{i^2}}{2p^+} + \frac{1}{p^+}\sum_{n=1}^{\infty} \alpha_{-n}^i \alpha_n^i \tag{2.50}$$

i.e.

$$p^2 = -2\sum_{n=1}^{\infty} \alpha_{-n}^i \alpha_n^i = -m^2 \quad . \tag{2.51}$$

Reintroduce the dimensionful constant as $\alpha' = \frac{1}{2\pi T}$, the Regge slope. Then the correct expression is

$$\alpha' m^2 = \sum_{n=1}^{\infty} \alpha_{-n}^i \alpha_n^i \quad . \tag{2.52}$$

The mass-squared is hence built up by an infinite set of "harmonic oscillator energies"[27]. For the quantum case each oscillator will contribute an energy due to zero-point fluctuations (which is due to the symmetrization above). The lowest mass level will then be

$$\alpha' m_0^2 = \frac{d-2}{2} \sum_{n=1}^{\infty} n \quad . \tag{2.53}$$

This is clearly a divergent sum which must be regularized. We do so by comparing the sum to the Riemann ζ-function[28]

$$\zeta(s) = \sum_{n=1}^{\infty} n^{-s} \quad . \quad \mathrm{Re}\ s > 1 \quad . \tag{2.54}$$

This is a function that can be analytically continued to s = -1 and

$$\zeta(-1) = -\frac{1}{12} \quad . \tag{2.55}$$

In this way the infinite series has been regularized into

$$\alpha' m_0^2 = -\frac{d-2}{24} \quad . \tag{2.56}$$

Hence this term should be included in p^-. It is also the reason for the -1 in (2.34).

There is also a more standard way[27] of obtaining this result by adding counterterms to the action (2.40), which amounts to renormalizing the speed of light, (which, of course, is another parameter of the theory). It is quite attractive that it is the quantum theory that demands a finite speed of light.

The really important consequence of Eq. (2.56) is that the lowest state is a tachyon. This really means that an interacting theory based on this bosonic string will not make sense. Also since the tachyon is the scalar state with no excitations of the higher modes, I find it hard to believe that there exists a consistent truncation in which the tachyon is left out.

Let us now redo the last analysis for closed strings. Inserting the solution (2.43) we find

$$\frac{\alpha'}{2} m^2 = \sum_{n=1}^{\infty} (\alpha_{-n}^{i}\alpha_{n}^{i} + \tilde{\alpha}_{-n}^{i}\tilde{\alpha}_{n}^{i}) \equiv N + \tilde{N} \quad . \tag{2.57}$$

Also in this sector we find a tachyon. For closed strings we get a further constraint. Consider again Eq. (2.38b). Integrating it between 0 and σ we find $x^-(\sigma)$. Although $x^-(\sigma)$ is depending on the x^i's we must demand that it represent a component of $x^\mu(\sigma)$ and hence must be periodic. Then

$$\int_0^\pi d\sigma\, x'^- = 0 = \frac{1}{p^+}\int_0^\pi d\sigma\, \dot{x}^i x'^i = \frac{\pi}{p^+}(N-\tilde{N}) \quad , \tag{2.58}$$

i.e. classically $N=\tilde{N}$ and quantum mechanically we impose this condition on the physical states.

The Poincaré algebra spanned by the generators (2.48)-(2.49) with the constraint (2.38b) contains all the information about the strings. This is typical for the light-cone gauge. By finding the non-linear representation we know the complete dynamics of the system, since the Hamiltonian, p^-, is one of the generators. The regrettable thing is that so far we have not found a deductive way to find the generators directly without deriving them from a covariant theory.

By describing the dynamics of free bosonic strings we have found that such a theory is much more constrained than a corresponding theory for point-particles. This is certainly a most wanted property, since one lesson we have learnt from modern gauge field theories for point particles is that seemingly whole classes of theories are theoretically consistent and only experiments can tell which theories Nature is using. In string theories we can entertain the hope that only one model is consistent and this then should be the theory of Nature!

3. Spinning strings

The representation of the Poincaré algebra found in the last section was found to lead to an inconsistent theory. To obtain a consistent one, we need to change the expression for p^- (2.50) so as to avoid tachyons. The most natural thing is to introduce a set of anti-commuting harmonic oscillators such that their zero-point fluctuations compensate the ones from the commuting oscillators.

The first problem to solve is to determine in which representation of the transverse symmetry group SO(d-2), the new oscillators should be chosen. The x-coordinates belong to the vector representation. We can always try this representation also for the new set, which we shall do first, but we should keep in mind, that for certain values of d-2, there are other representations with the same dimension as the vector one.

Consider hence a set of anticommuting harmonic oscillators d_n^i satisfying

$$\{d_m^i, d_n^j\} = \delta_{n+m,0}\delta^{ij} \quad . \tag{3.1}$$

If the relevant mass formula for open strings is

$$\alpha' p^2 = -\left(\sum_{n=1}^{\infty} \alpha_{-n}^i \alpha_n^i + \sum_{n=1}^{\infty} n\, d_{-n}^i d_n^i\right) \quad , \tag{3.2}$$

we can deduce p^- from this expression. To write it in a coordinate basis we introduce two normal-mode expansions

$$\lambda^{1i} = \sum_{n=-\infty}^{\infty} d_n^i\, e^{-in(\tau-\sigma)} \tag{3.3}$$

$$\lambda^{2i} = \sum_{n=-\infty}^{\infty} d_n^i\, e^{-in(\tau+\sigma)} \quad , \tag{3.4}$$

such that

$$\{\lambda^{Ai}(\sigma,\tau), \lambda^{Bj}(\sigma',\tau)\} = \pi\delta^{AB}\delta^{ij}\delta(\sigma-\sigma') \quad . \tag{3.5}$$

The generator p^- can then be written (classically) as

$$p^- = \frac{1}{2\pi p^+}\int_0^{\pi} d\sigma(\pi^2 p^{i^2} + x'^{i^2} - i\lambda^{1i}\lambda'^{1i} + i\lambda^{2i}\lambda'^{2i}) \quad . \tag{3.6}$$

Before trying to construct the remaining generators let us consider the dynamics following from (3.6). Since $p^- \sim i\frac{\partial}{\partial x^+} = \frac{1}{p^+}(i\frac{\partial}{\partial\tau}) \sim \frac{1}{p^+} H$, the corresponding action is

$$S = -\frac{1}{2\pi}\int d\tau \int_0^{\pi} d\sigma[\eta^{\alpha\beta}\partial_\alpha X^i \partial_\beta X^i + i\bar{\lambda}^i \rho^\alpha \partial_\alpha \lambda^i] \quad , \tag{3.7}$$

where we combine the two λ^i's into two-dimensional 2-component spinors and use the Majorana representation for the 2×2 Dirac matrices, here called ρ^α. To get a consistent theory the surface terms obtained upon variation of (3.7) must be zero. In the case of open strings the boundary conditions for the λ's are

$$\lambda^{1i}(0,\tau) = \lambda^{2i}(0,\tau) \tag{3.8}$$

$$\lambda^{1i}(\pi,\tau) = \begin{cases} \lambda^{2i}(\pi,\tau) & (3.9a) \\ -\lambda^{2i}(\pi,\tau) & (3.9b) \end{cases} \quad .$$

In fact we have two choices. The first choice together with equations of motion gives the solutions (3.3) and (3.4). In this sector we know that there are no tachyons. The other choice results in expansions

$$\lambda^{1i} = \sum_r b_r{}^i e^{-ir(\tau-\sigma)} \tag{3.10}$$

$$\lambda^{2i} = \sum_r b_r{}^i e^{-ir(\tau+\sigma)} \quad , \tag{3.11}$$

where the index r takes all half-integer values. The b's satisfy the anticommutators

$$\{b_r{}^i, b_s{}^j\} = \delta_{r+s,0}\, \delta^{ij} \quad . \tag{3.12}$$

The (classical) mass-shell condition now reads

$$\alpha' m^2 = \sum_{n=1}^{\infty} \alpha_{-n}{}^i \alpha_n{}^i + \sum_{r=1/2}^{\infty} r b_{-r}{}^i b_r{}^i \quad . \tag{3.13}$$

Computing the contributions from the zero-point fluctuations we find

$$\alpha' m_0^2 = \frac{d-2}{2}\left[\sum_{n=1}^{\infty} n - \sum_{n=1}^{\infty} (n-1/2)\right]$$

$$= \frac{d-2}{2} \sum_{n=1}^{\infty} n(1-\frac{1}{2}+1)$$

$$\to -\frac{d-2}{16} \quad , \tag{3.14}$$

when the sum is renormalized. Again we find tachyons! Hence this sector of the model is unphysical and basing an interacting theory upon it would lead to inconsistencies.

The states we have discovered spanned by the d- and b-oscillators together with the α's are in fact the spectrum of the Ramond[11)]-Neveu-Schwarz[12)] model. The states constructed out of d-oscillators all have to transform as fermions and constitute the states of the Ramond sector, while the ones constructed out of b-modes, which are bosonic, constitute the Neveu-Schwarz sector. Both sectors are needed in order to have a model with both fermions and bosons.

There is, in fact, a way to truncate the spectrum to avoid tachyons, which can be proven to be consistent with interactions[18)]. Consider the projector

$$P = \frac{1}{2}(1+(-1)^{\Sigma b_{-r} b_r}) \quad . \tag{3.15}$$

By demanding it be zero on physical states

$$P|phys\rangle = 0 \tag{3.16}$$

we obtain a tachyon-free spectrum. A consistent interaction can be set up, if also the spinors are chosen to be of Majorana-Weyl type (the critical dimension is 10 and such a choice is then possible). This leads to the superstrings, which we will describe in the next section.

The Poincaré generators can be constructed and in the quantum case the algebra only works in d=10. We will not give them here, but will discuss them in the next section.

In the case of closed strings there are also two sectors depending on what boundary conditions are chosen. One sector is obtained if the λ's are periodic in σ. This leads to 2 sets of integer moded oscillators $d_n^{\ i}$ and $\tilde{d}_n^{\ i}$. The other sector is obtained by choosing the λ's antiperiodic, which leads to 2 sets of half-integer moded oscillators $b_r^{\ i}$ and $\tilde{b}_r^{\ i}$. This sector has a tachyon. One can also have sectors with $d_n^{\ i}$ and $\tilde{b}_r^{\ i}$ or $\tilde{d}_n^{\ i}$ and $b_r^{\ i}$.

We have discussed the spinning string completely in the light-cone gauge. This is certainly enough as we have seen, but it is often advantageous to have a covariant formalism and in fact there is an action generalizing (2.3). It was done mimicking supergravity[14]. One considers two-dimensional supergravity (on the world-sheet σ,τ) with a "zweibein" $V_\alpha^{\ a}$, related to the metric $g_{\alpha\beta}$ in the usual way and a Majorana Rarita-Schwinger field ψ_a. Then the action is

$$S = -\frac{T}{2}\int d\sigma d\tau\ \eta_{\mu\nu}\ V\{g^{\alpha\beta}\partial_\alpha x^\mu \partial_\beta x^\nu + i\ V_a^{\ \alpha}\ \bar{\lambda}^\mu \rho^a \partial_\alpha \lambda^\nu$$

$$+ 2\ V_a^{\ \alpha} V_b^{\ \beta}\ \bar{\psi}_\alpha \rho^b \rho^a \lambda^\mu (\partial_\beta x^\nu + \frac{1}{2}\bar{\lambda}^\nu \psi_\beta)\} \quad . \tag{3.17}$$

This is an action with a lot of symmetry. Its physics is just the physics of free field theory although it contains interaction terms. It is not unlikely that this action could be used in other fields of physics if properly interpreted.

4. Superstrings

In the last section we added anticommuting degrees of freedom to cancel zero-point fluctuation. They transformed as the vector representation of SO(d-2). For d=3,4,6 and 10 we could also choose the lowest spinor representation, since it has the same dimension as the vector one. Since SO(d-2) is a compact group, the scalar product of two spinors is just the contracted sum as for vectors (which is the only product used in sect. 3). We can, then, take over all work in sect. 3. We only make the substitution

$$\lambda^{Ai} \rightarrow S^{Aa} \quad , \tag{4.1}$$

where A still is a 2-component spinor index and a is a d-2-component spinor index. This leads to the superstring theory, which was discussed in another formulation in the last section.

The action for the superstring theory is then[29)]

$$S = -\frac{1}{2\pi}\int d\tau \int_0^{\pi} d\sigma[\eta^{\alpha\beta}\partial_\alpha x^i \partial_\beta x^i + i\,\bar{S}^a \rho^\alpha \partial_\alpha S^a] \quad . \tag{4.2}$$

We know that this action leads to a sector (for open strings) which starts with massless particles as the lowest lying states. In this sector the solution to the equations of motion is

$$S_a^{\ 1} = \sum_{n=-\infty}^{\infty} S_n^{\ a}\, e^{-in(\tau-\sigma)} \tag{4.3}$$

$$S_a^{\ 2} = \sum_{n=-\infty}^{\infty} S_n^{\ a}\, e^{-in(\tau+\sigma)} \tag{4.4}$$

with the anticommutation rule

$$\{S_a^{\ A}(\sigma,\tau),\ S_b^{\ B}(\sigma',\tau)\} = \pi\ \delta_{ab}\delta^{AB}\ \delta(\sigma-\sigma') \tag{4.5}$$

$$\{S_n^{\ a},\ S_m^{\ b}\} = \delta_{n+m,0}\delta^{ab} \quad . \tag{4.6}$$

The operators $S_{-n}^{\ a}$ with n positive are creation operators. They will take a bosonic state that is acts on to a fermionic one. Hence this sector will contain both bosons and fermions, in fact equally many of each kind at each mass level, building up supermultiplets at each level (as we will soon prove).

The other sector which follows by using the other set of boundary conditions corresponding to (3.8) and (3.9b), we know has tachyons. Furthermore the fermionic oscillators will be half-integer moded and there will not be an equal number of bosons and fermions at each mass level, thus ruining the possibility to have a supersymmetry. Since the first sector contains all we want, we simply decree, that we only use the boundary conditions (3.8) and (3.9a).

Similarly for closed strings we decree that we only use periodic boundary conditions. This leads to the following solutions to the equations of motion

$$S_a^{\,1} = \sum_{n=-\infty}^{\infty} S_n^{\,1a}\, e^{-2in(\tau-\sigma)} \tag{4.7}$$

$$S_a^{\,2} = \sum_{n=-\infty}^{\infty} S_n^{\,2a}\, e^{-2in(\tau+\sigma)} \quad . \tag{4.8}$$

The arduous task now is to check if there is a representation of the Poincaré algebra spanned on this string theory. In fact there is[30)], and the marvellous fact is that it can also be extended to a super-Poincaré algebra! Since light-cone supersymmetry[31)] might not be too familiar, let me first review it. The supersymmetry charge Q in ten dimensions decomposes into two SO(8) light-cone spinors $Q_+^{\,a}$ and $Q_-^{\,\dot a}$, where the indices $a,\dot a = 1,\ldots,8$ denote the two inequivalent 8-component spinors of SO(8). The algebra is

$$\{Q_+^{\,a}, Q_+^{\,b}\} = 2p^+\delta^{ab} \tag{4.9a}$$

$$\{Q_-^{\,\dot a}, Q_-^{\,\dot b}\} = 2p^-\delta^{\dot a\dot b} \tag{4.9b}$$

$$\{Q_+^{\,a}, Q_-^{\,\dot b}\} = \sqrt{2}(\gamma_i)^{a\dot b} p^i \tag{4.9c}$$

For further notations, see Appendix.

Let me so write down the representation of the super-Poincaré algebra. Again I stress that there is some guesswork behind the construction of it. For the case of closed strings the algebra turns out to be an N=2 super-Poincaré algebra. The most general algebra is

$$p^+ = p^+ \tag{4.10a}$$

$$p^i = \int_0^{\pi} d\sigma\, p^i(\sigma,\tau) \tag{4.10b}$$

$$p^- = \frac{1}{2\pi p^+}\int_0^{\pi} d\sigma\, [\pi^2 p^{i\,2} + x'^{i\,2} - i(S^1 S'^1 - S^2 S'^2)] \tag{4.10c}$$

$$q_1^{+a} = \sqrt{\frac{2p^+}{\pi}} \int_0^\pi d\sigma\ S_1^{\ a} \tag{4.11a}$$

$$q_2^{+a} = \sqrt{\frac{2p^+}{\pi}} \int_0^\pi d\sigma\ S_2^{\ a} \tag{4.11b}$$

$$q_1^{-\dot{a}} = \frac{1}{\pi\sqrt{p^+}} \int_0^\pi d\sigma (\gamma^i S_1)^{\dot{a}} (\pi p^i - x'^i) \tag{4.11c}$$

$$q_2^{-\dot{a}} = \frac{1}{\pi\sqrt{p^+}} \int_0^\pi d\sigma (\gamma^i S_2)^{\dot{a}} (\pi p^i + x'^i) \tag{4.11d}$$

$$j^{ij} = \int_0^\pi d\sigma [x^i p^j - x^j p^i + \frac{1}{4\pi}(S^1\gamma^{ij}S^1 + S^2\gamma^{ij}S^2)] \tag{4.12a}$$

$$j^{+i} = \int_0^\pi d\sigma (x^+ p^i - x^i p^+) \tag{4.12b}$$

$$j^{+-} = x^+ p^- - x^- p^+ \tag{4.12c}$$

$$j^{-i} = \frac{1}{2} \int_0^\pi d\sigma [\{x^-(\sigma), p^i\} - \{x^i, p^-(\sigma)\} \tag{4.12d}$$

$$- \frac{i}{4\pi\sqrt{\pi p^+}} (S^1\gamma^{ij}S^1(\pi p^j - x'^j) + S^2\gamma^{ij}S^2(\pi p^j + x'^j)) + 4\frac{p^i}{p^+}] \quad , \tag{4.12e}$$

where

$$x'(\sigma) = \frac{\pi}{p^+} p^i x'^i + \frac{i}{2p^+}(S^1 S'^1 + S^2 S'^2) \quad . \tag{4.13}$$

In fact this algebra is enough to cover all known string models (apart from a d=2 model[32]).

(i) Type IIb superstrings: This is the full algebra (4.10)-(4.12) with periodic boundary conditions for the coordinates. Note that this is a chiral model, since the creation operators S^{1a}_{-n} and S^{2a}_{-n} create spinors of only one chirality.

(ii) Type IIa superstrings: The anticommuting coordinate $S_2{}^a$ can instead be chosen to transform as the other spinor representation, $S_2{}^{\dot{a}}$. Nothing is affected in the algebra since S_1 and S_2 are never contracted with each other. S_1 and S_2 cannot be put together as a 2-component spinor, but who cares? This model is not chiral since the spinor states can be combined to Majorana states.

(iii) Type I superstrings: For open strings we must use the boundary conditions corresponding to (2.8) and (3.8) and (3.9a). Then $q_1{}^+=q_2{}^+$ and $q_1{}^-=q_2{}^-$ and the supersymmetry is reduced to an N=1 one. One can also perform this truncation for closed strings.

(iv) The bosonic strings: Put $S^1=S^2=0$. No supersymmetry, of course.

(v) The heterotic string: This string model will be discussed in the next section.

(vi) The spinning string: By performing triality transformations back to vectors λ^i such as $S^{1a}S^{1a} \to \lambda^{1i}\lambda^{1i}$ and $S^1\gamma^{ij}S^1 \to \lambda^{1i}\lambda^{1j}$ and similarly for S^2 one can easily read off the representation for the spinning string. No supersymmetry survives, of course.

Considering the superstring theories I, IIa and IIb we know we have a quantum theory of free strings where the lowest lying states are massless. Next we like to know what is the spin content of these massless states. For the bosonic string one can choose a scalar vacuum state $|0\rangle$ as the ground state. Putting all fermionic oscillators in (4.12) to zero we find this state to be indeed a scalar one. However, in the superstring case there is an extra piece in the zero mode part of J^{ij} (4.12)

$$s_0{}^{ij} = \frac{1}{4}(S_0^1\gamma^{ij}S_0^1 + S_0^2\gamma^{ij}S_0^2) \quad . \tag{4.14}$$

Its effect on a vacuum state will be non-zero in general.

We should also realize that the massless level must be a supermultiplet. Such a one can be constructed from the anticommutator (4.9a) combined with the knowledge of (4.14) realizing that q_+ is linearly realized. The generator $q_+{}^a$ is real. In the 4-dimensional case it is customary to go from an SO(2) description to a U(1) one forming complex generators (with no Lorentz index) which then build up a Clifford algebra and we can define creation and annihilation operators from which we construct the supermultiplet. For d=10 we have so far used an SO(8) covariant notation. To decompose the generators into creation and annihilation operators we must break the covariance into SU(4) × U(1). This is the formalism we must use for the field theories for open strings. Here, however, I will describe an alternative formalism, which uses the full SO(8) covariance. This method can equally well be used in d=4. Consider the zero mode part of J^{ij} (4.12a) which is the relevant part on vacuum states and start with open strings

$$j_0{}^{ij} = \ell^{ij} + \frac{1}{2} S_0 \gamma^{ij} S_0 \equiv \ell^{ij} + s_0{}^{ij} \quad . \tag{4.15}$$

The last term in (4.15) is the spin contribution. If we try to start with a scalar vacuum $|0\rangle$ we need a constraint

$$s_0{}^{ij} \, |0\rangle = 0 \quad . \tag{4.16}$$

By multiplying (4.16) by another S_0 and using (4.16) and Fierz rearrangements we find that $S_0{}^a \, |0\rangle = 0$ and hence $|0\rangle = 0$ unless d=4.

For d=10 we have to try the next simplest thing. We take the vacuum to be a vector $|i\rangle$ with $\langle i|j\rangle = \delta^{ij}$. Then we must insist on a constraint

$$s_0{}^{ij} \, |k\rangle = \delta_{ik} \, |j\rangle - \delta_{jk} \, |i\rangle \quad . \tag{4.17}$$

This time we find when we bang an S_0 on (4.17) that for the general state

$$\psi^{ia} \equiv q_+{}^a \, |i\rangle \quad , \tag{4.18}$$

with the decomposition

$$\psi^{ia} = \tilde{\psi}^{ia} + \frac{1}{8}\gamma^i\gamma^j\psi^j \quad , \tag{4.19}$$

where $\gamma^i\tilde{\psi}^{ia} = 0$

that

$$\tilde{\psi}^{ia} = 0 \quad , \tag{4.20}$$

but that there is no constraint on

$$|\dot{a}> \equiv \frac{1}{8}(\gamma^i q_+)^{\dot{a}}\ |i> \quad . \tag{4.21}$$

with $<\dot{b}|\dot{a}> = p^+\delta_{\dot{a}\dot{b}}$. Checking the Lorentz properties of this state we find that

$$s_0^{ij}\ |\dot{a}> = -\frac{1}{2}(\gamma^{ij})^{\dot{a}\dot{b}}\ |\dot{b}> \quad . \tag{4.22}$$

Hence it transforms properly as a spinor. From the Fierz property (A.7)

$$q_+^a q_+^b = p^+\delta^{ab} + \frac{1}{16}(\gamma^{ij})^{ab}\ q_+\gamma^{ij}q_+ \quad , \tag{4.23}$$

we find that no other massless states can be constructed and the massless sector contains then one vector state and one spinor state which we recognize as the Yang-Mills multiplet in 10 dimensions. We can, of course, let both states transform according to some representation (such as the adjoint one) of some internal group. For type I theories, there is a standard way for including internal symmetry quantum numbers in scattering amplitudes introduced by Chan and Paton[33] a long time ago. What one does is to associate a matrix $(\lambda_i)_{ab}$ with the i'th external string state and multiply the N-point scattering amplitude with a group theory factor $tr(\lambda_1 \ldots \lambda_N)$. When checking that such factors factorize properly so as to not destroy the factorization properties of the scattering amplitudes one finds that the possible internal symmetry groups are SO(n), U(n) or Sp(2n)[34]. Note that the exceptional groups are not possible.

It may be instructive to also check the first excited level. Here one can form 128 boson states $\alpha_{-1}{}^{i}\,|j\rangle$ and $S_{-1}{}^{a}\,|\dot{b}\rangle$ and 128 fermionic states $\alpha_{-1}{}^{i}\,|\dot{a}\rangle$ and $S_{-1}{}^{a}\,|i\rangle$. These form various reducible SO(8) multiplets which can be reassembled into SO(9) representation (since they are massive) using the Lorentz generators. Doing this one finds for the bosons the SO(9) representations $\square\square$ and $\begin{smallmatrix}\square\\ \square\\ \square\end{smallmatrix}$ of dimensions 44 and 84 resp. The fermions form a single 128 dimensional Rarita-Schwinger SO(9) multiplet (like the $\tilde{\phi}^{ia}$ in (4.19)).

In the closed string case the massless spectrum is generated by two supersymmetry generators $q_+{}^1$ and $q_+{}^2$. If both belong to the same SO(8) representation (this means that they have the same chirality in 10 dimensions (type IIb), we can form complex generators

$$q_+{}^a = \frac{1}{\sqrt{2}}\,(q_+{}^1 + iq_+{}^2)^a \quad , \tag{4.24}$$

and hence use $q_+{}^a$ and q_+^{*a} as creation and annihilation operators. For this case we can introduce a scalar vacuum, (which has to be complex, which can be seen from the supersymmetry transformations). Checking $J_0{}^{ij}$ with the $s_0{}^{ij}$ term as in (4.14) we see that the vacuum is indeed a scalar state. Acting with $q_+{}^a$ we can form the following massless supermultiplet

$$\begin{aligned}
|\,0\rangle &\sim \phi \\
q_+{}^a\,|\,0\rangle &\sim \phi^a \\
q_+{}^a q_+{}^b\,|\,0\rangle &\sim \phi^{ab} \\
&\vdots \\
q_+{}^{a_1}\ldots q_+{}^{a_8}\,|\,0\rangle &\sim \phi^{a_1\ldots a_8} \quad .
\end{aligned}$$

It is easily seen that this is a reducible multiplet. To obtain an irreducible representation we impose the conditions

$$(\phi^{a_1a_2\cdots a_{2N}})^* = \frac{1}{(8-2N)!}\,\varepsilon^{a_1a_2\cdots a_8}\,\phi^{a_{2N+1}\cdots a_8} \qquad (4.25)$$

$$(\psi^{a_1a_2\cdots a_{2N+1}})^* = \frac{1}{(7-2N)!}\,\varepsilon^{a_1a_2\cdots a_8}\,\phi^{a_{2N+2}\cdots a_8} \quad . \qquad (4.26)$$

Because of the triality properties of the representations of SO(8) we can rewrite the states with vector indices. We now find the bosonic spectrum:

2 scalars ϕ

2 antisymmetric tensors $A^{ij} \sim (\gamma^{ij})^{ab}\phi^{ab}$

1 self-dual antisymmetric tensor $A^{ijk\ell}$

1 graviton g^{ij} ,

where the last two sets of states follow from $\phi^{a_1\cdots a_4}$. The fermionic spectrum is:

2 spinors ψ^a

2 Rarita-Schwinger states $\tilde{\psi}^{ia}$.

If q_+^1 and q_+^2 have opposite Weyl properties (type IIa) we cannot form creation operators in an SO(8) covariant way. We then have to follow the procedure of the N=1 case. We can start with a tensor state

$$|ij\rangle \sim |i\rangle \otimes |j\rangle \sim \phi^{ij}$$

and generate the states by stuttering the N=1 procedure

$$|a\rangle \otimes |i\rangle \sim \psi^{ai}$$

$$|i\rangle \otimes |\dot{a}\rangle \sim \chi^{\dot{a}i}$$

$$|a\rangle \otimes |\dot{a}\rangle \sim \phi^{a\dot{a}} \quad .$$

The bosonic spectrum is then

1 graviton $g^{ij} = \phi^{(ij)} - \frac{1}{8}\phi^{kk}\,\delta^{ij}$

1 antisymmetric tensor $A^{ij} = \phi^{[ij]}$

1 scalar $\phi = \phi^{ii}$

1 vector $A^{i} = (\gamma^{i})^{a\dot{a}}\,\phi^{a\dot{a}}$

1 antisymmetric tensor $A^{ijk} = (\gamma^{ijk})^{a\dot{a}}\phi^{a\dot{a}}$.

The fermionic spectrum is

2 spinors ψ^{a}, $\chi^{\dot{a}}$

2 Rarita-Schwinger states $\tilde{\psi}^{ai}$, $\tilde{\chi}^{\dot{a}i}$.

This technique could, of course, also have been used in the other case above.

We can impose constraints on the N=2 spectrum to obtain an N=1 spectrum. These closed strings (type I) are important since they will be able to couple to the open strings. To find this spectrum we linearly combine $Q^1+Q^2 = Q$ and use the N=1 technique. The spectrum is then clearly

$|i\rangle \otimes |j\rangle$

$|i\rangle \otimes |a\rangle$

i.e.

1 graviton	g^{ij}
1 antisymmetric tensor	A^{ij}
1 scalar	ϕ
1 spinor	ψ^{a}
1 Rarita-Schwinger state	$\tilde{\psi}^{ai}$.

So far we have only discussed free string states. We could anticipate that all closed string states also belong to some representation of some internal group. However, it will turn out that interaction is only possible if this representation is the trivial one.

Since the strings are extended objects in one dimension there is a possibility that they can carry an intrinsic orientation, like an "arrow" pointing in one direction along its length. When strings are oriented there are two distinct classical states for a single spatial configuration, corresponding to the two possible orientations. An open string is oriented, loosely speaking, if the end points are different, while a closed string is oriented, if one can distinguish a mode running one way around the string from a mode running the other way.

Hence the basic question is whether a string described by $x^{\mu}(\sigma)$, $S^{Aa}(\sigma)$ is the same as one described by $x^{\mu}(\pi-\sigma)$, $S^{Aa}(\pi-\sigma)$. Consider the closed string solution (2.43), (4.7) and (4.8). The replacement $\sigma \to \pi-\sigma$ corresponds to the interchanges $\alpha_n \leftrightarrow \tilde{\alpha}_n$, $S_n{}^1 \leftrightarrow S_n{}^2$. In the type II case the two strings connected by the interchanges are different, while for type I the constraints are just such as to make the two strings the same state. Hence we conclude that type II strings are oriented while type I closed strings are non-oriented. Heterotic strings which have different left-going and right-going modes are evidently oriented. The analysis here is classical but can be taken over to the quantum case, by considering matrix elements of the operators $x^{\mu}(\sigma)$ and $S^{Aa}(\sigma)$. The same conclusions are reached then.

For open strings we argued above that one can allow for an internal (global) symmetry (the remnant of the gauge symmetry in a covariant formalism). The interaction allows for letting the states transform as $\phi^a{}_b$ where the index a and b run over the fundamental representation and its complex conjugate resp. Hence $\phi^a{}_b$ transforms either as the adjoint representation or the singlet one. The intuitive way to interpret this is to say that each end carry a "quark" and an "antiquark". If, hence, the "quarks" are different from the "antiquarks" the string is oriented, otherwise not. Now these strings can join their ends to form closed strings, if they are singlets. Then the U(n) strings will form oriented closed strings, while the SO(n) and Sp(2n) strings will form non-oriented closed strings. But these strings must be of type I, since the open strings are, and hence should be non-oriented. Thus we conclude by this non-rigorous argument that only the gauge groups SO(n) or Sp(2n) are possible[35].

The notation of orientability will be important for perturbation expansions. For oriented strings the interaction must be such that the orientations match up. This will make the perturbation expansion of type I strings different from type II ones.

We have discussed the superstrings so far in the light-cone gauge. This is enough to understand the structure of the theory and to build an interacting theory. However, it would be advantageous to have a covariant formalism. Such an action has been found by Green and Schwarz[36] but it is not yet clear if that action can be quantized covariantly[37].

5. The heterotic string

We saw in the last section several representations embodied in the algebra (4.10)-(4.12). The most economic one is obtained in the bosonic string by writing $x^\mu(\tau,\sigma) = x^\mu(\tau-\sigma) + \tilde{x}^\mu(\tau+\sigma)$ and similarly for p^μ. The algebra then splits up into one piece from the right-moving coordinates $x^\mu(\tau-\sigma)$, $p^\mu(\tau-\sigma)$ and one from the left-moving part. The two parts do not mix and we can freely set one of them to zero. Similarly for the full algebra (4.10)-(4.12) it is straightforward to see that for the terms where S^A couple to x and p, S^1 couple to the right-moving part and S^2 to the left-moving one. Since they are right-moving and left-moving resp., again a consistent truncation can be made by putting, say the left-moving parts to zero. This reduces the algebra to an N=1 supersymmetry.

The heterotic string[21] is now constructed by putting together the algebra from one right-moving superstring constructed as above with the algebra from a 26-dimensional left-moving bosonic string. There is an obvious mismatch in dimensions! The solution to this dilemma is that we only add in the SO(1,9) subalgebra of the full bosonic algebra. Eventually we must interpret the extra coordinates $x^I(\tau,\sigma)$, I=1,...,16 and check what has happened to the SO(16) symmetry left out.

Let us now check the algebra closer. Consider the bosonic algebra. For a left-moving string

$$p^- = \frac{1}{4\pi p^+} \int_0^\pi d\sigma (\pi p^I + x'^I)^2, \qquad (5.1)$$

where we used (4.13). The combination $\pi p^I + x'^I$ is only left-moving. If we now <u>insist</u> that p^I and x'^I both <u>separately</u> are left-moving, we see in the mode expansion that they are the same (up to a π) and we have to change the canonical commutator to

$$[x^I(\sigma,\tau),\ p^J(\sigma',\tau)] = \frac{i}{2}\,\delta(\sigma-\sigma')\delta^{IJ} \tag{5.2}$$

and impose

$$[x^I(\sigma,\tau),\ x^J(\sigma',\tau)] = -\frac{i}{2}\,\pi\,\frac{1}{\partial_\sigma}\,\delta(\sigma-\sigma')\delta^{IJ} \quad , \tag{5.3}$$

where $\frac{1}{\partial_\sigma}\,\delta(\sigma-\sigma') = \varepsilon(\sigma-\sigma')$.

This change does not alter the closure of the rest of the algebra. Hence in the full algebra of the heterotic string

$$p^- = \frac{1}{2\pi p^+}\int_0^\pi d\sigma[\pi^2 p^{i\,2} + x'^{i\,2} - i\,S^a S'^a + 2\pi p^I x'^I] \tag{5.4}$$

and the constraint on $x^-(\sigma)$ is

$$x'^-(\sigma) = \frac{\pi}{p^+}\,p^i x'^i + \frac{i}{2p^+}\,S^a S'^a + \frac{\pi}{p^+}\,p^I x'^I \tag{5.5}$$

with $S \equiv S^1$. The remaining generators of the N=1 super-Poincaré algebra is obtained by putting $S^2=0$ in (4.10)-(4.12).

The action corresponding to the hamiltonian (5.4) is

$$S = -\frac{T}{2}\int_0^\pi d\sigma \int d\tau[\eta^{\alpha\beta}(\partial_\alpha x^i\partial_\beta x^i + \partial_\alpha x^I\partial_\beta x^I) + i\,S^a(\frac{\partial}{\partial\tau} + \frac{\partial}{\partial\sigma})S^a] \tag{5.6}$$

together with constraints

$$\Phi^I = (\frac{\partial}{\partial\tau} - \frac{\partial}{\partial\sigma})x^I = 0 \quad . \tag{5.7}$$

Alternatively a term $\lambda[(\frac{\partial}{\partial\tau} - \frac{\partial}{\partial\sigma})x^I]^2$ can be added to the action.

Siegel[38] has shown that the resulting action possesses a local gauge symmetry, which allows the Lagrange multiplier λ to be gauged away, leaving us with the constraints (5.7).

The canonical structure of the action can be analysed using Dirac's analysis. Taking into account the fact that the constraints (5.7) are second-class, the canonical commutators are found to be (2.44), (4.5) and (5.2).

Consider so a solution to the equation of motion for the x^I's

$$x^I(\tau+\sigma) = x^I + p^I(\tau+\sigma) + \frac{i}{2} \sum_{n\neq 0} \frac{\tilde{\alpha}_n^I}{n} e^{-2in(\tau+\sigma)} \quad . \tag{5.8}$$

However, it is not consistent with periodic boundary conditions in σ! To solve this dilemma we are forced to allow for more general boundary conditions. If the coordinates x^I lie on a hypertorus with radii R then the function $x^I(\sigma,\tau)$ maps the circle $0\leq\sigma\leq\pi$ onto the circle $0 \leq x^I \leq 2\pi R$, and such maps fall into homotopy classes characterized by a winding number L^I that counts how many times x^I wraps around the circle[39]. Then (5.8) is an allowed solution with $p^I=2L^IR$. We are hence forced to consider the extra 16 coordinates x^I to span a hypertorus. This is physically very appealing, since such a hypertorus can be small and we do not need to worry about the coordinates x^I on a macroscopic scale.

Before performing this compactification in detail, let us consider the effect of it on (5.4) and (5.5). Inserting the solution for x^i, S^a and x^I into (5.4) and (5.5) we obtain in analogy to (2.57) and (2.58) ($\alpha'=1/2$)

$$\frac{1}{4} m^2 = N + \tilde{N} -1 + \frac{1}{2} \sum_{I=1}^{16} (p^I)^2 \tag{5.9}$$

and

$$N = \tilde{N} - 1 + \frac{1}{2} \sum_{I=1}^{16} (p^I)^2 \quad , \tag{5.10}$$

where

$$N = \sum_{n=1}^{\infty} (\alpha_{-n}^{\ i}\alpha_n^{\ i} + n\, S_{-n}^{\ a} S_n^{\ a}) \tag{5.11}$$

$$\tilde{N} = \sum_{n=1}^{\infty} (\tilde{\alpha}_{-n}^{\ i}\tilde{\alpha}_n^{\ i} + \tilde{\alpha}_{-n}^{\ I}\tilde{\alpha}_n^{\ I}) \quad . \tag{5.12}$$

In (5.9) and (5.10) the subtraction of -1 is due to the regularization of the contribution from the zero-point fluctuations. Note, that although this contribution is not cancelled, there is no tachyon! It is enough to have the vacuum contributions cancel among the right-moving modes. Furthermore, we find from (5.10) that ${p^I}^2$ must be an even integer.

In the further study, the 16-dimensional hypertorus spanned by the x^I's may be thought of as R^{16} modulo a lattice Γ^{16}, generated by 16 independent basis vectors $e_i^{\ I}(i=1,\ldots,16)$. We choose an even lattice (to be justified shortly) with the length of the e_i's equal to $\sqrt{2}$.

$$x^I = x^I + \sqrt{2}\,\pi \sum_{i=1}^{16} n_i R_i e_i^{\ I} \quad , \tag{5.13}$$

n_i integers and R_i the radii.

For a torus R^{16}/Γ^{16}, the allowed momenta p^I lie on the dual lattice $\tilde{\Gamma}^{16}$ generated by

$$\sum_{I=1}^{16} e_i^{\ I} e_j^{*I} = \delta_{ij} \quad . \tag{5.14}$$

From (5.2) $2p^I$ generates translations, i.e.

$$p^I = \frac{1}{\sqrt{2}} \sum_{i=1}^{16} \frac{m_i}{R_i} e_i^{*I} \quad , \quad m_i \text{ integers} \quad . \tag{5.15}$$

However, we have also seen that $p^I = L^I$, a winding number, where

$$L^I = \frac{1}{\sqrt{2}} \sum_{i=1}^{16} R_i n_i e_i^{\ I} \quad . \tag{5.16}$$

From this we conclude that the allowed states must have momenta, which lie in the intersection of Γ^{16} and $\tilde{\Gamma}^{16}$. Furthermore, reintroducing the Regge slope parameter α' and comparing (5.15) and (5.16) we find

$$R_i \sim \sqrt{\alpha'} \quad . \tag{5.17}$$

This shows that the internal coordinates are small. (The parameter α' is so far a free parameter, which we might hope the theory determines eventually. It is certainly small, say $\sqrt{\alpha'}<10^{-17}$m, since we have not seen any Regge recurrencies, states with N=1 or higher, experimentally.) Finally, since $\tilde{\Gamma}\ \Gamma$, which is even, also $\tilde{\Gamma}$ is even and p^{I^2} will be even integers.

When the interacting case is investigated a further condition is derived. The (Euclidean) two-dimensional world-sheet which describes a one-loop closed string amplitude is a torus. This torus should be symmetric under the interchange of σ and τ, because of the global reparametrization invariance still left in the theory (duality). To achieve this symmetry the lattice has to be self-dual $\Gamma=\tilde{\Gamma}$.

Even self-dual lattices are extremely rare. In fact they only exist in 8n dimensions. In a pioneering work Goddard and Olive[40] showed that in 16 dimensions, there are only two such lattices, $\Gamma_8\times\Gamma_8$ and Γ_{16} which they proposed should be important in physics. Γ_8 is the root lattice of E_8. The construction of Γ_{16} is more complicated. The basis vectors for the SO(32) root lattice can be written in terms of a set of orthonormal vectors $\{u_i\}$, i=1,...,16 as $e_i = u_i - u_{i+1}$, i = 1,...,15, $e_{16}=u_{15}+u_{16}$. This root lattice is not self-dual. However, a self-dual lattice can be constructed from the root lattice by adding points that are multiplets of one of the spinor weights of Spin(32). We can choose the spinor weight to be $s_1 = \frac{1}{2}(u_1-u_2+u_3-u_4+\ldots+u_{15}-u_{16})$, $s_1^2=4$ and take the basis vectors for Γ_{16} to be s_1 and e_i, i=2,...16. Since e_1 is an integer linear combination of the basis vectors, Γ_{16} contains all the points of the root lattice of SO(32) plus additional points that correspond to spinor weights of Spin(32)/Z_2. This is not the same as SO(32). The center pf Spin(32) is $Z_2\times Z_2$. Removing the diagonal combination of the two Z_2 factors eliminates all spinor representations leaving SO(32). Removing one Z_2 factor eliminates one of the two spinor representations of Spin(32) and all representations in the same Z_2 conjugacy class as the spinor (including the vector).

Each of the two 16-dimensional self-dual lattice has 480 vectors of the minimal $\text{length}^2=2$.

Let us check the lowest lying states using (5.9) and (5.10). In the superstring sector the lowest lying states are in analogy to the analysis in sect. 4

$$|i\rangle_R$$

$$|\dot{a}\rangle_R \quad .$$

We must satisfy the condition (5.10). Hence the lowest lying states are

$$|i\rangle_R \otimes \tilde{\alpha}^j_{-1}\,|0\rangle_L \qquad g^{ij},\ A^{ij},\ \phi$$

$$|i\rangle_R \otimes \tilde{\alpha}^I_{-1}\,|0\rangle_L \qquad \text{16 vectors}$$

$$|i\rangle_R \otimes |p^I\rangle_L \qquad \text{480 vectors}$$

$$|\dot{a}\rangle_R \otimes \tilde{\alpha}^i_{-1}\,|0\rangle_L \qquad \tilde{\psi}^{\dot{a}i},\ \psi^{\dot{a}}$$

$$|\dot{a}\rangle_R \otimes \tilde{\alpha}^I_{-1}\,|0\rangle_L \qquad \text{16 spinors}$$

$$|\dot{a}\rangle_R \otimes |p^I\rangle_L \qquad \text{480 spinors}$$

in the notation of sect. 4.

This is the spectrum of N=1 supergravity coupled to a Yang-Mills theory with a gauge group with 496 generators.

Finally we need to investigate what has happened to the symmetry in the 16 compact dimensions. In fact, it is already known from the work of Frenkel and Kac, Segal and Goddard and Olive[41)] that connected to the lattices $\Gamma_8\times\Gamma_8$ and Γ_{16} one can construct representations of the groups $E_8\times E_8$ and $Spin(32)/Z_2$ resp. To this end we construct an operator $E(K^I)$, which acts on left-moving states. $E(K^I)$ represents the generator of G that translates states on the weight lattice by a root vector K^I. The construction is as follows. Consider

$$E(K) = \oint_0 \frac{dz}{2\pi iz} : e^{2iK^I x^I(z)} : C(K) \tag{5.18}$$

with ${K^I}^2=2$, $z=e^{2i(\tau+\sigma)}$.

These operators contain the usual translation operator $e^{2iK^I x^I}$, which translates the internal momenta (= winding numbers) by K^I. The additional term $C(K^I)$ can be viewed as an operator 1-cocycle, which will be chosen such that the 480 $E(K^I)$'s (5.18) together with the 16 operators p^I satisfy the Lie algebra of $E_8 \times E_8$ or $Spin(32)/Z_2$.

Using the properties of harmonic oscillator coherent states it is straigthforward to verify that

$$:e^{2iK^I x^I(z)}::e^{2iL^I x^I(w)}: =$$

$$= :e^{2i(K^I x^I(z)+L^I x^I(w))}: (z-w)^{K^I L^I} \text{ for } |w|<|z| \quad . \tag{5.19}$$

Consider now the commutator

$$[E(K),\ E(L)] \quad .$$

We may move the contour in z such that we only pick up contributions from the possible singularities at z=w. In order for the commutator to close properly we demand

$$C(K)C(L) = (-1)^{K\cdot L} C(L)C(K) \tag{5.20}$$

$$C(K)C(L) = \varepsilon(K,L)C(K+L) \quad . \tag{5.21}$$

The commutator will be non-zero if $K\cdot L = -1$. Then K+L is another root vector since $(K+L)^2 = 2$. The commutator is also non-zero if $K\cdot L = -2$ because of the normal ordering of the zero modes. In this case $(K+L)^2 = 0$. Hence K=-L. These are the only cases for which the commutator is non-zero. Therefore

$$[E(K),\ E(L)] = \begin{cases} \varepsilon(K,L)E(K+L) & K+L \text{ root} \\ K^I \cdot p^I & K = -L \\ 0 & \text{otherwise} \end{cases} \tag{5.22}$$

$$[p^I, E(K)] = K^I \quad . \tag{5.23}$$

This is precisely the commutators of the generators of $E_8 \times E_8$ or $Spin(32)/Z_2$ as long as $\varepsilon(K,L)$ are chosen to equal the structure constants (± 1 in this basis). It has been shown that it is possible to construct $C(K)$ and $\varepsilon(K,L)$ and hence we have the two explicit representations on the Fock space of the left-movers. It is remarkable that the SO(16) internal symmetry group we started with under the compactification turns into a 16-rank group. The extension of the string plays a crucial rôle here!

6. Conclusions

Consistent string model are, as we have seen, scarce. Among the ones discovered so far the $E_8 \times E_8$ heterotic string is quite a plausible candidate for a truly unified model. It contains a big enough gauge group, has chiral fermions without creating anomalies and seemingly can be compactified to four dimensions. We should ask, if further string models are possible? From the light-cone gauge analysis it seems quite likely that we have found the complete list. Now these models must be further checked. We must understand the quantum properties beyond one-loop graphs. Such an analysis could very well select just one model being completely consistent. If the consistency requirements also force the model to be compactified to four dimensions, then on purely theoretical grounds it would be the natural candidate for the ultimate model. Hopefully Nature reasoned the same way once upon a time.

Appendix

Some Notations and Conventions

The algebra of SO(8) has three inequivalent real eight-dimensional representations, one vector and two spinors. We use 8-valued indices $i,j, \ldots$ corresponding to the vector, $a,b, \ldots$ corresponding to one spinor and $\dot{a}, \dot{b}, \ldots,$ corresponding to the other spinor. Dirac matrices $\gamma^i_{a\dot{a}}$ may be regarded as Clebsch-Gordan coefficients for combining the three eights into a singlet. A second set of matrices $\tilde{\gamma}^i_{\dot{a}a}$ is also introduced. We choose

$$\tilde{\gamma} = \gamma^T \tag{A.1}$$

$$\{\gamma^i, \tilde{\gamma}^j\} = 2\,\delta^{ij} \quad . \tag{A.2}$$

The 16 × 16 matrices

$$\begin{bmatrix} 0 & \gamma^i{}_{a\dot{a}} \\ \tilde{\gamma}^i{}_{\dot{b}b} & 0 \end{bmatrix}$$

form a Clifford algebra. We also define

$$\gamma^{ij}_{ab} = \frac{1}{2}[\gamma^i_{a\dot{a}}\,\tilde{\gamma}^j_{\dot{a}b} - \gamma^j_{a\dot{a}}\,\tilde{\gamma}^i_{\dot{a}b}] \quad . \tag{A.3}$$

These matrices are seen to be antisymmetric in a and b using (A.1).

We can also define

$$\gamma^{ij}_{\dot{a}\dot{b}} = \frac{1}{2}[\tilde{\gamma}^i_{\dot{a}a}\,\gamma^j_{a\dot{b}} - \tilde{\gamma}^j_{\dot{a}a}\,\gamma^i_{a\dot{b}}] \tag{A.4}$$

which in a similar fashion is antisymmetric in $\dot{a}$ and $\dot{b}$.

To span the whole 8 × 8 dimensional matrix spaces we also define

$$\gamma^{ijk\ell}_{ab} \equiv (\gamma^{[i}\,\tilde{\gamma}^j\,\gamma^k\,\tilde{\gamma}^{\ell]})_{ab} \tag{A.5}$$

$$\gamma^{ijk\ell}_{\dot{a}\dot{b}} \equiv (\tilde{\gamma}^{[i}\,\gamma^j\,\tilde{\gamma}^k\,\gamma^{\ell]})_{\dot{a}\dot{b}} \quad . \tag{A.6}$$

These matrices are symmetric.

The general Fierz formula is

$$M_{ab} = \frac{1}{8}\,\delta_{ab}\,\mathrm{tr}\,M - \frac{1}{16}\,\gamma^{ij}_{ab}\,\mathrm{tr}(\gamma^{ij}M)$$
$$+ \frac{1}{384}\,\gamma^{ijk\ell}_{ab}\,\mathrm{tr}(\gamma^{ijk\ell}M) \quad . \tag{A.7}$$

In the case of SU(4), the six-vector can be obtained as the antisymmetric tensor product of two 4's or two $\overline{4}$'s. The corresponding Clebsch-Gordan coefficients (or Dirac matrices) are denoted $\rho^I{}_{AB}$ and

ρ^{IAB}. They are normalized as usual so that

$$\rho^{IAB}\,\rho^{J}{}_{BC} + \rho^{JAB}\,\rho^{I}{}_{BC} = 2\,\delta^{A}{}_{C}\,\delta^{IJ} \quad . \qquad (A.8)$$

We also define

$$\rho^{IJ}{}_{A}{}^{B} = \frac{1}{2}(\rho^{I}{}_{AC}\,\rho^{JCB} - \rho^{J}{}_{AC}\,\rho^{ICB}) \quad . \qquad (A.9)$$

References

1) G. Veneziano, Nuovo Cim. 57A (1968), 190.
2) Y. Nambu, Proc. Int. Conf. on Symm. and Quark Models, Wayne State Univ. (1969) (Gordon and Breach), 1970).
3) H.B. Nielsen, several talks, 1969;
15th Int. Conf. in High Energy Physics, Kiev (1970).
4) L. Susskind, Phys. Rev. D1 (1970), 1182.
5) M.A. Virasoro, Phys. Rev. D1 (1970), 2983.
6) R.C. Brower, Phys. Rev. D6 (1972), 1655.
P. Goddard and C.B. Thorn, Phys. Lett. 40B (1972), 235.
7) Y. Nambu, Lectures at Copenhagen Symposium, 1970 (unpublished).
8) O. Hara, Progr. Theor. Phys. 46 (1971), 1549.
9) T. Goto, Progr. Theor. Phys. 46 (1971), 1560.
10) P. Goddard, J. Goldstone, C. Rebbi and C.B. Thorn, Nucl. Phys. B56 (1973), 109.
11) P.M. Ramond, Phys. Rev. D3 (1971), 2415.
12) A. Neveu and J.H. Schwarz, Nucl. Phys. B31 (1971), 86; Phys. Rev. D4 (1971), 1109.
13) L. Brink, D.I. Olive, C. Rebbi and J. Scherk, Phys. Lett. 45B (1973), 379.
14) L. Brink, P. Di Vecchia and P.S. Howe, Phys. Lett. 65B (1976), 471
S. Deser and B. Zumino, Phys. Lett. 65B (1976), 369.
15) J.-L. Gervais and B. Sakita, Nucl. Phys. B34 (1971), 477.
16) A. Neveu and J. Scherk, Nucl. Phys. B36 (1972), 155.
17) J. Scherk and J.H. Schwarz, Nucl. Phys. B81 (1974), 118.
18) F. Gliozzi, J. Scherk and D.I. Olive, Phys. Lett. 65B (1976), 282; Nucl. Phys. B122 (1977), 253.
19) For reviews, see J.H. Schwarz, Phys. Rep. 69 (1982), 223
M.B. Green, Surveys in High Energy Physics 3 (1983), 127
L. Brink, CERN-TH 4006/84 in "Proceedings from Nato Advanced Study Institute, Supersymmetry." Bonn (1984).
20) M.B. Green and J.H. Schwarz, Phys. Lett. 149B (1984), 117.
21) D.J. Gross, J.A. Harvey, E. Martinec and R. Rohm, Phys. Rev. Lett. 54 (1985), 502;
Nucl. Phys. B256 (1985), 253; Princeton preprint (1985).
22) P. Candelas, G.T. Horowitz, A. Strominger and E. Witten, Nucl. Phys. B258 (1985), 46
E. Witten, Nucl. Phys. B258 (1985), 75.
23) the first of ref. 14.
24) A.M. Polyakov, Phys. Lett. 103B (1981), 207, 211.
25) P.A.M. Dirac, Can. J. Math. 2 (1950), 129.
26) J.H. Weis, unpublished.
27) L. Brink and H.B. Nielsen, Phys. Lett. 45B (1973), 332.

28) This method was suggested by F. Gliozzi, unpublished (1976).
29) M.B. Green and J.H. Schwarz, Phys. Lett. 109B (1982), 444.
30) M.B. Green and J.H. Schwarz, Nucl. Phys. B181 (1981), 502.
31) L. Brink, O. Lindgren and B.E.W. Nilsson, Nucl. Phys. B212 (1983), 401.
32) M. Ademollo, L. Brink, A. D'Adda, R. D'Auria, E. Napolitano, S. Sciuto, E. Del Giudice, P. Di Vecchia, S. Ferrara, F. Gliozzi, R. Musto, R. Pettorino, J.H. Schwarz, Nucl. Phys. B111 (1976), 77.
33) J. Paton and C. Hong-Mo, Nucl. Phys. B10 (1969), 519.
34. N. Marcus and A. Sagnotti, Phys. Lett. 119B (1982), 97.
35) J.H. Schwarz, Phys. Rep. 69 (1982), 223.
36) M.B. Green and J.H. Schwarz, Phys. Lett. 136B (1984), 367.
37) I. Bengtsson and M. Cederwall, Institute of Theoretical Physics, Göteborg 84-21 (1984).
38) W. Siegel, UCB-PTH-83/22 (1983).
39) E. Cremmer and J. Scherk, Nucl. Phys. B103 (1976), 399
M.B. Green, J.H. Schwarz and L. Brink, Nucl. Phys. B198 (1982), 474.
40) P. Goddard and D. Olive in "Vertex Operators in Mathematics and Physics" ed. J. Lepowsky et al. (MSRI) publ. No 3 Springer Verlag p. 419 (1984).
41) I.B. Frenkel and V.G. Kac, Inv. Math 62 (1980), 23
G. Segal, Comm. Math. Phys. 80 (1982), 301
P. Goddard and D. Olive see ref. 39
P. Goddard, this volume.

LECTURES ON SUPERSTRINGS

Michael B. Green,

Dept. of Physics, Queen Mary College, University of London, U.K.

Mile End Road, London E1 4NS

Superstring theories have not yet been formulated in terms of a single compelling principle such as that of general relativity. However, enough is now known about the structure of these theories to justify the optimism that certain of them might be consistent quantum theories that unify gravity and the other forces. The fact that the quantum consistency of superstring theories restricts the possible ten-dimensional unifying symmetry groups to be $E_8 \times E_8$ or SO(32) (or (Spin 32)/Z_2 which has the same algebra as SO(32)) is a novel development in particle physics. The case of $E_8 \times E_8$ is particularly interesting since, in the process of compactification from ten to four dimensions, it can break to a realistic chiral symmetry group describing all the observed interactions and the spectrum of the known particles.

The subject is still in a somewhat primitive state by comparison with the sophistication of our understanding of conventional "point" field theory. Results have been arrived at by a variety of techniques which I shall survey in these lectures.

In the first lecture I will start with the description of the dynamics of a free classical superstring moving in a flat super-space-time background. In order to avoid problems associated with covariant quantization in this formulation I shall describe the first-quantized superstring theory in the light-cone gauge. This provides the basis for calculating the spectra of the various types of superstring theories and is adequate for most perturbation theory calculations around a background space with no Riemann curvature.

The second lecture will survey the formulation of the second-quantized interacting field theory of superstrings in the light-cone gauge. This involves fields which create and destroy complete strings and which are therefore functionals of the string configurations. The use of the light-cone gauge is presumably an undesirable feature since it obscures much of the geometric structure of the theory. However, for the moment this is the only more or less complete understanding of interacting super string field theory we have.

The third lecture will summarize the present status of the one-loop calculations. I will also present, in some detail, the calculation of the Yang-Mills anomaly in the open-string theory and demonstrate its cancellation for the group SO(32).

I COVARIANT SUPERSTRING DYNAMICS

(a) The Covariant Action

The most geometrically appealing formulation of superstring theories begins with a generalization of the Nambu-Goto action[1] of bosonic string theory (namely, the area of the world-sheet swept out as the string moves through space-time) to super-space-time[2]. The other formulation, based on the observation that the spectrum of the "spinning string" theory[3] can be truncated to be supersymmetric in ten-dimensional space-time[4], does not incorporate space-time supersymmetry manifestly.

As a string moves through space-time it sweeps out a world-sheet that is parametrized by a time-like parameter τ and a spacelike parameter σ. The coordinates of the superstring map the world-sheet into superspace . These are the space-time coordinates $X^\mu(\sigma,\tau)$ (where μ = 0,1,..,D-1 is a space-time index in D dimensions) and, in general, two Grassmann (anticommuting) coordinates $\theta^{Aa}(\sigma,\tau)$ which are space-time spinors (A=1,2) and the spinor index a = $1,2,...2^{D/2}$). In ten dimensions these spinors will be taken to be both Majorana (i.e. real in the Majorana representation of the Dirac gamma matrices) and to satisfy the chirality (Weyl) constraints

$$(1 + \eta^A \gamma_{11})^{ab} \theta^{Ab} = 0 \tag{1.1}$$

where $\eta^A = \pm 1$ and $\gamma_{11} = \gamma^0 \gamma^1 \dots \gamma^9$. The Dirac gamma matrices γ^μ satisfy $\{\gamma^\mu,\gamma^\nu\} = -2\eta^{\mu\nu}$ where the ten-dimensional Minkowski metric $\eta^{\mu\nu} = \mathrm{diag}(-1,1,..,1)$. When $\eta^1 = -\eta^2$ the theory has no net chirality whereas when $\eta^1 = \eta^2$ the theory is chiral. The fact that both the Majorana and the chirality conditions can be imposed simultaneously is a special property of ten dimensions (more generally of 2 mod 8 dimensions). More generally, the classical theory will also make sense in dimensions D = 3 (with Majorana spinors), 4 (with Majorana or Weyl spinors) and 6 (with Weyl spinors) but since the case D = 10 is of particular interest in the quantum theory I will use notation appropriate to that dimension. The super-Poincaré transformations on which these theories are based are (suppressing spinor indices)

$$\delta\theta^A = \frac{1}{4}\, \omega_{\mu\nu}\, \gamma^{\mu\nu}\, \theta^A + \epsilon^A \tag{1.2}$$

$$\delta X^\mu = \omega^\mu{}_\nu\, X^\nu + a^\mu + i\bar{\epsilon}^A\, \gamma^\mu\, \theta^A \tag{1.3}$$

where $\omega_{\mu\nu}$ and a_μ are infinitessimal parameters of the Poincaré group while $\epsilon^A \equiv \epsilon^{Aa}$ are the two 32-component Grassmann spinor parameters which also satisfy Majorana and Weyl conditions. [In six dimensions θ^A is chiral but not Majorana and the term $i\bar{\epsilon}^A\gamma^\mu\theta^A$ is replaced by $i(\bar{\epsilon}^A\gamma^\mu\theta^A - \bar{\theta}^A\gamma^\mu\epsilon^A)$.] For the case of the heterotic string[5] there is only one (Majorana-Weyl) spinor coordinate.

The natural covariant action for a relativistic point particle is the length of the world-line traversed as it moves in space-time. In the bosonic string theory this action is generalized[1] to the area of the world-sheet. The fact that this is invariant under arbitrary reparametrizations of the world-sheet ($\sigma \to \tilde{\sigma}(\sigma,\tau)$ and $\tau \to \tilde{\tau}(\sigma,\tau)$) is crucial for the consistency of the theory. The action for the superstring consists of several terms

$$S = S_1 + S_2 \quad (+\ S_3 \text{ in the case of the heterotic string}). \tag{1.4}$$

The first term is the obvious guess for a supersymmetric generalization of the bosonic action. It is convenient to write this in the form that invokes a two-dimensional metric tensor $g^{\alpha\beta}(\sigma,\tau)$ so that the action looks like two-dimensional general relativity on the world-sheet (with the coordinates X^μ and θ^{Aa} being scalars under two-dimensional reparametrizations of the world-sheet)

$$S_1 = -\frac{T}{2}\int \eta_{\mu\nu}\,\sqrt{-g}g^{\alpha\beta}\,\Pi^\mu_\alpha\Pi^\nu_\beta\,d^2\xi \tag{1.5}$$

(where $d^2\xi = d\sigma\,d\tau$ and the indices $\alpha,\beta = \sigma,\tau$) and

$$\Pi^\mu_\alpha = \partial_\alpha X^\mu - i\bar{\theta}^A\gamma^\mu\partial_\alpha\theta^A \tag{1.6}$$

when θ^A is a Majorana spinor (otherwise the term $i\bar{\theta}^A\gamma^\mu\partial_\alpha\theta^A$ is replaced by $\frac{1}{2}\,i(\bar{\theta}^A\gamma^\mu\partial_\alpha\theta^A - \partial_\alpha\bar{\theta}^A\gamma^\mu\theta^A)$). The string has been taken to be moving in flat Minkowski space-time ($\eta^{\mu\nu}$ is the D-dimensional Minkowski metric). Recent considerations of strings moving in curved space-time backgrounds involve the replacement of $\eta^{\mu\nu}$ by the background metric $G^{\mu\nu}(X)$ (as well as the addition of other terms) in which case the resulting non-linear σ model is only a consistent string theory for very special spaces. S_1 is manifestly invariant under the global transformations of eqs.(1.2) and (1.3) (since Π^μ_α is manifestly supersymmetric) as well as under arbitrary reparametrizations of σ and τ. The metric, $g^{\alpha\beta}(\sigma,\tau)$, is an auxiliary field which (at least in the classical theory) can be eliminated by replacing it in the action by the solution of its equations of motion ($g_{\alpha\beta} = f(\sigma,\tau)\, \Pi^\mu_\alpha \Pi_{\mu\beta}$ where $f(\sigma,\tau)$ is an arbitrary function). The resulting expression is the "area" of the world-sheet in superspace. The form of the action S_1 in eq. (1.5) is a generalization to superstrings of the action of ref. 6 and used by Polyakov[7] in discussing the quantization of the bosonic and spinning string theories.

By itself S_1 does not define a conformally-invariant quantum theory, presumably because of the terms cubic and quartic in the coordinates. However, it is possible to add another term, S_2, to the action to remedy this where

$$S_2 = -iT \int \eta_{\mu\nu}\, \epsilon^{\alpha\beta} \Big\{ \partial_\alpha X^\mu \, (\bar{\theta}^1 \gamma^\nu \partial_\beta \theta^1 - \bar{\theta}^2 \gamma^\nu \partial_\beta \theta^2) - i\bar{\theta}^1 \gamma^\mu \partial_\alpha \theta^1 \, \bar{\theta}^2 \gamma^\nu \partial_\beta \theta^2 \Big\}\, d^2\xi \quad . \tag{1.7}$$

This term is also manifestly invariant under reparametrizations of σ and τ due to the presence of the two-dimensional Levi-Cevita tensor density, $\epsilon^{\alpha\beta}$. The fact that S_2 is also invariant under the global supersymmetry transformations is not so manifest. To verify this consider, for simplicity, the case of the heterotic superstring obtained by setting $\theta^2 = 0$ in eq.(1.7). The variation under supersymmetry transformations is then given by substituting the ϵ variation from eqs. (1.2) and (1.3)

$$\delta_\epsilon S_2 = \int (\bar{\epsilon} \gamma^\mu \partial_\alpha \theta \; \bar{\theta} \gamma_\mu \partial_\beta \theta + \text{total derivatives})\, d\sigma\, d\tau \; . \tag{1.8}$$

It is straightforward to verify that this vanishes by using the identity (proved by using Fierz transformations)

$$\bar{\epsilon}\gamma^{\mu}\lambda_{[1}\bar{\lambda}_{2}\gamma^{\mu}\lambda_{3]} = 0 \tag{1.9}$$

where $\lambda_1 = \theta$, $\lambda_2 = \partial_\tau\theta$, $\lambda_3 = \partial_\sigma\theta$ and [ ] denotes antisymmetrization. This is the same identity that is used in proving the supersymmetry of supersymmetric Yang-Mills theories in dimensions D = 3, 4, 6 and 10 (with the appropriate kind of spinors in each of these dimensions). There is therefore a restriction on the possible space-time dimensionality in the classical theory. We shall see that the critical dimension in the quantum theory is D = 10 which means that the excitations of the string are purely transverse in ten dimensions. [If it is possible to define consistent quantum theories in 3, 4 or 6 dimensions it is presumably necessary to account for new longitudinal modes in the manner suggested by Polyakov[7].]

The relative coefficient of S_1 and S_2 in eq.(1.5) and (1.7) is uniquely determined by requiring that the total action have extra local symmetries which lead to it describing a conformally invariant two-dimensional theory. In particular there is a local fermionic invariance (analogous to that for the superparticle discussed in ref. 8). It is useful to introduce projection operators

$$P_{\pm} = \frac{1}{2}\left[g^{\alpha\beta} \pm \frac{\epsilon^{\alpha\beta}}{\sqrt{-g}}\right] \tag{1.10}$$

which project onto the self-dual and anti-self-dual pieces of two-dimensional vectors. If any two-vector V_α satisfies $V^\alpha \equiv V^\alpha_\pm = P^{\alpha\beta}_\pm V_\beta$ then the two components are equal (for + sign) or equal in magnitude but opposite in sign (for - sign). The action $S_1 + S_2$ is invariant under fermionic transformations with Grassmann parameters $\kappa^{1\alpha}$ and $\kappa^{2\alpha}$,

$$\delta_\kappa\theta^A = 2i\gamma.\Pi_\alpha\kappa^{A\alpha} \quad , \qquad \delta_\kappa X^\mu = i\bar{\theta}^A\gamma^\mu\delta_\kappa\theta^A ,$$
$$\delta_\kappa(\sqrt{-g}g^{\alpha\beta}) = -16\sqrt{-g}(\bar{\kappa}^{1\alpha}\partial^\beta_-\theta^1 + \bar{\kappa}^{2\alpha}\partial^\beta_+\theta^2) \tag{1.11}$$

The parameters $\kappa^{1\alpha}$ and $\kappa^{2\alpha}$ have suppressed space-time spinor indices and are self-dual ($\kappa^{1\alpha} \equiv \kappa_{+}^{1\alpha}$) and anti-self-dual ($\kappa^{2\alpha} \equiv \kappa_{-}^{2\alpha}$) respectively. The proof of the κ invariance of S involves the same Fierz identity used in the proof of its ϵ invariance. These supersymmetries are reminiscent of two-dimensional supersymmetry although the parameters are not two-dimensional spinors and there is no two-dimensional gravitino field.

The action S possesses further local invariance under bosonic transformations which can be discovered by trying to close the algebra of the κ transformations. These bosonic transformations are

$$\delta_{\lambda}\theta^{A} = \sqrt{-g}\partial_{\alpha}\theta^{A}\lambda^{A\alpha} , \qquad \delta_{\lambda}X^{\mu} = i\bar{\theta}^{A}\gamma^{\mu}\delta_{\lambda}\theta^{A} ,$$

$$\delta_{\lambda}(\sqrt{-g}g^{\alpha\beta}) \quad . \tag{1.12}$$

The parameters satisfy $\lambda^{1\alpha} = \lambda_{+}^{1\alpha}$, $\lambda^{2\alpha} = \lambda_{-}^{2\alpha}$ respectively. For the heterotic string where there is just one superspace spinor $\Theta(\sigma,\tau)$ the action contains, in addition to S_1 and S_2, the third term S_3 that incorporates the internal quantum numbers of (Spin 32)/Z_2 (which has the same Lie algebra as SO(32)) or $E_8 \times E_8$. This term may be written in terms of a two-dimensional chiral fermion field Ψ^{I} (I = 1,2,..,32) as

$$S_3 = \frac{1}{2}\, i\, T \int \bar{\Psi}^{I}(1-\rho_3)e_{A}^{\alpha}\rho^{A}\partial_{\alpha}\Psi^{I}\, d^2\xi \tag{1.13}$$

where ρ^{A} are the two-dimensional Dirac matrices and $\rho_3 = \rho_1\rho_2$. The zweibein e_{A}^{α} defines the metric by $g^{\alpha\beta} = e_{A}^{\alpha}\, e^{\beta A}$. The consistency of the heterotic superstring when one-loop corrections are incorporated requires that SO(32) or $E_8 \times E_8$ be symmetries of S_3. This is acheived by arranging that Ψ^{I} transform as a 32-component vector of SO(32) or as the (16,16) representation of the SO(16) × SO(16) subgroup of $E_8 \times E_8$. Since this very appealing idea is discussed in great detail elsewhere in this workshop I will not describe it in any detail in these lectures.

The geometrical interpretation of the term S_2 has been clarified[9] by writing it in terms of the supersymmetric one-form (for example, for the case of the heterotic string)

$$\Omega^M \equiv (\Omega^\mu, \Omega^a) \tag{1.14}$$

where

$$\Omega^\mu = dX^\mu - i\bar{\Theta}\gamma^\mu d\Theta \equiv \Omega^\mu_\alpha \, d\xi^\alpha \tag{1.15}$$

$$\Omega^a = d\Theta^a \equiv \Omega^a_\alpha \, d\xi^\alpha \tag{1.16}$$

and $d\xi^\alpha \equiv (d\tau, d\sigma)$. In this notation the action S_1 is written as

$$S_1 = -\frac{1}{2} T \int \sqrt{-g} \, g^{\alpha\beta} \, \eta_{\mu\nu} \Omega^\mu_\alpha \, \Omega^\nu_\beta \tag{1.17}$$

S_2 can be written as a kind of Wess-Zumino term by formally extending the dimension of the world-sheet to a three-dimensional space (in the style of Witten[10]) and introducing the manifestly super-Poincaré invariant three-form

$$\Omega^3 = -i(C\gamma^\mu)_{ab} \, \Omega_\mu \, \Omega^a \, \Omega^b \tag{1.18}$$

so that

$$S_2 = -\frac{1}{2} T \int \Omega^3 \tag{1.19}$$

The earlier expression for S_2 (eq.(1.7)) is recovered by using the fact that $\Omega^3 = d\Omega^2$ where $\Omega^2 = -idX^\mu \wedge \bar{\Theta}\gamma_\mu d\Theta$ in the case of the heterotic superstring and a generalization involving two Θ's in the case of type II superstring theories.

The situation is reminiscent of a non-linear σ-model defined on a group manifold in which the addition of a Wess-Zumino term leads to a free theory for a certain value of the relative couplings of the two terms[11]. In that case the action $S = S_1/g + S_2$ has a beta function which has a zero at a special value of g at which the theory is conformally invariant. The Wess-Zumino term can also be interpreted in terms of a torsion which parallelizes the curvature at the point at which the beta function vanishes[12]. A similar interpretation is possible in the case of the superstring theories (where the σ-model

$$S_2 = -\frac{1}{2} T \int \Omega^3 \tag{1.19}$$

The earlier expression for S_2 (eq.(1.7)) is recovered by using the fact that $\Omega^3 = d\Omega^2$ where $\Omega^2 = -idX^\mu \wedge \bar{\theta}\gamma_\mu d\theta$ in the case of the heterotic superstring and a generalization involving two θ's in the case of type II superstring theories.

The situation is reminiscent of a non-linear σ-model defined on a group manifold in which the addition of a Wess-Zumino term leads to a free theory for a certain value of the relative couplings of the two terms[11]. In that case the action $S = S_1/g + S_2$ has a beta function which has a zero at a special value of g at which the theory is conformally invariant. The Wess-Zumino term can also be interpreted in terms of a torsion which parallelizes the curvature at the point at which the beta function vanishes[12]. A similar interpretation is possible in the case of the superstring theories (where the σ-model defines a mapping of the two-dimensional world-sheet into $N = 1$ or $N = 2$ superspace).

The covariant action S has recently been generalized (for $N = 1$ theories) to describe a curved gravitational background[13]. This involves using a curved metric instead of $\eta^{\mu\nu}$ in S_1 and coupling the using a three-form that arises in ten-dimensional supergravity[14] to generalize Ω^3 in S_2. A similar construction has also been carried out for $N = 2$ theories[15].

(b) Equations of Motion

In type I or type II theories the equation of motion for $g^{\alpha\beta}$ arises only from the S_1 term in the action and gives the two-dimensional Einstein equation

$$\Pi^{\mu}_{\alpha}\,\Pi_{\mu\beta} - \frac{1}{2}\, g_{\alpha\beta}\, g^{\gamma\delta}\, \Pi^{\mu}_{\gamma}\,\Pi_{\mu\delta} = 0 \ . \tag{1.20}$$

[For the heterotic string there is an additional term arising from the variation of the zweibein in S_3.] This equation expresses the vanishing of the two-dimensional energy-momentum tensor which is traceless and symmetric and has two independent components. In a conformal gauge, defined by

$$g^{\alpha\beta} = e^{\Phi}\, \eta^{\alpha\beta} \tag{1.21}$$

where $\eta^{\alpha\beta} = \begin{bmatrix} -1 & 0 \\ 0 & 1 \end{bmatrix}$, eq. (1.20) describes two independent constraints on the coordinates. In terms of the quantities $\Pi_{\pm}{}^{\mu}$

$$\Pi^{\mu}_{\pm} \equiv (\, \Pi^{\mu}_{\tau} \pm \Pi^{\mu}_{\sigma}\,)/\sqrt{2} \tag{1.22}$$

(the components of $P_{\mp\alpha\beta}\Pi^{\mu\beta}$) these "Virasoro" constraints are

$$\Pi_{+}\cdot\, \Pi_{+} = 0 = \Pi_{-}.\Pi_{-} \quad . \tag{1.23}$$

The equations for $X^{\mu}(\sigma,\tau)$, determined from the action, can be written in a conformal gauge as

$$\partial_{\alpha}(\partial^{\alpha}X^{\mu} - 2i\bar{\theta}^{1}\gamma^{\mu}\partial^{\alpha}_{+}\theta^{1} - 2i\bar{\theta}^{2}\gamma^{\mu}\partial^{\alpha}_{-}\theta^{2}) = 0 \tag{1.24}$$

(where $\partial^{\alpha}_{\pm} \equiv P^{\alpha\beta}_{\mp}\, \partial_{\beta}$). The θ^{Aa} equations are

$$\gamma.\Pi_{-}\partial_{+}\theta^{1} = 0 = \gamma.\Pi_{+}\partial_{-}\theta^{2} \tag{1.25}$$

where $\partial_{\pm} \equiv (\partial_{\tau}\pm\partial_{\sigma})/\sqrt{2}$. In deriving these equations from the action various boundary conditions have been imposed which eliminate surface terms. For closed strings these conditions impose periodicity of the coordinates in σ. For open strings the conditions require

$$\theta^1 = \theta^2 , \qquad \Pi^{\mu}_{\sigma} = 0 \text{ at the endpoints } \sigma = 0,\pi. \tag{1.26}$$

From the transformations in eq.(1.2) it is clear that the open-string supersymmetry is truncated to N = 1 since eq.(1.26) requires $\epsilon^1 = \epsilon^2$.

The covariant quantization of this action is hampered by the fact that there are additional phase-space constraints involving the momenta, p^A_θ, conjugate to θ^A. Defining

$$p^A_\theta = \frac{\delta S}{\delta \theta^A} \tag{1.27}$$

it is easy to see that p_θ is related to functions of X^μ, P^μ and θ^A. These constraints are mixtures of first and second class constraints which must be separated before quantization. Despite some progress[16] it seems probable that covariant quantization will involve introducing extra variables into the action to relax the constraints[17]. I will take the pragmatic route of passing to the light-cone gauge in which the super-Poincaré invariance is not manifest but the quantum theory is easy to formulate.

(c) **The Light-Cone Gauge**

The parameters, $\kappa^{A\alpha}$ have the same number of independent components as θ^A (after allowing for the fact that they are (anti) self-dual). However, it follows from the transformation laws in eq.(1.11) that only half the components of the θ's can be gauged away because the operators $\gamma.\Pi_\pm$ are nilpotent (since $(\gamma.\Pi_\pm)^2 = (\Pi_\pm)^2 = 0$ by use of eq.(1.23)). By a suitable choice of κ^A the fermionic gauge invariance can be used to choose

$$\frac{1}{2} (\gamma^- \gamma^+)^{ab} \theta^{Ab} = 0 \tag{1.28}$$

where $\gamma^{\pm}$ are the light-cone Dirac matrices which satisfy $(\gamma^{+})^2 = 0 = (\gamma^{-})^2$ so that $\frac{1}{2}(\gamma^{+}\gamma^{-})$ and $\frac{1}{2}(\gamma^{-}\gamma^{+})$ are projection operators. The ± components of any 10-vector, V^{μ}, are defined by $V^{\pm} = (V^0 \pm V^9)/\sqrt{2}$. Each spinor, θ, satisfying eq.(1.28) has eight independent components, half as many as a general Majorana-Weyl spinor. With this choice of κ gauge the equation of motion for $X^{+}(\sigma,\tau)$ in the conformal gauge, eq. (1.24) becomes $\partial^2 X^{+}(\sigma,\tau) = 0$. Just as in the original development of the light-cone gauge treatment of the bosonic theory in ref. 18 this allows the choice of a special parametrization, known as the light-cone gauge, in which the "time" coordinate, $X^{+}(\sigma,\tau)$, takes a common value for all values of σ

$$X^{+}(\sigma,\tau) = x^{+} + \frac{1}{\pi T} p^{+}\tau \quad . \tag{1.29}$$

[The tension T will often be set equal to $1/\pi$ from now on.] The equations of motion are particularly simple in this gauge. From eqs.(1.24) and (1.25) we have

$$\partial^2 X^I = 0 \qquad \text{with } I = 1,2,..,8 \quad , \tag{1.30}$$

$$\partial_+\theta^1 \equiv \frac{1}{\sqrt{2}}(\partial_\tau + \partial_\sigma)\,\theta^1 = 0 = \partial_-\theta^2 \equiv \frac{1}{\sqrt{2}}(\partial_\tau - \partial_\sigma)\,\theta^2 \quad . \tag{1.31}$$

The last two equations are the components of the two-dimensional Dirac equation

$$\rho.\partial\theta = 0 \tag{1.31'}$$

where the two-component spinor θ is defined by

$$\theta = \begin{bmatrix} \theta^{1a} \\ \theta^{2a} \end{bmatrix} \tag{1.32}$$

In passing to the light-cone gauge the ten-dimensional scalars, θ^1 and θ^2, which were independent world-sheet scalars in the covariant action have become the two components of a world-sheet spinor.

This is a consequence of the way in which super-Poincaré transformations act in the light-cone gauge. Since the light-cone gauge conditions (eqs.(1.28) and (1.29)) are not covariant they are altered by certain super-Poincaré transformations. In order to ensure that the gauge conditions are unaltered in the transformed frame the transformations must be supplemented by compensating (super) reparametrizations. The net effect is that under these compensated transformations the θ's transform as two-dimensional spinors. The Virasoro constraint equations $\Pi_{+}^{2} = 0 = \Pi_{-}^{2}$ can be explicitly solved to express X^- in terms of the coordinates X^I and θ^A by substituting eqs.(1.28) and (1.29) into these constraint equations. The result is

$$\dot{X}^- = \frac{1}{p^+}(\dot{\underline{X}}^2 + \underline{X}'^{\,2}) + \frac{i}{\sqrt{2}}\bar{\theta}^2\gamma^-\partial_+\theta^2 + \frac{i}{\sqrt{2}}\bar{\theta}^1\gamma^-\partial_-\theta^1 \qquad (1.33)$$

$$\underline{X}'^- = \frac{1}{p^+}(\dot{\underline{X}}.\underline{X}') + \frac{i}{\sqrt{2}}\bar{\theta}^2\gamma^-\partial_+\theta^2 - \frac{i}{\sqrt{2}}\bar{\theta}^1\gamma^-\partial_-\theta^1 \qquad (1.34)$$

where $\underline{X} \equiv X^I$ and $\dot{\underline{X}} \equiv \partial_\tau\underline{X}$ and $\underline{X}' \equiv \partial_\sigma\underline{X}$.

It is convenient to introduce an SO(8) spinor notation in which any sixteen-component Majorana-Weyl spinor ψ^a $(a = 1,2,..,16)$ is written as the sum of two inequivalent SO(8) spinors

$$\psi = \tfrac{1}{2}\gamma^-\gamma^+\psi + \tfrac{1}{2}\gamma^+\gamma^-\psi \qquad (1.35a)$$

$$\equiv \psi^a + \psi^{\dot{a}} \qquad (1.35b)$$

where the superscripts a and $\dot{a}$ in the last line take the values 1,2,..,8. The physical modes of the type I or type II theories are now represented by the SO(8) vector

$$X^I(\sigma,\tau) \qquad (1.36a)$$

and two SO(8) spinors, written in the two-dimensional spinor notation as

$$\theta^a(\sigma,\tau) \equiv \begin{bmatrix} \theta^{1a}(\sigma,\tau) \\ \theta^{2a}(\sigma,\tau) \end{bmatrix} \tag{1.36b}$$

(while the heterotic string has only a single SO(8) spinor that can be thought of as eight chiral two-dimensional spinors). In the non-chiral type IIa theory one of the two SO(8) spinors is a dotted spinor while the other is an undotted spinor.

The light-cone gauge equations can be deduced from a light-cone gauge action[19]

$$S = -\frac{1}{2}T \int_0^\pi d\sigma \int d\tau \left\{ \eta^{\alpha\beta}\partial_\alpha X^I \partial_\beta X^I + \frac{1}{2} i p^+ \bar{\theta}^a \rho . \partial \theta^a \right\} . \tag{1.37}$$

A bar on top of a spinor now denotes *two-dimensional* conjugation (i.e. $\bar{\theta} \equiv \theta\rho^0$). This action is invariant under the supersymmetry transformations

$$\delta\theta^a = \eta^a + \gamma^I_{a\dot{b}} \rho . \partial X^I \epsilon^{\dot{b}} \tag{1.38}$$

$$\delta X^I = \frac{i}{p^+} \bar{\epsilon}^{\dot{a}} \tilde{\gamma}^I_{\dot{a}b} \theta^b \tag{1.39}$$

where the Grassmann parameters η^a and $\epsilon^{\dot{a}}$ are both SO(8) spinors and world-sheet spinors

$$\eta^a \equiv \begin{bmatrix} \eta^a_1 \\ \eta^a_2 \end{bmatrix} \qquad \epsilon^{\dot{a}} \equiv \begin{bmatrix} \epsilon^{\dot{a}}_1 \\ \epsilon^{\dot{a}}_2 \end{bmatrix} \tag{1.40}$$

Together they build up the 32 components of the two supercharges.

The matrices $\gamma^I_{a\dot{b}}$ and $\tilde{\gamma}^I_{\dot{b}a}$, defined by

$$\gamma^I_{a\dot{b}} \tilde{\gamma}^J_{\dot{b}c} + \gamma^J_{a\dot{b}} \tilde{\gamma}^I_{\dot{b}c} = 2\delta^{IJ}\delta_{ac} \tag{1.41a}$$

$$\tilde{\gamma}^I_{\dot{a}b} \gamma^J_{b\dot{c}} + \tilde{\gamma}^J_{\dot{a}b} \gamma^I_{b\dot{c}} = 2\delta^{IJ}\delta_{\dot{a}\dot{c}} \tag{1.41b}$$

are related to the 16 × 16 Dirac matrices of the chiral ten-dimensional theory by

$$\gamma^I = \begin{bmatrix} 0 & \gamma^I_{a\dot{b}} \\ \tilde{\gamma}^I_{\dot{b}a} & 0^{ab} \end{bmatrix} \quad . \tag{1.42}$$

The solutions of the equations of motion (eqs.(1.30) and (1.31)) can be written as Fourier expansions so that, for example, the open string coordinates satisfying the boundary conditions of eq. (1.26) (i.e. $\theta^1 = \theta^2$ and $\partial_\sigma X^I = 0$ at the endpoints) are given by

$$\theta^{1a}(\sigma,\tau) = \sum_{-\infty}^{\infty} \theta^a_n \, e^{-in(\tau-\sigma)} \qquad \theta^{2a}(\sigma,\tau) = \sum_{-\infty}^{\infty} \theta^a_n \, e^{-in(\tau+\sigma)} \tag{1.43}$$

$$X^I(\sigma,\tau) = x^I + p^I\tau + i\sum_{n=1}^{\infty} \frac{1}{n}(\alpha^I_n e^{-in\tau} - \alpha^I_{-n} e^{in\tau}) \cos n\sigma \tag{1.44}$$

(where $\alpha^I_{-n} \equiv \alpha^{I*}_n$ and $\theta^a_{-n} \equiv \theta^{a*}_n$). The momentum conjugate to $X^I(\sigma,\tau)$ is given by

$$P^I(\sigma,\tau) = \frac{\delta S}{\delta \dot{X}^I} = \frac{1}{\pi} \sum_{-\infty}^{\infty} \alpha^I_n \, e^{-in\tau} \cos n\sigma \quad . \tag{1.45}$$

Poisson brackets can be obtained from the action in the usual way and I will immediately transcribe them to (anti)-commutator brackets (setting $\hbar = 1$) which gives

$$[\, P^I(\sigma,\tau) \;,\; X^J(\sigma',\tau)] = -i \, \delta^{IJ} \, \delta(\sigma - \sigma') \tag{1.46}$$

$$\{\, \theta^a(\sigma,\tau) \;,\; \theta^b(\sigma',\tau) \,\} = \frac{2\pi}{p^+} \, \delta^{ab} \, \delta(\sigma-\sigma') \quad . \tag{1.47}$$

Substituting the mode expansions into these expressions gives the relations

$$[\, \alpha^I_m \;,\; \alpha^J_n \,] = m \, \delta^{IJ} \, \delta_{m+n,0} \tag{1.48}$$

$$\{\, \theta^a_m \;,\; \theta^b_n \,\} = \frac{2}{p^+} \delta^{ab} \, \delta_{m+n,0} \quad . \tag{1.49}$$

The generators of all the super-Poincaré transformations can be represented in terms of the bosonic oscillator modes $\underline{\alpha}_n$ and their fermionic partners, Θ_n. The modes Θ_n are related to the S_n's of ref. 20 by $S_n = \sqrt{p^+}\,\Theta_n$. These factors of $\sqrt{p^+}$ slightly alter the representation of the generators J^{+-} and J^{-i} from those in ref. 20 (where $J^{\mu\nu}$ are the Lorentz generators). I will not describe this in detail as I shall want to use a slightly different formalism in the treatment of superstring field theory in the next lecture. Note that the hamiltonian operator, h (the operator conjugate to X^+ defined by $h \equiv \int_0^\pi d\sigma\, P^-(\sigma,\tau)$) is given by $h = \frac{1}{\pi}\int_0^\pi d\sigma\, \dot{X}^-$ which can be obtained from the expression for X^- (eq. (1.34)) and is given in modes by

$$h \equiv p^- = \frac{1}{p^+}(N + \frac{\underline{p}^2}{2}) \quad . \tag{1.50}$$

where

$$N = \sum_{n=1}^{\infty}(\underline{\alpha}_n^\dagger \cdot \underline{\alpha}_n + \frac{1}{2}p^+ n\Theta_n^{\dagger a}\,\Theta_n^a) \tag{1.51}$$

Notice that the normal ordering constants cancel between Bose and Fermi modes at each value of n. This is not the case in the heterotic theory. The mass of any open-string state is given by

$$(\text{Mass})^2 = 2p^+p^- - \underline{p}^2 = 2N \quad . \tag{1.52}$$

The massless ground states in the open-string sector form the Yang-Mills supermultiplet. These are

an SO(8) vector $|i\rangle$ of gauge bosons

an SO(8) spinor $|a\rangle$ of fermions

Similar arguments for the closed-string theories lead to two independent sets of modes α_n^I, Θ_n^a and $\tilde{\alpha}_n^I$, $\tilde{\Theta}_n^a$ corresponding to waves running around the string in either direction. The states of type II closed string theories have masses determined by

$$(\text{Mass})^2 = 4\ (N+\tilde{N}) \tag{1.53}$$

subject to the constraint

$$N = \tilde{N} \tag{1.54}$$

which follows by requiring that $X^-(\sigma,\tau)$ is periodic (by integrating eq. (1.34) from $\sigma = 0$ to $\sigma = \pi$). The ground states of the type IIb theory form the massless D = 10 chiral N = 2 supergravity multiplet consisting of

$|i\rangle \otimes |\tilde{j}\rangle$, $|a\rangle \otimes |\tilde{b}\rangle$ 128 boson states

$|i\rangle \otimes |\tilde{a}\rangle$, $|a\rangle \otimes |\tilde{i}\rangle$ 128 fermion states

where $|\ \rangle$ and $|\tilde{\ }\rangle$ indicate the open-string states in each oscillator space. [In the IIa theory the two spaces have fermion ground states of opposite type, $|\dot{a}\rangle$ and $|\tilde{a}\rangle$.]

The type I closed-string states are obtained as a truncation of the type IIb theory by symmetrizing the states between the two types of spaces. This halves the number of ground states which now form the massless supermultiplet of N=1 supergravity in ten dimensions.

More details of the SO(8) formalism are given by Brink in his contribution to this workshop.

(d) SU(4) × U(1) Formalism[21].

The fact that the Θ's are self-conjugate (and therefore do not anti-commute with themselves) means that they are simultaneously "position" and "momentum" variables. This will not be satisfactory for formulating a field theory of superstrings in which the fields are functions of the coordinates. For the type IIb theory it is possible to take the complex combinations $\theta^{1a} + i\theta^{2a}$ and $\theta^{1a} - i\theta^{2a}$, etc. to be position and momentum variables[22] but this does not adapt to N = 1 theories (type I or heterotic). A more satisfactory resolution is to form complex combinations of the components of a single spinor which amounts to breaking the manifest SO(8) symmetry down to SU(4) × U(1) by the identifications (for the type IIb theory)

$$\begin{aligned} \theta^{1a} &\rightarrow \theta_A \,,\ \lambda^A \\ \theta^{2a} &\rightarrow \tilde{\theta}_A \,,\ \tilde{\lambda}^A \\ 8 &\rightarrow \bar{4}_{1/2} \,,\ 4_{-1/2} \end{aligned} \tag{1.55}$$

where the last line indicates the SU(4) × U(1) content and an upper index indicates an SU(4) spinor whereas a lower index indicates an SU(4) anti-spinor (and $\theta^{\bar{A}} \equiv \theta_A$). [For the type IIa theory the SU(4) indices on the spinors is changed to θ_A, $\tilde{\theta}^A$, λ^A and $\tilde{\lambda}_A$.] Similarly

$$\begin{aligned} X^I &\rightarrow X^i \,,\ X^L \,,\ X^R \\ 8 &\rightarrow 6_0 \,,\ 1_1 \,,\ 1_{-1} \end{aligned} \tag{1.56}$$

where $i = 1,2,..,6$ labels the 6 of SU(4) and $X^L \equiv (X^7+iX^8)/\sqrt{2}$, $X^R \equiv (X^7-iX^8)/\sqrt{2}$ are SU(4) singlets. The fact that an SU(4) subgroup is picked out amounts to treating the six transverse dimensions differently from the other two. Since we hope that six dimensions will compactify in the end such a formalism is certainly adequate for our purposes.

In writing the mode expansions of the coordinates in the interacting theory it will prove useful to use a normalization of the parameter σ such that it spans the region

$$0 \leqslant \sigma \leqslant 2\pi|p^+| \equiv \pi|\alpha| \tag{1.57}$$

where $\alpha = 2p^+$ as introduced by Mandelstam[23]. The anticommutation relations in eq. (1.47) are now given by

$$\{ \lambda^A(\sigma,\tau) \, , \, \theta_B(\sigma',\tau) \} = \delta^A{}_B \delta(\sigma-\sigma') = \{ \tilde{\lambda}^A(\sigma,\tau) \, , \, \tilde{\theta}_B(\sigma,\tau) \} \qquad (1.58)$$

For diversity, I will consider the mode expansions for the type IIb theory in this subsection. These are

$$X^I = x^I + p^I\tau + \frac{1}{2}i\sum_{n\neq 0} \frac{1}{n}(\alpha^I_n e^{2in(\tau+\sigma)/|\alpha|} + \tilde{\alpha}^I_n e^{2in(\tau-\sigma)/|\alpha|}) \qquad (1.59a)$$

$$\theta_A = \frac{1}{\alpha} \sum_{-\infty}^{\infty} Q_{An} e^{-2in(\tau-\sigma)/|\alpha|} \qquad (1.59b)$$

$$\tilde{\theta}_A = \frac{1}{\alpha} \sum_{-\infty}^{\infty} \tilde{Q}_{An} e^{-2in(\tau+\sigma)/|\alpha|} \qquad (1.59c)$$

$$\lambda^A = \frac{1}{\pi|\alpha|} \sum_{-\infty}^{\infty} Q^A_n \, e^{-2in(\tau-\sigma)/|\alpha|} \qquad (1.59d)$$

$$\tilde{\lambda}^A = \frac{1}{\pi|\alpha|} \sum_{-\infty}^{\infty} \tilde{Q}^A_n \, e^{-2in(\tau+\sigma)/|\alpha|} \qquad (1.59e)$$

(and $P^I(\sigma,\tau) = \frac{\dot{X}^I}{\pi}(\sigma,\tau)$ as before). The α_n and $\tilde{\alpha}_n$ modes satisfy the commutation relations of eq. (1.48) whereas the Q_n's and $\tilde{Q}_n$'s satisfy

$$\{ Q^A_m \, , \, Q_{Bn} \} = \alpha \, \delta_{m+n,0} \, \delta^A{}_B = \{ \tilde{Q}^A_m \, , \, \tilde{Q}_{Bn} \} \qquad (1.60)$$

with the other anticommutators vanishing.

Each sixteen-component supercharge generator breaks into two SO(8) spinors, q^a and $q^{\dot{a}}$, where the undotted piece generates a linear transformation of the coordinates whereas the dotted piece is more complicated since it also performs a compensating κ transformation to restore the gauge condition. Each undotted SO(8) supercharge splits into two SU(4) spinors which are integrals of charge densities which may be determined by the Noether method to be given by

$$q_{1A}(\sigma) = \epsilon(\alpha) \, \theta_A(\sigma) \qquad\qquad q_{2A}(\sigma) = \epsilon(\alpha) \, \tilde{\theta}_A(\sigma)$$

$$q^A_1(\sigma) = \lambda^A(\sigma) \qquad\qquad q^A_2(\sigma) = \tilde{\lambda}(\sigma) \qquad (1.61)$$

(where $\epsilon(\alpha) = \mathrm{sign}(\alpha)$) and hence

$$\{ q_1^A(\sigma) , q_{1B}(\sigma') \} = \delta^A_B \, \delta(\sigma-\sigma') = \{ q_2^A(\sigma) , q_{2B}(\sigma') \} . \qquad (1.62)$$

When integrated these relations give the piece of the $N = 2$ supercharge algebra associated with the linearly realized supersymmetries. The dotted SO(8) spinors split into SU(4) spinors which will be denoted by $\bar{q}_{1A}$, $\bar{q}_1^{\,A}$, $\bar{q}_{2A}$ and $\bar{q}_2^{\,A}$. These are represented by integrals of quadratic functions of the coordinates. For example,

$$\bar{q}_1^{\,A} = \int_0^{\pi|\alpha|} \{ \sqrt{2}\rho_i^{AB} (P^i - \frac{\acute{X}^i}{\pi}) \, \theta_B + 2\pi\epsilon(\alpha)(P^L - \frac{\acute{X}^L}{\pi})\tilde{\lambda}^A \} \, d\sigma \qquad (1.63)$$

where the matrices ρ_i^{AB} are Clebsch-Gordon coefficients for SU(4) normalized so that

$$\rho_i^{AB} \rho_{jBC} + \rho_j^{AB} \rho_{iBC} = \delta_{ij} \, \delta^A_C \, . \qquad (1.64)$$

The formulae for the other q^-'s can be found in ref. 21. The rest of the anticommutation relations of the supercharge algebra can be obtained from them. These include

$$\{ \bar{q}_1^{\,A} , \bar{q}_{1B} \} = 2\delta^A_B \, h_{cl} = \{ \bar{q}_2^{\,A} , \bar{q}_{2B} \} \qquad (1.65)$$

$$\{ q_1^A , \bar{q}_1^{\,B} \} = \sqrt{2} \, \rho_i^{AB} \, P^i = \{ q_2^A , \bar{q}_2^{\,B} \} \qquad (1.66)$$

$$\{ q_{1A} , \bar{q}_{1B} \} = \sqrt{2} \, \rho_{iAB} \, P^i = \{ q_{2A} , \bar{q}_{2B} \} \qquad (1.67)$$

The subsidiary condition $N = \tilde{N}$ must be used in verifying the closure of the algebra. The hamiltonian, h_{cl}, which appears in eq.(1.65) is given by

$$h_{cl} = \frac{1}{\pi} \int_0^{\pi|\alpha|} d\sigma \, \{\epsilon(\alpha)(\pi^2 \underline{P}^2 + \underline{\acute{X}}^{\,2}) - 2\pi i(\Theta_A \acute{\lambda}^A + \tilde{\Theta}_A \acute{\tilde{\lambda}}^A)\} \tag{1.68}$$

$$= \frac{4}{\alpha}(N + \tilde{N}) + \frac{\underline{P}^2}{\alpha} \tag{1.69}$$

where

$$N = \sum_{n\neq 0} \left(\frac{1}{2} \alpha^I_{-n} \cdot \alpha^I_n + \frac{n}{\alpha} Q_{-nA} Q^A_n\right) \tag{1.70}$$

$$\tilde{N} = \sum_{n\neq 0} \left(\frac{1}{2} \tilde{\alpha}^I_{-n} \cdot \tilde{\alpha}^I_n + \frac{n}{\alpha} \tilde{Q}_{-nA} \tilde{Q}^A_n\right) \quad . \tag{1.71}$$

The Lorentz generators $J^{\mu\nu}$ may be represented in analogous fashion and the closure of the Lorentz algebra verified, including the notorious term $[J^{I-}, J^{J-}]$. This term only vanishes for the ten-dimensional theory obtained from the covariant action. If the superstring theories in D = 3, 4, or 6 dimensions are assumed to have purely transverse excitations in the light-cone gauge this commutator is found not to vanish[24] (at least with the usual definitions of the Lorentz generators). This reinforces the observation that these theories are not in their critical dimensions.

In this SU(4) formalism the massless ground states are described by wave functions that depend on Θ_0 and $\tilde{\Theta}_0$ which are the wave functions of the corresponding super-Yang-Mills or supergravity point field theories. The excited states are constructed as before by operating on the ground states with the creation operators.

The states of the type II theories are tensor products of states in the untilde and tilde Fock spaces obtained by applying α_{-n}, $\tilde{\alpha}_{-n}$, Q_{-n} and $\tilde{Q}_{-n}$ to the massless ground state N = 2 supergravity multiplet. The subsidiary condition $N = \tilde{N}$ must also be enforced.

The type I closed-string states are obtained from the type IIb states by symmetrizing them in the oscillators α and $\tilde{\alpha}$ as well as Q and $\tilde{Q}$. This truncates the supersymmetry to N = 1.

The open-string algebra has one ten-dimensional supercharge (since $\theta_0 = \tilde{\theta}_0$ due to the boundary conditions) given by averaging the N = 2 supercharges so that any of the various SU(4) components of the supercharges is given by

$$q = \frac{1}{2}(q_1 + q_2) \quad . \tag{1.72}$$

The open-string states are created by just one set of α_n and Q_n modes operating on the ground-state Yang-Mills supermultiplet.

Rules for calculating scattering amplitudes can be formulated in the first-quantized formalism by inventing vertices for emitting on-shell ground states which are consistent with the constraints of supersymmetry. This was how many of the calculations were initially performed both for tree diagrams[20] and for one-loop diagrams[42].

II LIGHT-CONE-GAUGE FIELD THEORY OF SUPERSTRINGS

The first-quantized superstring theory is adequate for calculations in perturbation theory around flat space-time in the critical dimension. The amplitudes for the bosonic and spinning string theories have been constructed in the light-cone gauge[23] by a generalization of Feynman's path-integral formalism, as sums over all possible connected world-sheets that join the incoming and outgoing strings. The interactions occur at the ponts where the world-sheet splits. The "Born" (or "tree") diagrams are given by summing over those surfaces which have no handles attached and no holes cut out (in the case of an open string theory where the boundary of a hole corresponds to the world-line of a string endpoint). Adding handles or holes corresponds to the higher order corrections (associated with loop diagrams, for example). The functional treatment of the covariant formulation of string theory was suggested[25] some time ago but not studied in depth until recently[7]. The functional formulation of the perturbation theory diagrams for superstring theories is the subject of intensive study both in the light-cone gauge[26] and in a covariant formulation.

Just as in any point particle theory this first-quantized formulation is probably not appropriate for understanding non-perturbative aspects, such as the compactification of the theory. In order to develop our understanding of these theories it will be necessary to study them in the language of string field theory in which string fields create and destroy complete strings. There is, presumably, something like a geometrical formulation of the theory that generalizes the familiar description of general relativity in terms of the geometry of space-time. Although there has been progress towards a supercovariant, gauge-invariant interacting string field theory[27], this is only incompletely understood for the moment.

The interacting quantum field theory of strings has, however, been formulated in the light-cone gauge, both for the older string theories[28] and superstring theories[21,22,29]. This is not completely satisfying since it is likely that any deep geometric structure in the theory will be obscured by the choice of a special gauge. Nevertheless it may be adequate for certain purposes, such as studying compactified solutions in which four dimensions are flat (including the + and - directions). This field theory formalism also provides a systematic way of generating the light-cone-gauge superstring perturbation theory diagrams including certain local operators that have to be inserted at the interaction points where the world-sheet splits. [Such operators do not occur in the bosonic theory.]

In this lecture I will outline the way in which the superstring fields are defined and the construction of the free field representation of the super-Poincaré algebra. I will then show how the interaction terms are almost uniquely specified by demanding that the algebra be represented non-linearly on the fields in the interacting theory. The whole discussion is a generalization of conventional point field theory. However, one particularly striking general feature of the light-cone gauge field theory of closed strings is the fact that the interactions are simply cubic in the string fields (in contrast with point field theories containing the Einstein-Hilbert action which involve infinite orders of interactions between the fluctuations of the metric). This is intuitively plausible from the fact that the string interactions correspond to critical points on the world-sheet, which generically involve the splitting of one string into two or the joining of two strings into one. I will show how this is required by the consistency of the supersymmetry algebra in the interacting field theory. Most of this lecture is based on material contained in ref. (21).

(a) Light-Cone-Gauge Superstring Fields

A string field is a scalar *functional* of the light-cone string superspace coordinates. In the light-cone gauge it is often useful to use a Fourier transform with respect to X^- so that the momentum p^+ is a variable ($p+ = i\partial/\partial X^-$). A string field is then a function of $\Theta(\sigma)$, $\tilde{\Theta}(\sigma)$, $\underset{\sim}{X}(\sigma)$, X^+ and p^+ (or α) which means that it is a function of the configuration of the whole string and is not an explicit function of σ. By convention a field operator with $p^+ > 0$ will be a creation operator whereas one with $p^+ < 0$ will be an annihilation operator. A closed-string field will be denoted by

$$\Psi[\underset{\sim}{X}(\sigma),\Theta_A(\sigma),\tilde{\Theta}_A(\sigma),\alpha,X^+]$$

for a type IIb (i.e. chiral N=2) field (the non-chiral type IIa field is $\Psi[\underset{\sim}{X}(\sigma),\Theta_A(\sigma),\tilde{\Theta}^A(\sigma),\alpha,X^+]$ i.e. it is a function of $\tilde{\Theta}$ in a 4 instead of one in a $\bar{4}$). An open-string field is denoted by

$\Phi[\mathbf{X}(\sigma),\theta_A(\sigma),\tilde{\theta}_A(\sigma),\alpha,X^+]$

These fields describe an infinite set of ordinary point fields, one for every state of excitation of the string. The open-string field is also a matrix in an internal symmetry group. By direct analogy with ordinary scalar field theory in light-cone coordinates[30] the action S is related to the field-theory hamiltonian H by

$$S = \int D^{16}Z\ dX^+\ dX^-\ (\partial_+\Psi\partial_-\Psi + \mathrm{tr}\partial_+\Phi\partial_-\Phi) - \int dX^+\ H \qquad (2.1)$$

Note that $\partial_\pm = \partial/\partial X^\pm$ so that the action is linear in time derivatives (if H is independent of them). In eq. (2.1) $D^{16}Z$ denotes the functional integration over the eight transverse coordinates $\mathbf{X}(\sigma)$ and eight Grassmann coordinates $\theta_A(\sigma)$, $\tilde{\theta}_A(\sigma)$ i.e.

$$D^{16}Z \equiv D^8X^I(\sigma)\ D^4\theta_A(\sigma)\ D^4\tilde{\theta}_A(\sigma) \qquad (2.2)$$

which will be interpreted as the infinite product of the differentials of the modes of the coordinates.

For closed strings it is important to impose the condition that the superfield Ψ does not depend on the origin of the σ parameter

$$\Psi[X^I(\sigma+\sigma_0),\theta_A(\sigma+\sigma_0),\tilde{\theta}_A(\sigma+\sigma_0)] = \Psi[X^I(\sigma),\theta_A(\sigma),\tilde{\theta}_A(\sigma)] \qquad (2.3)$$

where σ_0 is an arbitrary constant. For infinitessimal σ_0 this equation becomes

$$\int_0^{\pi\alpha} d\sigma \left\{ X'^I\frac{\delta}{\delta X^I} + \theta'_A\frac{\delta}{\delta\theta_A} + \tilde{\theta}'_A\frac{\delta}{\delta\tilde{\theta}_A} \right\} \Psi = 0 \qquad (2.4)$$

This is exactly the same condition as $N-\tilde{N} = 0$ obtained earlier (eq. (1.54)) and in this context is to be imposed as a (first class) constraint on the string fields in the action. The massless type IIb states, in particular, are described by the ground state component superfield $\Psi(\mathbf{x},\theta_{oA},\theta_{oA},\alpha,x^+)$ which has 2^8 terms when expanded in powers of Θ, corresponding to the states of type IIb supergravity.

Type I closed string fields satisfy the additional constraint that expresses the fact that they are unoriented

$$\Psi[X^I(\sigma),\Theta_A(\sigma),\tilde{\Theta}_A(\sigma)] = \Psi[X^I(-\sigma),\tilde{\Theta}_A(-\sigma),\Theta_A(-\sigma)] \tag{2.5}$$

The symmetry under $\sigma \leftrightarrow -\sigma$ and $\Theta \leftrightarrow \tilde{\Theta}$ expressed by eq. (2.5) implies (from the mode expansions in eqs. (1.59)) that the physical states are symmetric under the interchange of (α_n,Q_n) and $(\tilde{\alpha}_n,\tilde{Q}_n)$ as required for type I closed strings.

Open strings carry quantum numbers at their endpoints associated with the defining representation of one of the classical groups SO(n), U(n) or Sp(2n) which appear to be the only ones that can be incorporated in this manner[31] in the classical theory. In the quantum theory anomalies rule out all groups except SO(32). In this ("Chan-Paton") scheme[32] the indices a and b of Φ_{ab} are associated with either endpoint of the string and the particular group G represented in this way depends on the conditions imposed on the Φ. These conditions are designed so that the massless ground states lie in the adjoint representation of G. For example, the group SO(n) is obtained by requiring Φ_{ab} to be real with a,b = 1,2,...,n and imposing the condition

$$\Phi_{ab}[X^I(\sigma),\Theta_A(\sigma),\tilde{\Theta}_A(\sigma)] = -\Phi_{ba}[X^I(\pi|\alpha|-\sigma),\tilde{\Theta}_A(\pi|\alpha|-\sigma),\Theta_A(\pi|\alpha|-\sigma)] \tag{2.6}$$

This relates a superfield to one with reversed orientation which (as follows from the mode expansions of eqs. (1.59)) identifies states which differ by the substitutions $\alpha_n^I \rightarrow (-1)^n\alpha_n^I$ and $\Theta_{An} \rightarrow (-1)^n\tilde{\Theta}_{An}$. This means that the even mass levels (with masses satisfying $(\text{mass})^2/2\pi T = 0,2,4\ldots$) have component fields satisfying $\Phi_{ab} = -\Phi_{ba}$ and therefore lie in the adjoint representation. The odd levels (with $(\text{mass})^2/2\pi T = 1,3,5,\ldots$) have components satisfying $\Phi_{ab} = \Phi_{ba}$.

(b) Free Field Theory of Superstrings

For every generator of the super-Poincaré algebra g, which was expressed in terms of the coordinates and momenta in the first lecture, we now associate a field theory generator G made out of the string superfields. In general this generator will be a complicated function of the fields which I shall assume can be expanded as a series

$$G = G_2 + G_3 + \ldots . \tag{2.7}$$

where G_2 is quadratic in fields, G_3 is cubic and so on. Actually, this counting works for purely closed-string theories or for the purely open-string part of the type I theories. However, in the terms involving mixtures of open and closed strings a closed string field counts as two powers of an open string field. The G's are constructed to satisfy the same algebra as the g's.

In the free theory all the G's are quadratic and have the form

$$G_2 = \int_0^\infty \alpha d\alpha \int D^{16}Z \; (\Psi_{-\alpha}\, g\, \Psi_\alpha + \mathrm{tr}\Phi_{-\alpha}\, g\, \Phi_\alpha) \tag{2.8}$$

where the dependence of the fields on α (i.e. on p^+) is explicitly displayed (and reality of the fields implies $\Psi_{-\alpha} = \Psi_\alpha^*$ and $\Phi_{-\alpha} = \Phi_\alpha^\dagger$). The form of eq. (2.8) generalizes that of conventional scalar field theory expressed in light-cone coordinates. [Note that the normalization of the string fields implicit in this equation differs by a power of $\sqrt{\alpha}$ from that adopted by some authors.] The G's satisfy an algebra that is isomorphic to the algebra of the g's as will be seen by use of Poisson brackets.

The momenta conjugate to the fields, Π_Ψ and Π_Φ, are defined by

$$\Pi_\Psi \equiv \frac{\delta S}{\delta \partial_+ \Psi} = \partial_- \Psi \qquad \Pi_\Phi \equiv \frac{\delta S}{\delta \partial_+ \Phi} = \partial_- \Phi \tag{2.9}$$

These are phase-space constraints since the momenta are proportional to "spatial" derivatives of the fields. These constraints lead to a simple modification of the canonical Poisson brackets as is well-known in scalar light-cone field theory[30]. Even though all the arguments of this lecture are at the level of the classical string field theory (i.e. I will not treat the fields as quantum operators so that the ordering of the fields will never matter) I shall describe these constrained Poisson brackets as if they were equal time quantum commutation relations by including appropriate powers of i. These relations, analogous to those of scalar point field theory can be deduced simply from the requirement that the G's defined by eq. (2.8) form a closed algebra. For closed strings this gives

$$[\Psi[1]\ ,\ \Psi[2]] = \frac{\delta(\alpha_1+\alpha_2)}{\alpha_2}\int_0^{\pi|\alpha_2|}\frac{d\sigma_0}{\pi|\alpha_2|}\,\Delta^{16}[Z_1(\sigma)-Z_2(\sigma+\sigma_0)] \tag{2.10}$$

where the integral over σ_0 takes into account the constraint of eq. (2.3) and $\Delta^{16}(Z)$ represents the product of an infinite number of Dirac delta functions, one for each mode of X^I, θ_A and $\tilde{\theta}_A$. For type I closed strings there is also a projection on the right-hand side which imposes the constraint of eq. (2.5). For open strings the commutation relations which are consistent with the group theory constraints are (when the internal group is SO(32)),

$$\begin{aligned}[\Phi_{ab}[1]\ ,\ \Phi_{cd}[2]] = \frac{\delta(\alpha_1+\alpha_2)}{2\alpha_2}\Big[&\delta_{ab}\delta_{cd}\Delta^{16}(Z_1(\sigma)-Z_2(\sigma)) \\ &- \delta_{ac}\delta_{bd}\Delta^{16}(Z_1(\sigma)-Z_2(\pi|\alpha_2|-\sigma))\Big]\end{aligned} \tag{2.11}$$

with analogous expressions for the other possible groups.

From these commutation relations it is easy to check that the G_2's (defined in eq. (2.8) generate the appropriate changes in the fields,

$$[G_2,\Psi] = g\Psi \qquad \text{and} \qquad [G_2,\Phi] = g\Phi \tag{2.12}$$

For example, the time dependence or the fields is generated by the hamiltonian H_2, with h being given (eq. (1.69)) by the sum of harmonic oscillator hamiltonians (and their Grassmann equivalents). The time (i.e. X^+) dependence of the free field theory is therefore determined by the equation of motion

$$i \frac{\partial \Psi}{\partial X^+} = h \Psi \tag{2.13}$$

with a similar equation for the open string field. A string field can be expanded as a sum over a complete set of eigenfunctions of h in the form (for open strings)

$$\Phi[X(\sigma), \theta_A(\sigma), \tilde{\theta}_A(\sigma)] =$$

$$\sum_{\{n_k^I, m_{As}\}} \Phi_{\{n_k^I, m_{As}\}}(x, \theta_0) \prod_k \prod_I H_{n_k^I}(x_k^I) \prod_s \prod_A \Omega_{m_{As}}(\theta_{As}) \tag{2.14}$$

where the H_n's are Hermite functions (eigenfunctions of the piece of h involving bosonic modes) whereas the Ω_n's are the Grassmann eigenfuntions. The numbers n_k^I and m_{As} are the occupation numbers of the levels labelled by (k,I) for the Bose modes, x_k^I, and (s,A) for the Fermi modes, θ_{As}. The coefficients $\Phi_{\{n\}}$ are ordinary point superfields which are either fermionic or bosonic depending on whether the level {n} has an even or odd number of θ excitations. By substituting this expansion into the free field action it can be written as an infinite sum of ordinary point field free field actions

$$S = \int_0^\infty d\alpha \int dx\, d\theta_0 \sum_{\{n\}} \left\{ \frac{1}{2} \Phi_{-\alpha\{n\}} (\Box - N_{\{n\}}) \Phi_{\alpha\{n\}} + \frac{1}{\alpha} \Psi_{-\alpha\{n\}} (\Box - N_{\{n\}}) \Psi_{\alpha\{n\}} \right\} \tag{2.15}$$

where the fermionic superfields have been denoted $\Psi_{\{n\}}$ to distinguish them from the bosonic ones.

(c) Interacting Superstring Fields.

The choice of the light-cone gauge spoils the manifest super-Poincaré invariance of a theory. As a result, in the interacting light-cone-gauge field theory certain of the super-Poincaré generators act non-linearly on the fields. These are the generators that incorporate compensating gauge transformations which restore the light-cone gauge conditions, namely, J^{+-}, J^{I-}, $Q^{-}{}_{A}$, $\tilde{Q}^{-}{}_{A}$, Q^{-A}, $\tilde{Q}^{-A}$ and P^{-} (which is the hamiltonian, H). The form of these generators can be determined uniquely by demanding that they satisfy the super-Poincaré algebra (and assuming that they contain no more than two spatial derivatives). In fact, requiring the supersymmetry sub-algebra to be satisfied already determines the form of the generators almost uniquely. The appropriate equations are

$$\{ Q^{-A} , Q^{-}_{B} \} = 2 \, \delta^{A}{}_{B} \, H \tag{2.16}$$

$$\{ Q^{-A} , Q^{-B} \} = 0 = \{ Q^{-}{}_{A} , Q^{-}{}_{B} \} \tag{2.17}$$
$$\vdots$$

By substituting the series expansions (eq. (2.7)) for the Q's and H into these equations we can satisfy them order by order in the string fields (allowing for the comment after eq. (2.7) concerning terms involving mixtures of open and closed string fields).

(d) The Cubic Interaction of Open Superstrings.

In order to illustrate the method I will begin with the term in the interaction hamiltonian which describes the splitting of one open string into two or the joining of two open strings into one as illustrated in fig. 2.1a.

Fig. 2.1a

We want to solve the equations like (2.16) and (2.17) by substituting the series expansions (eq. (2.7)) for the generators. The lowest order involves only the quadratic pieces of the generators and the equations are satisfied by the free field theory expressions. At the next order eqs. (2.16) and (2.17) give (for terms involving purely open-string or purely closed-string fields)

$$\{ Q_3^{-A}, Q_{2B}^{-} \} + \{ Q_2^{-A}, Q_{3B}^{-} \} = 2\, \delta^A_B\, H_3 \qquad (2.18)$$

$$\vdots$$

Needless to say the rest of the superalgebra, which involves those generators that generate linear transformations and do not get interaction corrections, provides strong constraints on the form of H_3, and the Q_3^-'s. Together with eqs. (2.18) these constraints almost completely determine these generators. The small remaining ambiguity involves overall powers of α (i.e. p^+) in the expressions for the interactions. In principle, these could be fixed by using the Lorentz generators J^{I-} and J^{+-} (but in practice it is easier to fix them by treating a special case with external ground-state particles).

The interaction can best be represented on a plot of the σ–τ space first introduced by Mandelstam[23] (with the interaction taken at $\tau = 0$).

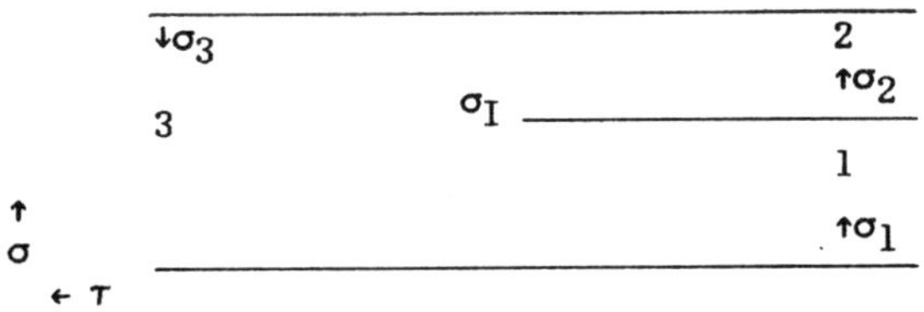

Fig. 2.1b

The boundaries of the open strings are represented by solid lines. Each string has its own parameter σ_r (where $r = 1,2,3$ labels the string) which runs between 0 and $\pi\alpha$ in a sense that depends on whether the string is incoming or outgoing as indicated in the figure.. The α's are chosen to be positive for incoming strings (1 and 2) and negative for outgoing ones (3). The coordinates on each string, $X^{(r)}(\sigma_r)$, $\theta^{(r)}(\sigma_r)$ and $\tilde{\theta}^{(r)}(\sigma_r)$ (setting $\tau=0$), have mode expansions like eqs. (1.59) on each string. It is useful to use one parameter, $\sigma = -\pi\alpha_3 - \sigma_3 = \sigma_1 = \pi\alpha_1 + \sigma_2$, and to define $X_r(\sigma) \equiv X^{(r)}(\sigma_r)$ (with the understanding that X_r vanishes outside of the region in σ in which string r is defined) with similar definitions for the Grassmann coordinates.

One physical ingredient that is built into the form of the interaction hamiltonian is that the coordinates should be continuous at the interaction time. This means that at $\tau = 0$ we must enforce the conditions

$$\sum_{r=1}^{3} \epsilon_r Z_r(\sigma) = 0 \tag{2.19}$$

where $\epsilon_r = \mathrm{sign}(\alpha_r)$ and $Z_r(\sigma) \equiv (X_r(\sigma), \theta_r(\sigma), \tilde{\theta}(\sigma))$. This is incorporated into the expression for any of the generators by a delta functional. The general form of a generator is therefore

$$G_3 = g\int (\prod_{r=1}^{3} d\alpha_r D^{16}Z_r)\delta(\Sigma\alpha_r)\Delta^{16}[\Sigma\epsilon_r Z_r]\hat{G}(\sigma_I)\,\mathrm{tr}(\Phi[1]\Phi[2]\Phi[3]) \tag{2.20}$$

where $\hat{G}$ is an operator that acts on the fields at the interaction point σ_I. The notation $\Phi[r]$ denotes the field as a functional of the coordinates of string r. The expression for G_3 can also be written in terms of the Fourier transform of the field $\tilde{\Phi}$ in which case the coordinate *continuity* Δ funtional is replaced by one expressing the *conservation* of momenta, $\underset{\sim}{P}_r(\sigma)$, $\lambda_r^A(\sigma)$, $\tilde{\lambda}_r^A(\sigma)$ at the interaction time.

This form of the interaction terms is not only motivated by the physical picture of the strings interacting at their endpoints but also guarantees that the linearly realized symmetries are correctly incorporated. These are the symmetries generated by J^{IJ}, J^{I+}, p^I, Q^A and Q_A which do not get interaction corrections.

In the case of the bosonic string theory considered in ref. 28 the form of the interaction was guessed to be like eq. (2.20) without the operator $\hat{G}$ (and obviously with no fermionic variables) but with certain extra factors in the integration measure. This expression was shown to agree with that obtained by Mandelstam in ref.23 (who had obtained the vertex by considering the Feynman path integral for the infinite time process shown in the fig. 2.1 and then factored off the external legs). The major difference in the case of the superstring theories (as well as in the spinning string theory) is the occurrence of the operator $\hat{G}$ at the interaction point.

The expression for G_3 can be written in terms of the component fields by substituting the expansion of eq. (2.14) into eq. (2.20) and integrating over the non-zero modes of the coordinates. This gives an expression of the form

$$G_3 = \int \left(\prod_{r=1}^{3} d\alpha_r d^8 x_r d^4 \theta_{0r}\right) \sum_{\{n_1, n_2, n_3\}} C_{\{n_1 n_2 n_3\}} \; \Phi_{\{n_1\}} \Phi_{\{n_2\}} \Phi_{\{n_3\}} \tag{2.21}$$

where the coefficients $C_{\{n_1 n_2 n_3\}}$ determine the interactions between states of arbitrary occupation numbers, denoted $\{n_r\}$ for string r. These coefficients can be written in terms of states of definite occupation numbers for each string as

$$C_{\{n_1 n_2 n_3\}} = \langle n_1 | \otimes \langle n_2 | \otimes \langle n_3 | \; G \rangle \tag{2.22}$$

where $|G\rangle$ is a "vertex" ket vector defined in the tensor product of the Fock spaces of the three strings (and $\langle n_r |$ is the bra vector for string r in a state with occupation numbers $\{n_r\}$). This relation can be inverted to give

$$| \; G \; \rangle = \sum_{\{n_1 n_2 n_3\}} C_{\{n_1 n_2 n_3\}} \; | n_1 \rangle \otimes | n_2 \rangle \otimes | n_3 \rangle \tag{2.23}$$

For all the generators this can be written in the form

$$|G\rangle = \hat{G}\ |V\rangle \tag{2.24}$$

where $\hat{G}$ is the oscillator-basis representation of the operator $\hat{G}(\sigma_I)$ that occurred in eq. (2.20) and the vertex $|V\rangle$ is the oscillator-basis representation of the Δ functional in eq. (2.20). It is given by the exponential of a quadratic form in the creation operators for each string, α^r_{-n} and θ^r_{-n} acting on the product of the three ground states. The explicit expression for $|V\rangle$ is given in ref. 21 (and generalizes that of the bosonic string theory which only involved the α modes).

The representation of the theory in the oscillator basis gives well-defined expressions with no singular operators. This makes it convenient for deriving the explicit expressions for the $\hat{G}$'s by solving the equations like eq. (2.18). In the position (or momentum) basis there are delicacies in defining various operators (such as the momentum, $\underline{P}(\sigma)$) in the vicinity of the interaction point since they are singular and must be regulated. This singular behaviour is essential for there to be a non-trivial interaction. However, (as stressed in ref. (36)) in the oscillator basis it is not at all obvious that the theory is local on the world-sheet. This locality is an essential ingredient in understanding duality which therefore appears to be an accidental miracle in the oscillator basis. The position-space representation of the interactions can be reconstructed in the form given in eq. (2.20) where the operator $\hat{G}$ is given as a polynomial in the functional derivative

$$Z^J = -\, i\ \sqrt{\pi}\ \lim_{\epsilon\to 0} \sqrt{\epsilon}\ \frac{\delta}{\delta X^J(\sigma_I-\epsilon)} \tag{2.25}$$

and a Grassmann coordinate

$$Y_A = \lim_{\epsilon\to 0} \sqrt{\epsilon}\ [\theta_A(\sigma_I-\epsilon) + \tilde{\theta}_A(\sigma_I-\epsilon)]/\sqrt{2} \tag{2.26}$$

where the factors of $\sqrt{\epsilon}$ cancel the singular behaviour of the operators near the interaction point. It does not matter on which string these operators are defined - they all give the same limiting answer. In terms of these operators the generators, $\hat{G}$, appearing in the supercharges are given by

$$\hat{Q}^{-}_{3A} = Y_A \tag{2.27}$$

$$\hat{Q}^{-A}_{3} = \frac{2}{3} \epsilon^{ABCD} Y_B Y_C Y_D \tag{2.28}$$

$$H_3 = \frac{1}{\sqrt{2}} Z^L + \frac{1}{2} (\rho^i)^{AB} Z^i Y_A Y_B + \frac{1}{6\sqrt{2}} Z^R \epsilon^{ABCD} Y_A Y_B Y_C Y_D \tag{2.29}$$

The fact that the expression for the interaction term in the hamiltonian is linear in Z (i.e. in the momentum $P(\sigma_I)$) is reminiscent of the usual Yang-Mills cubic interaction. In fact, it is easy to isolate the term involving just the zero modes (the usual light-cone gauge Yang-Mills superfields) in the interaction from eq. (2.20). This term has the form of the familiar light-cone gauge cubic super-Yang-Mills interaction but smeared out by a gaussian factor (due to the fact that the centres of masses of the three strings do not coincide). In the limit of infinite string tension the interaction reduces to that of usual point super-Yang-Mills field theory.

(e) Open - Closed Superstring Interaction.

The fact that the interaction of fig. 1 is local on the string means that the same joining or splitting process can take a single open string into a closed string and vice versa.

Fig. 2.2

The expression for this interaction, which involves $g\Psi \mathrm{tr}\Phi$, is again deduced by determining its contribution to the supersymmetry generators[21]. [It was also discussed in detail in the bosonic theory using a lattice regulator in the σ-τ plane[34].] As expected on the grounds of locality the expressions for the generators involve the same local operator $\hat{G}(\sigma_I)$ acting at the joining or splitting point as in the cubic open-string interaction.

(f) **Closed Supertring Interactions.**

Two oriented (type II) closed strings can interact by touching at a point

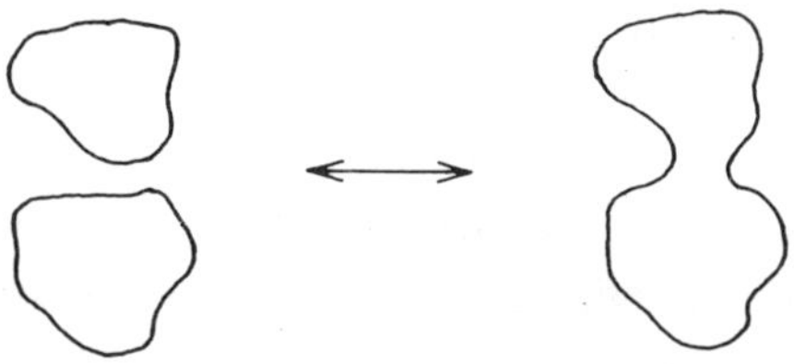

Fig. 2.3

This is a local interaction which is cubic in the closed-string fields (i.e. contains Φ^3) and is of gravitational strength κ (since it reduces to the cubic gravitational coupling in the low energy limit). It has a form that is very simply related to the cubic open-string interaction described above. The operator acting on the fields in the interaction terms for any of the generators is the product of two open-string factors. In the oscillator basis these generators are given by tensor products of two factors where one factor is made of the untilde modes and the other of the tilde ones.

$$| Q_1^{-A} \rangle = | Q^{-A} \rangle \otimes | \tilde{H} \rangle \tag{2.30}$$

$$| Q_2^{-A} \rangle = | H \rangle \otimes | \tilde{Q}^{-A} \rangle \tag{2.31}$$

$$| Q^-_{1A}\rangle = |Q^-_A\rangle \otimes | \tilde{H} \rangle \tag{2.32}$$

$$| Q^-_{2A}\rangle = | H \rangle \otimes | \tilde{Q}^-_A\rangle \tag{2.33}$$

$$| H \rangle_{closed} = | H \rangle \otimes | \tilde{H} \rangle \tag{2.34}$$

where the right-hand sides of these equations involve the open-string expressions of eqs. (2.27) - (2.29). The closure of the closed-string algebra follows simply from the fact that the open-string algebra closes. The expressions for the generators in terms of the fields can be deduced from these expressions in a similar manner to the open string case.

The type I cubic closed-string interaction is obtained by symmetrizing these expressions between the two types of oscillator space.

The generators for the heterotic string (which has only one supercharge) are obtained by substituting the (twenty-six dimensional) bosonic string vertex $|\tilde{V}\rangle$ instead of the right-hand factors in eqs.(2.30)-(2.34) and compactifying the zero modes in sixteen of the dimensions on a suitable hypertorus[35]. This involves a subtle issue concerning the fact that these internal bosonic coordinates are constrained to be right-moving (or left-moving) on the world-sheet and therefore do not commute.

(g) Other Interactions

The interaction illustrated in fig. 2.3 is the only contribution to the oriented closed-superstring theories (types II or heterotic). For the (unoriented) type I theories there are a number of other interaction terms in which two internal points touch. The existence of all these terms is obviously necessary to ensure the closure of the superalgebra.

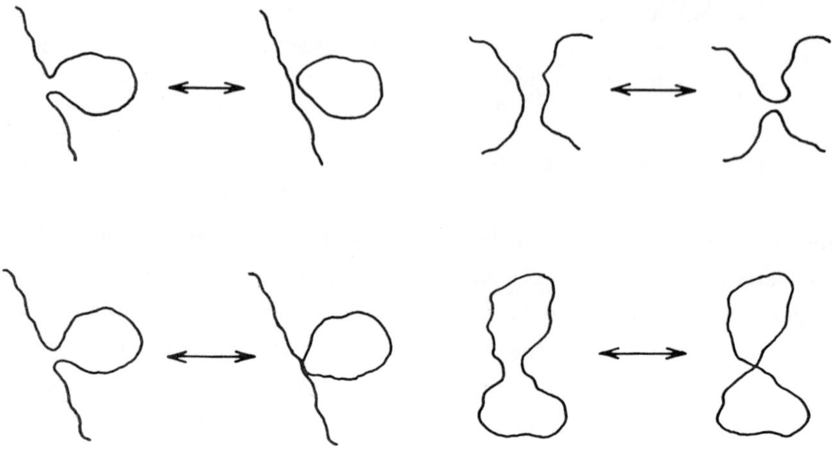

Fig. 2.4

All these terms are of strength κ and they all involve the *same* local touching interaction. [The last two were inadvertently omitted from ref. 21 but included in ref. 36.] Since the interaction is local along the string the form of the expressions for each of these terms is determined by the cubic closed-string coupling which involves the same touching interaction. This should be checked explicitly by verifying the closure of the super-Poincaré algebra including these terms.

(h) The Absence of Higher Order Interactions

The seven interaction terms depicted in the figures are the only ones needed to satisfy the closure of the algebra. None of these interactions is of higher order than cubic in closed-string fields and there is just one term, the $(\Phi)^4$ term, which is quartic in open-string fields. Notice that this only contributes to the scattering of two incoming strings into two outgoing ones (and not to the $1\to3$ or $3\to1$ processes) so that it does not contain the whole Yang-Mills contact term. This quartic term also only contributes to one of the possible group theory trace factors in a four-particle amplitude. The fact that these are the only interactions conforms with intuition based on the picture of splitting and joining of oriented and non-oriented world-sheets.

The absence of higher-order terms can be deduced by considering the next terms in the series expansion of the superalgebra. For simplicity, I shall consider the pieces of the generators containing only open-string fields (which is a subsector of the type I theory). For example, eq. (2.16) implies the relation

$$\{ Q_3^{-A}, Q_{3B}^- \} + \{ Q_2^{-A}, Q_{4B}^- \} + \{ Q_4^{-A}, Q_{2B}^- \} = 2\, \delta^A_B \; H_4 \tag{2.35}$$

Notice that in the *quantum* field theory of strings there could, in principle, also be corrections to the right-hand side arising from effects of normal ordering the string fields. The arguments presented here all treat the string fields *classically*. In order to show that the quartic terms are absent we would have to show that

$$\{ Q_3^{-A}, Q_{3B}^- \} \tag{2.36}$$

vanishes. Since we know that there is a quartic open-string interaction that contributes to a particular group theory factor the easiest way to investigate this anticommutator is to consider four-particle matrix elements between states of definite quantum numbers which exclude that factor. Fig. 2.5 depicts the diagrams that contribute with a ("Chan-Paton") factor $tr(\lambda_1\lambda_2\lambda_3\lambda_4)$ (where λ_r is a matrix in the fundamental representation of the internal symmetry group representing the quantum numbers of the particle r) in which case the possible intermediate states are a one-string state and a three-string state. The internal lines in these diagrams represent the complete sums over intermediate states (and not propagators). The labelling of the vertices indicates the occurrence of the local fermionic operators ($\hat{Q}^{-A}$ or $\hat{Q}_{\bar{B}}$) acting at the interaction points ($\sigma = \sigma_A$ and $\sigma = \sigma_B$).

+ A ↔ $\bar{B}$

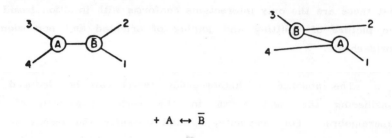

Fig. 2.5

In the σ-τ plane the two interactions occur at the same value of τ, so *both* the diagrams shown explicitly in fig. 2.5 are represented in parameter space by

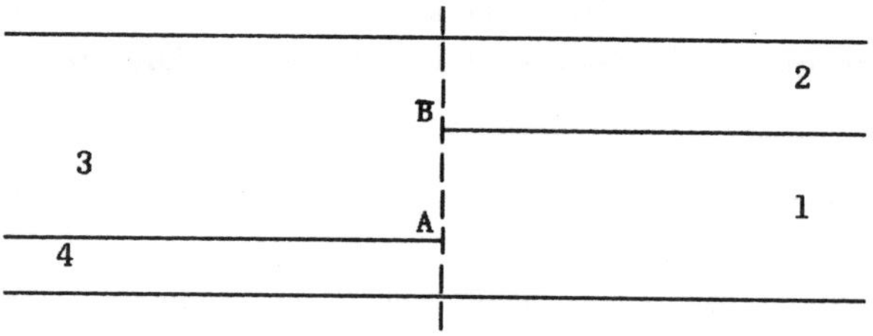

Fig. 2.6

The two cases in fig. 2.5 differ only in the order of the fermionic factors in the vertices. However,

$$\{ \hat{Q}^{-A}(\sigma_A) \ , \ \hat{Q}^{-}_{B}(\sigma_B) \} = 0 \qquad . \tag{2.37}$$

Therefore the matrix element of the anticommutator vanishes.

This reasoning breaks down if the points A and B coincide. This

only happens in the forward direction ($p_1^\mu = -p_4^\mu$ and $p_2^\mu = -p_3^\mu$) since for non-forward scattering the relative values of σ_A and σ_B can be changed by a Lorentz transformation. The result is proportional to

$$\delta^{10}(p_1 + p_4)\ \delta^4(\alpha_1\theta_1 + \alpha_4\theta_4)\ \delta^{10}(p_2 + p_3)\ \delta^4(\alpha_2\theta_2 + \alpha_3\theta_3) \qquad (2.38)$$

which suggests that the anticommutator (2.36) is proportional to the identity operator. It seems to imply an extra term in H_4 of a rather peculiar nature that only contributes to the disconnected piece of the S matrix and hence does not affect tree diagrams. Similar considerations should also apply yo the closed-string theories.

The full Yang-Mills contact term is obtained as an effective interaction at low energy due to the exchange of the massive string states between cubic vertices. This can be derived from the string field theory action by explicitly[37] integrating the massive fields in the generating functional in the limit $T \to \infty$.

It has been argued that the form of the cubic interaction between either open or closed strings derived by the methods used in this section is unique (up to powers of p+ which can only be determined by using the full Lorentz algebra). This suggests that there are no possible counterterms other than the action itself which are consistent with super-Poincaré invariance. This in turn is suggestive of finiteness or renormalizability[38]. However, the analysis has only been carried out at the level of the classical superstring theory and it ignores quantum effects that we know can give rise to anomalies in the super-Poincaré algebra.

III LOOP AMPLITUDES AND ANOMALIES

The string perturbation expansion is the generalization to string theories of the Feynman diagram expansion of point field theory. The diagrams can be thought of as representing functional integrals over all possible world-sheets joining initial and final strings including surfaces of non-trivial topology which have handles attached or holes cut out. In this lecture I will first review in outline the divergence structure of the one-loop calculations. I will then show that superstring theories are gauge invariant at the level of tree diagrams as a prelude to describing the calculation of the one-loop Yang-Mills anomaly in the type I theory. Requiring the absence of this anomaly[39] restricts the gauge group to SO(32). The one-loop diagrams are also finite for this choice of group[36].

The question of checking gauge invariance in string theories and the rules for setting up the calculation using the covariant operator formalism are well known and straightforward. The difficulties in the evaluation of the anomaly involve subtleties that have apparently caused some confusion in other covariant formulations of the problem (such as the functional formalism).

(a) Loops of Open Strings (Type I theories)

The one-loop open-string amplitude is given by a sum of contributions from path integrals over orientable and non-orientable surfaces. The contributions from planar oriented surfaces to the amplitude can be represented by any of the diagrams in fig. 3.1. The first two diagrams look like a box diagram and a self-energy diagram respectively. The equivalence of these ways of representing the amplitude illustrates the duality property of string theories. The box diagram is divergent due to the infinite sum over massive states circulating around the loop and due to the high loop momentum. It is often very interesting to represent the amplitude in the third way shown in fig. 3.1(c) which is obtained by another distortion of the world-sheet. The diagram has the form of an open-string tree diagram with a closed string emitted into the vacuum at zero momentum k^μ.

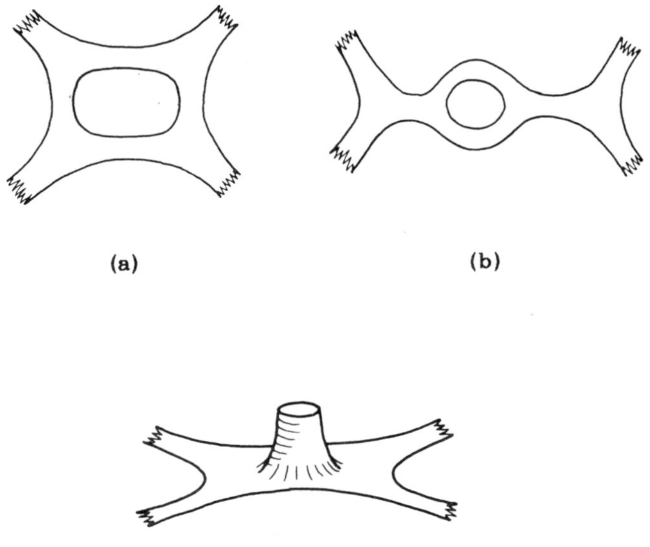

(a) (b)

(c)

Fig. 3.1

Since the closed-string sector contains a massless scalar "dilaton" state the fig. 3.1c is divergent due to the dilaton propagator, $1/k^2$, evaluated at $k^\mu = 0$. In the old string theories (the bosonic and spinning string theories) there is also a tachyon state in the emitted closed-string tadpole (which makes the divergence appear even worse in the usual representation of the loop due to the use of an invalid integral representation). The form of the divergence, first discussed in detail in ref. 40 for the bosonic theory, is that of the derivative of a tree diagram with respect to the string tension[41].

The amplitude of fig. 3.1 can also be represented by an annulus in parameter-space by an appropriate choice of the two-dimensional metric as illustrated in fig. 3.2. In the expression for the amplitude the inner radius, r, of the annulus is to be integrated from 0 to 1 (with the outer radius fixed equal to 1). The residual Möbius invariance of the integrand can then be used to fix one of the external particles at z = 1 on the outer boundary with the positions, z_r, of the N-1 other ones being integrated around the outer boundary.

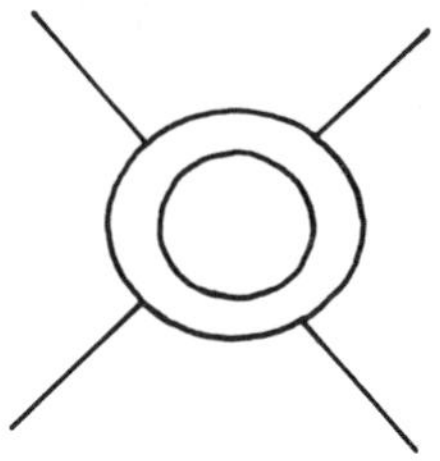

Fig. 3.2

The other divergent diagrams are represented by non-orientable surfaces (Möbius strips).

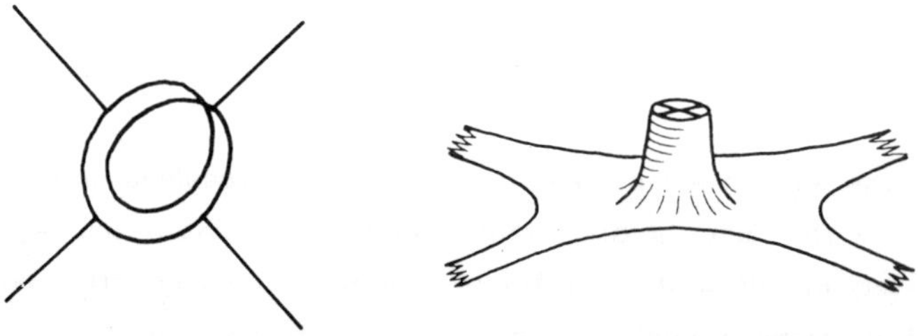

(a) (b)

Fig. 3.3

Fig. 3.3(b) illustrates a divergence due to the emission of a closed string into the vacuum at zero momentum, this time via a non-orientable cross cap (a boundary with diametrically opposite points identified).

Open strings carry quantum numbers of an internal symmetry group at their endpoints in the manner first proposed by Chan and Paton[32]. The only groups that can be included in this way at the level of the open-string tree diagrams are the classical groups SO(n), Sp(2n) and U(n)[31] (with U(n) excluded when considering the coupling of open to closed strings).

The open-string one-loop amplitudes in superstring theories (with four external ground state bosons) were considered in ref. 36,42 using the light-cone-gauge ground state emission vertices obtained in ref. 20. Certain of these amplitudes vanish due to supersymmetry trace identities. These identities are expressed in terms of supertraces of functions of the fermionic SO(8) spinor zero modes, θ_0^a, (or the S_0's of ref. 20) which enter the light-cone treatment and satisfy (eq. 1.49)

$$\{ \theta_0^a, \theta_0^b \} = \frac{2}{p^+}\delta^{ab} \quad . \tag{3.1}$$

The vertices for emitting ground-state on-shell bosons in the light-cone gauge involve operators $K^{ij} = \frac{1}{4}\,\theta_0\gamma^{ij}\theta_0$ which act like spin operators on the ground state super-Yang-Mills vector and spinor indices. The identities

$$\mathrm{Tr}\, K^{ij} = \mathrm{Tr}\, K^{ij} K^{kl} = \mathrm{Tr}\, K^{ij} K^{kl} K^{mn} = 0 \tag{3.2}$$

can easily be checked[20]. These identities lead to the vanishing of the open-string diagrams with one, two or three external on-shell ground-state bosons. Generally, the four-particle amplitude has divergences from both the planar orientable loop (fig. 3.1) and the non-orientable loop (fig. 3.3) but these cancel when the symmetry group is SO(32)[36]. This result has recently been generalized to certain one-loop N-particle amplitudes for $N > 4$[43]. [There is an ambiguity in the way the separate divergences are regularized in ref. 36 that raises doubts about the generalization of the results to higher loop diagrams.]

The last class of one-loop diagrams built from open strings has orientable world-sheets with external particles attached to both boundaries as, for example, in fig. 3.4.

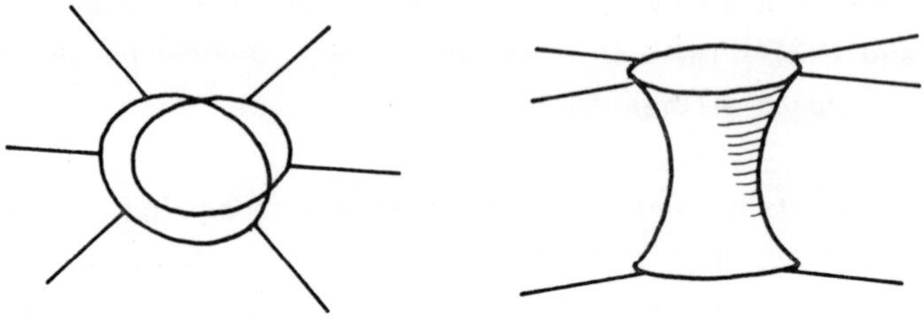

(a) (b)

Fig. 3.4

This contribution is not divergent. However, it generally has a branch cut in the Mandelstam invariant of the channel with vacuum quantum numbers that leads to a violation of unitarity. In 1971 it was realized by Lovelace[44] that this cut became a pole in D = 26 dimensions (for the bosonic theory) if he made the (correct) assumption that in that dimension an extra set of states decoupled by virtue of the Virasoro gauge conditions. [This corresponds to the condition for transversality of the physical Hilbert space.] The pole is interpreted as a closed string bound state formed by the joining of the ends of a single open string. [This was originally proposed as the origin of the Pomeranchuk Regge pole in hadronic physics.]

(b) Closed-String Loop Amplitude

In the case of the type IIa, type IIb and heterotic superstring theories the perturbation series only involves orientable closed surfaces with handles but no boundaries. [The type I closed-string amplitudes are represented by non-orientable as well as orientable surfaces with holes cut out as well as handles attached. None of these have yet been calculated.] An N-particle one-loop amplitude can be represented by

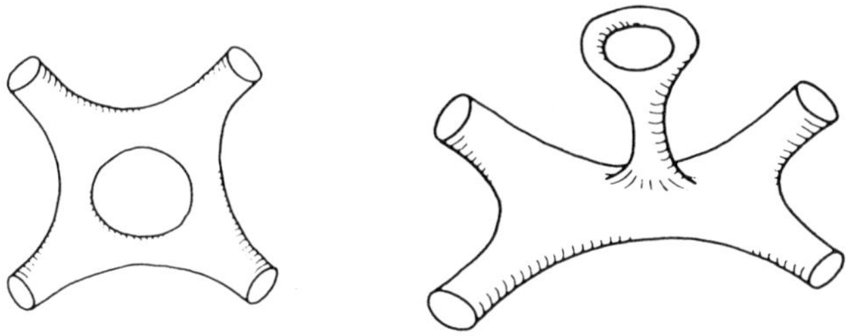

(a) (b)

Fig. 3.5

In the bosonic closed-string theory[45] there is a divergence in this one-loop diagram. This can once again be associated with the emission into the vacuum of a tadpole which disappears into the vacuum via a closed-string loop as in fig. 3.5(b). By an appropriate choice of world-sheet metric the σ-τ parameter space can be mapped into a torus. The expression for the amplitude involves complex integrations over the modulus of the torus, τ, as well as the positions, ν_r, of N-1 of the external particles which are integrated over the whole surface of the torus (with ν_N held fixed). In $D = 26$ dimensions the integrand is a modular function (which means that it is invariant under the transformations $\tau \to \tau + 1$ and $\tau \to -1/\tau$) and the τ integration must be restricted to the fundamental region of the modular group (in other dimensions there is a violation of unitarity).

The divergence of the expression for the amplitude arises from the endpoint of the integrations at which all the external particles come together on the surface of the torus (i.e. all the ν_r are equal) irrespective of the value of τ. In this limit the tadpole neck in fig. 3.3(b) has vanishingly small radius. The divergence is again associated with the emission of a dilaton at zero momentum. The residue of the divergence is proportional to the coupling of an on-shell dilaton to a torus. This residue is also divergent in the bosonic string theory due to the tachyon circulating around the loop (which is *not* an ultra-violet divergence).

In the case of type II superstring theories the supersymmetry identities in eq. (3.2) cause the one, two and three-particle amplitudes to vanish[42]. This suggests that the N-particle amplitude should be finite in these theories since the residue of the dilaton tadpole now vanishes. The four-particle amplitude was found to be finite by explicit calculation[19]. This has also now shown to be true for the heterotic superstring theories[5].

(c) Multi-Loop Amplitudes

There are no firm results yet about the divergences of multi-loop diagrams but the subject is under intensive study. The oriented closed-string theories are easiest to consider because there is only one diagram at any order in perturbation theory (at L loops its world-sheet is topologically a sphere with L handles). Assuming that the only possible divergences arise from the emission of dilatons into the vacuum there is a new potential divergence at each order. For example, fig. 3.6 shows two representations of the two-loop diagram.

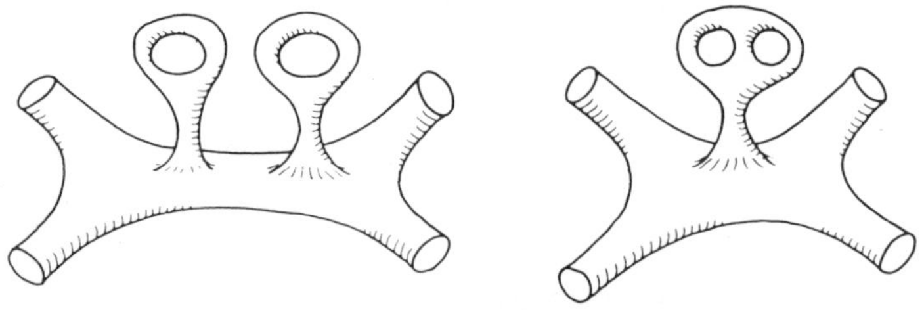

(a) (b)

Fig. 3.6

The two-loop tadpole of fig. 3.6(b) (the "E.T." diagram) contains a dilaton in its neck which can give a new divergence while at L loops there is an L-loop tadpole. Denoting the coupling of an on-shell dilaton to an L-loop torus by  the condition for finiteness is therefore

 $= 0$ (3.3)

This equation is also the condition that supersymmetry be unbroken. In perturbation theory around ten-dimensional Minkowski space supersymmetry cannot be broken in perturbation theory unless there are Lorentz anomalies (since the gravitini do not have the right partners with which to form a massive states). This means that

$$\text{absence of anomalies} \leftrightarrow \text{finiteness}$$

at any order in perturbation theory[46]. The question of whether the theories are actually free of infinities (or of anomalies) requires explicit calculations of L-loop diagrams which is at present the subject of extensive study[26].

(d) One-Loop Chiral Gauge Anomalies (in type I theories).

The occurrence of gauge anomalies in any theory is a disaster since they result in the coupling of unphysical gauge modes to physical ones with a consequent breakdown of unitarity. In ten dimensions there can be anomalies both in the Yang-Mills and in the gravitational currents[47]. These arise at one loop from hexagon diagrams[48] with circulating chiral particles and external gauge particles. In string theories there is an infinite set of unphysical modes associated with the infinity of covariant oscillators from which the states are constructed. This suggests that there may therefore be an infinite number of anomalies one of which is described by the hexagon string diagrams with massless external particles.

Neither the gravitational anomalies for the type IIb (closed and chiral) superstring theory northe Yang-Mills and gravitational anomalies for the heterotic superstring theory have yet been evaluated explicitly. However, the finiteness of the one-loop amplitudes makes it very plausible that the anomalies vanish by the arguments mentioned above. Further evidence for the absence of anomalies in these theories is that they reduce to anomaly-free point field theories at low energy.

I will now describe the calculation of the Yang-Mills anomaly in type I theories in some detail. Even though the type I theories may not, in the end, be phenomenologically interesting or even perhaps, consistent, the calculation illustrates techniques that may be of interest. To begin with I shall review the proof of the gauge invariance of the open-string tree diagrams.

Gauge Invariance of Tree Diagrams.

Consider an N-particle tree diagram in the covariant approach to the theory. For example consider a fermionic string emitting ground-state vector particles with momenta k_r and polarizations ζ_r satisfying the mass-shell condition

$$k_r^2 = 0 \tag{3.4}$$

and the transversality condition

$$\zeta_r \cdot k_r = 0 \quad . \tag{3.5}$$

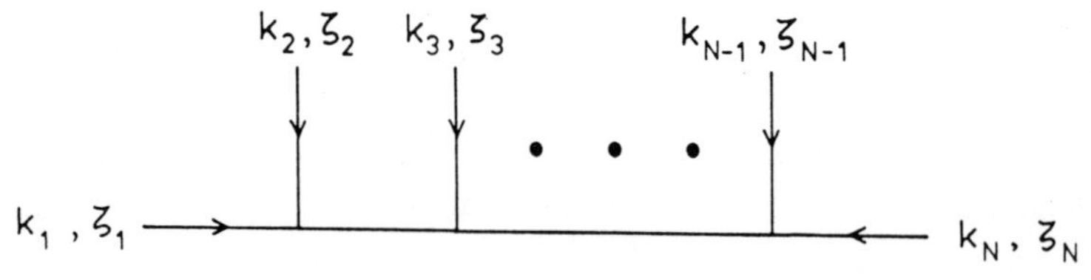

Fig. 3.7

The states are defined in terms of covariant bosonic and fermionic oscillator modes, α_n^μ and d_n^μ where

$$[\alpha_m^\mu , \alpha_n^\nu] = m \, \delta_{m+n,0} \, \eta^{\mu\nu} \tag{3.6}$$

$$\{ d_m^\mu , d_n^\nu \} = \delta_{m+n,0} \, \eta^{\mu\nu} \quad . \tag{3.7}$$

The amplitude is given by the expression (ignoring the group theory factor)

$$T_N = \langle k_1 | V(k_2,\zeta_2,1)\ \Delta \ldots V(k_{N-1},\zeta_{N-1},1)\ \frac{1}{2}(1+\Gamma_{11})|k_N\rangle \tag{3.8}$$

where the fermionic string propagator, Δ, and the vertex for emitting an on-shell vector particle of momentuom k_r and polarization ζ_r, $V(k_r,\zeta_r,z)$, are constructed from the momentum operator, $P^\mu(z)$, and the Ramond field[3], $\Gamma^\mu(z)$. The variable $\ln z/2\pi i$ is the proper time, τ, (i.e. $z = 1$ is $\tau = 0$) and the particles are attached to the end of the string $\sigma = 0$. These operators are defined by

$$P^\mu(z) = \sum_{-\infty}^{\infty} \alpha_n z^n \equiv \frac{dX^\mu(z)}{dz} \tag{3.9}$$

$$\Gamma^\mu(z) = \gamma^\mu + i\sqrt{2}\,\gamma_{11} \sum_{1}^{\infty}(d_n^\mu z^{-n} + d_n^\mu z^n) \tag{3.10}$$

The propagator is defined by

$$\Delta = (F_0)^{-1} = \frac{F_0}{L_0} \tag{3.11}$$

where

$$F_0 = \frac{1}{i\sqrt{2}} \int \frac{dz}{2\pi iz}\, \Gamma(z).P(z) \tag{3.12}$$

and

$$L_0 = F_0^2 = \frac{1}{2}P^2 + N \tag{3.13}$$

where the number operator N is defined by

$$N = \sum_{n=1}^{\infty} (\alpha_n^\dagger.\alpha_n + n\, d_n^\dagger.d_n) \quad . \tag{3.14}$$

The vertex for emitting string r at $\tau = \ln z/2\pi i$ is defined by

$$V(k_r,\zeta_r,1) = g\zeta_r.\Gamma(1)\ V_0(k_r,1) \tag{3.15}$$

where $V_0(k_r,1)$ is the usual factor of $\exp\{ik_r.X(1)\}$. The factor of

$\frac{1}{2}(1+\Gamma_{11})$ in eq. (3.8) is the generalized chirality projection operator introduced[4] to obtain a space-time supersymmetric theory. Γ_{11} is the string generalization of γ_{11} and defined by $\Gamma_{11} = \gamma_{11}\,(-1)^{\Sigma d_{-n}^{11} \cdot d_n}$ so that Γ_{11} anti-commutes with the matrices $\Gamma^{\mu}(z)$. It also anti-commutes with the propagator and with the vertex so that the factors of $\frac{1}{2}(1+\Gamma_{11})$ associated with each internal line can all be moved to the right-hand end of the tree as in eq. (3.8).

Gauge invariance can be checked by coupling the logitudinal mode of any given external gauge particle by substituting its momentum, k_r, for its polarization, ζ_r, in the vertex. The rth vertex becomes

$$V(k_r, k_r, 1) \equiv k_r . \Gamma(1)\, e^{ik_r . X(1)}$$

$$= i\sqrt{2}\left[F_0 \,,\, e^{ik_r . X(1)} \right] \qquad . \qquad (3.16)$$

The two terms in the commutator have a factor of F_0 which cancels the propagator one side or the other of the vertex, leaving the product of two factors of V_0 from adjacent vertices. In string theories the product of two vertices in a tree with no propagator between them is singular and this causes the tree to vanish (essentially because of a factor $\exp\{-k_r . k_{r+1}\ \Sigma 1/n\}$ which vanishes). As a result of this "cancelled propagator argument" the tree amplitude is proved to be gauge invariant. In usual point field theory the terms with cancelled propagators do not vanish and need to be compensated by the presence of contact terms in the action.

The cancelled propagator argument must be used with extreme caution in loop amplitudes since a zero may be compensated by an infinity from the loop. This is what happens in the loop anomaly calculation where the result is finite.

The One-Loop Anomaly.

Further details of this section may be found in ref. 39. The first anomalous diagrams in open-string theories are hexagon diagrams with external massless gauge particles (open-string ground states) and circulating chiral fermions.

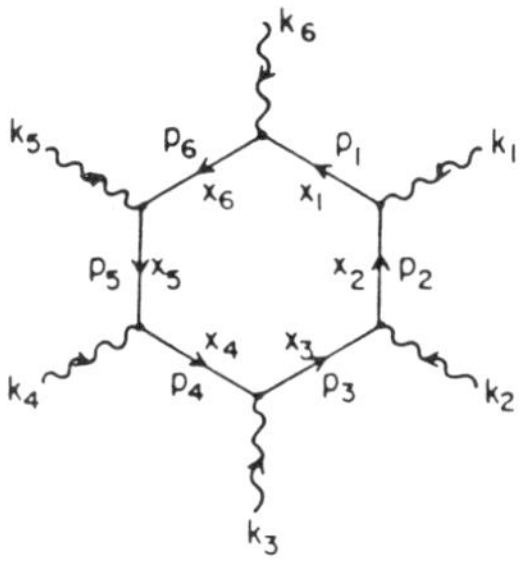

Fig. 3.8

The planar diagram (which is like fig. 3.1 but with six external states) has a group theory Chan-Paton factor

$$G^{\mathrm{planar}} = n\ \mathrm{tr}(\lambda_1 \ldots \lambda_6) \quad . \tag{3.17}$$

The λ's are matrices in the fundamental representation of SO(n), Sp(2n) or U(n) and the factor of n comes from the trace around the inner boundary which gives tr(1). The full amplitude must also be Bose symmetrized.

Only the components in the amplitude with an odd number of Γ_{11} factors will contribute to the anomaly. These give

$$T = \int d^{10}p\ \mathrm{Tr}\{\Delta V(k_1,\zeta_1,1)\Delta \ldots \Delta V(k_6,\zeta_6,1)\Gamma_{11}\} \tag{3.18}$$

where the propagator Δ is defined in eq. (3.11) and the vertex $V(k_r,\zeta_r,1)$ is defined in eq. (3.16). An overall constant and the group theory factor have been omitted from eq. (3.18). In writing this expression factors of $(1+\Gamma_{11})$ have been moved around the loop so that only one power of Γ_{11} remains and is arbitrarily placed next to the vertex for particle 6. Just as in ordinary point field theory anomaly calculations this is only a formal manoevre since T is divergent. However, the calculation is easier in this asymmetric configuration with only one power of Γ_{11} and the answer can easily be related to the result of the symmetric calculation.

The next step is to regulate T. In ref. 39 two different regulators were considered one of which was analogous to a Pauli–Villars regulator while the other was analogous to a Fujikawa regulator. They gave different expressions for the anomaly although they reduced to the same low-energy limit. This may merely reflect the fact that the theory is sick if there is any non-zero anomaly. I will describe the "Pauli–Villars" (P.V.) method here. Introducing a P.V. propagator

$$\Delta_m = (F_0 - im)^{-1} = \frac{F_0 + im}{L_0 + m^2} \tag{3.19}$$

in T instead of Δ defines T_m where $T-T_m$ is the regulated amplitude. The regulator mass, m, will be taken to infinity at the end of the calculation. Gauge invariance can be checked for the regulated amplitude by substituting k_r for ζ_r successively on each line. Due to the asymmetric position of Γ_{11} the only non-trivial case is that of line 6 (since the cancelled propagator argument works on the other legs in the finite, regulated expression). The vertex for line 6 can be rewritten using

$$k_6.\Gamma(1)V_0(k_6,1)\Gamma_{11} = i\sqrt{2}\left\{F_0, V_0(k_6,1)\,\Gamma_{11}\right\} \tag{3.20}$$

in T and

$$k_6 \cdot \Gamma(1) V_0(k_6,1)\Gamma_{11} = i\sqrt{2}\left\{ F_0 - im,\ V_0(k_6,1)\Gamma_{11}\right\}$$
$$-2\sqrt{2}\, m\, V_0(k_6,1)\Gamma_{11} \tag{3.21}$$

in T_m. In each case the term in { } involves an inverse propagator that leads to a vanishing contribution (recalling that $T-T_m$ is finite). The only non-zero contribution to the anomaly therefore comes from the last term in eq. (3.21) which leads to the anomaly A of the form

$$A = m \int d^{10}p \,\mathrm{Tr}\left\{ \frac{F_0+im}{L_0+m^2} V(k_1,\varsigma_1,1)\ldots V(k_5,\varsigma_5,1) \frac{F_0+im}{L_0+m^2} V_0(k_6,1)\Gamma_{11}\right\} \tag{3.22}$$

The Dirac trace can be performed easily due to the γ_{11} which eats up ten γ^μ factors out of the eleven which are present in the integrand of eq. (3.22) (one in each of the six F_0's and one in each of the five V's). The result is

$$A = im^2\, \epsilon(\varsigma,k) \int d^{10}p \,\mathrm{Tr}\left\{\frac{1}{L_0+m^2} V_0(k_1,1)\ \ldots \frac{1}{L_0+m^2}(-1)^{\Sigma d_{-n}\cdot d_n}\right\} \tag{3.23}$$

where

$$\epsilon(\varsigma,k) \equiv \epsilon_{\mu_1\ldots\mu_5\nu_1\ldots\nu_5}\, \varsigma_1^{\mu_1}\ldots\ \varsigma_5^{\mu_5}\, k_1^{\nu_1}\ldots\ k_5^{\nu_5} \tag{3.24}$$

is proportional to the usual anomaly in point field theory in ten dimensions. The integral is analogous to standard bosonic open-string loop integrals. It is usual to represent each propagator by

$$(L_0 + m^2)^{-1} = \int_0^1 dx\ x^{L_0+m^2-1} \tag{3.25}$$

where the x's are labelled as in fig. 3.8. The trace in eq. (3.23) is reexpressed by using

$$x^{L_0}\, V_0(k,1) = V_0(k,x)\ x^{L_0} \tag{3.26}$$

Using the variables

$$\rho_i = x_1 \ldots x_i \quad , \qquad \rho_6 = \omega \tag{3.27}$$

gives

$$A = m^2 \,\epsilon(\zeta,k) \int \Pi \frac{dx^i}{x^i}\, \omega^{m^2} \int d^{10}p \;\mathrm{Tr}\left\{ \omega^{L_0}\, V_0(k_1,\rho_1) \ldots \right.$$
$$\left. \ldots V_0(k_6,\rho_6)\, (-1)^{\Sigma d_{-n}\cdot d_n} \right\} \tag{3.28}$$

In the limit of large m this expression vanishes exponentially fast unless ω is close to 1. The region near $\omega = 1$ can be examined by a change of variables (a Jacobi imaginary transformation) to

$$z_i = \exp\left\{ 2\pi i \frac{\ln \rho_i}{\ln \omega} \right\} \tag{3.29}$$

where $z_6 = 1$ and the z_i's are ordered around the unit circle in the complex plane. The oscillator arithmetic in eq. (3.28) can be elegantly described by using an identity for $\int d^{10}p\ \mathrm{Tr}\{\ldots\}$ (proved in ref. 49) which recasts this expression into a form involving vertices which are functions of z_r giving

$$A = m^2\, \epsilon(\zeta,k) \int \Pi \frac{dz_i}{z_i} \int_0^1 d\omega\, \omega^{m^2-1}$$

$$\times \langle p = 0|\,\mathrm{Tr}\left\{ \exp\left(\frac{4\pi^2 N}{\ln \omega}\right) V_0(z_1,k_1) \ldots V_0(z_6,k_6) \right\} |p = 0\rangle \tag{3.30}$$

where the trace is now over non-zero modes only. Recalling that the number operator N has integer eigenvalues we see that as m gets large only the term in the trace with N = 0 survives. Furthermore, since

$$\int_0^1 d\omega\, \omega^{m^2-1} = m^{-2} \tag{3.31}$$

the result is finite and is given by

$$A = \epsilon(\zeta,k) \int \prod_{i=1}^{5} \frac{dz_i}{z_i} \langle 0 | V_0(z_1) \ldots V_0(z_6) | 0 \rangle$$

$$= \epsilon(\zeta,k) \int \prod_{i=1}^{5} \frac{dz_i}{z_i} \prod_{i<j} (z_j - z_i)^{k_i \cdot k_j} \tag{3.32}$$

At low momenta (i.e. when $k_i.k_j \ll 1$ in the units with $T = 1/\pi$) the integral in eq. (3.32) is a constant and the result reduces to the usual hexagon diagram anomaly of point field theory. The integral contains all the "stringiness" which includes non-trivial dynamical structure, the significance of which is not clear.

In addition to the planar diagrams we must add the diagrams with twists on internal propagators. Those with an odd number of twists are Möbius strips and only have one boundary. They are associated with group theory factors like

$$G^{\text{Möbius}} = \pm \operatorname{Tr} (\lambda_1 \ldots \lambda_6) \tag{3.33}$$

with the + sign for USp(2n) and the - sign for SO(n) (for U(n) there is no contribution from these non-orientable diagrams - this case is, in any case, known to be inconsistent at the level of loop amplitudes and need not be considered here[31]). These signs, associated with a twist, can be understood from the symmetry properties of the group generators expressed in terms of the fundamental representation. In contrast to eq.(3.17) there is no factor of n in this expression because there is only one boundary on a Möbius strip. However, there are 32 diagrams with an odd number of twists on internal propagators and apart from the group theory factor, they contribute with the same weight to the anomaly as in the planar case (after Bose symmetrization). The calculation is very similar to the earlier one and will not be repeated here. Once again, the low energy limit of the contribution to the anomaly from these terms coincides with the calculation of a hexagon diagram in the massless field theory.

After adding the planar and Möbius strip contributions, the anomaly has a factor of

$$(n + 32) \quad \text{for} \quad USp(2n)$$
$$(n - 32) \quad \text{for} \quad SO(n) \tag{3.34}$$

The last class of diagrams is those with an even number of twists such that the world sheet is an annulus with particles attached to both the inner and the outer boundaries such as that shown in fig. 3.4(a). Elementary group theory shows that for the groups SO(n) or USp(n) there must be an even number of particles attached to each boundary of the world-sheet. The contribution shown in fig. 3.8 has a group theory factor of $tr(\lambda_1\lambda_2\lambda_3\lambda_4)tr(\lambda_5\lambda_6)$. The anomaly calculation is similar to the earlier one with the important distinction that the exponential factor inside eq. (3.30) now becomes

$$\exp\left\{\frac{4\pi^2(8N + k_5.k_6)}{8 \ln \omega}\right\} \tag{3.35}$$

which vanishes for any value of $N \geqslant 0$ (with $k_5.k_6 \geqslant 0$ by a suitable continuation, if necessary). This result at first seems rather surprising since the low-energy limit of zero is zero! This does not coincide with the (non-zero) contribution to the anomaly arising from the corresponding hexagon diagrams in the massless effective field theory. The resolution to this paradox lies in the fact that in the low energy limit fig. 3.4(a) gives another anomalous contribution in addition to the usual hexagon loop. This arises from the closed-string bound states in the 5-6 channel which can be seen by distorting the diagram into the cylinder of fig. 3.4(b). These states include the massless supergravity multiplet which survives the low energy limit in which massive states decouple. At low energy it is clear from fig. 3.4(b) that there is a contribution from a *tree* diagram with the exchange of the supergravity multiplet. This tree contains an anomalous piece arising from the exchange of the antisymmetric tensor field, $B_{\mu\nu}$.

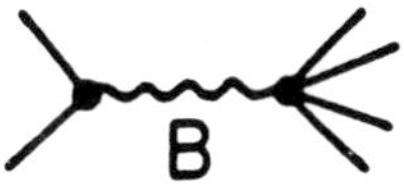

Fig. 3.9

The coupling between $B_{\mu\nu}$ and the pair of gauge particles at one end of the tree is the usual coupling present in the much-studied system of D = 10 super-Yang-Mills coupled to supergravity[50] (which I shall refer to as the "minimal" theory). However, there is, in addition, an *anomalous* coupling between the $B_{\mu\nu}$ field and four gauge particles that is not included in the "minimal" theory.

The total anomaly is therefore given entirely by the planar and non-orientable diagrams. From eq. (3.34) we see that the anomaly cancels for the group SO(32). Furthermore, the cancellation can be understood in the language of the low-energy point field theory as a cancellation between the usual one-loop quantum effect and a classical effect due to the presence of an anomalous term in the action.

(f) Anomaly Cancellations in the Low-Energy Theory

As yet there has not been an explicit calculation of gravitational anomalies in any string theory or of the Yang-Mills anomaly in the heterotic string theory. However the argument for the cancellation of the Yang-Mills anomaly based on the low-energy field theory generalizes to gravitational and mixed anomalies. The requirement that all these anomalies should cancel by the addition of local terms to the action is very restrictive. Details of how these restrictions arise are given in ref. 39 and I shall only quote the results here. The gravitational anomalies can only be cancelled if the Yang-Mills group has dimension 496 (i.e. there are 496 species of chiral fermions in the Yang-Mills sector). The Yang-Mills anomaly and the mixed Yang-Mills gravitational anomalies can only be cancelled if the gauge group has the property that for an arbitrary matrix, F, in the adjoint representation

$$\mathrm{Tr}\, F^6 = \frac{1}{48}\, \mathrm{Tr}\, F^2\left\{(\mathrm{Tr}\, F^4 - \frac{1}{300}\, (\mathrm{Tr}\, F^2)^2\right\} \tag{3.36}$$

These conditions only hold for the algebras of SO(32) and $E_8 \times E_8$ (apart from the apparently trivial cases of $U(1)^{496}$ and $E_8 \times U(1)^{248}$). This suggested that there ought to be a superstring theory with the gauge algebra of $E_8 \times E_8$ as well as SO(32). The heterotic superstring is just such a theory. It actually has (Spin 32)/Z_2 instead of SO(32) but these have the same Lie algebra.

The absence of the gravitational and Yang-Mills anomalies does not imply the vanishing of anomalies in the divergence of the supercurrent[51]. For completeness it would be satisfying to calculate these anomalies explicitly in superstring theories or merely in the low energy point field theory limits.

The absence of infinitessimal gauge anomalies is a pre-requisite for the consistency of the theory. It is also necessary for anomalies associated with "large" ten-dimensional general coordinate transformations, which are not continuously connected to the identity, to be absent. This has been shown[52] to be the case, at least when space-time is taken to be a ten-sphere, S^{10}.

The restriction to these two groups in the heterotic theory can be seen by considering an arbitrary loop diagram. In a closed-string theory consistency requires that the expression for a loopdiagram is the integral of a modular function. This restricts the possible groups of the heterotic string to the ones associated with rank 16, self-dual, even lattices[5]. Requiring modular invariance of the loop amplitudes for closed strings is equivalent to requiring invariance of the theory under "large" reparametrizations of the world-sheet i.e. reparametrizations not continuously connected to the identity[53].

REFERENCES

(1) Y. Nambu, Lectures at the Copenhagen Symposium, 1970; T. Goto, Progr. Theor. Phys. **46** (1971) 1560.

(2) M.B. Green and J.H. Schwarz, Phys. Lett. **136B** (1984) 367; Nucl. Phys. **B243** (1984) 285.

(3) P.M.Ramond, Phys. Rev. **D3** (1971) 2415; A. Neveu and J.H. Schwarz, Nucl. Phys. **B31** (1971) 86; Phys. Rev. **D4** (1971) 1109.

(4) F. Gliozzi, J. Scherk and D.I. Olive, Nucl. Phys. **B122** (1977) 253

(5) D.J. Gross, J.A. Harvey, E. Martinec, and R. Rohm, Phys. Rev. Lett. **54** (1985) 502; Nucl. Phys. **B256** (1985) 253; "Heterotic String Theory II. The Interacting Heterotic String", Princeton preprint (June 1985).

(6) S. Deser and B. Zumino, Phys. Lett. **65B** (1976) 369. L. Brink, P. Di Vecchia and P.S. Howe, Phys. Lett. **65B** (1976) 471

(7) A.M. Polyakov, Phys. Lett. **103B** (1981) 207; Phys. Lett. **103B** (1981) 211.

(8) W. Siegel, Phys. Lett. **128B** (1983) 397.

(9) M. Henneaux and L. Mezincescu, Phys. Lett. **152B** (1985) 340; T.L. Curtright, L. Mezincescu, C.K. Zachos, Argonne preprint ANL-HEP-PR-85-28 (1985).

(10) E. Witten, Nucl. Phys. **B223** (1983) 422.

(11) E. Witten, Comm. Math. Phys. **92** (1984) 455.

(12) T. Curtright and C. Zakhos, Phys. Rev. Lett. **53** (1984) 1799.

(13) E. Witten, "Twistor-Like Transform in Ten Dimensions", Princeton preprint (May, 1985).

(14) B. Nielsen, Nucl. Phys. **B188** (1981) 176.

(15) M.T. Grisaru, P. Howe, L. Mezincescu, B. Nilsson, P.K. Townsend, D.A.M.T.P. Cambridge preprint (1985).

(16) I. Bengtsson and M. Cederwall, Göteborg preprint (1984). T. Hori and K. Kamimura, U. of Tokyo preprint (1984).

(17) W. Siegel, Berkeley preprint UCB-PTH-85/23 (May, 1985).

(18) P. Goddard, J. Goldstone, C. Rebbi and C.B. Thorn, Nucl. Phys.**B56** (1973) 109.

(19) M.B. Green and J.H. Schwarz, Phys. Lett. **109B** (1982) 444.

(20) M.B. Green and J.H. Schwarz, Nucl. Phys. **B198** (1982) 252.

(21) M.B. Green and J.H. Schwarz, Phys. Lett. **140B** (1984) 33; Nucl. Phys. **B243** (1984) 475.

(22) M.B. Green, J.H. Schwarz and L. Brink, Nucl. Phys. B219 (1983) 437.
(23) S. Mandelstam, Nucl. Phys. B64 (1973) 205; B69 (1974) 77.
(24) L. Brink and M.B. Green (unpublished).
(25) C.S. Hsue, B. Sakita and M.A. Virasoro, Phys. Rev. D2 (1970) 2857 J.- L. Gervais and B. Sakita, Phys. Rev. D4 (1971) 2291
(26) S. Mandelstam, Proceedings of the Niels Bohr Centennial Conference, Copenhagen, Denmark (May, 1985); Proceedings of the Workshop on Unified String Theories, ITP Santa Barbara, 1985. A. Restuccia and J. G. Taylor, King's College, London preprints (June, 1985).
(27) W. Siegel, Phys. Lett. 149B (1984) 157, 162; Banks and M. Peskin, Proceedings of the Argonne-Chicago Symposium on Anomalies, Geometry and Topology (World Scientific, 1985); SLAC preprint SLAC-PUB-3740 (July,1985); D. Friedan, E.F.I. preprint EFI 85-27 (1985); A. Neveu and P.C. West, CERN preprint CERN-TH 4000/85; K. Itoh, T. Kugo, H. Kunitomo and H. Ooguri, Kyoto preprint(1985) W. Siegel and B. Zweibach, Berkeley preprint UCB-PTH-85/30(1985); M. Kaku and Lykken, CUNY preprint (1985); M. Kaku, Osaka University preprints OU-HET 79 and 80 (1985); S. Raby, R. Slansky and G. West, Los Alamos preprint (1985).
(28) M. Kaku and K. Kikkawa, Phys. Rev. D10 (1974) 1110, 1823; E. Cremmer and J.-L. Gervais, Nucl. Phys. B76 (1974) 209; J.F.L. Hopkinson R.W. Tucker and P.A. Collins, Phys. Rev. D12 (1975) 1653.
(29) M.B. Green and J.H. Schwarz, Nucl. Phys. B218 (1983) 43.
(30) E. Tomboulis, Phys. Rev. D8 (1973) 2736.
(31) J.H. Schwarz, Proceedings of Johns Hopkins Workshop on Current Problems in Particle Theory (Florence, 1982) 233; N. Marcus and A. Sagnotti, Phys. Lett. 119B (1982) 97.
(32) H.M. Chan and J. Paton, Nucl. Phys. B10 (1969) 519.
(33) C.B. Thorn, University of Florida UFTP-85-8.
(34) R. Giles and C.B. Thorn, Phys. Rev. D16 (1976) 366.
(35) L. Brink, M. Cederwall and M.B. Green, Göteborg preprint, (1985).
(36) M.B. Green and J.H. Schwarz, Phys. Lett. 151B (1985) 21.
(37) M.B. Green, Proceedings of the Argonne-Chicago Symposium on Algebra, Geometry and Topology (1985).
(38) A.K.H. Bengtsson, L. Brink, M. Cederwall and M. Ogren, Nucl. Phys. B254 (1985) 625.
(39) M.B. Green and J.H. Schwarz, Phys. Lett., 149B (1984) 117; M.B. Green and J.H. Schwarz, Nucl. Phys. B225 (1985) 93.
(40) A. Neveu and J. Scherk, Phys. Rev. D1 (1970) 2355.
(41) J.A. Shapiro, Phys. Rev. D11 (1975) 2937; M. Ademollo et al., Nucl. Phys. B124 (1975) 461.
(42) M.B. Green and J.H. Schwarz, Nucl. Phys.B198 (1982) 441;
(43) P.H. Frampton P. Moxhay and Y.J. Ng, Harvard preprint HUTP-85/A059 (1985); L. Clavelli, U. of Alabama preprint (1985).
(44) C. Lovelace, Phys. Lett. 34 (1971) 500.

(45) J. A. Shapiro, Phys. Rev. D5 (1972) 1945.
(46) M.B. Green, Caltech preprint CALT-68-1219 (1984), to be published in the volume in honour of the 60th birthday of E.S. Fradkin.
(47) L. Alvarez-Gaume and E. Witten, Nucl. Phys. **B234** (1983) 269.
(48) P.H.Frampton and T.W.Kephart,Phys. Rev. Lett.,50(1983)1343;1347; P.K. Townsend and G. Sierra, Nucl. Phys. **B222** (1983) 493.
(49) L. Clavelli and J.A.Shapiro, Nucl. Phys. **B57** (1973) 490.
(50) A.H. Chamseddine, Phys. Rev. **D24** (1981) 3065
E. Bergshoeff, M. de Roo, B. de Wit and P. van Nieuwenhuisen, Nucl. Phys. **B195** (1982) 97;
G.F. Chapline and N.S. Manton, Phys. Lett. **120B** (1983) 105.
(51) R. Kallosh, Phys. Lett. **159B** (1985) 111.
(52) E. Witten, "Global Gravitational Anomalies", Princeton preprint (1985).
(53) E. Witten, Proceedings of the Argonne-Chicago Symposium on Anomalies, Geometry and Topology (1985).

HETEROTIC STRING THEORY

David J. Gross*
Joseph Henry Laboratories
Princeton University
Princeton, New Jersey 08544

I. INTRODUCTION

High energy physics is, at present, in an unusual state. It has been clear for some time that we have succeeded in achieving many of the original goals of particle physics. We have constructed theories of the strong, weak and elecromagnetic interactions and have understood the basic constituents of matter and their interactions. The "standard model" has been remarkably successful and seems to be an accurate and complete description of physics, at least at energies below a Tev. Indeed, as we have heard in the experimental talks at this meeting, there are at the moment no significant experimental results that cannot be explained by the color gauge theory of the strong interactions (QCD) and the electroweak gauge theory. New experiments continue to confirm the predictions of these theories and no new phenomenon have appeared.

This success has not left us sanguine. Our present theories contain too many arbitrary parameters and unexplained patterns to be complete. They do not satisfactorily explain the dynamics of chiral symmetry breaking or of CP symmetry breaking. The strong and electroweak interactions cry out for unification. Finally we must ultimately face up to including quantum gravity within the thoery. However, we theorists are in the unfortunate situation of having to address these

*Research supported in part by NSF Grant PHY80-19754.

questions without the aid of experimental clues. Furthermore extrapolation of present theory and early attempts at unification suggest that the natural scale of unification is 10^{16} Gev or greater, tantalizingly close to the Planck mass scale of 10^{19} Gev. It seems very likely that the next major advance in unification will include gravity. I do not mean to suggest that new physics will not appear in the range of Tev energies. Almost all attempts at unification do, in fact, predict a multitude of new particles and effects that could show up in the Tev domain (Higgs particles, Supersymmetric partners, etc.), whose discovery and exploration is of the utmost importance. But the truly new threshold might lie in the totally inaccessible Planckian domain.

In this unfortunate circumstance, when theorists are not provided with new experimental clues and paradoxes, they are forced to adopt new strategies. Given the lessons of the past decades it is no surprise that much of exploratory particle theory is devoted to the search for new symmetries. However it is not enough simply to dream up new symmetries, one must also explain why these symmetries are not apparent, why they have been heretofore hidden from our view. This often requires both the discovery of new and hidden degrees of freedom as well as mechanisms for the dynamical breaking of the symmetry.

Some of this effort is based on straightforward extrapolations of established symmetries and dynamics, as in the search for grand unified theories (SU_5, SO_{10}, E_6,...), or in the development of a predictive theory of dynamical chiral gauge symmetry breaking (technicolor, preons, ...). Ultimately more promising, however, are the suggestions for radically new symmetries and degrees of freedom.

First there is supersymmetry, a radical and beautiful extension of space-time symmetries to include fermionic charges. This symmetry principle has the potential to drastically reduce the number of free parameters. Most of all it offers an explanation for the existence of fermionic matter, quarks and leptons, as compelling as the argument that the existence of gauge mesons follows from local gauge symmetry.

An even greater enlargement of symmetry, and of hidden degrees of freedom is envisaged in the attempts to revive the idea of Kaluza and Klein, wherein space itself contains new, hidden, dimensions. These new degrees of freedom are hidden from us due to the spontaneous compactification of the new spatial dimensions, which partially breaks many of the space-time symmetries of the larger manifold. Although strange at first, the notion of extra spatial dimensions is quite reasonable when viewed this way. The number of spatial dimensions is clearly an experimental question. Since we would expect the compact dimensions to have sizes of order the Planck length there clearly would be no way to directly observe many (say six) extra dimensions. The existence of such extra dimensions is not without consequence. The unbroken isometries of the hidden, compact, dimensions can yield a gravitational explanation for the emergence of gauge symmetries (and, in supergravity theories, the existence of fermionic matter). A combination of supergravity and Kaluza-Klein thus has the potential of providing a truly unified theory of gravity and matter, which can provide an explanation of the known low energy gauge theory of matter and predict its full particle content.

Attempts to utilize these new symmetries in the context of ordinary QFT, however, have reached an impasse. The problems one encounters are most severe if one attempts to be very ambitious and contemplate a unified theory of pure supergravity (in, say, 11 dimensions), which would yield the observed low energy gauge group and fermionic spectrum upon compactification. First of all we do not have a satisfactory quantum theory of gravity, even at the perturbative level. Einstein's theory of gravity, as well as its supersymmetric extensions, is nonrenormalizable. We know that that means that there must be new physics at the Planck length. We are clearly treading on thin ice if we attempt to use this potentially inconsistent theory as the basis for unification.

Even if we ignore this issue, and focus on the low energy structure of such theories, it appears to be impossible to construct realistic theories without a great loss of predictive power. The primary obstacle is the existence of chiral fermions (i.e. the fact that the weak interactions are V-A in structure). In order to generate the observed spectrum of chiral quarks and leptons it appears to be necessary to retreat from the most ambitious Kaluza-Klein program, which would uniquely determine the low energy gauge group as isometries of some compact space and introduce gauge fields by hand. Furthermore the supergravity theories ubiquitously produce a world which would have an intolerably large cosmological constant. Finally no realistic and compelling model has emerged. This brings us to string theories which offer a way out of this impasse.

II. STRING THEORIES

String theories offer a way of realizing the potential of supersymmetry, Kaluza-Klein and much more. They represent a radical departure from ordinary quantum field theory, but in the direction of increased symmetry and structure. They are based on an enormous increase in the number of degrees of freedom, since in addition to fermionic coordinates and extra dimensions, the basic entities are extended one dimensional objects instead of points. Correspondingly the symmetry group is greatly enlarged, in a way that we are only beginning to comprehend. At the very least this extended symmetry contains the largest group of symmetries that can be contemplated within the framework of point field theories--those of ten-dimensional supergravity and super Yang-Mills theory.

The origin of these symmetries can be traced back to the geometrical invariance of the dynamics of propagating strings. Traditionally string theories are constructed by the first quantization of a classical relativistic one dimensional object, whose motion is determined by requiring that the invariant area of the world sheet it sweeps out in space-time is extremized. In this picture the dynamical degrees of freedom of the string are its coordinates, $X_\mu(\sigma,\tau)$ (plus fermionic coordinates in the superstring), which describe its position in space time. The symmetries of the resulting theory are all consequences of the reparametrization invariance of σ,τ parameters which label the world sheet. As a consequence of these symmetries one finds that the free string contains massless gauge bosons. The closed string automatically contains a massless spin two meson, which can be identified as the graviton whereas the open string, which has ends to which charges can be attached, yields massless vector mesons which can be identified as Yang-Mills gauge bosons.

String theories are inherently theories of gravity. Unlike ordinary quantum field theory we do not have the option of turning off gravity. The gravitational, or closed string, sector of the theory must always be present for consistency even if one starts by considering only open strings, since these can join at their ends to form closed strings. One could even imagine discovering the graviton in the attempt to construct string theories of matter. In fact this was the course of events for the dual resonance models where the graviton (then called the Pomeron) was discovered as a bound state of open strings. Most exciting is that string theories provide for the first time a consistent, even finite, theory of gravity. The problem of ultraviolet divergences is bypassed in string theories which contain no short distance infinities. This is not too surprising considering the extended nature of strings, which softens their interactions. Alternatively one notes that interactions are introduced into string theory by allowing the string coordinates, which are two dimensional fields, to propagate on world sheets with nontrivial topology that describe strings splitting and joining. From this first quantized point of view one does not introduce an interaction at all, one just adds handles or holes to the world sheet of the free string. As long as reparametrization invariance is maintained there are simply no possible counterterms. In fact all the divergences that have ever appeared in string theories can be traced to infrared divergences that are a consequence of vacuum instability. All string theories contain a massless partner of the graviton called the dilaton. If one constructs a string theory about a trial vacuum state in which the dilaton has a nonvanishing vacuum expectation value, then infrared infinities will occur due to massless dilaton tadpoles. These divergences however are just a sign of the instability of the original trial vacuum. This is the source of the divergences that occur in one loop diagrams in the old bosonic string theories (the Veneziano model). Superstring theories have vanishing dilaton tadpoles, at least to one-loop order. Therefore both the superstring and the heterotic string are explicitly finite to one loop order and there are strong arguments that this persists to all orders!

String theories, as befits unified theories of physics, are incredibly unique. In principle they contain no freely adjustable parameters and all physical quantities should be calculable in terms of h, c, and m_{planck}. In practice we are not yet in the position to exploit this enormous predictive power. The fine structure constant α, for example, appears in the theory in the form α exp(-D), where D is the aforementioned dilaton field. Now the value of this field is undetermined to all orders in perturbation theory (it has a "flat potential"). Thus we are free to choose its value, thereby choosing one of an infinite number of degenerate vacuum states, and thus to adjust α as desired. Ultimately we might believe that string dynamics will determine the value of D uniquely, presumably by a nonperturbative mechanism, and thereby eliminate the nonuniqueness of the choice of vacuum state. In that case all dimensionless parameters will be calculable. Even more, string theories determine in a rather unique fashion the gauge group of the world and fix the number of space-time dimensions to be ten.

Finally and most importantly, string theories lead to phenomenologically attractive unified theories, which could very well describe the real world.

III. CONSISTENT STRING THEORIES

The number of consistent string theories is extremely small, the number of phenomenologically attractive theories even smaller. First there are the closed superstrings, of which there are two consistent versions. These are theories which contain only closed strings which have no ends to which to attach charges and are thus inherently neutral objects. At low energies, compared to the mass scale of the theory which we can identify as the Planck mass, we only see the massless states of the theory which are those of ten dimensional supergravity. One version of this theory is non-chiral and of no interest since it could never reproduce the observed chiral nature of low energy physics. The other version is chiral. One might then worry that it would suffer from anomalies, which is indeed the fate of almost all chiral supergravity theories in ten dimensions. Remarkably the particular supergravity theory contained within the chiral superstring is the unique anomaly free theory in ten dimensins. It however contains no gauge interactions in ten dimensions and could only produce such as a consequence of compactification. This approach raises the same problems of reproducing chiral fermions that plagued field theoretic Kaluza-Klein models and has not attracted much attention.

Open string theories, on the other hand, allow the introduction of gauge groups by the time honored method of attaching charges to the ends of the strings. String theories of this type can be constructed which yield, at low energies, N=1 supergravity with any SO(N) or Sp(2N) Yang-Mills group. These however, in addition to being somewhat arbitrary, were suspected to be anomalous. The discovery by Green and Schwarz, last summer, that for a particular gauge group--SO_{32}--the would be anomalies cancel, greatly increased the phenomenological prospects of unified string theories.

The anomaly cancellation mechanism of Green and Schwarz can be understood in terms of the low energy field theory that emerges from the superstring, which is a slightly modified form of d=10 supergravity. One finds that the dangerous Lorentz and gauge anomalies cancel, if and only if, the gauge group is SO_{32} or $E_8 \times E_8$. The ordinary superstring theory cannot incorporate $E_8 \times E_8$. The apparent correspondence between the low energy limit of anomaly free superstring theories and anomaly free supergravity theories provided the motivation that led to the discovery of a new string theory, by J. Harvey, E. Martinec, R. Rohm and myself, whose low energy limit contained an $E_8 \times E_8$ gauge group --the heterotic string. The heterotic string is a closed string theory that produced by a stringy generalization of the Kaluza-Klein mechanism of compactification, gauge interactions. These are determined by consistency to be $E_8 \times E_8$ or Spin $32/Z_2$. It is of more than academic interest to construct this theory since its phenomenological prospects are much brighter.

IV. THE FREE HETEROTIC STRING

Free string theories are constructed by the first quantization of an action which is given by the invariant area of the world sheet swept out by the string or by its supersymmetric generalization. The fermionic coordinates of superstrings can be described either as ten-dimensional spinors or by ten-dimensional fermionic vector fields. Similarly the sixteen, left-moving coordinates of the heterotic string can be described by 32 real fermionic coordinates, by the local coordinates on the group manifolds of $E_8 \times E_8$ or spin $(32)/Z_2$, or by sixteen bosonic coordinates. Here we shall take the right-moving fermionic coordinates to be described by a ten-dimensional spinor and the left-moving internal coordinates to be sixteen bosonic fields. This will have the advantage of making the ten-dimensional supersymmetry manifest and of yielding a rather physical picture of the left-moving internal space. The price we pay is that this formulism is only tractable in light cone gauge, so that we must relinquish manifest Lorentz invariance.

The manifestly supersymmetric action for the heterotic string is given by the Green-Schwarz action for the right-movers plus the Nambu action for the left-movers. The superstring action is that of non-linear sigma model on superspace (the super-translation group manifold) with a Wess-Zumino term. Consider an element of the supertranslation group, $h = \exp i(X\cdot P + \theta\cdot Q)$, where $X_\mu (\mu=0,1,\ldots,9)$ and θ^a (a ten-dimensional Majorana-Weyl spinor) are the superspace coordinates, P and Q are the generators of translations and supersymmetry translations. Then

$$\Pi_\alpha = h^{-1}(x)\, \partial_\alpha h(x) = (\partial_\alpha X^\mu - i\bar{\theta}\gamma^\mu \partial_\alpha \theta)\, P_\mu + \partial_\alpha\, \theta\cdot Q, \qquad (1)$$

and defining $tr(P_\mu P_\nu) = \eta_{\mu\nu}$, $tr(Q_a Q_b) = 0$, $tr(Q_a Q_b P^\mu) = (\gamma^\mu C^{-1})_{ab}$, we have

$$S_R = -T/2\{\int d^2\xi\; e g^{\alpha\beta}\; tr\; \Pi_\alpha \Pi_\beta + \int_M d^3\xi\; \varepsilon^{\alpha\beta\gamma}\; tr\; \Pi_\alpha \Pi_\beta \Pi_\gamma$$

$$+ \int d^2\xi e\; \lambda^{++}\; (e_+^\alpha\; \Pi_\alpha)^2\}. \qquad (2)$$

Here $\xi^{\pm} = \tau\pm\sigma$ are two dimensional light cone coordinates, e^{α}_{a} is the two-bein $(e^{\alpha}_{a}e^{\beta a}=g^{\alpha\beta})$. The second term is the Wess-Zumino term, where M is a three-dimensional manifold whose boundary coincides with the world sheet. The last term enforces the constraint that the coordinates are only right moving.

The left movers are given by a similar action

$$S_L = -T/2 \int d^2\xi \; e[\tfrac{1}{2} g^{\alpha\beta}\partial_{\alpha}X^A\partial_{\beta}X^A + \lambda^{--} (e^{\alpha}_{-} \partial_{\alpha}X^A)^2], \tag{3}$$

where A=0,...25.

When expanded, the full action is recognizable as the Green-Schwarz action for right movers alone plus the Nambu action for 26 left movers. It is therefore invariant under ten-dimensional Poincare transformations, N=1 supersymmetry, which acts on the right movers alone, and 16-dimensional Poincare transformations of X^I (I=A-9=1,2,...16). In addition it is invariant under local two-dimensional reparametrizations and the local fermionic transformation of Green and Schwatrz. These local symmetries enable us to choose a gauge--"light cone gauge"--where $g^{\alpha\beta} \sim \eta^{\alpha\beta}$, $X^+(\sigma,\tau) = \pi/T\, P^+\tau+x^+$, and $\gamma^+\theta = 0$. The resulting dynamics is given by the light cone action

$$S_{HeT} = -\frac{1}{2\pi} \int d^2\xi \; [(\partial_{\alpha}X^i)^2 + \frac{i}{2} \bar{S}(\partial_{\tau}+\partial_{\sigma})S + (\partial_{\alpha}X^I)^2 + \lambda[(\partial_{\tau}-\partial_{\sigma})X^I]^2], \tag{4}$$

where we have chosen units in which the string tension $T = \frac{1}{2\pi\alpha'} = \frac{1}{\pi}$. The physical degrees of freedom are now manifest--eight transverse coordinates $X^i(i=1...8)$, a Majorana-Weyl-light cone right moving spinor S^a $((1+\gamma^{11})S^a = \gamma^+S^a = 0)$, and sixteen left-moving coordinates $X^I(I=1,...,16)$. The equations of motions are then

$$\partial^2X^i = 0; \qquad (\partial_{\tau}+\partial_{\sigma})S^a = 0; \qquad (\partial_{\tau}-\partial_{\sigma})X^I = 0; \tag{5}$$

to which we must append the constraints that follow from the gauge fixing.

Canonical quantization of the transverse and fermonic coordinates is straightforward. Since we are dealing with closed strings, the fields are periodic functions of $0 \leqslant \sigma \leqslant \pi$ and can be expanded as

$$X^i(\tau-\sigma) = \frac{1}{2} x^i + \frac{1}{2} p^i(\tau-\sigma) + \frac{i}{2} \sum_{n\neq o} \frac{\alpha_n^i}{n} e^{-2in(\tau-\sigma)}$$

$$X^i(\tau+\sigma) = \frac{1}{2} x^i + \frac{1}{2} p^i(\tau+\sigma) + \frac{i}{2} \sum_{n\neq o} \frac{\tilde{\alpha}_n^i}{n} e^{-2in(\tau+\sigma)} \tag{6}$$

$$S^a(\tau-\sigma) = \sum_{n=-\infty}^{+\infty} S_n^a \, e^{-2in(\tau-\sigma)}, \tag{7}$$

where

$$[x^i, p^j] = i\delta^{ij}, \; [\alpha_n^i, \alpha_m^j] = [\tilde{\alpha}_n^i, \tilde{\alpha}_m^j] = n\delta_{n+m,o} \, \delta^{ij}$$

$$[\alpha_i^n, \tilde{\alpha}_j^m] = 0$$

$$\{S_m^a, S_n^b\} = (\gamma^+ h)^{ab} \, \delta_{n+m,o}. \tag{8}$$

The quantization of the left moving coordinates, X^I, is more subtle. We must impose the second class constraint (which is actually the equation of motion) that

$$\phi(\sigma,\tau) = (\partial_\tau - \partial_\sigma) X^I = 0. \tag{9}$$

But this is not consistent with the cannonical commutator of X^I and its conjugate momentum $P^I = \frac{1}{\pi} \partial_\tau X^I$

$$[X^I(\sigma,\tau), P^J(\sigma',\tau)] = i\delta^{IJ} \, \delta(\sigma-\sigma'). \tag{10}$$

We therefore modify the commutation relations, ala Dirac, defining

$$[A,B]_{DIRAC} = [A,B] - \sum_{\phi_i,\phi_j} [A,\phi_i] \, C_{ij} \, [\phi_j B],$$

where the sum runs over all constraints ϕ_i whose commutator is $[\phi_i,\phi_j] = (C^{-1})_{ij}$. The Dirac bracket is then consistent with the constraint, $[A,\phi_i] = 0$, which can be imposed as an operator identity. In our case we must modify Eq. (10) to read

$$[X^I(\sigma,\tau),\ P^J(\sigma',\tau)] = \frac{i}{2}\,\delta^{IJ}\,\delta(\sigma-\sigma').$$

$$[X^I(\sigma,\tau),\ X^J(\sigma',\tau)] = -\frac{i}{4}\,\delta^{IJ}\,\mathrm{Sgn}(\sigma-\sigma'). \tag{11}$$

Therefore $X^I = X^I(\tau+\sigma)$ has the expansion

$$X^I(\tau+\sigma) = X^I + P^I(\tau+\sigma) + \frac{i}{2}\sum_{n\neq 0}\frac{\tilde{\alpha}_n^I}{n}\,e^{-2in(\tau+\sigma)}, \tag{12}$$

where upon quantization

$$[\tilde{\alpha}_n^I,\ \tilde{\alpha}_m^J] = n\delta_{n+m,o}\,\delta^{IJ}$$

$$[X^I,\ P^I] = \frac{i}{2}\,\delta^{IJ}. \tag{13}$$

Note the factor of 1/2 in the commutator of X^I and P^I, which implies that $2P^I$ is the generator of translations in the internal space. The allowed values of P^I must be restricted since X^I must be a periodic function of σ. They will be determined by the structure of the internal sixteen dimensional space.

In light-cone gauge $X^+(\tau,\sigma) = x^+ + p^+\tau$ and X^- is determined by solving the constraints that result from choosing a conformally flat metric on the world sheet. If we expand $X^-(\tau,\sigma)$ as

$$X^-(\tau,\sigma) = x^- + p^-\tau + \frac{i}{2}\sum_{n\neq o}\frac{1}{n}\left(\alpha_n^-\,e^{-2in(\tau-\sigma)} + \tilde{\alpha}_n^-\,e^{-2in(\tau+\sigma)}\right) \tag{14}$$

then α_n^- is given, as in the fermionic string, by $(n\neq o)$

$$\alpha_n^- = \frac{1}{p^+}\sum_n \alpha_m^i\,\alpha_{n-m}^i + \frac{1}{2p^+}\sum_m \left(m - \frac{n}{2}\right)\bar{S}_{n-m}\,\gamma^-\,S_m \tag{15}$$

and $\tilde{\alpha}_n^-$ is constructed, as in the bosonic string, as ($n \neq o$)

$$\tilde{\alpha}_n^- = \frac{1}{p^+} \sum_m (\tilde{\alpha}_m^i \tilde{\alpha}_{n-m}^i + \tilde{\alpha}_m^I \tilde{\alpha}_{n-m}^I). \tag{16}$$

(Note, in these formulas $\alpha_0^i = \tilde{\alpha}_0^i = \frac{1}{2} p^i$, $\tilde{\alpha}_0^I = p^I$.)
These same constraints determine p^-, the generator of τ translations conjugate to X^+, and thereby the mass operator $m^2 = 2p^+p^- - (p^i)^2$ of the string

$$\frac{1}{4} (\text{mass})^2 = N + (\tilde{N}-1) + \frac{1}{2} \sum_{I=1}^{16} (p^I)^2, \tag{17}$$

where $N(\tilde{N})$ are the normal ordered number operators for the right (left) movers

$$N = \sum_{n=1}^{\infty} (\alpha_{-n}^i \alpha_n^i + \frac{1}{2} n \bar{S}_{-n} \gamma^- S_n)$$

$$\tilde{N} = \sum_{n=1}^{\infty} (\tilde{\alpha}_{-n}^i \tilde{\alpha}_n^i + \tilde{\alpha}_{-n}^I \tilde{\alpha}_n^I). \tag{18}$$

The subtraction of -1 in (2.17) is due to the normal ordering of $\tilde{N}$, which is unnecessary in the case of the right movers due to fermion-boson cancellations. This subtraction can also be seen to be necessary to ensure ten dimensional Lorentz invariance. Finally the factor of $(p^I)^2/2$ comes from the internal momentum and winding number (see below) of the left movers.

In addition there is a further constraint that requires

$$N = \tilde{N} - 1 + \frac{1}{2} \sum_{I=1}^{16} (p^I)^2. \tag{19}$$

This constraint has a simple physical explanation. Since there is no distinguished point on a closed string, we are free to shift the origin of the σ coordinate by an arbitrary amount Δ. This is achieved by the unitary operator

$$U(\Delta) \equiv e^{2i\Delta(N-\tilde{N}+1-1/2\sum_I (P^I)^2)} , \qquad (20)$$

which (recall that $[X^I, P^I] = \frac{i}{2}\delta^{IJ}$) satisfies $U(\Delta)\ F(\tau,\sigma)\ U^+(\Delta) = F(\tau,\sigma+\Delta)$, where F can be X^i, X^I or S^a. The operator $U(\Delta)$ must therefore equal the identity operator on the space of physical states, which must therefore satisfy eq. (19). Again the subtraction constant, -1, is due to the normal ordering of $\tilde{N}$ and is necessary for Lorentz invariance.

Now let us consider the internal left-moving coordinates X^I whose properties are responsible for the new features of the heterotic string. These coordinates can be thought of as parametrizing an internal space T. It is unlikely that T could be curved without producing an inconsistent theory, since the resulting two-dimensional field theory of X^I would be an interacting nonlinear σ model with conformal anomalies. Therefore we consider only flat internal manifolds, taking T to be a sixteen-dimensional torus.

Since closed strings contain gravity in their low energy limit one expects that a compactified closed string theory will contain massless vectors associated with the isometries of the compact space. In the case of a flat 16-dimensional torus this would yield the gauge bosons of $[U(1)]^{16}$. A remarkable feature of closed string theories is that for special choices of the compact space there will exist additional massless vector mesons. These are in fact massless solitons of the closed string theory. They combine with the Kaluza-Klein gauge bosons to fill out the adjoint representation of a simple Lie group whose rank equals the dimension of T. In the case of the heterotic string the structure of T is so severely limited that only two choices are consistent. These produce the gauge mesons of $Spin(32)/Z_2$ or $E_8 \times E_8$!

To see this take T to be the most general torus, which may be thought of as R^{16} modulo a lattice Γ generated by 16 basis vectors e_i^I (i=1...16). We identify the center of mass coordinate X^I with its translation by $\pi\ e_i^I$ [$\sqrt{e_i{}^I \cdot e_i{}^I}$ is the diameter of the torus in the ith direction]

$$X^I \equiv X^I + \pi \sum_{i=1}^{16} n_i \, e_i{}^I \equiv X^I + \pi \, L^I \tag{21}$$

with n_i integers. On such a torus the allowed center of mass momenta P^I lie on the dual lattice, Γ^*, generated by $e_i{}^{*I}$ (i=1...16), defined by

$$\sum_{I=1}^{16} e_i{}^I \cdot e_j{}^{*I} = \delta_{ij} \, . \tag{22}$$

In other words $\exp(2\pi i \, P^I \cdot L^I)$, which represents a translation around T by the element πL^I (recall that $2P^I$ generates translations), must equal one, so that

$$P^I = \sum_{i=1}^{16} m_i \, e_i^{*I} \quad (m_i = \text{integer}). \tag{23}$$

For the heterotic string, which is a periodic function of σ on T,

$$X^I = X^I + P^I\tau + L^I\sigma + \ldots = X^I(\tau+\sigma) = X^I + P^I(\tau+\sigma) + \text{oscillators} \tag{24}$$

represents a string configuration which winds around the torus n_i times in the ith direction, with momenta P^I which equals the winding number L^I. This is in fact a soliton--a classical solution of the equations of motion which is characterized by topological quantum numbers (n_i) that classify the maps of the one sphere $(0 \leq \sigma \leq \pi)$ onto T $(\pi_1(T) = Z^{16})$. Such solitons must be included in the spectrum of the interacting closed string, since a string with $L^I = 0$ can split into one with $+L^I$ and $-L^I$.

Next we recall the constraint of Eq. (19), which requires that the allowed values of $(P^I = L^I)^2$ are <u>even</u> integers. Therefore, since all $(L^I + L^J)$ are allowed windings, $L^I \cdot L^J$ must be an integer (since $(L^I + L^J)^2$, $(L^I)^2$ and $(L^J)^2$ must all be even numbers). The winding numbers L^I must therefore lie on an <u>integer</u>, <u>even</u> lattice, so that the "metric" $g_{ij} = \sum_{I=1}^{16} e_i{}^I \, e_j{}^I$ is integer valued, and g_{ii} = even. The momenta P^I lie on the dual lattice Γ^*, which for integer Γ, contains Γ. [Since every vector in Γ, $V^I = \sum_{i=1}^{16} n_i e_i^{\ I}$, can be written as a vector in Γ^*, $V_i = \Sigma m_i e_i^{*I}$, with $m_i = \sum_{j=1}^{16} n_j g_{ji}$ = integer.] In general Γ^* contains more points than Γ, including some that are not of even length squared. For example if Γ = hypercubic lattice with spacing $\sqrt{2}$, Γ^* = hypercubic lattice with spacing $1/\sqrt{2}$. From a geometrical point of view we would, however, expect that all vectors in Γ^* are allowed momenta, in which case Γ^* must also be integer and even. But then $\Gamma > \Gamma^*$ and since $\Gamma^* > \Gamma$ the lattice must be self dual, $\Gamma = \Gamma^*$.

In that case

$$L^I = P^I + \sum_{i=1}^{16} n_i e_i^{\ I} \ (n_i = \text{integer}), \tag{25}$$

and the diameters of the torus are all equal to $\sqrt{2} = \sqrt{\frac{2}{\pi T}}$.

Actually, for the free string we need not take Γ^* to be integer and even. But if we do so, the non geometric nature of the momenta would produce trouble at the interacting level where we would find two-dimensional (global) diffeomorphism anomalies that render string loops nonunitary or sick.

Even self dual lattices are extremely rare. They only exist in 8n dimensions. In sixteen dimensions there are two such lattices, $\Gamma_8 \times \Gamma_8$ and Γ_{16}. The first is the direct product of two Γ_8's, where Γ_8 is the root lattice of the exceptional Lie algebra E_8; the second is the weight lattice of $Spin(32)/Z_2$, generated by the weights of the adjoint of SO(32) plus one of the spinor representations. Let us consider $\Gamma_8 \times \Gamma_8$, generated by $e_i^{\ I}$, the roots of $E_8 \times E_8$. In this case the torus T is the "maximal torus" of $E_8 \times E_8$, generated by $\exp iH_i$, where H_i constitute the Cartan subalgebra of $E_8 \times E_8$.

The appearance of these special tori and the consequent emergence of the gauge group $G(G=E_8 \times E_8$ or $Spin(32)/Z\ )$ might seem mysterious. To those familiar with the theory of affine Lie (Kac-Moody) algebras the mathematical framework is familiar; nonetheless let us take a physicist's approach. Consider the massless states of the theory. These must satisfy $N = \tilde{N} - 1 + 1/2\ \Sigma (P_I)^2 = 0$. The bosonic states with $\tilde{N}=1$,

$P^I=0$, consist of gravitons $(\tilde{\alpha}^i_{-1}|0\rangle)$ and 16 corresponding gauge bosons $(\tilde{\alpha}^I_{-1}|0\rangle)$. These massless vectors are expected--they arise from the $U(1)^{16}$ isometry of the torus. The corresponding conserved charges are simply the components of the internal (left-handed) momenta P^I. However there are additional massless vectors, solitons of the string theory, with winding numbers $L^I=P^I$ such that $(P^I)^2=2$. For the allowed self-dual lattices there are precisely 480 such vectors. The solitons have nonvanishing $U(1)^{16}$ charges, which can be identified with the non-zero weights of the adjoint representation of G, and they are massless vector mesons. It is well known that the only consisent theory of massless vector bosons with nonvanishing charges is a local gauge theory. But why couldn't the gauge group be $U(1)^{496}$? The reason is that the interactions will allow a soliton of charge, which equals winding number, P^I to break up into two solitons of charge $P^I+K^I$and $-K^I$ respectively. Thus each gauge boson couples to every other one, and the corresponding gauge group must be G. To show this explicitly we should construct the generators of G in the Fock space of the string oscillators. This we shall do below when we discuss interactions, since these generators are simply the vertex operators for a string to emit a massless gauge boson of vanishing momenta.

Let us now consider the full spectrum of the heterotic string. The physical states are simply direct products of the Fock spaces $|\rangle_R \times |\rangle_L$ of the right moving fermionic string and the left moving bosonic string, subject to the constraint, which ensures that the masses are non negative,

$$(\text{mass})^2 = 8N \geq 0$$
$$N = \tilde{N} + \frac{1}{2}(P^I)^2 - 1. \tag{26}$$

The right-handed ground state is annihilated by α^i_n and S^i_n $(n>0)$ and N. It consists of 8 bosonic states $|i\rangle_R$ and 8 fermionic states $|a\rangle_R$, which form a massless vector and spinor supermultiplet. The left-moving ground state consists of $\tilde{\alpha}^i_{-1}|0\rangle_L$ $(\tilde{N}=1,\ P^I=0)$ and $\tilde{\alpha}^I_{-1}|0\rangle_L$ $(\tilde{N}=1,\ P^I=0)$ and $|P^I,\ (P^I)^2=2\rangle_L$. The most general state

$$\prod_i \alpha^i_{-n_i} \prod_j S^{a_j}_{-m_j} \prod_k \tilde{\alpha}^k_{-n_k} \,|\text{Ground State}\rangle,$$

can be decomposed into irreducible representations of D=10, N=1 supersymmetry and of the group G.

The demonstration that these states are indeed Lorentz invariant and supersymmetric is straightforward. It is a consequence of the fact that the generators of Lorentz and supersymmetry transformations act separately on the right and left movers, and that the left movers are Lorentz invariant in 26 dimensions and the right movers Lorentz invariant and N=1 supersymmetric in 10 dimensions.

The ground state is a direct product of $|i\rangle_R + |a\rangle_R$ with $\tilde{\alpha}^i_{-1}|0\rangle_L$ (which yields the N=1 D=10 supergravity multiplet), and with $\tilde{\alpha}^I_{-1}|0\rangle_L$ and $|(P^I)^2=2\rangle_L$ (which yields the N=1 Yang Mills supermultiplet in the adjoint of $E_8\times E_8$). The higher mass states are easily assembled into $SO(9)\times G$ multiplets. For example the first massive level, with $(\text{mass})^2 = 8$, has N=1 and $(\tilde{N}, (P^I)^2) = (2,0)$, $(1,2)$ or $(0,4)$. We separately assemble the right and left moves into SO(9) multiplets and take direct products. The right movers contain $\alpha^i_{-1}|j\rangle_R$ and $S^a_-\,(b)_R$ which fill out the SO(9) representations $\underline{44}$ = 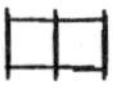 and $\underline{84}$ =  , as well as the fermion states $\alpha^i_{-1}|a\rangle_R$ and $S^a_{-1}|i\rangle_R$ which form the 128 of SO(a). In the left-moving sector we have the $(E_8\times E_8,\ SO(9))$ representations: $((1,1), \underline{44})$ which contain $\tilde{\alpha}^i_{-1}\,\tilde{\alpha}^j_{-1}|0\rangle_L$, $((248,1) + (1,248), \underline{9})$ which contains $\tilde{\alpha}^i_{-1}|(P^I)^2=2\rangle_L$, and $((3875,1) + (1,2875) + (248,248) + 2(1,1), \underline{1})$ which contains $|(P^I)^2=4\rangle$. Altogether, at this level, we have 18,883,584 physical degrees of freedom!

At higher mass levels the number of states will increase rapidly, not only will we have the usual proliferation of higher spin states but also we will get even larger representations of $E_8\times E_8$. The number of states of mass M in the heterotic string increases as $d(M) \underset{M\to\infty}{\sim} \exp(\beta_H M)$, with β_H given by $(2+\sqrt{2})\,\pi\,\sqrt{\alpha'}$, the mean of corresponding factors for the fermionic superstring and the 26 dimensional bosonic string.

The heterotic string theory has, by now, been developed to the same stage as other superstring theories. Interactions have been introduced and shown to preserve the symmetries and consistency of the theory, radiative corrections calculated and shown to be finite.

V. STRING PHENOMENOLOGY

In order to make contact beween the string theories and the real world one is faced with a formidable task. These theories are formulated in ten flat space-time dimensions, have no candidates for fermionic matter multiplets, are supersymmetric and contain an unbroken large gauge group--say $E_8 \times E_8$. These are not characteristic features of the physics we observe at energies below a Tev. If the theory is to describe the real world one must understand how six of the spatial dimensions compactify to a small manifold leaving four flat dimensions, how the gauge group is broken down to $Su_3 \times SU_2 \times U_1$, how supersymmetry is broken, how families of light quarks and leptons emerge, etc. Much of the recent excitement concerning string theories has been generated by the discovery of a host of mechanisms, due to the work of Witten and of Candelas, Horowitz and Strominger, and of Dine, Kaplonovsky, Nappi, Seiberg, Rohm, Breit, Ovrut, Segre, and others, which indicate how all of this could occur. The resulting phenomenology, in the case of the $E_8 \times E_8$ heterotic string theory is quite promising.

The first issue that must be addressed is that of the compactification of six of the dimensions of space. The heterotic string, as described above, was formulated in ten dimensional flat spacetime. This however is not neccessary. Since the theory contains gravity within it the issue of which spacetime the string can be embedded in is one of the string dynamics. That the theory can consistently be constructed in perturbation theory about flat space is equivalent to the statement that ten dimensional Minkowski spacetime is a solution of the classical string equations of motion. Such a solution yields the background expectation values of the quantum degrees of freedom. We can then ask are there other solutions of the string equations of motion that describe the string embedded in, say, four dimensional Minkowski spacetime times a small compact six dimensional manifold?

At the moment we do not possess the full string functional equations of motion, however one can attack this problem in an indirect fashion. One method is to deduce from the scattering amplitudes that describe the string fluctuations in ten dimensional Minkowski space an effective Lagrangian for local fields that describe the string modes. Restricting one's attention to the masseless modes, the resulting Lagrangian yields equations which reduce to Einstein's equations at low energies, and can be explored for compactified solutions. Another method is to proceed directly to construct the first quantized string about a trial vacuum in which the metric η_{ab} (as well as other string modes) have assumed background values. In this approach one starts with the action of Equation (2), or its supersymmetric generalization, but allows $\eta_{ab}(x)$ to be the metric of a curved manifold. A consistent string theory can be developed as long as the two dimensional field theory of the coordinates $X^{\alpha}(\sigma,\tau)$ is conformally invariant. This is a nontrivial requirement, since the theory described by (2) is an interacting nonlinear σ-model. The condition that the two dimensional theory be conformal invariant is equivalent to demanding that the string equations of motion are satisfied. Thus one can search for alternative vacuum states by looking for σ-models (actually supersymmetric σ-models), for which the relevant β functions (which are local functions of the metric $\eta_{ab}(x)$ and its derivatives) vanish. In addition one must check that the anomaly in the commutators of the stress energy tensor is not modified. Given such a theory one can construct a consistent string theory and if $\eta_{ab}(x)$ describes a curved manifold the string will effectively be embedded in this manifold.

Remarkably there do exist a very large class of conformally invariant supersymmetric σ-models, that yield solutions of the string classical equations of motion to all orders and describe the compactification of ten dimensions to a product of four dimensional Minkowski space times a compact internal six dimenional manifold. These compact manifolds are rather exotic mathematical constructs (they are Kahler and admits a Ricci flat metric--i.e. they have SU_3 holonomy) and are called "Calabi-Yau" manifolds. In general they have many free parameters (moduli) which, among the rest, determine their size. Once again, this is an indication of the enormous vacuum degeneracy of the string theory, at least when treated perturbatively, and leads to many (at the present stage of our understanding) free parameters. This abundance of riches should not displease us, at the moment we would like to know whether there are any solutions of the theory which resemble the real world, later we can try to understand why the dynamics picks out a particular solution.

In the case of the heterotic string it is not sufficient to simply embed the string in a Calabi-Yau manifold. One must also turn on an SU_3 subgroup of $E_8 \times E_8$ gauge group of the string. This is because the internal degree of freedom of the heterotic string consist of right-moving fermions, which feel the curvature of space-time, and left-moving coordinates which know nothing of the space-time curvature but are sensitive to background gauge fields. Unless there is a relation between the curvature of space and the curvature (field strength) of the gauge group there is a right left mismatch which gives rise to anomalies. Therefore one must identify the space-time curvature with the gauge curvature (embed the spin connection in the gauge group). One does this by turning on background gauge fields in an SU_3 subgroup of one of the E_8's, thereby breaking it down to E_6 (or possibly O_{10} or SU_5).

These Calabi-Yau compactifications, produce for each manifold K, a consistent string vacuum, for which the gauge group is no larger than $E_6 \times E_8$ and N=1 supersymmetry is preserved. Furthermore there now exist massless fermions which naturally form families of quarks and leptons. Recall that after Kaluza-Klein compactification the spectrum of massless chiral fermions is determined by the zero modes of the Dirac operator on the internal space. Since, for heterotic string, the gauge and spin connections are forced to be equal one can count the number of chiral fermions by geometrical arguments. The massless fermions fall into $\underline{27}$'s of E_6. This is good, E_6 is an attractive grand unified model and each $\underline{27}$ can incorporate one generation of quarks and leptons. The number of generations is equal to half the Euler character of the manifold (which counts the number of "handles" it has), and is normally quite large. If there exists a discrete symmetry group, Z, which acts freely on K, one can consider the smaller manifold K/Z, whose Euler character is reduced by the dimension of Z. By this trick, and after some searching, manifolds have been constructed with 1,2,3,4,... generations. It seems that to be realistic we must restrict attention to manifolds with three, or perhaps four, generations.

The compactification scheme also produces a natural mechanism for the breaking of E_6 down to the observed low energy gauge group. If K/Z is multiply connected one can allow flux of the unbroken E_6 (or of the E_8, for that matter) or to run through it, with no change in the vacuum energy. The net effect is that when we go around a hole in the manifold through which some flux runs we must perform a nontrivial gauge transformation on the charged degrees of freedom. These noncontractible Wilson loops act like Higgs bosons, breaking E_6 down to the largest subgroup that commutes with all of them. By this mechanism one can, without generating a cosmological constant, find vacua whose unbroken low energy gauge group is, say $Su_3 \times SU_2 \times U_1 \times$ (typically, an extra U_1 or two). Moreover there exists a natural reason for the existence of massless Higgs bosons which are weak isospin doublets (and could be responsible for the electroweak breaking at a Tev), without accompanying color triplets. Many of the successful features of grand unified models, such as the prediction of the weak mixing angle, carry over, and many of the unsuccessful predictions, such as quark lepton mass ratios, do not.

Of course it is also necessary to break the remaining N=1 supersymmetry. For this purpose the extra E_8 gauge group might be useful. Below the compactification scale it yields a strong, confining gauge theory like QCD, but without light matter fields. In general this sector would be totally unobservable to us, consisting of very heavy glueballs, which would only interact with our sector with gravitational strength at low energies. However there could very well exist in this sector a gluino condensate which can serve as source for supersymmetry breaking.

Thus the heterotic string theory appears to contain, in a rather natural context, many of the ingredients necessary to produce the observed low energy physics. I do not mean to suggest that there are not many problems and unexplained mysteries. There exists the danger (common to many grand unified models, especially suprrsymmetric ones) of too rapid proton decay, there is no deep understanding of why the cosmological constant, so far zero, remains zero to all orders, and when supersymmetry is broken, at least by the mechanism discussed above, the theory tends to relax back to ten dimensional flat space. Nonetheless, the early successes are very reassuring and they give one the feeling that there are no insuperable obstacles to deriving all of low energy physics from the $E_8 \times E_8$ heterotic string theory.

VI. OUTLOOK

I do not want to leave the impression that string theory has brought us close to the end of particle physics. Quite the opposite is the case. Not only are there many unsolved problems and deep mysteries that need to be understood before one can claim success, in addition we have only begun to probe the structure of these new theories. I prefer, therefore, to conclude with a list of open problems.

VI.1 What is String Theory?

We do not fully understand the deep symmetry principles and symmetries that underly string theories. To date these theories have been constructed in a somewhat adhoc fashion and often the formulism has produced, for reasons that are not totally understood, structures that appear miraculous.

VI.2 How Many String Theories Are There?

Do there exist more consistent theories than the known five? Do there exist fewer, in the sense that some of the ones we know already are perhaps different manifestations (different vacua?) of the same theory?

VI.3 String Technology

This is not a question but a program of development of the techniques for performing calculations within string theory, including control of multiloop perturbation theory and the construction of manifestly covariant and supersymmetric methods of calculation. In addition one needs to develop, in a manifestly covariant approach, a useful second quantized formulation of the theory--string field theory.

VI.4 What is the Nature of String Perturbation Theory

Does the perturbative expansion of the string theory converge? If not, when does it give a reliable asymptotic expansion? How can one go beyond perturbation theory?

VI.5 String Phenomenology

Here there are many issues that remain to be resolved. They can all be included in the question--can one construct a totally realistic model which agrees with observation and why is it picked out?

VI.6 What Picks the Correct Vacuum?

This is one of the greatest mysteries of the theory, which seems to have an enormous number of acceptable vacuum states. Why then don't we live in ten dimensional flat space? How does the value of the dilaton field get fixed and thereby the dilaton acquire a mass? Does the vanishing of the cosmological constant survive the physical mechanism that lifts the vacuum degeneracy?

VI.7 What Is the Nature of High Energy Physics?

By this I mean what does physics look like at energies well above the Planck mass scale? This is a question that is addressable, in principle, for the first time and might be of more than academic interest for cosmology. Does the string undergo a transition to a new phase at high temperatures and densities? Can one avoid in string theory the ubiquitous singularities that plague ordinary general relativity?

VI.8 Is There a Measurable, Qualitatively Distinctive, Prediction of String Theory

String theories can make many "postdictions" (such as the calculation of mass ratios of quarks and leptons, Higgs masses, gauge couplings, etc.). They can also make many new predictions (such as the masses of the various supersymmetric partners). These would be sufficient to establish the validity of the theory, however one could imagine conventional field theories coming up with similar pre or post dictions. It would be nice to predict a phenomenon which might be accessible at observable energies and is uniquely characteristic of string theory.

REFERENCES

I have made no attempt to give detailed references to the papers in this rapidly growing field. An incomplete set of references to recent work is given below.

Reviews:

Unified String Theories (Proceedings of the String Workshop at Santa Barbara, World Scientific, 1986).

Superstrings (Reprint Volume, edited by J. Schwarz, World Scientific, 1985).

Superstring Anomalies

M.B. Green and J.H. Schwarz, Phys. Lett. 149B (1984) 117;
L. Alvarez-Gaume and E. Witten, Nucl. Phys. B234 (1983) 269.

Heterotic String

D.J. Gross, J.A. Harvey, E. Martinec and R. Rohm, Phys. Rev. Lett. 54 (1985) 502, Nucl. Phys. B256 (1985) 253, and to be published.

String Phenomenology

P. Candelas, G. Horowitz, A. Strominger and E. Witten, Nucl. Phys. B258 (1985) 46; E. Witten, Nucl. Phys. B258 (1985) 75.

KALUZA-KLEIN AND SUPERSTRING THEORIES

P. H. Frampton
Institute of Field Physics, Department of Physics and Astronomy
University of North Carolina, Chapel Hill, NC 27514, USA

These lectures are in three sections with the subheadings

I. Strings for Strong Interactions

II. Non-string Kaluza-Klein Theories

III. Superstring Theories

I. Strings for Strong Interactions

This covers the period 1968-75 and will amount to a synopsis of an ancient book[1] published in 1974, out-of-print in 1979, and to be reprinted some time soon. The history starts with the Veneziano[2] model for the scattering of two particles into two particles (Fig. 1). The amplitude A(s, t) = A(t, s) is a crossing-symmetric function of the two Mandelstam invariants s, t. Veneziano actually chose the convenient process $\pi^a\pi^b \to \pi^c\omega$ with unique spin structure $\varepsilon_{abc}\varepsilon_{k\lambda\mu\nu}p_1^k p_2^\lambda p_3^\mu A(s, t)$. The question is what is A(s, t) as a function of s and t? This could be answered by a quantum field theory but in 1968 no such theory existed. The guess for A(s, t) was based on four phenomenological clues:

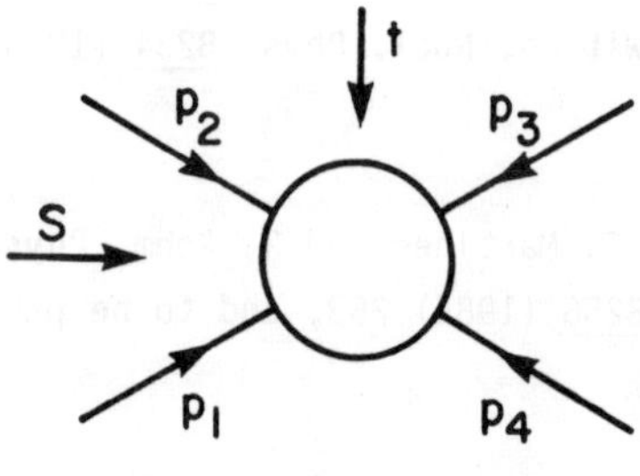

Figure 1.

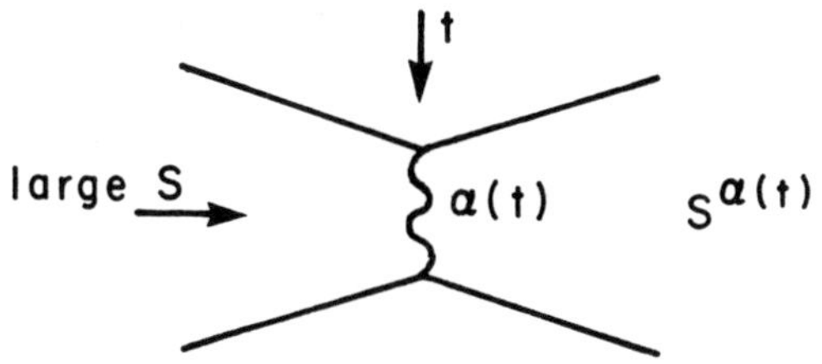

Figure 2.

(i) Regge behavior. For $s \to \infty$ with t fixed and spacelike (Fig. 2) we have

$$A(s,\ t) \sim s^{\alpha(t)} \tag{1}$$

where the Regge trajectory $\alpha(t) \simeq \alpha(o) + \alpha' t$ is approximately linear with Regge slope $\alpha' \simeq 0.9\ \mathrm{GeV}^{-2}$. The mesons lie quite accurately on straight-line trajectories (e.g. Fig. 3.)

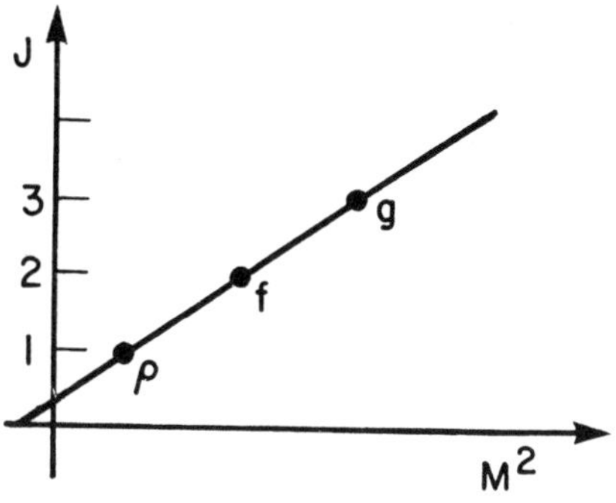

Figure 3.

(ii) Resonances. At low s, there are resonances with width << mass (Fig. 4) so that in this region we may write

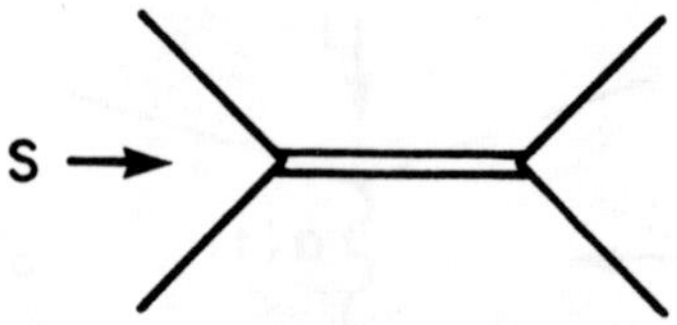

Figure 4.

$$A(s,\ t) = \sum_i \frac{a_i(t)}{s - s_i} \tag{2}$$

with residues $a_i(t)$ polynomial in t of order equal to the resonance spin.

(iii) Analyticity. Because of Regge behavior and dispersion relations (analyticity), the absorptive part of A(s, t) satisfies[3] the Finite Energy Sum Rule (FESR)

$$\int^{\bar{s}} ds\ \mathrm{Im}A(s,\ t) = \left(\frac{\bar{s}}{s_o}\right)^{\alpha(t)+1} \frac{\beta(t)}{\alpha(t) + 1} \tag{3}$$

Varying $\bar{s}$, the left hand side varies as $\bar{s}^{-1}$ (resonances) while the right side goes as $\bar{s}^{\alpha(t)+1}$. How is this possible while keeping A(s, t) = A(t, s)? was one conundrum before Veneziano. The relation of analyticity to linearly-rising Regge trajectories in a zero-width resonance approximation was discussed.[4]

(iv) Duality. A phenomenological question is: should one add the low-energy resonances to the high-energy Regge poles? Looking at the pion-nucleon charge exchange forward amplitude A(s, o), Dolen, Horn and Schmid showed that, phenomenologically, the Regge pole contribution, extrapolated back to low energy, gives an average description of the low-energy resonances (Fig. 5). Thus, addition of Regge plus resonances would be double-counting and instead there is a "duality" of the form

$$\langle \text{Resonances} \rangle_{\text{average}} = \text{Regge} \tag{4}$$

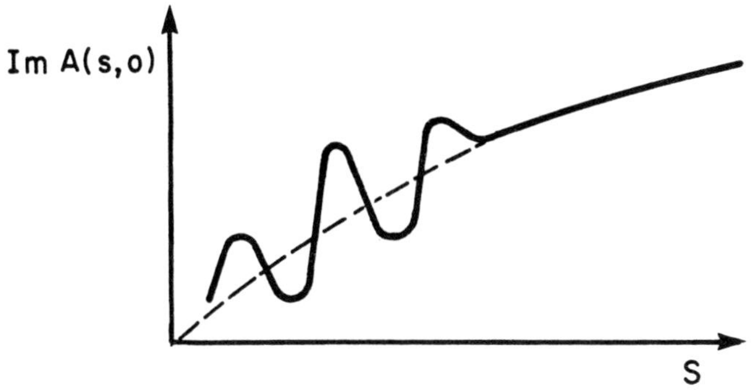

Figure 5.

One representation of this is through the quark diagrams of Harari[5] and Rosner[6], e.g. for meson-meson scattering (Fig. 6).

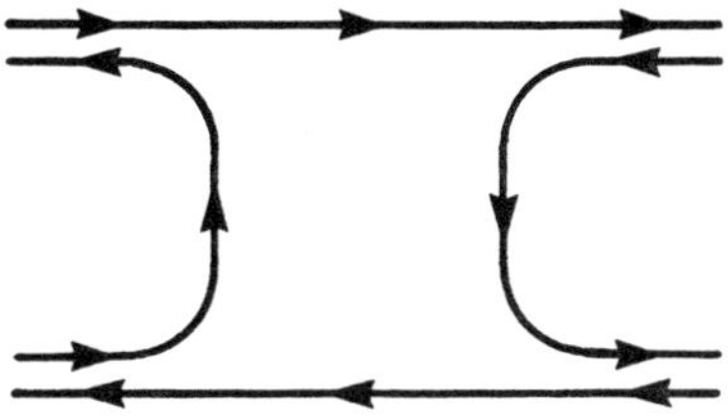

Figure 6

In elastic scattering there is a diffractive component leading to approximately constant total cross sections, through the optical theorem, and this is associated with the Pomeron singularity in the angular momentum plane. Just as the normal meson Regge trajectories are dual to resonances, the Pomeron appears phenomenologically[7] to be dual to non-resonant background, the relevant quark diagram being Fig. 7.

Based on these four clues Veneziano[2] made the guess that

$$A(s, t) = B\big(-\alpha(s), -\alpha(t)\big) \tag{5}$$

where

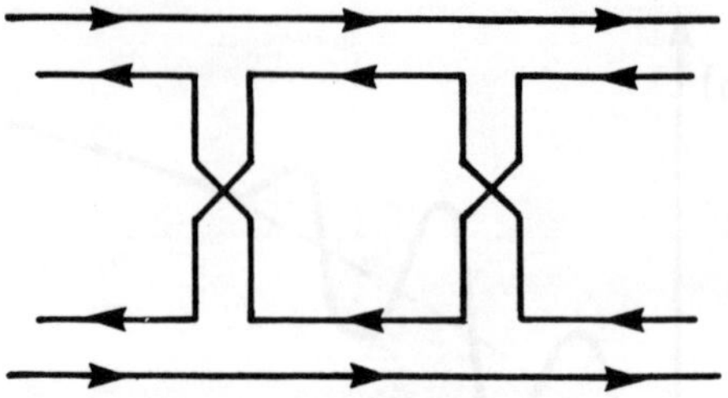

Figure 7

$$B(X, Y) = \frac{\Gamma(X)\Gamma(Y)}{\Gamma(X + Y)} \tag{6}$$

is the Euler B-function and $\alpha(s)$ is a linear function $\alpha(s) = \alpha(o) + \alpha' s$. This model has remarkable properties: manifest crossing-symmetry, expandable completely into poles in s (meromorphic), Regge behavior in all complex directions except the positive real axis, analyticity and duality. The question then was: what is the physics underlying Eq. (5)?

The first step was the generalization[8] to N-particle amplitudes (Fig. 8) incorporating similar duality properties. These amplitudes

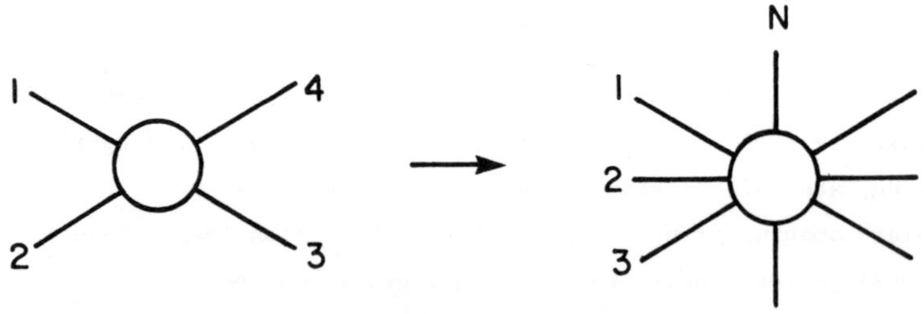

Figure 8

were then factorized[9] on a finite spectrum with degeneracy growing as $\sim e^{am}$, where m = mass, a = numerical constant. This then led to a covenient operator formalism[10] with Hamiltonian

$$H = \sum_{n=1}^{\infty} n a_\mu^{(n)+} a_\mu^{(n)} \tag{7}$$

with simple harmonic oscillator modes which led, in turn, to a physical string picture.[11]

Calculation[12] of the nonplanar orientable loop amplitude (Fig. 7) remarkably led to a new singularity, naturally (then) identified with the Pomeron.

The algebra of the 4-vector modes $\left(g_{\mu\nu} = \text{diag.}(+\,---)\right)$

$$[a_\mu^{(m)}, a_\nu^{(u)+}] = -\,\delta_{mn} g_{\mu\nu} \tag{8}$$

means that the timelike excitations are ghosts. For spacetime dimension $d \leqslant 26$, however, all such ghosts decouple.[13] Examination of the Pomeron diagram which is redrawn in two different ways in Fig. 9 reveals a spectrum of closed string states with slope $\frac{1}{2}\alpha'$ and

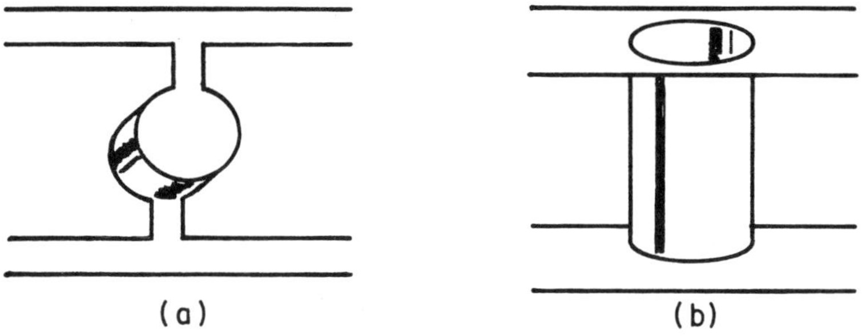

Figure 9

leading intercept $\alpha(o) = 2$. This sector appears consistently only for $d = 26$ exactly. When we scatter such closed strings we find a new type of singularity structure (Fig. 10) with a four-point function of the form[14]

$$A(s, t, u) = \frac{\Gamma(-\alpha_s/2)\Gamma(-\alpha_t/2)\Gamma(-\alpha_u/2)}{\Gamma\left(\frac{-\alpha_s-\alpha_t}{2}\right)\Gamma\left(\frac{-\alpha_t-\alpha_u}{2}\right)\Gamma\left(\frac{-\alpha_u-\alpha_s}{2}\right)} \tag{9}$$

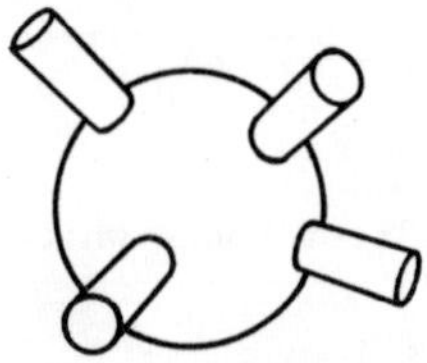

Figure 10

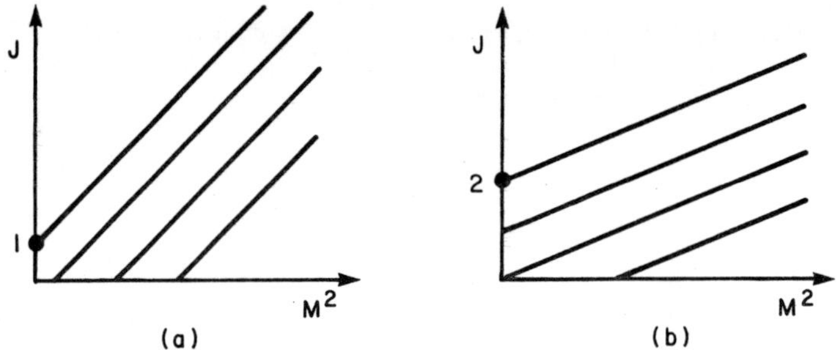

Figure 11

The spectra for open and closed strings are as indicated in Fig. 11. The intercepts required for strong interaction physics are $\alpha(o) \simeq 1/2\ (\rho)$ and $\alpha(o) \simeq 1$ (Pomeron), off by about a factor two from the dual resonance model prediction.

In 1971, a new dual model incorporating fermions was constructed[15] with the same intercepts but a critical spacetime dimension now $d = 10$. The odd G parity sector of this old model is precisely the O(32) superstring discussed in section III below.

Clearly the string does not provide a fundamental theory of strong interactions (alone). Aside from the serious difficulties of the spectrum and the dimension already given, the amplitudes for off-shell currents as well as the "granularity" needed for the parton model were other shortcomings. Finally, the success of QCD drove another nail into the coffin of the old string model which essentially died in 1975.

Nevertheless, we should bear in mind that hadrons do lie on linear Regge trajectories, there is phenomenological duality for hadron scattering, etc., so strings may approximate QCD in certain regimes, e.g., the linear quarkonium potential at large interquark separation.

The current rebirth of strings as applied to gravitational, and possibly all, interactions requires the identification[16] of the misplaced massless spin-two Pomeron with the graviton and decreases α' from about 1 GeV^{-2} to about 10^{-38} GeV^{-2}. This has led to five superstring candidates for possibly finite quantum gravity theories in $d = 10$:

(1) O(32) open and closed strings[15]

(2) O(32) heterotic string[17]

(3) E(8) x E(8) heterotic string[17]

(4) Nonchiral N=2 closed strings[18]

(5) Chiral N=2 closed strings[18]

But the phenomenological motivations from strong interactions have disappeared for quantum gravity. There are precedents, e.g. spontaneous symmetry breaking was used with some success in strong interactions before its crucial use in electroweak interactions. It may be that the old string people guessed a unique answer to the S-matrix problem and only now is it being used in the correct way.

II. Non-string Kaluza-Klein Theories

II.a. Anomalies

A principal difficulty of, and restriction on, all Kaluza-Klein theories is to obtain chiral fermions on M_4, four dimensional Minkowski space. This leads to two simple, but important, remarks:

(i) The initial dimension must be even $d = 2n$, with a spatial rotation group $O(2n - 1)$, n = integer. In $O(2n)$ rotational invariance implies that parity (P) is conserved because P is not a discrete operation but merely part of the rotation group e.g. for $d = 3$ and O(2) parity is a 180° rotation (Fig. 12). In such a case, no chirality can be defined.

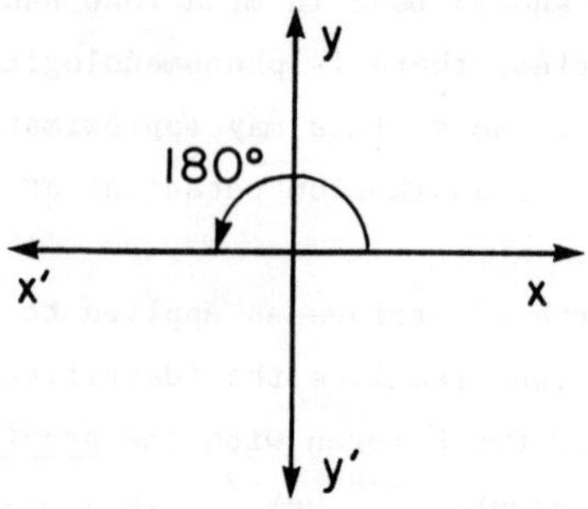

Figure 12

Equivalently, for d = 4, the four gamma matrices γ_μ can be combined to give the pseudoscalar γ^5 which allows the decomposition

$$\psi = \left(\frac{1+\gamma_5}{2}\right)\psi + \left(\frac{1-\gamma_5}{2}\right)\psi = \psi_R + \psi_L \tag{10}$$

In d = 5, by contrast, $\gamma^0, \gamma^1, \gamma^2, \gamma^3, \gamma^4$ form a complete set and there is no "γ^6".

(ii) There must be explicit gauge fields in the starting dimension d = 2n > 4. If we start with a chiral fermion ψ_L in d = 2n > 4 (eigenvalue of γ_{2n+1} equal to −1) then compactification to d = 4 with a compact manifold C_{2n-4} gives chiralities as follows

$$M_{2n} \to M_4 \times C_{2n-4}$$

$$\begin{aligned} \gamma_{2n+1} &= \gamma_5 \times \gamma_{2n-3} \\ &= + \;\times\; - \\ &\; - \;\times\; + \end{aligned} \tag{11}$$

so that both chiralities appear symmetrically on M_4. Thus the original Kaluza-Klein program starting with just higher dimensional gravity fails. One must introduce an asymmetry between the chiralities by arranging explicit gauge fields in a topologically nontrivial vacuum configuration on C_{2n-4}; this then allows survival of chirality on M_4.

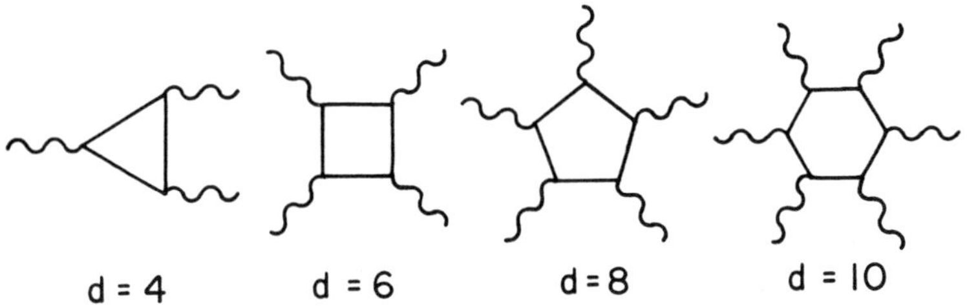

Figure 13

The necessity of explicit Yang-Mills fields in these theories was disappointing from the viewpoint of unification. One did, however, still hope to explain fermion quantum numbers, including the existence of families, by new anomaly cancellation conditions. This is because such gauge theories with chiral fermions have anomalies which must be zero in order to define the quantum theory and keep unitarity. The nonrenormalizability of such higher dimensional theories is irrelevant to the necessity for anomaly cancellation.

The leading contribution to the gauge anomaly in dimension $d = 2n$ is from a polygonal Feynman graph with $(n + 1)$ sides e.g. the triangle in $d = 4$ and the hexagon in $d = 10$ (Fig. 13). These anomalies may be computed using Fujikawa's method, or by differential geometry. We used[19] a brute force method of studying the linear divergence of the Feynman diagram (Fig. 14). Shifting integration variables appropriately one finds that the $5! = 120$ crossed diagrams combine in 60 pairs and enable one to compute

$$2q_k V_{k\lambda\mu\nu\rho\sigma}(p_a p_b p_c p_d p_e)$$

$$= 2^6 X_6 (ST)\ \varepsilon_{\lambda\mu\nu\rho\sigma\alpha\beta\gamma\delta\varepsilon} p_a^\alpha p_b^\beta p_c^\gamma p_d^\delta p_e^\varepsilon \tag{12}$$

with

$$X_6 = \frac{-1}{2^{10}\pi^5 5!} \tag{13}$$

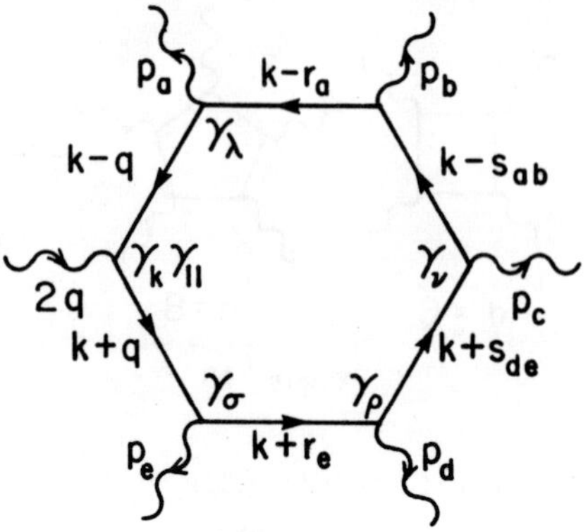

Figure 14

and

$$(ST) = STr(\Lambda^a\Lambda^b\Lambda^c\Lambda^d\Lambda^e) \tag{14}$$

is the symmetrized trace of generators in the fermion representation. We must decompose this trace into irreducible tensors formed from the defining representation

$$\begin{aligned} STr(\Lambda^a\Lambda^b\Lambda^c\Lambda^d\Lambda^e\Lambda^f) &= A_6 STr(\lambda^a\lambda^b\lambda^c\lambda^d\lambda^e\lambda^f) \\ &+ A_6^{4,2} S\big(tr(\lambda^a\lambda^b\lambda^c\lambda^d)tr(\lambda^e\lambda^f)\big) \\ &+ A_6^{3,3} S\big(tr(\lambda^a\lambda^b\lambda^c)tr(\lambda^d\lambda^e\lambda^f)\big) \\ &+ A_6^{2,2,2} S\big(tr(\lambda^a\lambda^b)tr(\lambda^c\lambda^d)tr(\lambda^e\lambda^f)\big) \ . \end{aligned} \tag{15}$$

For SU(N) and k-rank antisymmetric tensors the relevant formulas for these hexagon anomalies are

$$A_6(N,k) = \sum_{p=0}^{k=1} (-1)^{k-p+1}(k-p)^5\binom{N}{p} \tag{16}$$

$$A_6^{4,2}(N,k) = 15\ A_4(N-2,\ k-1) \tag{17}$$

where

$$A_4(N, k) = \sum_{p=0}^{k=1} (-1)^{k-p+1}(k - p)^3 \binom{N}{p} \tag{18}$$

$$A_6^{2,2,2}(N, k) = 15 \binom{N-6}{k-3} \tag{19}$$

$$A_6^{3,3}(N,k) = \frac{2}{3} A_6^{4,2}(N, k) + \frac{4}{3} A_6^{2,2,2}(N, k) \tag{20}$$

For example:

$$A_6(N, 0) = 1 \tag{21}$$

$$A_6(N, 2) = (N - 32) \tag{22}$$

The $k = 2$ representation of SU(N) behaves like the adjoint of O(N). Only the disconnected (non-leading) anomalies may be compensated by Chern-Simons terms. O(32) is the <u>only</u> nonexceptional group for which the adjoint has leading hexagon anomaly $A_6 = 0$.

One immediate point is that if we are in $d = 10$ with only left-handed, and no right-handed, Weyl spinors then, without adding Chern-Simons classes, there cannot be anomaly cancellation for any nontrivial representation of any gauge group G. This can be seen by going into the Cartan subalgebra (diagonal) with real eigenvalues λ_i and noticing that

$$\sum_i \lambda_i^6 > 0 \tag{23}$$

The only possible cancellation in these nonstring cases is between left and right helicities (in $d = 4k + 2$). Thus, all supersymmetric Yang-Mills theories in $d = 10$ (or any $d = 4k + 2$) are anomalous. Because of the zero-slope limit it appeared that open superstrings were anomalous although we did say[20] in a footnote: "The only possible loophole would be if the infinite sum over massive fermions gives a nonvanishing anomaly precisely cancelling the anomaly of the massless fermions. We regard this as extremely unlikely." For O(32), however, this is precisely what happens from this viewpoint.

We now discuss those combinations of the representations [k] (K-rank antisymmetric) of SU(N) which have, in d = 2n, no gauge, gravity, or mixed chiral anomalies. For d = 6, the first non-trivial solution[21] ("non-trivial" means able to give chirality in d = 4 after compactification) is for SU(6) where there is only one possibility. For higher SU(N) there is one solution for SU(7), two for SU(8,9), three for SU(10,11), and so on. In general, for d = 2n and SU(N) with only antisymmetric representations the number of such independent solutions is

$$\left[\frac{1}{2}(N - n - 1)\right] \tag{24}$$

where [x] is the integer part of x.

There is a systematic procedure[22] for constructing these solutions: consider the superalgebra SU(N/M) with $(N - M) \geq (n + 1)$. Choose first a combination of $[k]^{(N-M)}$ which is real (n = even) or pure imaginary (n = odd). Now promote these Young tableaus $[k]^{(N-M)}$ to super Young tableaus $[k]^{(N/M)}$ defined by

$$[k]^{(N/M)} = \sum_{p=0}^{m} (-1)^p \left([k-p]^{(N)}, \{p\}^{(M)}\right) \tag{25}$$

where $\{p\}^{(M)}$ is a symmetric p-rank representation of SU(M) and the sign $(-1)^p$ determines the helicity of the state. The combination arrived at has no chiral anomalies of gauge or gravitational type.

It is convenient to introduce the principal representations[23,24]

$$P_{(2n)}(N, k) = [k]^{(N/N-2k+1)} + (-1)^n [k-1]^{(N/N-2k+1)} \tag{26}$$

and by taking

$$\left[\frac{1}{2}(n + 3)\right] \leq k \leq k_{max} = \begin{cases} [N/2] & n = \text{odd} \\ [\frac{1}{2}(N-1)] & n = \text{even} \end{cases} \tag{27}$$

we obtain the full number $[\frac{1}{2}(N - n - 1)]$ of independent solutions as conjectured by others. Note, in particular, that the number of independent choices is small.

II. b. Compactification

We seek solutions of the Einstein-Yang-Mills equations in $d = 2n$ with less than the maximal symmetry (M_{2n}). Actually, we shall discuss the manifolds of the form $M_4 \times S_{2n-4}$ (sphere).

Such classical solutions require that the energy-momentum tensor T_{MN} receive contributions from a nonzero vacuum value of the gauge field e.g. a monopole configuration for $n = 3$ (S_2), an instanton configuration for $n = 4$ (S_4), etc. This drives the compactification. At the same time, and quite attractively, this precisely enables survival of chiral fermions on M_4.

We shall require classical stability[25] of the vacuum under small perturbations--surely a necessary condition for stability in the complete quantum theory.

Consider first the case $d = 6$ with manifold $M_4 \times S_2$ and monopole-induced compactification.[26] The solution sought is therefore of the form

$$g_{MN} dz^M dz^N = g_{mn}(x) dx^m\, dx^n + g_{\mu\nu}(y) dy^\mu\, dy^\nu \tag{28}$$

$$g_{\mu\nu}(y) dy^\mu\, dy^\nu = a^2 (d\theta^2 + \sin^2\theta\, d\phi^2) \tag{29}$$

$$A_M dz^M = A_\mu(y) dy^\mu \tag{30}$$

$$= \frac{n}{2g} (\cos\theta \mp 1) d\phi \qquad \begin{matrix} 0 \leqslant \theta < \pi \\ 0 < \theta \leqslant \pi \end{matrix} \tag{31}$$

Here, n = monopole charge. It is worth pointing out that the monopole is not in the space: the field on S_2 is as if there were a monopole at its center, see Eq. (31).

To investigate which fields remain massless on M_4, we expand in monopole harmonics $Y_{q,LM}(\theta,\phi)$ which are eigenfunctions of $\underline{J}^2 = (\underline{L}^2 + \underline{q}^2)$ with $q = ny/2$ where y is the U(1) charge of the field considered. For example, a massless (in $d = 6$) scalar field is expanded as

$$\Phi(x, y) = \sum_{JM} \phi_{JM}(x) Y_{q,JM}(\theta, \phi) \tag{32}$$

The only massless component (recall $J \geq |q|$) is for $q = 0$. If $n \neq 0$, this means $y = 0$. Recall that

$$\Box_z = \Box_x + \Box_y \tag{33}$$

so the eigenvalue of $\Box_y$ must be zero (s-wave, independent of θ and ϕ).

More interesting is a 4-component Weyl spinor Ψ_L in $d = 6$ with its expansion

$$\begin{pmatrix} \Psi_{L1} \\ \Psi_{L2} \\ \Psi_{L3} \\ \Psi_{L4} \end{pmatrix} = \sum_{JM} \begin{pmatrix} \Psi^1_{L,JM}(x)\ Y_{q-1/2,JM}(\theta, \phi) \\ \Psi^2_{L,JM}(x)\ Y_{-q+1/2,JM}(\theta, \phi) \\ \Psi^1_{R,JM}(x)\ Y_{q+1/2,JM}(\theta, \phi) \\ \Psi^2_{R,JM}(x)\ Y_{-q-1/2,JM}(\theta, \phi) \end{pmatrix} \tag{34}$$

For a given q, consider $J = |q| - 1/2$. Only the upper two ($q > 0$) or lower two ($q < 0$) components are nonzero. The corresponding multiplicity is $(2J + 1) = 2|q| = |ny|$. Hence we obtain $|ny|$ Weyl spinors massless on M_4 with the same ($q > 0$) or opposite ($q < 0$) helicity compared to the helicity in $d = 6$.

Using this rule, we can easily analyze the anomaly-free combinations.

It turns out that these $d = 6$ examples are not very successful phenomenologically ($d = 6$ supergravity may work better?) so I will go immediately to instanton-induced compactification in $d = 8$ with manifold $M_4 \times S_4$. This is similar to the $d = 6$ case except that now we write[27]

$$A^a_\mu(y) dy^\mu = \frac{2}{g} \frac{\eta_{a\mu\nu} y^2}{y^2 + b^2} dy^\mu . \tag{35}$$

where μ = 5, 6, 7, 8 and a = 1, 2, 3 for the generators of SU(2)g. The full gauge group is SU(N) $\supset$ SU(2)g. Again we take Weyl spinors in antisymmetric representations $[k]^{(N)}$ of SU(N). These contain only t = 0 and t = 1/2 representations of SU(2)g. Writing

$$S_4 = \frac{SO(5)}{SO(4)} = \frac{SO(5)}{SU(2)_A \times SU(2)_B} \tag{36}$$

the invariance group of the instanton is $SU(2)_{B+g}$ and we must analyze an 8-component chiral spinor Ψ_L in d = 8 by expansion into irreducible representations of $SO(5) \supset SU(2)_A \times SU(2)_{B+g}$. We find, by a generalization of the procedure done explicitly above for d = 6, that

$$[k]_L^{(N)} \rightarrow \underset{SO(5)\ \text{singlet}}{[k-1]_L^{(N-2)}} \tag{37}$$

This is a very simple rule. In fact, it extends to complete superalgebra tableaus

$$[k]_L^{(N/M)} \rightarrow [k-1]^{(n-2/M)} \tag{38}$$

Now we ask: can we find completely anomaly-free fermions in d = 8 that give interesting chiral fermions on M_4? This is not a trivial question since the anomaly conditions are so restrictive. Nevertheless, the answer is affirmative.[23,24]

To arrive at SU(N) in d = 4 we start from SU(N + 2) in d = 8. Examples are:

$$N = 11 \qquad [5]_L^{(13/3)} + 2[4]_L^{(13/5)} + 3[3]_L^{(13/7)}$$

$$\rightarrow \quad (11^4 + \overline{11}^3 + \overline{11}^2 + \overline{11}) \text{ of } SU(11) \tag{39}$$

which is a three-family model.[28]

$$N = 9 \qquad [4]_L^{(11/3)} + 3[3]_L^{(11/5)}$$

$$\rightarrow \quad 9^3 + 9(\bar{9}) \text{ of } SU(9). \tag{40}$$

This also has three families[29] since on reduction to SU(5)

$$SU(9) \to SU(5)$$

$$[3]^{(9)} : \quad 84 = \overline{10} + 4(10) + 6(5) + 4(1) \tag{41}$$

$$9[8]^{(9)} : \quad 9(\bar{9}) = 9(\bar{5} + 4(1)) \tag{42}$$

so that adding, and dropping real representations that are expected to develop superheavy Dirac masses, leaves

$$3(10 + \bar{5}) \text{ of } SU(5). \tag{43}$$

The SO(5) isometry group is passive: all the massless chiral fermions on M_4 are SO(5) singlets. SO(5) nonsinglets are superheavy, near the Planck scale.

Finally, concerning nonstring Kaluza-Klein theory we should remark that the quantum fluctuations are expected to be large (the theory is non-renormalizable) and the only features of the classical solutions expected to survive are those massless particles protected by an unbroken symmetry like gauge vectors and chiral fermions. We should not take seriously any massless scalars which, since unprotected here by supersymmetry, will become superheavy due to quantum corrections.

Rather, we have shown how to obtain phenomenologically an acceptable mass spectrum for spin 1 and spin 1/2 in the effective theory on M_4.

Two objections to this approach are

1) the theory is non-unified; the gauge coupling is independent of the gravitational coupling.

2) the quantum corrections are out of control.

Both objections may be overcome by superstrings.

III. Superstring Theories

III.a. Anomaly Cancellation

Let us first focus on the anomaly cancellation[30] which occurs with the gauge group O(32). The d = 10 chiral fermions are in the 496 dimensional adjoint representation of O(32).

For the adjoint of O(N), put k = 2 in the SU(N) formulae already given to find

$$A_6(N, 2) = N - 32 \tag{44}$$

$$A_6^{4,2}(N, 2) = 15 \tag{45}$$

$$A_6^{2,2,2}(N, 2) = 0 \tag{46}$$

$$A_6^{3,3}(N,2) = 10 \tag{47}$$

But now we must realize that $tr(\lambda\lambda\lambda) = 0$ in O(N) if $N \neq 6$ and hence we have

$$STr(\lambda\lambda\lambda\lambda\lambda\lambda) = (N - 32)Str(\lambda\lambda\lambda\lambda\lambda\lambda) + 15S\big(tr(\lambda\lambda\lambda\lambda)tr(\lambda\lambda)\big) \tag{48}$$

for Λ of O(N) in the adjoint representation. We see that O(32) supersymmetric Yang-Mills theory is not enough.

The field theory limit of the superstring has the bosonic terms

$$S = \int d^{10}x \sqrt{g} \, [- \frac{1}{2\kappa^2} R - \frac{1}{\kappa^2} \frac{1}{\phi^2} \partial_\mu\phi\partial_\mu\phi$$

$$- \frac{1}{4g^2\phi} F^2 - \frac{3\kappa^2}{2g^4} \frac{1}{\phi^2} H^2] \tag{49}$$

where

$$F = dA + A^2 \tag{50}$$

$$H = dB - \omega^{o}_{3Y} - \omega^{o}_{3L} \tag{51}$$

Note that the ten dimensional spacetime dictates that the dimensions are $[g] \sim L^3$, $[\kappa] \sim L^4$ so

$$\frac{\kappa}{g^2} \sim \frac{1}{L^2} \sim \frac{1}{\alpha'} \sim T \tag{52}$$

where α' = Regge slope and T = string tension. In fact, g^2 is absorbable into $\langle\phi\rangle$ leaving no dimensionless parameters at all, just T.

The presence of the Yang-Mills Chern-Simons form ω^{o}_{3Y} in Eq. (51) was discovered in N = 1 supergravity plus N = 1 super-Yang-Mills in d = 10 by two different groups[31] in 1982. One has

$$\omega^{o}_{3Y} = \mathrm{tr}(AF - \frac{1}{3} A^3) \tag{53}$$

$$d\omega^{o}_{3Y} = \mathrm{tr}\ F^2 \tag{54}$$

$$\omega^{o}_{3L} = \mathrm{tr}(\omega R - \frac{1}{3} \omega^3) \tag{55}$$

$$d\omega^{o}_{3L} = \mathrm{tr}\ R^2 \tag{56}$$

where the last two will be important in the gravitational anomalies.

Although $B_{\mu\nu}$ is in the supergravity multiplet, and hence a "gauge singlet," it does transform under Yang-Mills gauge transformation as dictated by local supergravity. Under

$$\delta A = d\Lambda + [A, \Lambda] \tag{57}$$

where Λ = gauge function, then

$$\delta\omega^{o}_{3Y} = d\omega^{1}_{2Y} \tag{58}$$

Corresponding to

$$\mathrm{TrF}^6 \doteq (N - 32)\mathrm{trF}^6 + 15\ \mathrm{trF}^4\mathrm{trF}^2 \tag{59}$$

the consistent anomaly is

$$G_1 \sim \int (\frac{1}{3}\ \omega^{1}_{2Y}\mathrm{trF}^4 + \frac{2}{3}\ \omega^{1}_{6Y}\mathrm{trF}^2) \tag{60}$$

If we add to the effective action

$$S_1 = \int (B\mathrm{trF}^4 + \frac{2}{3}\ \omega^{o}_{3Y}\omega^{o}_{7Y}) \tag{61}$$

with $\delta B = (-\ \omega^{1}_{2Y} - \omega^{1}_{2L})$ we find that

$$\delta S_1 = \int \delta B\ \mathrm{tr}F^4 - \frac{2}{3}\left(d\omega^1_{2Y})\omega^o_{7Y} - \frac{2}{3}\ \omega^o_{3Y}(d\omega^1_{6Y})\right) \tag{62}$$

$$= \int (\delta B\ \mathrm{tr}F^4 + \frac{2}{3}\ \omega^1_{2Y}\mathrm{tr}F^4 - \frac{2}{3}\ \omega^1_{6Y}\mathrm{tr}F^2) \tag{63}$$

$$= - G_1 - \int \omega^1_{2L}\mathrm{tr}F^4 \tag{64}$$

The miracle continues for 0(32) since one finds that one $(3/2)_L$ gravitino plus one $(1/2)_R$--the content of the supergravity multiplet--plus n $(1/2)_L$ gives a gravitational anomaly[32] of the form

$$\frac{n - 496}{7560}\ \mathrm{tr}R^6 + \left[\frac{1}{8} + \frac{n - 496}{5760}\right]\ \mathrm{tr}R^4\mathrm{tr}R^2$$
$$+ \left[\frac{1}{32} + \frac{n - 496}{13824}\right]\ (\mathrm{tr}R^2)^3 \tag{65}$$

so we must have n = 496 to cancel the leading gravitational anomaly = dimension 0(32). By adding

$$S_2 = - \int \left(\frac{1}{32}\ B(\mathrm{tr}R^2)^2 + \frac{1}{8}\ B\ \mathrm{tr}R^4 + \frac{1}{12}\ \omega^o_{3L}\omega^o_{7L}\right) \tag{66}$$

and using $\delta\omega^o_{3L} = - d\omega^1_{2L},\ \delta\omega^o_{7L} = - d\omega^1_{6L}$ we find

$$\delta S_2 = - G_2 + \int \left(\frac{1}{32}\ \omega^1_{2Y}(\mathrm{tr}R^2)^2 + \frac{1}{8}\ \omega^1_{2Y}\mathrm{tr}R^4\right) \tag{67}$$

where

$$G_2 = - \int \left(\frac{1}{32}\ \omega^1_{2L}(\mathrm{tr}R^2)^2 + \frac{1}{24}\ \omega^1_{2L}\mathrm{tr}R^4 + \frac{1}{12}\ \omega^1_{6L}\mathrm{tr}R^2\right) \tag{68}$$

is the consistent anomaly corresponding to Eq. (65).

The mixed gauge-gravity anomalies require two further numerical coincidences, making a total of four, in order that 0(32) be an anomaly-free superstring.

For 0(32) the mixed anomaly is

$$- \mathrm{tr}R^2\left[\mathrm{tr}F^4 + \frac{\hat{A}}{8}\ (\mathrm{tr}F^2)^2\right] + \hat{B}\mathrm{tr}F^2\left[\frac{1}{8}\ \mathrm{tr}R^4 + \frac{5}{32}\ (\mathrm{tr}R)^2\right] \tag{69}$$

where $\hat{A} = \hat{B} = 1$. For general 0(N) one would have in this formula

$$\hat{A} = \frac{24}{N - 8} \quad \text{and} \quad B = \frac{\hat{N} - 2}{32} \tag{70}$$

The corresponding consistent anomaly is

$$\begin{aligned} G_3 = \int \Big(& \frac{1}{3}\, \omega^1_{2L} \mathrm{tr}F^4 + \frac{2}{3}\, \omega^1_{6Y} \mathrm{tr}R^2 \\ & + \frac{\hat{A}}{24}\, \omega^1_{2L} (\mathrm{tr}F^2)^2 + \frac{\hat{A}}{12}\, \omega^1_{2Y} \mathrm{tr}R^2 \mathrm{tr}F^2 \\ & - \frac{\hat{B}}{24}\, \omega^1_{2Y} \mathrm{tr}R^4 - \frac{\hat{B}}{12}\, \omega^1_{6L} \mathrm{tr}F^2 \\ & - \frac{5\hat{B}}{96}\, \omega^1_{2Y} (\mathrm{tr}R^2) + \frac{5\hat{B}}{48}\, \omega^1_{2L} \mathrm{tr}R^2 \mathrm{tr}F^2 \Big) . \end{aligned} \tag{71}$$

We now add to the effective action

$$\begin{aligned} S_3 = \int \Big(& \frac{K}{8}\, B \mathrm{tr}F^2 \mathrm{tr}R^2 + \frac{L}{48}\, \omega^{o}_{3L} \omega^{o}_{3Y} \mathrm{tr}R^2 \\ & - \frac{M}{24}\, \omega^{o}_{3Y} \omega^{o}_{3L} \mathrm{tr}F^2 - \frac{2N}{3}\, \omega^{o}_{3L} \omega^{o}_{7Y} + \frac{P}{12}\, \omega^{o}_{3Y} \omega^{o}_{7Y} \Big) \end{aligned} \tag{72}$$

where K, L, M, N, P are initially at our disposal. Now we wish to arrange that

$$\delta S_3 = - G_3 - (\delta S_1 + G_1) - (\delta S_2 + G_2) \tag{73}$$

and by looking at coefficients of particular terms we find:

$$\omega^1_{2L} (\mathrm{tr}F^2)^2 : \qquad M = \hat{A} \tag{74}$$

$$\omega^1_{2Y} \mathrm{tr}F^2 \mathrm{tr}R^2 : \qquad M = 3 - 2\hat{A} \text{ hence } \hat{A} = 1 \tag{75}$$

$$\omega^1_{2L} \mathrm{tr}F^2 \mathrm{tr}R^2 : \qquad L = 6 - 5\hat{B} \tag{76}$$

$$\omega^1_{2Y} (\mathrm{tr}R^2)^2 : \qquad 2L = 5\hat{B} - 3 \text{ and hence } \hat{B} = 1. \tag{77}$$

Hence we soon find that $\hat{A} = \hat{B} = 1$ both requiring 0(32) uniquely. Of

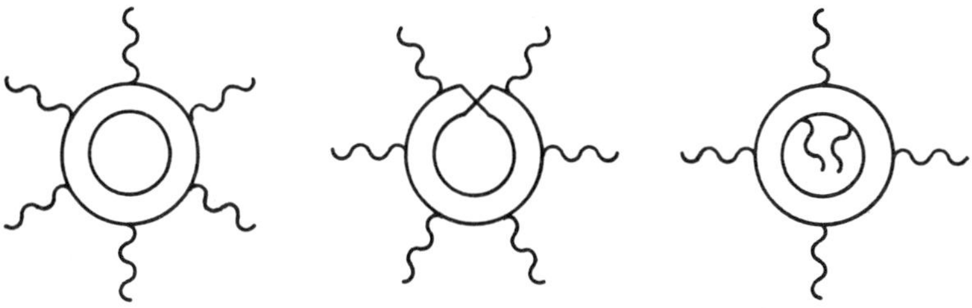

Figure 15

course, ones inevitable reaction is that one is discovering anomaly cancellations at too superficial a level and that all this must come from automatic anomaly-freedom in a higher-dimensional spacetime.

A similar series of anomaly cancellations works for the group E(8) x E(8) and the student is encouraged to work this out in similar detail as an exercise.

In O(32), consider the amplitude with six external O(32) massless gauge bosons. The three contributions are depicted in Fig. 15. The nonplanar orientable diagram (the final figure of Fig. 15) is non-anomalous because the hexagon anomaly corresponding to the first figure in Fig. 16 is cancelled by the massless tensor $(B_{\mu\nu})$ exchange shown also in Fig. 16. The $B_{\mu\nu}$ is in the same supermultiplet as the graviton and is a massless closed-string state. This diagram is really a "tree" rather than a "one-loop" diagram. The appearance of Planck's constant is complicated by the fact that closed strings have an intrinsic "h". Hence, the prejudice from quantum field theory that quantum corrections are loop effects is not true for the open super string loop expansion.

In any case, one has arrived at an anti-Kaluza-Klein theory of precisely the interesting type in terms of producing chiral fermions on M_4. Also, it is fully unified with gravity and may be a finite quantum theory (see below).

One may pause and ask how this new type of anomaly cancellation

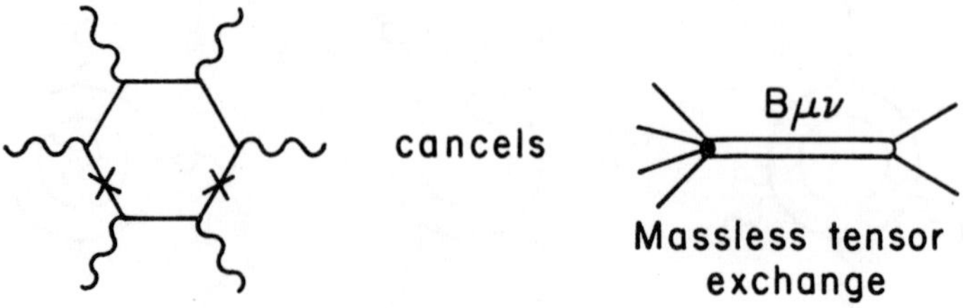

Figure 16

impacts on nonstring Kaluza-Klein theory, in case we ever become disenchanted by superstrings. The answer is[33] that the non-leading gauge anomalies, the non-leading pure gravitational anomalies, and all the mixed gauge-gravitational anomalies can always be cancelled (at least for d = 6, 8, 10) by adding antisymmetric tensor fields transforming as suitable Chern-Simons classes. So one need cancel only the leading (connected) gauge anomaly in nonstring theories. The leading gravitational anomaly may be cancelled by gauge singlets that do not effect the light fermion content on M_4.

III.b. One Loop Finiteness of Open 0(32) Superstring Theory

We shall consider one loop diagrams with some number M of external massless ground states of the superstring. This will imply all one loop graphs are finite by factorization if we can do any M. We shall use light-cone gauge. For example, for the bosonic modes the action in light-cone variables is just

$$S = -\frac{1}{4\pi\alpha'}\int_0^\pi d\sigma \int d\tau\, \partial_\alpha X^i \partial^\alpha X^i \tag{78}$$

The corresponding equations of motion are

$$\left(\frac{\partial^2}{\partial\sigma^2} - \frac{\partial^2}{\partial\tau}\right) X^i(\sigma, \tau) = 0 \tag{79}$$

with boundary conditions for an open string

$$\frac{\partial}{\partial\sigma} X^i = 0 \quad \text{at } \sigma = 0 \text{ and } \sigma = \pi \,. \tag{80}$$

The general solution is

$$X^i(\sigma, \tau) = x^i + p^i\tau + i \sum_{n\neq 0} \frac{1}{n} \alpha_n^i \cos n\sigma \, e^{-in\tau} \tag{81}$$

Upon quantization, one imposes

$$[\alpha_m^i, \alpha_n^j] = m\delta_{m+n,o}\delta^{ij} \tag{82}$$

Or defining, for $n \geqslant 1$, $\alpha_n^i = \sqrt{n}\, a_n^i$ and $\alpha_{-n}^i = \sqrt{n}\, a_n^{i+}$

$$[a_m^i, a_n^{j+}] = \delta_{mn}\delta^{ij} \tag{83}$$

Constraints give a mass-shell condition

$$\alpha'(\text{mass})^2 = N - 1 \tag{84}$$

$$N = \sum_{n=1}^{\infty} \alpha_{-n}^i \alpha_n^i = \sum_{n=1}^{\infty} n a_n^{i+} a_n^i \tag{85}$$

For superstrings, the $X^i(\sigma, \tau)$ are joined by spinor variables S^a which satisfy the subsidiary conditions in light-cone gauge

$$(h^-)^{ab} S^b = (\gamma^+)^{ab} S^b = 0 \tag{86}$$

where $h^- = \frac{1}{2}(1 - \gamma^{11})$. As discussed in Ref. 34, these conditions together with a Majorana condition and the Dirac equation reduce S^a to only <u>eight</u> real components. Ab initio there are 64 complex components--2-spinors on the 2-spacetime of the world sheet and 32-spinors on the 10-spacetime manifold; these 128 real components are halved four times by the Weyl condition, the Majorana condition, the light-cone (γ^+) constraint and finally the Dirac equation. At the endpoint $\sigma = 0$ we expand

$$S^a = \sum_{n=-\infty}^{\infty} S_n^a \, e^{-in\tau} \tag{87}$$

and quantize by

$$\{S^a_m, \bar{S}^b_n\} = (\gamma^+ h^-)^{ab} \delta_{m+n,o} \tag{88}$$

$$[\alpha^i_m, S^a_n] = 0 \ . \tag{89}$$

The superstring mass shell condition is

$$\alpha'(\text{mass})^2 = N = \sum_{n=1}^{\infty} (n a^{i\dagger}_n a^i_n + \frac{n}{2} \bar{S}_{-n} \gamma^- S_n) \tag{90}$$

The ground state of the open superstring is comprised of 16 massless states: 8 form a vector

$$|i\rangle \qquad 1 \leqslant i \leqslant 8 \tag{91}$$

$$\langle i\,|j\rangle = \delta_{ij} \tag{92}$$

Since $[S_o, N] = 0$ we may act with S_o to obtain a massless Majorana-Weyl spinor

$$|a\rangle = \frac{i}{8} (\gamma_i S_o)^a \,|i\rangle \quad 1 \leqslant a \leqslant 8 \tag{93}$$

The S^a_o satisfy

$$S^a_o \bar{S}^b_o = \frac{1}{2} (\gamma^+ h^-)^{ab} + \frac{1}{4} (\gamma^{ij+} h^-)^{ab} R^{ij}_o \tag{94}$$

where γ^{ij+} is the fully antisymmetrized produce of $\gamma^i \gamma^j \gamma^+$ and

$$R^{ij}_o = \frac{1}{8} \bar{S}_o \gamma^{ij-} S_o \tag{95}$$

are generators of O(8) in the transverse space.

The excited massive superstring states may be assembled by acting on the ground state with the creation operators α^i_{-n} and S^a_{-n}. For example, at N = 1, there are 128 boson states

$$\alpha^i_{-1}|j\rangle \ \text{ and } \ S^a_{-1}|b\rangle \tag{96}$$

and 128 fermion states

$$\alpha^i_{-1}|a\rangle \ \text{ and } \ S^a_{-1}|i\rangle \tag{97}$$

The student should verify there are 1152 of each at N = 2.

It is sometimes useful in computations to define

$$\psi_n^a = \frac{1}{2^{3/4}} \gamma^o \gamma^- S_n^a = \frac{1}{2^{1/4}} S_n^a \tag{98}$$

satisfying the simple anticommutation relations

$$\{\psi_n^a, \psi_n^{b+}\} = \delta^{ab} \tag{99}$$

Since

$$\frac{1}{2} \bar{S}_{-n} \gamma^- S_n = \psi_n^+ \psi_n \tag{100}$$

it follows that e.g.

$$\mathrm{tr}\, w^{1/2n\bar{S}_{-n}\gamma^- S_n} = \mathrm{tr}\, w^{n\psi_n^+\psi_n} = (1 - w^n)^8 \tag{101}$$

$$\mathrm{tr}\, w^{N_F} = \prod_{n=1}^{\infty} (1 - w^n)^8 \tag{102}$$

where

$$N_F = \sum_{n=1}^{\infty} \frac{1}{2} n\bar{S}_{-n}\gamma^- S_n \; . \tag{103}$$

In order to calculate our one-loop superstring graphs with external massless gauge bosons we shall need only the vertex for emitting such a state. In the bosonic string, the massless vector state is

$$|i\rangle = \alpha_{-1}^i |0\rangle \tag{104}$$

If $\zeta^\mu(k)$ is the polarization vector with $k \cdot k = k \cdot \zeta = 0$ then the appropriate vertex is

$$V = g\zeta \cdot P(\tau) e^{ik \cdot X(\tau)} \tag{105}$$

$$P^\mu(\tau) = \frac{\partial}{\partial \tau} X^\mu(\tau) \Big|_{\sigma=0} \tag{106}$$

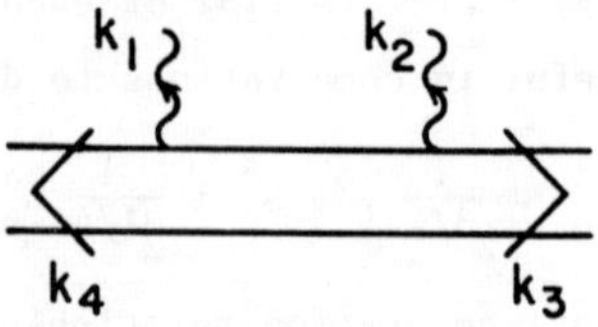

Figure 17

In light cone gauge, with $k^+ = 0$ and $\zeta^+ = 0$ this is just $g\zeta^i P^i V_o(k)$. For the superstring the relevant vertex is

$$g\zeta^i(P^i + k^j R^{ij})V_o(k) \tag{107}$$

where $R^{ij} = \frac{1}{8}\bar{S}\gamma^{ij-}S$. Here there is no ordering ambiguity because $k \cdot k = k \cdot \zeta = 0$. Although light-cone gauge itself is not restrictive, the fact that we are putting $k_+ = 0$ means that we may deal only with $M \leq 10$ external lines.

For a beginning, consider the bosonic string tree amplitudes with external spinless tachyon ground states. The vertex is

$$V_o(k) = g: e^{ik \cdot X(\tau)} : \tag{108}$$

$$= g \exp\Big(\sum_{n=1}^{\infty} \frac{k \cdot a_n^+}{\sqrt{n}}\Big)\exp\Big(-\sum_{n=1}^{\infty} \frac{k \cdot a_n}{\sqrt{n}}\Big) \tag{109}$$

The propagator is

$$\Delta = \frac{1}{2}(L - 1)^{-1} = \int_0^1 dx\, x^{L-2} \tag{110}$$

$$L = N + \frac{1}{2}p^2 \tag{111}$$

$$N = \sum_{n=1}^{\infty} \alpha_{-n}^i \alpha_n^i = \sum_{n=1}^{\infty} n a_n^{i+} a_n^i \tag{112}$$

For the four-point function of Fig. 17 the prototypical string calculation is just a "vacuum" value; we regard the k_3 and k_4 tachyons as unexcited Fock space vacua and compute

$$\langle 0| \exp\left(- k_1 \cdot \sum_{n=1}^{\infty} \frac{\alpha_n}{n}\right) x^N \exp\left(k_2 \cdot \sum_{m=1}^{\infty} \frac{\alpha_{-m}}{m}\right)|0\rangle$$

$$\exp\left(- k_1 \cdot k_2 \sum_{n=1}^{\infty} \frac{x^n}{n}\right) = (1 - x)^{k_1 \cdot k_2} \tag{113}$$

and we may write

$$A_4 = \int_0^1 dx\ x^{-\frac{1}{2} s-2} \ (1-x)^{-\frac{1}{2} t-2} \tag{114}$$

for which the crossing symmetry between s and t was the first "string miracle." The M point function is obtained by an extension of this, and is most conveniently written with M complex variables z_i on a unit circle (Fig. 18) in the form

$$A_M = \int \frac{dz_i}{dV_3} \prod_{1 \leqslant I < J \leqslant M} (z_I - z_J)^{k_I \cdot k_j} \tag{115}$$

with

$$dV_3 = \frac{dz_a dz_b dz_c}{(z_a - z_b)(z_b - z_c)(z_c - z_a)} \tag{116}$$

with the z_a, z_b, z_c arbitrarily chosen, reflecting the 3-parameter Moebius invariance of an SO(2,1) group.

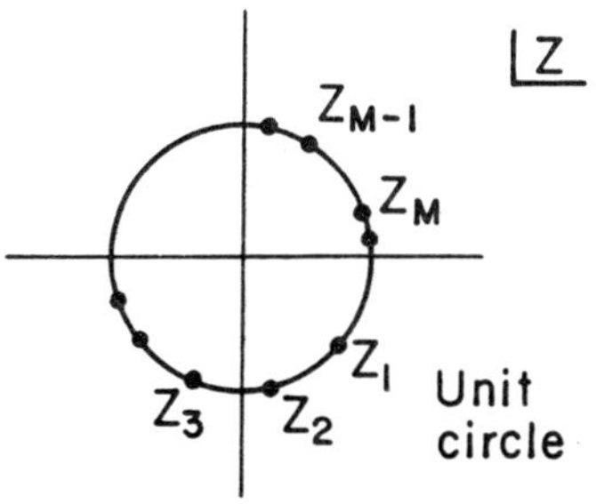

Figure 18

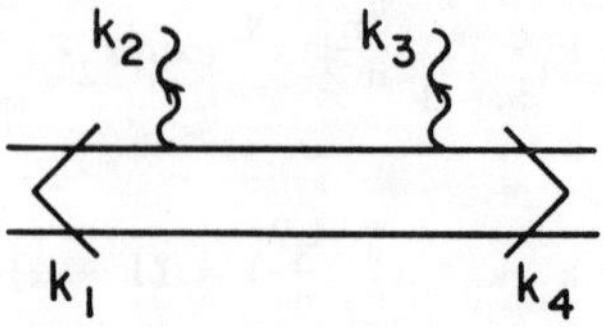

Figure 19

For the open superstring amplitude of Fig. 19 the S_o modes are very crucial. The relevant matrix element for four external massless gauge bosons is

$$\langle k_1, \zeta_1 | V(k_2, \zeta_2)\ \Delta\ V(k_3, \zeta_3) | k_4, \zeta_4 \rangle \tag{117}$$

with $V(k, \zeta)$ given by Eq. (107). The result is the crossing symmetric amplitude

$$A_4 = K_4 \frac{\Gamma(-\tfrac{1}{2}s)\Gamma(-\tfrac{1}{2}t)}{\Gamma(-\tfrac{1}{2}s - \tfrac{1}{2}t)} \tag{118}$$

with the kinematic factor given by

$$K_4 = \zeta_1^{i_1} \zeta_2^{i_2} \zeta_3^{i_3} \zeta_4^{i_4}\, k_1^{j_1} k_2^{j_2} k_3^{j_3} k_4^{j_4}\, \mathrm{tr}\left(R_o^{i_1 j_1} R_o^{i_2 j_2} R_o^{i_3 j_3} R_o^{i_4 j_4}\right) \tag{119}$$

Derivation of Eq. (118) is a good exercise.

We shall need the twisting operator in our discussion of the Moebius strip superstring loops. In the tree diagrams Fig. 20

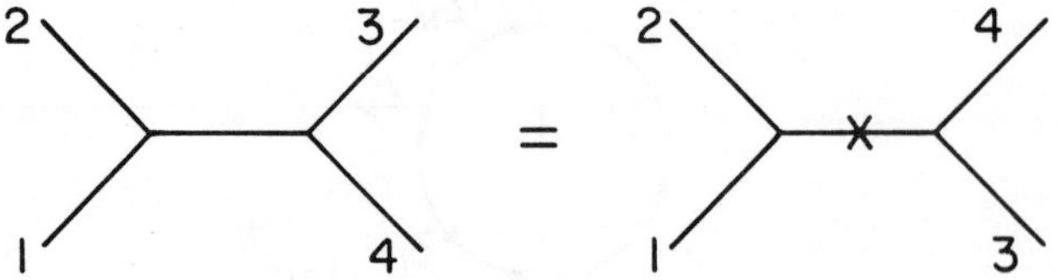

Figure 20

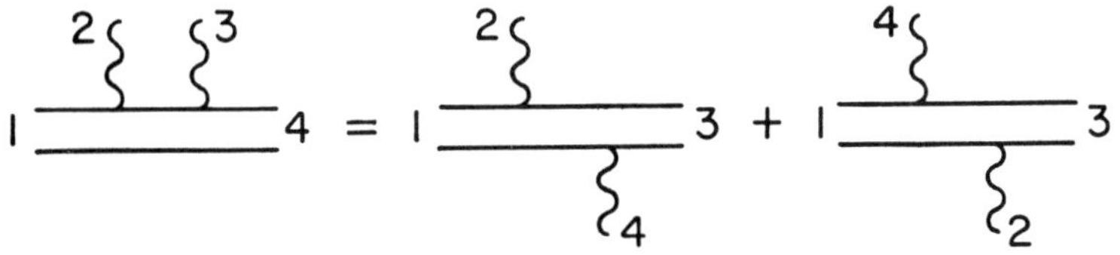

Figure 21

provides a simplistic notion. It has potential pit-falls; for example, if Ω is the twisting operator we have for the four-point tree amplitude

$$\langle 1|V(2)\Delta V(3)|4\rangle = \langle 1|V(2)\Delta\Omega V(4)|3\rangle + \langle 1|V(4)\Delta\Omega V(2)|3\rangle \tag{120}$$

as suggested in Fig. 21. The point here is that the full amplitude is given by Eq. (114) while the pieces on the right hand side of Eq. (120) represent

$$\int_0^1 = \int_0^{1/2} + \int_{1/2}^1 \tag{121}$$

In practice, we may evaluate only one twisted graph, then adjust the region of integration to account for the number of twisted graphs. This is an awkwardness of the operator formalism which can be rendered unique in path integrals.

The operators R_o^{ij} of Eqs. (94), (95) play an important role in superstring loop diagrams since one can show that in the S_o^a space

$$\mathrm{Tr}(1) = \mathrm{Tr}(R_o) = \mathrm{Tr}(R_oR_o) = \mathrm{Tr}(R_oR_oR_o) = 0. \tag{122}$$

These vanishing traces imply immediately the superstring renormalizatin theorems that the loop diagrams with two (Fig. 22) or three (Fig. 23) external gauge vectors vanish. As indicated in the Figures, this holds for both the annulus and Moebius strip diagrams.

On the other hand, since $\mathrm{tr}(R_oR_oR_oR_o)$ does not vanish, the diagrams with four external gauge bosons do not vanish (Fig. 24). Actually these diagrams diverge due to soft dilaton emission into the

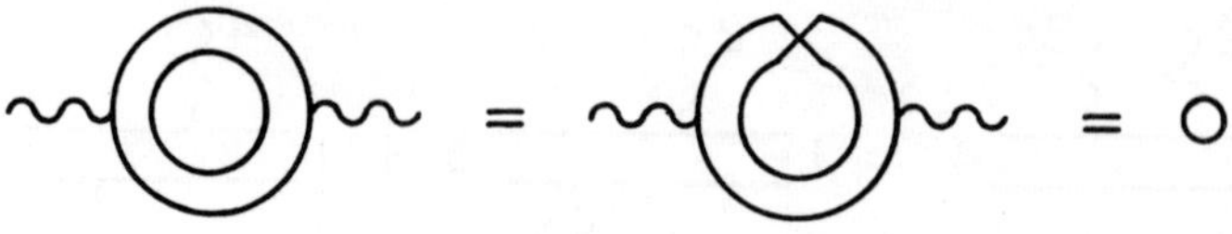

Figure 22

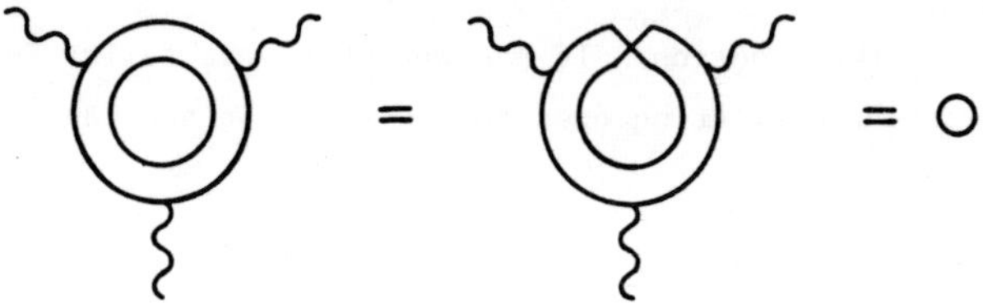

Figure 23

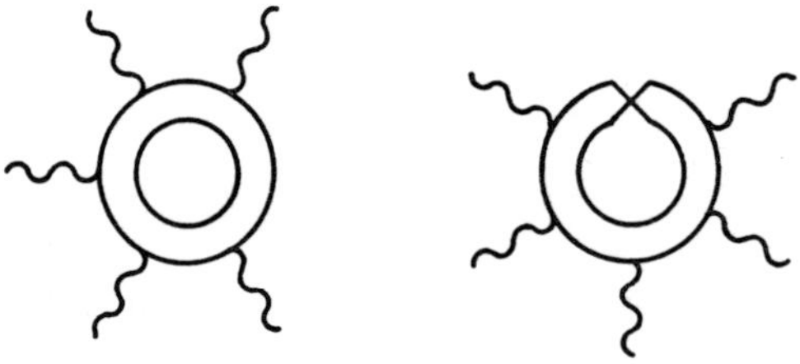

Figure 24

vacuum.

We are now ready to compute string loop diagrams. The technique is old,[35] dating from 1969, but unfamiliar to those of you who have computed one loop graphs only in quantum field theory. But once one has computed a one-loop graph for the bosonic-string, the essential stringiness of the loop has been done and one-loop superstring graphs are only slightly different. We shall spend the rest of these lectures discussing one-loop string graphs.

Consider the bosonic string in $d = 26$ with M external ground state tachyons emitted from the outer boundary of an annular string loop. We must compute the amplitude

$$\int d^{26}p \; \mathrm{tr}\left(\Delta V(k_1)\Delta V(k_2)\cdots\Delta V(k_M)\right) \tag{123}$$

and hence we need the trace

$$\mathrm{Tr}\left[x_1^N V(k_1) x_2^N V(k_2)\cdots x_M^N V(k_M)\right] \tag{124}$$

where

$$V(k) = g{:}\; e^{ik\cdot x(0)}{:} \tag{125}$$

$$N = \sum_{n=1}^{\infty} n a_n^{i+} a_n^i \tag{126}$$

To evaluate Eq. (124) one uses coherent states which are eigenstates of the annihilation operator. The basic formulas are

$$|z\rangle = e^{a^+ z}|0\rangle \tag{127}$$

$$a|z\rangle = z|z\rangle \tag{128}$$

$$e^{a^+ y}|z\rangle = |z + y\rangle \tag{129}$$

$$y^{a^+ a}|z\rangle = |yz\rangle \tag{130}$$

$$1 = \frac{1}{\pi}\int d^2\lambda \; e^{-|\lambda|^2} \; |\lambda\rangle\langle\lambda| \tag{131}$$

$$\mathrm{tr}A = \frac{1}{\pi}\int d^2\lambda \; e^{-|\lambda|^2} \; \langle\lambda|A|\lambda\rangle \tag{132}$$

where a, y, λ are complex numbers. Hence the trace of Eq. (124) can be written

$$g^M \prod_{n=1}^{\infty} \prod_{i=1}^{D-2} T_{ni} \tag{133}$$

where

$$T_{ni} = \int \frac{d^2\lambda}{\pi} e^{-|\lambda|^2} \langle\lambda| x_1^{na^+a} \exp(\frac{k_1^i a^+}{\sqrt{n}}) \exp(-\frac{k_1^i a}{\sqrt{n}}) x_2^{na^+a} \cdots$$

$$\cdots\cdot x_M^{na^+a} \exp(\frac{k_M^i a^+}{\sqrt{n}}) \exp(-\frac{k_M^i a}{\sqrt{n}}) |\lambda\rangle \qquad (134)$$

where $a = a_n^i$. Now define $q_I = k_I/\sqrt{n}$, $u_I = x_I^n$ and

$$O_I = u_I^{a^+a} e^{q_I a^+} e^{-q_I a} \qquad (135)$$

then

$$O_I|z\rangle = e^{-q_I z} |u_I(z + q_I)\rangle \qquad (136)$$

and

$$O_L O_{L+1} \cdots O_M |z\rangle = e^{-\beta_L z} |z_L\rangle \qquad (137)$$

defines Z_L and β_L. These satisfy the recursion relations

$$z_L = u_L(z_{L+1} + q_L) \qquad (138)$$

$$\beta_L = q_L z_{L+1} + \beta_{L+1} \qquad (139)$$

In particular, defining

$$z_1 = Wz + C \qquad (140)$$

$$\beta_1 = -Bz - D \qquad (141)$$

the required trace is

$$\frac{1}{\pi} \int d^2z\, e^{-|z|^2} e^{-\beta_1} \langle z|z_1\rangle =$$

$$= \frac{1}{\pi} \int d^2z \exp(-|z|^2 - \beta_1 + z^* z_1) \qquad (142)$$

$$= \frac{1}{\pi} \int d^2z \exp[(1 - W)|z|^2 + Bz + Cz^* + D] \,. \qquad (143)$$

Putting $z = x + iy$ the Gaussian integrals result in

$$\frac{1}{1 - W} \exp\left(\frac{- D(W - 1) + BC}{1 - W}\right)$$

$$= \frac{1}{1 - W} \exp\left(\frac{- \sum_{I,J=1}^{M} q_I^i q_J^i C_{IJ}}{1 - W}\right) \tag{144}$$

where

$$C_{IJ} = \begin{cases} (u_1 u_2 \cdots u_I)(u_{J+1} \cdots\cdots u_M) & (I \leqslant J) \\ u_{J+1} u_{j+2} \cdots u_I & (I > J) \end{cases} \tag{145}$$

Here we used the solutions of Eqs. (138) and (139), namely

$$W = \prod_{I=1}^{M} u_I \tag{146}$$

$$C = \sum_{I=1}^{M} q_I (u_1 u_2 \cdots u_I) \tag{147}$$

$$B = - \sum_{k=1}^{M} q_k \sum_{I=k+1}^{M+1} u_I \tag{148}$$

$$D = - \sum_{1 \leqslant K < J \leqslant M} q_K q_J (u_{K+1} \cdots u_J) \tag{149}$$

with $u_{M+1} \equiv 1$. Taking the product over modes and dimensions our trace becomes

$$g^M (f(w))^{2-D} \prod_{n=1}^{\infty} \prod_{I<J} \exp\left(\frac{-k_I \cdot k_J (C_{JI}^n + (w/C_{JI})^n - 2w^n)}{n(1 - w^n)}\right) \tag{150}$$

where $w = x_1 x_2 \cdots x_m$ and

$$f(w) = \prod_{p=1}^{\infty} (1 - w^p) \,. \tag{151}$$

We may further massage Eq. (150) by doing the logarithmic sums over n inside the exponent giving

$$g^M f(w)^{2-D} \prod_{I<J} \Big[\prod_{p=1}^{\infty} \frac{(1 - w^{p-1}C_{JI})(1 - w^p/C_{JI})}{(1 - w^p)^2} \Big]^{k_I \cdot k_J} \tag{152}$$

The momentum integral is

$$\int d^D p \prod_{i=1}^{M} x_I^{p_I^2/2} \tag{153}$$

where

$$p_I = p - \sum_{J=1}^{I-1} k_J = p + \sum_{J=I}^{M} k_J \tag{154}$$

The integrand may be written

$$\exp\Big[\sum_{I=1}^{M} \ln x_I \big(\frac{1}{2} p^2 + p \sum_{J=1}^{M} k_J\big) - \frac{1}{2} \sum_{I<J} k_I \cdot k_J \ln C_{JI} \Big] \tag{155}$$

and gaussian integration then gives, for Eq. (153),

$$\Big(- \frac{2\pi}{\ln w}\Big)^{D/2} \prod_{I<J} \Big[C_{IJ}^{-1/2} \exp\Big(\frac{\ln^2 C_{IJ}}{2 \ln w}\Big) \Big]^{k_I \cdot k_J} \tag{156}$$

Incidentally, the gaussian integral we keep using is

$$\int_{-\infty}^{\infty} dx\, e^{-a^2 x^2 \pm bx} = \frac{\sqrt{\pi}}{a} \exp\Big(\frac{b^2}{4a^2}\Big) \tag{157}$$

Collecting results, the bosonic annulus with M external spinless tachyons has amplitude

$$A_M = g^M \int \prod_{I=1}^{M} dx_I\, w^{-2} f(w)^{-24} \Big(- \frac{2\pi}{\ln w}\Big) \prod_{I<K} (\psi_{IJ})^{k_I \cdot k_J} \tag{158}$$

where

$$\psi_{IJ} = \psi(C_{JI}, w) \tag{159}$$

$$\psi(x, w) = \frac{1}{\sqrt{x}} \exp\Big(\frac{\ln^2 x}{2 \ln w}\Big) \prod_{n=1}^{\infty} \frac{(1 - w^{n-1}x)(1 - w^n/x)}{(1 - w^n)^2} \tag{160}$$

It is now useful to transform to the disc variables

$$\nu_I = \frac{\ln(x_1 x_2 \cdots x_I)}{\ln w} \tag{161}$$

$$q = \exp\left(\frac{2\pi^2}{\ln w}\right) \tag{162}$$

so

$$d\nu_I = \frac{1}{\ln w} \sum_{J=1}^{I} \frac{dx_J}{x_J} + \frac{\ln(x_1 \cdots x_J)}{2\pi^2} \frac{dq}{q} \tag{163}$$

$$\frac{dq}{q} = - \frac{2\pi^2}{(\ln w)^2} \sum_{I=1}^{M} \frac{dx_I}{x_I} \tag{164}$$

giving

$$\sum_{I=1}^{M} dx_I = \frac{1}{2\pi^2} w(-\ln w)^{M+1} \frac{dq}{q} \prod_{I=1}^{M-1} d\nu_I \theta(\nu_{I+1} - \nu_J) \tag{165}$$

In these disc variables the amplitude looks like

$$A_M = g^M \int_0^1 \prod_{I=1}^{M-1} d\nu_J \; \theta(\nu_{I+1} - \nu_I)$$

$$\int \frac{dq}{q^3} \left(- \frac{2\pi^2}{\ln q}\right)^M f(q^2)^{-24} \prod_{I<J} (\psi_{IJ})^{k_I \cdot k_J} \tag{166}$$

As $q \to 0$, the hole at the center of the annulus shrinks to a point and one picks up the closed string pole corresponding to the massless scalar dilaton coupling to the vacuum. Using

$$\psi_{IJ} \sim (\ln q)^{-1} \tag{167}$$

$$\sum_{I<J} k_I \cdot k_J = + M \tag{168}$$

(since $k_I^2 = + 2$) we see the integrand of Eq. (166) is meromorphic at $q = 0$. Actually it is a higher order pole here because of tachyons; also, the $q^2 \to 1$ singularity of $f(q^2)^{-24}$ is severe.

Rather than continuing to discuss the unpleasant singularities of Eq. (166) we go straight to the open superstring. Consider the annulus with M = 4. Firstly we find a factor (the propagator is $(\frac{1}{2}p^2 + N_B + N_F)^{-1}$)

$$\mathrm{tr}\,(w^{N_F}) = f(w)^{+8} \tag{169}$$

which precisely cancels the $f(w)^{-8}$ from the bosonic modes. Hence supersymmetry removes the worst divergence immediately. Taking the gauge group O(N) the result is[36]

$$A_4^{(\mathrm{Annulus})} = NK_4 \int_0^1 \frac{dq}{q} \int_0^1 \prod_{I=1}^{3} d\nu_I\, \theta(\nu_{I+1} - \nu_I) \prod_{I<J} (\psi_{IJ})^{k_I \cdot k_J} \tag{170}$$

where N is the group factor coming from the empty boundary and

$$K_4 = 16\pi^3 g^4\, \mathrm{Tr}(\lambda^{a_1}\lambda^{a_2}\lambda^{a_3}\lambda^{a_4})\, t^{i_1 j_1 i_2 j_2 i_3 j_3 i_4 j_4}$$
$$\zeta_1^{i_1}\zeta_2^{i_2}\zeta_3^{i_3}\zeta_4^{i_4} k_1^{j_1} k_2^{j_2} k_3^{j_3} k_4^{j_4} \tag{171}$$

$$t^{i_1 j_1 i_2 j_2 i_3 j_3 i_4 j_4} = \mathrm{tr}(R_0^{i_1 j_1} R_0^{i_2 j_2} R_0^{i_3 j_3} R_0^{i_4 j_4}) \tag{172}$$

Note especially that the $(\ell nq)^{-M-1}$ from the Jacobian cancels the $(\ell nq)^5$ from the momentum integral $\int d^{10}p$. Also, since $\sum_{I<J} k_I \cdot k_J = 0$ for massless vector external states, the integrand is meromorphic at q = 0. It is now a simple dilaton pole since tachyons are also eliminated.

The M = 4 Moebius strip (Fig. 25) has no N since there is no empty boundary. The twisting operator on the propagator is

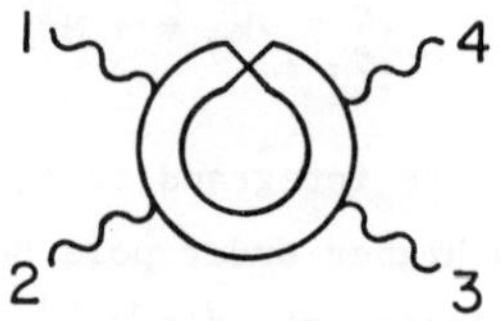

Figure 25

$$\Omega = (-1)^{N_B+N_F} \tag{173}$$

so it makes $x_1 \to -x_1$ and $w \to -w$. The integration range of ν_1 becomes (0, 2) instead of (0, 1) (c.f. Eq. (121)) and there is an overall − sign for O(N) (it would be + for Sp(2N) and 0 for SU(N)). Thus

$$A_4^{(\text{Moebius})} = -K_4 \int_0^1 \frac{dq}{q} \int_0^2 \prod_{I=1}^{3} d\nu_I \, \theta(\nu_{I+1} - \nu_I) \prod_{I<J} (\psi_{IJ,N})^{k_I \cdot k_J} \tag{174}$$

In fact, the annulus and the Moebius strip are so similar that we may write[36]

$$A_4^{(\text{Annulus})} + A_4^{(\text{Moebius})} = \int_0^1 \frac{dq}{q} \left[NF(q^2) - 8F(-\sqrt{q}) \right] \tag{175}$$

and changing variables to $\lambda = q^2$ and $\lambda = \sqrt{q}$ respectively in the two terms, this can be written for O(32) only as

$$16 \int_{-1}^{+1} \frac{d\lambda}{\lambda} F(\lambda) \qquad \text{(principal part)} \tag{176}$$

We should say that no consistent regularization has been worked through in detail for superstrings. Dimensional regularization is presumably not useful since d = 10 seems fixed. Thus, the above procedure is ambiguous: for example, if we put $\lambda = q$ instead of q^2 for the annulus the result would diverge. Nevertheless, the principal part prescription is the way to isolate O(32) and will be used in what follows.

Consider now the M = 5 pentagon (Fig. 26). The divergence cancellation for $M \geq 6$ will become clear from M = 5. Really the M = 4 case is too close to the nonrenormalization theorems for the general case to be clear.

In the vector emission vertex, Eq. (107), both P^i and R^{ij} have both zero modes (n = 0) and nonzero modes ($n \neq 0$). P is linear in α_n while R is bilinear in S_n. We need at least eight S_o factors for

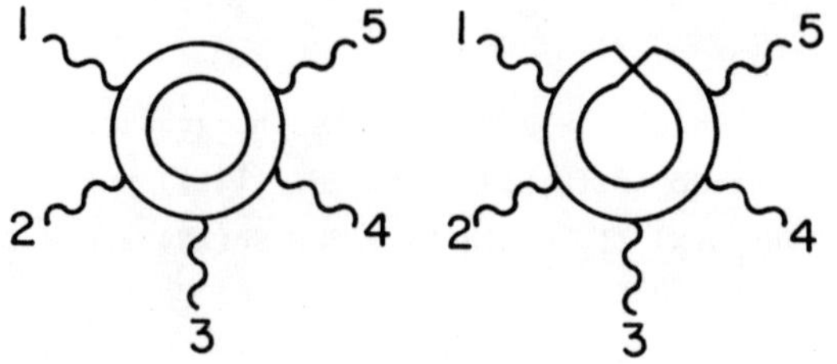

Figure 26

the trace to be nonvanishing.

Already for M = 4 we needed

$$t^{i_1j_1i_1j_2i_3j_3i_4j_4} = \mathrm{tr}\,(R_o^{i_1j_1} R_o^{i_2j_2} R_o^{i_3j_3} R_o^{i_4j_4}) \tag{177}$$

$$\begin{aligned} &= a_1 \varepsilon^{i_1j_1i_2j_2i_3j_3i_4j_4} \\ &+ a_2[(\delta^{i_1i_2}\delta^{j_1j_2} - \delta^{i_1j_2}\delta^{i_2j_1}) \\ &\quad (\delta^{i_3i_4}\delta^{j_3j_4} - \delta^{i_3j_4}\delta^{i_4j_3}) \\ &+ (13)(24) + (14)(23)] \\ &+ a_3[\delta^{j_1i_2}\delta^{j_2i_3}\delta^{j_3i_4}\delta^{j_4i_1} + \cdots]_{48 \text{ terms}} \end{aligned} \tag{178}$$

The coefficients a_i of these three irreducible tensors can be found by special choices of indices e.g.

$$a_1 = \mathrm{tr}(R_o^{12} R_o^{24} R_o^{56} R_o^{78}) = -\frac{1}{32}\,\mathrm{tr}(\gamma^- \gamma^{78} \gamma^{56} \gamma^{34} \gamma^{12} h_- \gamma^+) \tag{179}$$

$$= -\frac{1}{2} \tag{180}$$

$$a_2 = \mathrm{tr}(R_o^{12} R_o^{12} R_o^{34} R_o^{34}) = -\frac{1}{32}\,\mathrm{tr}(\gamma^- \gamma^{34} \gamma^{34} \gamma^{12} \gamma^{12} h_- \gamma^+) \tag{181}$$

$$= -\frac{1}{2} \tag{182}$$

$$a_3 = \mathrm{tr}(R_o^{12} R_o^{23} R_o^{34} R_o^{41}) = (+1)_{\mathrm{bosons}} + (-\tfrac{1}{2})_{\mathrm{fermions}} = +\tfrac{1}{2} \quad (183)$$

For M = 5 we consider the parity-violating and parity-conserving pieces of

$$\mathrm{tr}\left(R_o^{i_1 j_1} R_o^{i_2 j_2} R_o^{i_3 j_3} R_o^{i_4 j_4} R_o^{i_5 j_5}\right) \quad (184)$$

The parity-violating part can be written

$$c_1[\delta^{i_1 j_2} \varepsilon^{j_1 i_2 j_3 i_4 j_4 i_5 j_5} + \cdots]_{20 \text{ terms}}$$

$$+ c_2[\delta^{i_1 j_3} \varepsilon^{j_1 i_3 i_2 j_2 i_4 j_4 i_5 j_5} +]_{20 \text{ terms}} \quad (185)$$

where the terms group into those with nearest neighbor and next-nearest neighbor vertices on ϕ respectively. By considering particular indices one finds $c_1 = 3/20$ and $c_2 = 1/20$. Actually, there are Gram determinental constraints which reduces the 40 tensors in (185) to 31. These constraints are the analog of

$$\delta^{\mu\alpha}\varepsilon^{\beta\gamma\delta\varepsilon} + \delta^{\mu\beta}\varepsilon^{\gamma\delta\varepsilon\alpha} + \delta^{\mu\gamma}\varepsilon^{\delta\varepsilon\alpha\beta} + \delta^{\mu\delta}\varepsilon^{\varepsilon\alpha\beta\gamma} + \delta^{\mu\varepsilon}\varepsilon^{\alpha\beta\gamma\delta} = 0 \quad (186)$$

in d = 4. But we may keep an overcomplete basis if we assign special indices and do not use tensor identities (i.e. do not compare the non-unique coefficients of specific tensors).

The parity-conserving pieces has 524 terms of the types

$$[(\delta\delta)(\delta\delta\delta)]_{160 \text{ terms}} + [(\delta\delta\delta\delta\delta)]_{384 \text{ terms}} \quad (187)$$

We may separate these further into irreducible parts by giving typical groupings:

$$a_1[(12)(345)]_{80 \text{ terms}} + a_2[(13)(245)]_{80 \text{ terms}}$$

$$+ \; b_1[(12345)]_{32 \text{ terms}} \; + \; b_2[(13524)]_{32 \text{ terms}}$$

$$+ \; b_3[(13254)]_{160 \text{ terms}} + b_4[(12354)]_{160 \text{ terms}} \quad (188)$$

Looking at particular values of the indices gives $a_1 = a_2 =$

$-b_2 = b_3 = -b_4 = -\frac{1}{4}$ and $b_1 = \frac{3}{4}$. Note that the only bosonic contribution to this supertrace is in b_1.

Thus the contribution to A_P (P = planar) is [37]

$$A_P^{(R_o^5)} = \zeta_1^{i_1}\zeta_2^{i_2}\zeta_3^{i_3}\zeta_4^{i_4}\zeta_5^{i_5} k_1^{j_1}k_2^{j_2}k_3^{j_3}k_4^{j_4}k_5^{j_5}$$

$$\mathrm{tr}(R_o^{i_1j_1} R_o^{i_2j_2} R_o^{i_3j_3} R_o^{i_4j_4} R_o^{i_5j_5})$$

$$\int \prod_{I=1}^{5} dx_I \frac{1}{w}\left(-\frac{2\pi}{\ell n w}\right) \prod_{1\leq I<J\leq 5} (\psi_{IJ})^{k_I\cdot k_J} \tag{189}$$

To go from A_P to A_N (N = non-orientable Moebius strip) we replace w by $-w$ and ψ_{IJ} by $\psi_{N,IJ}$. This is the only piece of A_5 with ten S_o factors, the rest have eight S_o factors. Consider first the terms with one S_n and one S_{-n} ($n \neq 0$). There are eight nonvanishing terms in the factor e.g.

$$\frac{1}{64}\mathrm{Tr}\{[S_b^a + \sum_{k\geq 1}(\bar{S}_k^a + \bar{S}_{-k}^a)](\gamma^{i_1j_1-})^{ab}\,[S_o^b + \sum_{\ell\geq 1}(S_\ell^b + S_{-\ell}^b)]$$

$$[\bar{S}_o^c + \sum_{m\geq 1}\bar{S}_m^c x_2^{-m} + \bar{S}_{-m}^c x_2^m)](\gamma^{i_2j_2-})^{cd}\,[S_o^d + \sum_{p\geq 1}(S_p^d x_2^{-p} + S_{-p}^d x_2^p)]$$

$$W^{N_F} R_o^{i_3j_3} R_o^{i_4j_4} R_o^{i_5j_5}\}\ . \tag{190}$$

Note that $\mathrm{tr}(\bar{S}_k\gamma^{i_1j_1-} S_{-k}) = 0$ because there is no second rank antisymmetric tensor in O(8). The eight nonvanishing terms in (190) are all equal. We must then add the ten inequivalent pairs of vertices and obtain[37]

$$A_p^{(S_o^8 S_n S_{-n})} = \zeta_1^{i_1}\zeta_2^{i_2}\zeta_3^{i_3}\zeta_4^{i_4}\zeta_5^{i_5}\, k_1^{j_1}k_2^{j_2}\, k_3^{j_3}k_4^{j_4}k_5^{j_5}\ \frac{1}{64}$$

$$\int \prod_{I=1}^{5} dx_I w^{-1}\left(-\frac{2\pi}{\ell n w}\right)^5 \prod_{1\leqslant I<J\leqslant 5} (\psi_{IJ})^{k_I\cdot k_J}$$

$$\sum_{\ell\geqslant 1} \frac{1}{1-w^\ell}\{[(\frac{w}{x_2})^\ell - x_2^\ell]$$

$$t^{i_1j_1i_2j_2ij}\,\mathrm{tr}(R_o^{ij} R_o^{i_3j_3} R_o^{i_4j_4} R_o^{i_5j_5})$$

$$+ [(\frac{w}{x_2x_3})^\ell - (x_2x_3)^\ell]t^{i_1j_1i_3j_3ij}\,\mathrm{tr}(R_o^{ij} R_o^{i_2j_2} R_o^{i_4j_4} R_o^{i_5j_5})$$

$$+ \text{cyclic permutations}\}_{2\text{x}5 \text{ terms}} \tag{191}$$

where

$$t^{i_1j_1i_2j_2ij} = \mathrm{tr}(\gamma^{i_1j_1}\gamma^{i_2j_2}\gamma^{ij}) \tag{192}$$

$$= 32[\delta_{ij_1}\delta_{i_2j}\delta_{i_1j_2} - \delta_{ij_1}\delta_{i_2j}\delta_{j_ij_2}$$

$$- \delta_{ij_1}\delta_{i_1i_2}\delta_{jj_2} + \delta_{ii_1}\delta_{i_2j_1}\delta_{jj_2} - \delta_{ii_2}\delta_{i_1j_2}\delta_{jj_1}$$

$$+ \delta_{ii_2}\delta_{i_1j}\delta_{j_1j_2} + \delta_{ij_2}\delta_{i_1i_2}\delta_{jj_1} - \delta_{ij_2}\delta_{i_2j_1}\delta_{i_1j}] \tag{193}$$

We obtain $A_N^{(S_o^8 S_n S_{-n})}$ from $A_P^{(S_o^8 S_n S_{-n})}$ by replacing ψ_{IJ} by $\psi_{N,IJ}$, w by − w and x_1 by − x_1.

The remaining terms are of the form $(R_o^4 P)$ with

$$P^i = p^i + \sum_{n\geqslant 1} \sqrt{n}\,(a_n^i + a_n^{i+}) \tag{194}$$

The zero mode piece (p^i) modifies the gaussian in the momentum integral already computed in Eqs. (153) and (156) above. The nonzero modes change the bosonic trace in Eq. (144) above. Modifying the operator O_I of Eq. (135) to

$$\hat{O}_I^i = P^{i_I}(u_I)O_I = \sqrt{n}\,(a u_I^{-1} + a^+ u_I)O_I \tag{195}$$

giving a modified trace

$$\hat{T} = \frac{1}{\pi}\int d^2z\,\sqrt{n}\,(z_1 u_1^{-1} + z^* u_1)e^{-|z|^2}\,e^{-\beta_1}\,e^{z^* z_1} \tag{196}$$

$$= \sqrt{n}\,\left(W u_i^{-1}\frac{\partial}{\partial B} + C u_1^{-1} + u_1\frac{\partial}{\partial C}\right)T \tag{197}$$

where T is the old trace (c.f Eq. (144))

$$T = \frac{1}{1-W}\exp\left(-\frac{D(W-1)-BC}{1-W}\right) \tag{198}$$

It follows that

$$\hat{T} = \frac{1}{1-W}\left(C u_1^{-1} + B u_1\right)T \tag{199}$$

Collecting results we find that[37]

$$A_p^{(R_o^4P)} = \int \prod_{I=1}^{5} dx_I\, w^{-1}\left(-\frac{2\pi}{\ell n w}\right)^5 \prod_{1\leq I<J\leq 5}(\psi_{IJ})^{k_I\cdot k_J}$$

$$[(\zeta_1^{i_1}\sum_{L=1}^{5} k_L^{i_1}\frac{\ell n(x_1\cdots x_L)}{-\ell n w}K_4(2345)$$

$$+\ \zeta_2^{i_2}\sum_{L=2}^{1} k_L^{i_2}\frac{\ell n(x_2\cdots x_L)}{-\ell n w}K_4(3451)$$

$$+\ \text{cyclic permutations})_{5\ \text{terms}}$$

$$+\ \{K_4(2345)\zeta_1^{i_1}\sum_{\ell\geq 1}\frac{1}{1-w^\ell}\,(k_2^{i_1}[x_2^\ell - (\frac{w}{x_2})^\ell]$$

$$+\ k_3^{i_1}[(x_2x_3)^\ell - (\frac{w}{x_2x_3})^\ell]$$

$$+\ k_4^{i_1}[(\frac{w}{x_5x_1})^\ell - (x_5x_1)^\ell] + k_5^{i_1}[(\frac{w}{x_1}) - x_1^\ell])$$

$$+ \text{ cyclic permutations}\}_{5 \text{ terms}}] \tag{200}$$

where e.g.

$$K_4(2345) = \zeta_2^{i_2}\zeta_3^{i_3}\zeta_4^{i_4}\zeta_5^{i_5}\; k_2^{j_2}k_3^{j_3}k_4^{j_4}k_5^{j_5}\; \mathrm{tr}(R_o^{i_2j_2}R_o^{i_3j_3}R_o^{i_4j_4}R_o^{i_5j_5}) \tag{201}$$

As usual, for the nonorientable Moebius strip we replace ψ_{IJ}, w, x_1 by $\psi_{N,IJ}$, $-$ w, $-$ x_1 respectively.

This completes the M = 5 results. Higher point functions are computed similarly. We now make an asymptotic expansion for $q \to 0$

$$A_p = \int_0^1 \frac{dq}{q}\left[a_o + \frac{a_1}{\ell nq} + \frac{a_2}{(\ell nq)^2} + \cdots\right] \tag{202}$$

The infra-red divergence $(q \to 0)$ is from the a_o and a_1 terms only. From the physical picture of the dilaton pole, presumably all $a_k = 0$ for $k \geqslant 1$. We should note the (ℓnq) factors are $(\ell nq)^5$ and $(\ell nq)^{-M-1}$ from the momentum integrand and Jacobian respectively. for M = 4 it is trivially meromorphic and the annulus and Moebius strip diagrams (Fig. 27) cancelled. Recall that we put $q = q'^4$ to relate the q variables and used the change

$$\int_0^2 \prod_{I=1}^{M-1} d\nu_I' = 2^{M-1}\int_0^1 \prod_{I=1}^{M-1} d\nu_I \tag{203}$$

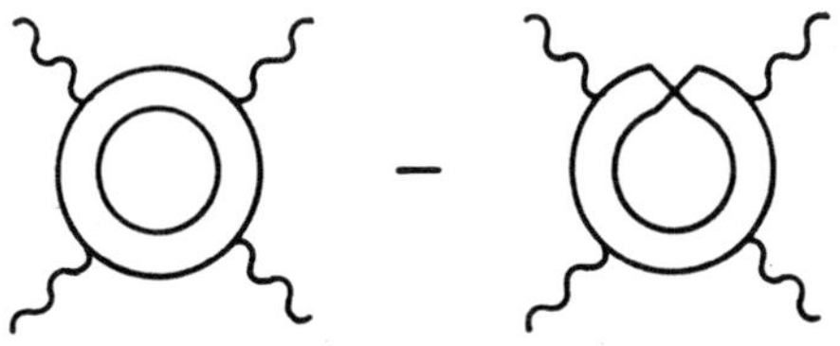

Figure 27

together with

$$\frac{dq}{q} = 4\,\frac{dq'}{q'} \tag{204}$$

to obtain (for M = 4) 4 times 8 = 32. But how can the 2^{M-1} be independent of M, the number of external lines? The answer lies in the mode sum

$$F_p(\nu, w) = \sum_{n=1}^{\infty} \frac{c^n - (w/c)^n}{1 - w^n} \quad \text{and } c = w^\nu \tag{205}$$

$$= \sum_{n=1}^{\infty} \sum_{m=o}^{\infty} \left(e^{n\nu\ell nw} - e^{n(1-\nu)\ell nw}\right)e^{mn\ell n\nu} \tag{206}$$

$$= \sum_{m=0}^{\infty} \left(\frac{e^{(\nu+m)\ell nw}}{1 - e^{(\nu+m)\ell nw}} - \frac{e^{(1-\nu+m)\ell nw}}{1 - e^{(1-\nu+m)\ell nw}}\right) \tag{207}$$

$$= \frac{1}{(-\ell nw)}\left(A_o + A_1(-\ell nw) + A_2(-\ell nw)^2 + \cdots\right) \tag{208}$$

Now we find easily that

$$A_o = \frac{1}{\nu} + 2\nu \sum_{m=1}^{\infty} \frac{1}{\nu^2 - m^2} = \pi \cot \pi\nu \tag{209}$$

The nonorientable Moebius strip contains

$$F_N(\nu, w) = \sum_{\ell=1}^{\infty} \frac{1}{1 - (-w)^\ell}\left[c^\ell - (-w/c)^\ell\right] \tag{210}$$

$$\sim -\left(\frac{1}{2}\right)\frac{\ell nq}{2\pi} \cot\left(\frac{\pi\nu}{2}\right)\left[1 + 0\left(\frac{1}{\ell nq}\right)\right] \tag{211}$$

as $q \to 0$. Note that

$$F_p(\nu, w) \sim -\frac{\ell nq}{2\pi} \cot \pi\nu \left[1 + 0\left(\frac{1}{\ell nq}\right)\right] \tag{212}$$

the extra (1/2) in Eq. (211) is the essential point. When we generate the extra $(\ell nq)^{M-4}$ from the mode sums, necessary to identify the coefficient a_o in Eq. (202), we also generate a relative factor 2^{4-M}

in the Moebius strip relative to the annulus. Thus the terms may be combined as

$$\frac{1}{2}\int_0^1 \frac{d\lambda}{\lambda}\left[NF(\lambda) - \frac{4 \cdot 2^{M-1}}{2^{M-4}} F(-\lambda)\right] = 16 \int_{-1}^{+1} \frac{d\lambda}{\lambda} F(\lambda) \qquad (213)$$

and the principal part prescription generalizes to all M for the gauge group O(32).

References

1. P. H. Frampton, Dual Resonance Models, Benjamin (1974).

2. G. Veneziano, Nuovo Cim. 57A, 190 (1968).

3. K. Igi and S. Matsuda, Phys. Rev. Lett. 18, 625 (1967); Phys. Rev. 163, 1622 (1967).

4. S. Mandelstam, Phys. Rev. 166, 1539 (1968).

5. H. Harari, Phys. Rev. Lett. 22, 562 (1962).

6. J. L. Rosner, Phys. Rev. Lett. 22, 689 (1969).

7. P. G. O. Freund, Phys. Rev. Lett. 20, 235 (1968); H. Harari, Phys. Rev. Lett. 20. 1395 (1968).

8. H. M. Chan and T. S. Tsun, Phys. Lett. 28B, 485 (1969); K. Bardakci and H. Ruegg, Phys. Rev. 181, 1884 (1969); Z. Koba and H. B. Nielsen, Nucl. Phys. B10, 633 (1969); ibid. B12, 517 (1969).

9. S. Fubini and G. Veneziano, Nuovo Cim. 64A, 811 (1969); K. Bardakci and S. Mandelstam, Phys. Ref. 184, 1640 (1969).

10. Y. Nambu, in Symmetries and Quark Models, editor R. Chand, p. 269 (1969). S. Fubini, D. Gordon and G. Veneziano, Phys. Lett. 29B, 679 (1969).

11. Y. Nambu, lecture notes prepared for the Summer Institute of the Niels Bohr Institute, Copenhagen (1970); T. Goto, Prog. Theor. Phys. 46, 1560 (1971); P. Goddard, J. Goldstone, C. Rebbi and C. B. Thorn, Nucl. Phys. B56, 109 (1973).

12. D. J. Gross, A. Neveu, J. Scherk and J. H. Schwarz, Phys. Rev. D2, 697 (1970).

13. R. C. Brower, Phys. Rev. D6, 1655 (1972).
E. Del Guidice, P. Di Vecchia and S. Fubini, Ann. Phys. 70, 378 (1972);
P. H. Frampton and H. B. Nielsen, Nucl. Phys. B45, 318 (1972);
P. Goddard and C. B. Thorn, Phys. Lett. 40B, 235 (1972).

14. M. A. Virasoro, Phys. Rev. 177, 2309 (1969).

15. P. Ramond, Phys. Rev. D3, 2415 (1971);
A. Neveu and J. H. Schwarz, Nucl. Phys. B31, 86 (1971).

16. T. Yoneya, Prog. Theor. Phys. 51, 1907 (1974);
J. Scherk and J. H. Schwarz, Nucl. Phys. B81, 118 (1974); Phys. Lett. 57B, 463 (1975).

17. D. J. Gross, J. A. Harvey, E. Martinec and R. Rohm, Phys. Rev. Lett. 54, 502 (1985); Nucl. Phys. B256, 253 (1985) and to be published.

18. M. B. Green and J. H. Schwarz, Phys. Lett. 109B, 444 (1982).

19. P. H. Frampton, Phys. Lett. 122B, 351 (1983);
P. H. Frampton and T. W. Kephart, Phys. Rev. Lett. 50, 1343, 1347 (1983); Phys. Rev. D28, 1010 (1983).

20. P. H. Frampton and T. W. Kephart, Phys. Lett. 131B, 80 (1983).

21. P. H. Frampton, Phys. Lett. 140B, 313 (1984).

22. L. Caneschi, G. Farrar and A. Schwimmer, Phys. Lett. 138B, 386 (1984).

23. P. H. Frampton and K. Yamamoto, Phys. Rev. Lett. 52, 2016 (1984).

24. P. H. Frampton and K. Yamamoto, Nucl. Phys. 254B, 349 (1985).

25. S. Randjbar-Daemi, A. Salam and J. Strathdee, Phys. Lett 124B, 345 (1983).
P. H. Frampton, P. J. Moxhay and K. Yamamoto, Phys. Lett. 144B, 354 (1984).

26. S. Randjbar-Daemi, A. Salam and J. Strathdee, Nucl. Phys. B214, 491 (1983).

27. S. Randjbar-Daemi, A. Salam and J. Strathdee, Phys. Lett. 132B, 386 (1984); Nucl. Phys. B242, 447 (1984).

28. H. Georgi, Nucl. Phys. B156, 126 (1979).

29. P. H. Frampton, Phys. Lett. 89B, 352 (1980).

30. M. B. Green and J. H. Schwarz, Phys. Lett. 149B, 117 (1984).

31. G. F. Chapline and N. S. Manton, Phys. Lett. 120B, 105 (1983). E. Bergshoeff, M. De Roo, B. De Wit and P. Van Nieuwenhuizen, Nucl. Phys. B195, 97 (1982).

32. L. Alvarez-Gaumé and E. Witten, Nucl. Phys. B234, 269 (1983).

33. P. H. Frampton and K. Yamamoto, Phys. Lett. 156B, 345 (1985).

34. J. H. Schwarz, Phys. Reports 89, 223 (1982). This review is more current than Reference 1 and is a good source for more background material.

35. D. Amati, C. Bouchiat and J. L. Gervais, Nuovo Cim. Lett. 2, 399 (1969).

36. M. B. Green and J. H. Schwarz, Phys. Lett 151B, 21 (1985).

37. P. H. Frampton, P. Moxhay and Y. J. Ng, UNC-Chapel Hill Reports IFP-255-UNC, to be published in Phys. Rev. Lett. and IFP-256-UNC.

PATH INTEGRAL AND ANOMALIES

Kazuo Fujikawa

Research Institute for Theoretical Physics
Hiroshima University, Takehara, Hiroshima 725

ABSTRACT

The path integral approach to the anomalies in quantum field theory is briefly summarized. It is emphasized that the study of anomalies is the study of quantum symmetries. After giving a quick derivation of Feynman path integral formula, we discuss the chiral, conformal and gravitational anomalies. The path integral of relativistic strings and the Virasoro condition are also briefly discussed.

1. Introduction

The anomalies in quantum field theory have been discussed by many authors from various view points in the past.[1)] In the early days, it was emphasized that the anomalies arise from the breaking of classical symmetries by the regularization procedure. Although this view point is still correct, it is sometimes more convenient to regard anomalies as the breakdown of certain classical symmetries in the quantization procedure. In other words, the classical symmetry properties and the quantum symmetry properties are different in general, and the latter quantum symmetry properties are more restrictive. This interpretation of anomalies is quite natural in the path integral approach, where the anomalies are identified with the Jacobian factors under the classical symmetry transformations.[2)] In the following we discuss the path integral approach to various anomalous identities.

2. Feynman Path Integral Formula

We first briefly summarize the basic properties of the Feynman path integral formula. We start with the Lagrangian

$$\mathcal{L}_J = \frac{1}{2}\partial_\mu\phi\partial_\mu\phi - \frac{1}{2}m^2\phi^2 - V(\phi) + \phi(x)J(x) \quad . \tag{2.1}$$

The last term in (2.1) stands for the source term, which creats or destroys the field $\phi(x)$ at space-time point x^μ. Here we assume that $J(x)$ is localized in the sence

$$\lim_{t\to\pm\infty} J(x) = 0 \quad . \tag{2.2}$$

The equation of motion from (2.1) is given by

$$-\Box\phi - m^2\phi - V'(\phi) + J(x) = 0 \quad . \tag{2.3}$$

The action principle of Schwinger,[3] which is equivalent to the Schrödinger equation,[4] is stated as

$$\delta\langle\infty|-\infty\rangle_J = i\int d^4x\langle\infty|\delta\mathcal{L}(x)|-\infty\rangle_J \tag{2.4}$$

where the left-hand side stands for the variation in the vacuum-to-vacuum applitude caused by the infinitesimal variation of the Lagrangian $\delta\mathcal{L}(x)$. In particular, under the change of the source $J(x)$, we have

$$\frac{\delta}{i\,\delta J(x)}\langle\infty|-\infty\rangle_J = \langle\infty|\phi(x)|-\infty\rangle_J \quad . \tag{2.5}$$

From (2.3) and (2.5), we obtain the constraint

$$\left[-\Box\frac{\delta}{i\delta J(x)} - m^2\frac{\delta}{i\delta J(x)} - V'\left(\frac{\delta}{i\delta J(x)}\right) + J(x)\right]\langle\infty|-\infty\rangle_J = 0 \quad . \tag{2.6}$$

To evaluate the transition amplitude is then reduced to integrating the equation (2.6). Performing the functional Fourier analysis,[5] one obtains the solution

$$\langle\infty|-\infty\rangle_J = \frac{1}{N}\int\prod_x\mathcal{D}\phi(x)\ \exp[iS_J] \tag{2.7}$$

with $S_J = \int\mathcal{L}_J(x)dx$. To be precise, one obtains the condition from (2.6)

$$\int\prod_x\mathcal{D}\phi(x)\frac{\delta}{\delta\phi(y)}iS_J e^{iS_J} = \int\prod_x\mathcal{D}\phi(x)\frac{\delta}{\delta\phi(y)}\ e^{iS_J} = 0 \quad . \tag{2.8}$$

The basic requirement of the path integral measure is then the "translation" invariance in the functional space in the sense

$$\prod_x \mathcal{D}\phi(x) = \prod_x \mathcal{D}[\phi(x) + \varepsilon(x)] \tag{2.9}$$

for an "arbitrary" infinitesimal function $\varepsilon(x)$. If (2.9) is valid, one obtains

$$\int \prod_x \mathcal{D}\phi(x) e^{iS_J(\phi)} \mp \int \prod_x \mathcal{D}[\phi(x) + \varepsilon(x)] e^{iS_J(\phi+\varepsilon)}$$
$$= \int \prod_x \mathcal{D}\phi(x) e^{iS_J(\phi+\varepsilon)} \tag{2.10}$$

By expanding (2.10) in $\varepsilon(x)$, one can confirm that the right-hand side of (2.8) vanishes, as is required by (2.6).

There are two simple ways to satisfy (2.9):

(i) $\phi(x)$ = ordinary number and $\int \mathcal{D}\phi(x)$ the conventional integration (Bosonic case),

(ii) $\phi(x)$ = Grassmann number (i.e., the totally anti-commuting number) and $\mathcal{D}\phi(x) = \frac{\delta}{\delta\phi(x)}$, the left-derivative (Fermionic case).[6)]

One has thus recovered the conventional path integral formula, although the Lagrangian (2.1) should be replaced by the Dirac Lagrangian for fermions. The discussion of the anomaly in the path integral then becomes a more careful definition of the path integral measure.

3. Chiral U(1) Anomaly

The chiral anomaly associated with the global U(1) symmetry plays an important role in the general discussion of the chiral anomaly. We thus study the chiral U(1) anomaly on the basis of the QCD Lagrangian

$$\mathcal{L} = \bar{\psi}(i\not{D} - m)\psi - \frac{1}{2g^2}\mathrm{Tr}F^{\mu\nu}F_{\mu\nu} \tag{3.1}$$

$$\not{D} = \gamma^\mu(\partial_\mu - iA_\mu)$$

$$A_\mu \equiv A_\mu^a(x)T^a$$

$$[T^a, T^b] = if^{abc}T^c, \quad \mathrm{Tr}T^aT^b = \frac{1}{2}\delta^{ab}.$$

We note that the γ-matrices in our convention become anti-hermitean after Wick-rotation to the Enclidean theory

$$\gamma^{\mu\dagger} = - \gamma^{\mu} . \tag{3.2}$$

The path integral is then defined by

$$\int \prod_x \mathcal{D}\overline{\psi}(x)\mathcal{D}\psi(x)[\mathcal{D}A_\mu(x)]\exp\left[\int \mathcal{L}(x)dx\right] \tag{3.3}$$

where $[\mathcal{D}A_\mu(x)]$ includes the gauge fixing and compensating terms. For the moment, we forget $[\mathcal{D}A_\mu]$ and define

$$d\mu(\psi) = \prod_x \mathcal{D}\overline{\psi}(x)\mathcal{D}\psi(x) . \tag{3.4}$$

Under the infinitesimal chiral U(1) transformation specified by a <u>localized</u> parameter $\alpha(x)$

$$\begin{aligned} \psi(x) &\to \psi'(x) = e^{i\alpha(x)\gamma_5}\psi(x) \\ \overline{\psi}(x) &\to \overline{\psi}(x)' = \overline{\psi}(x)e^{i\alpha(x)\gamma_5} \end{aligned} \tag{3.5}$$

the Lagrangian changes as

$$\mathcal{L} \to \mathcal{L} - \partial_\mu\alpha(x)\overline{\psi}\gamma^\mu\gamma_5\psi(x) - \alpha(x)2mi\overline{\psi}\gamma_5\psi(x) \tag{3.6}$$

where the coefficient of $\partial_\mu\alpha(x)$ defines the symmetry current. If the measure (3.4) should remain invariant under (3.5), we would have

$$\int d\mu(\psi)e^{S(\psi)} \equiv \int d\mu(\psi')e^{S(\psi')} = \int d\mu(\psi)e^{S(\psi')} \tag{3.7}$$

with $S(\psi) = \int \mathcal{L}(x)dx$. By expanding (3.7) in powers in $\alpha(x)$, one obtains the naive identity

$$\partial_\mu\langle\overline{\psi}\gamma^\mu\gamma_5\psi(x)\rangle - 2im\langle\overline{\psi}\gamma_5\psi(x)\rangle = 0 \tag{3.8}$$

where $\langle O(x)\rangle$ stands for the average of $O(x)$ in the path integral formula.

A careful estimate of the Jacobian factor for the transformation (3.5) in fact gives

$$d\mu(\psi') = d\mu(\psi)\exp\left[-i\int \alpha(x)A(x)dx\right] \tag{3.9}$$

and the relation (3.8) is modified to

$$\partial_\mu\langle\overline{\psi}\gamma^\mu\gamma_5\psi(x)\rangle - 2im\langle\overline{\psi}\gamma_5\psi(x)\rangle = iA(x) \tag{3.10}$$

the well-known anomalous chiral U(1) identity.[1)]

To evaluate the Jacobian factor in (3.9), it becomes more transparent if one expands $\psi(x)$ and $\overline{\psi}(x)$ as

$$\begin{aligned}\psi(x) &= \sum_n a_n \varphi_n(x) = \sum_n a_n \langle x|\varphi_n\rangle \\ \overline{\psi}(x) &= \sum_n \overline{b}_n \varphi_n(x)^\dagger = \sum_n \overline{b}_n \langle\varphi_n|x\rangle\end{aligned} \tag{3.11}$$

with

$$\not{D}\varphi_n(x) = \lambda_n \varphi_n(x) \ , \qquad \int \varphi_m(x)^\dagger \varphi_n(x)dx = \delta_{m,n} \ . \tag{3.12}$$

The coefficients a_n and $\overline{b}_n$ in (3.11) are the elements of the Grassmann algebra. The expansion such as (3.11) is well-known in the path integral formalism.[7)] In effect one can make the short-distance behavior more manageable by means of the expansion (3.11). The path integral measure is then given by

$$d\mu(\psi) = \frac{1}{\det[\langle x|\varphi_n\rangle]\det[\langle\varphi_n|x\rangle]} \prod_n da_n d\overline{b}_n = \prod_n da_n d\overline{b}_n \tag{3.13}$$

In the present case, where the basic operator $\not{D}$ in (3.12) is hermitean, the Jacobian for the transformation from $\psi(x)$ and $\overline{\psi}(x)$ to a_n and $\overline{b}_n$ in (3.13) becomes unity. In terms of (3.11), the fermionic action in (3.1) becomes

$$\int \overline{\psi}(i\not{D} - m)\psi dx = \sum_n (i\lambda_n - m)\overline{b}_n a_n \tag{3.14}$$

Namely, the fermionic action is formally diagonalized, which may justify the use of the particular basis set in (3.12).

Under the chiral transformation (3.5), the coefficient a_n, for example, is transformed as

$$\psi'(x) = \sum_n a_n' \varphi_n(x) = \sum_m e^{i\alpha(x)\gamma_5} a_m \varphi_m(x) \tag{3.15}$$

Namely,

$$a_m' = \sum_n \int \varphi_m(x)^\dagger e^{i\alpha(x)\gamma_5} \varphi_n(x)dx \ a_n \tag{3.16}$$

and

$$\prod_m da_m{}' = \det\left[\int \varphi_m(x)^\dagger e^{i\alpha(x)\gamma_5} \varphi_n(x)dx\right]^{-1} \prod_n da_n \tag{3.17}$$
$$= \exp\left[-i\sum_n \int \varphi_n(x)^\dagger \alpha(x)\gamma_5 \varphi_n(x)dx\right] \prod_n da_n \, .$$

One may sum the series in (3.17) starting from small eigenvalues ($|\lambda_n| \leq M$ and $M \to \infty$ later)

$$\sum_n \varphi_n(x)^\dagger \gamma_5 \varphi_n(x) \equiv \lim_{M\to\infty} \varphi_n(x)^\dagger \gamma_5 e^{-\lambda_n{}^2/M^2} \varphi_n(x)$$
$$\equiv \lim_{M\to\infty} \varphi_n(x)\gamma_5 e^{-\not{D}^2/M^2} \varphi_n(x) \, . \tag{3.18}$$

In this regularized form, one may change the basis set to plane waves (interaction picture), and one obtains

$$\lim_{M\to\infty} \mathrm{Tr}\int \frac{d^4k}{(2\pi)^4} e^{-ikx}\gamma_5\, e^{-\not{D}^2/M^2} e^{ikx} \tag{3.19}$$

$$= \lim_{M\to\infty} \mathrm{Tr}\int \frac{d^4k}{(2\pi)^4} \gamma_5 \exp\left[-\frac{(ik_\mu+D_\mu)(ik^\mu+D^\mu)}{M^2} + \frac{i}{4M^2}[\gamma^\mu, \gamma^\nu]F_{\mu\nu}\right]$$

since

$$\not{D}^2 = D_\mu D^\mu - \frac{i}{4}[\gamma^\mu, \gamma^\nu]F_{\mu\nu} \, . \tag{3.20}$$

The trace in (3.19) runs over the Dirac and internal symmetry indices.

After re-scaling $k_\mu \to Mk_\mu$, one may perform the $1/M$ expansion in (3.19). If one takes the trace with γ_5 into account, (3.19) can be rewritten as

$$\lim_{M\to\infty} \mathrm{Tr}M^4 \int \frac{d^4k}{(2\pi)^4} \gamma_5 \exp\left[-k_\mu k_\mu + \frac{i}{4M^2}[\gamma^\mu, \gamma^\nu]F_{\mu\nu}\right]$$

$$= \lim_{M\to\infty} \mathrm{Tr}\gamma_5 \frac{1}{2!}\left(\frac{i}{4}[\gamma^\mu, \gamma^\nu]F_{\mu\nu}\right)^2 \int \frac{d^4k}{(2\pi)^4} e^{-k_\mu k_\mu}$$

$$= \frac{1}{32\pi^2} \mathrm{Tr}\, \epsilon^{\mu\nu\alpha\beta} F_{\mu\nu} F_{\alpha\beta} \, . \tag{3.21}$$

In 2n-dimensions, the calculation proceeds in an identical manner, and one obtains

$$\mathrm{Tr}\, \frac{1}{n!} \gamma_5 \left(\frac{i}{4}[\gamma^\mu, \gamma^\nu]F_{\mu\nu}\right)^n \int \frac{d^{2n}k}{(2\pi)^{2n}} e^{-k_\mu k_\mu} = \frac{1}{(2\pi)^n} \frac{i^n}{n!} \mathrm{Tr}F^n \tag{3.22}$$

with the form notation

$$F \equiv \frac{1}{2} F_{\mu\nu} dx^{\mu} dx^{\nu} \quad . \tag{3.23}$$

The γ-matrices in (3.22) play the role of dx^{μ} in (3.23) in the precence of $Tr\gamma_5$; the precise sign factor in (3.22) depends on the definition of γ_5. We note that the final result (3.21) or (3.22) is independent[2] of the particular form of the regulator $\exp[-\lambda_n^2/M^2]$ in (3.18).

Coming back to 4-dimensions, we have equal contributions from $\mathcal{D}\psi$ and $\mathcal{D}\bar{\psi}$, and

$$d\mu \to d\mu \exp[-2iTr\int dx\alpha(x)\frac{\varepsilon^{\mu\nu\alpha\beta}}{32\pi^2} F_{\mu\nu}F_{\alpha\beta}] \tag{3.24}$$

and the anomalous chiral U(1) identity

$$\partial_{\mu}\langle\bar{\psi}\gamma^{\mu}\gamma_5\psi(x)\rangle - 2im\langle\bar{\psi}\gamma_5\psi(x)\rangle = \langle\frac{i}{16\pi^2} Tr\, \varepsilon^{\mu\nu\alpha\beta}F_{\mu\nu}F_{\alpha\beta}\rangle \ . \tag{3.25}$$

We note that the calculation (3.18) corresponds to a local version of the Atiyah-Singer index theorem.[8] If one assumes that a global limit $\alpha(x) \to$ constant can be taken in (3.24), one obtains

$$d\mu \to \sum_{\nu} d\mu_{(\nu)}\exp[-2i\alpha\nu] \tag{3.26}$$

with ν the Pontryagin index, and the θ-vacuum structure.[9]

4. Non-Abelian Anomaly

We now discuss the non-Abelian anomaly, i.e., the possible anomaly in non-Abelian transformations.[10] For this purpose we start with

$$\mathcal{L} = \bar{\psi}i\not{D}\psi$$

$$\not{D} = \gamma^{\mu}(\partial_{\mu} - iV_{\mu}^{a}T^{a} - iA_{\mu}^{a}T^{a}\gamma_5) \tag{4.1}$$

$$[T^a, T^b] = if^{abc}T^c, \quad TrT^aT^b = \frac{1}{2}\delta^{ab} \quad .$$

A salient feature of (4.1) is that the basic operator $\not{D}$ is <u>not</u> hermitean in the Euclidean sence

$$(\Phi, \not{D}\Psi) \equiv (\not{D}^{\dagger}\Phi, \Psi) \tag{4.2}$$

for

$$(\Phi,\ \Psi) = \int \Phi(x)^{\dagger}\Psi(x)dx \quad . \tag{4.3}$$

To be precise

$$\not{D}^{\dagger} = \gamma^{\mu}(\partial_{\mu} - iV_{\mu} + iA_{\mu}\gamma_5) \neq \not{D} \ . \tag{4.4}$$

There are basically two distinct ways to handle this situation:

(i) "Analytic" continuation[11] in A_{μ}, $A_{\mu} \to iA_{\mu}$, and

$$\not{D} \equiv \gamma^{\mu}(\partial_{\mu} - iV_{\mu} + A_{\mu}\gamma_5) = \not{D}^{\dagger} \ . \tag{4.5}$$

This gives rise to the integrable (or consistent) anomaly, if one uses the basis set

$$\not{D}\varphi_n(x) = \lambda_n\varphi_n(x) \tag{4.6}$$

and the corresponding regulator with a hermitean $\not{D}$ in (4.5) as

$$e^{-\not{D}^2/M^2} \quad . \tag{4.7}$$

For example, one obtains the Jacobian factor

$$-2i\mathrm{Tr}\int \frac{d^4k}{(2\pi)^4}\, e^{-ikx}\gamma_5 T^a e^{-\not{D}^2/M^2} e^{ikx} \tag{4.8}$$

for the chiral transformation

$$\begin{aligned} \psi(x) &\to \psi'(x) = e^{i\alpha^a(x)T^a\gamma_5}\psi(x) \\ \overline{\psi}(x) &\to \overline{\psi}(x)' = \overline{\psi}(x)e^{i\alpha^a(x)T^a\gamma_5} \ . \end{aligned} \tag{4.9}$$

The calculation (4.8) is straightforward but tedious, and one obtains the result[11]

$$\frac{1}{24\pi^2}\,\mathrm{Tr}T^a\,\varepsilon^{\mu\nu\alpha\beta}\partial_{\mu}(W_{\nu}\partial_{\alpha}W_{\beta} - \tfrac{i}{2}W_{\nu}W_{\alpha}W_{\beta}) \tag{4.10}$$

if one sets $W_{\mu} = V_{\mu} = -A_{\mu}$ after the calculation of (4.8).

A notable feature of the prescription (4.5) is that the vector transformation

$$\psi(x) \to \psi'(x) = e^{i\alpha^a(x)T^a}\psi(x)$$
$$\bar{\psi}(x) \to \bar{\psi}'(x) = \bar{\psi}(x)\, e^{-i\alpha^a(x)T^a} \qquad (4.11)$$

is always anomaly-free.

(ii) The second method is to use the polar decomposition of the original $\not{D}$ in (4.1) by means of the eigenvalue equations,[12)]

$$\not{D}^\dagger\not{D}\,\varphi_n(x) = \lambda_n{}^2\,\varphi_n(x)$$
$$\not{D}\not{D}^\dagger\phi_n(x) = \lambda_n{}^2\phi_n(x)$$
$$\psi(x) = \sum_n a_n\varphi_n(x) \equiv \sum_n a_n\langle x|\varphi_n\rangle \qquad (4.12)$$
$$\bar{\psi}(x) = \sum_n \bar{b}_n\phi_n(x)^\dagger = \sum_n \bar{b}_n\langle\phi_n|x\rangle$$

and

$$d\mu = \Pi\mathcal{D}\bar{\psi}\mathcal{D}\psi = \det[\langle x|\varphi_n\rangle]^{-1}\det[\langle\phi_n|x\rangle]^{-1}\prod_n d\bar{b}_n da_n$$
$$= \det[\langle x|\phi_n\rangle]\det[\langle\varphi_n|x\rangle]\prod_n d\bar{b}_n da_n\ .$$

One thus obtains

$$\det\not{D} \equiv \int d\mu\, e^{\int\mathcal{L}dx}$$
$$= \int d\mu\,\exp[\sum_n i\lambda_n\bar{b}_n a_n]$$
$$= \det[\langle x|\phi_n\rangle]\prod_n\lambda_n\det[\langle\varphi_n|x\rangle] \qquad (4.13)$$

if one notes

$$\not{D}\,\varphi_n(x) = \lambda_n\phi_n(x) \qquad (4.14)$$

for a suitable choice of the phase factor. Eq.(4.13) corresponds to the polar decomposition of a finite dimensional matrix M and det M = $\det[U\Lambda V^\dagger]$ with two unitary matrices U and V and a diagonal Λ.

The Jacobian factor for the transformation (4.9) is thus given by

$$
\begin{aligned}
&(-i) \sum_n \mathrm{Tr}[\varphi_n(x)^\dagger \gamma_5 T^a \varphi_n(x) + \phi_n(x)^\dagger \gamma_5 T^a \phi_n(x)] \\
&\equiv (-i)\lim_{M\to\infty} \sum_n \mathrm{Tr}[\varphi_n(x)^\dagger \gamma_5 T^a e^{-\not{D}^\dagger \not{D}/M^2} \varphi_n(x) \\
&\qquad\qquad + \phi_n(x)^\dagger \gamma_5 T^a e^{-\not{D}\not{D}^\dagger/M^2} \phi_n(x)] \qquad (4.15)\\
&= (-i)\lim_{M\to\infty} \mathrm{Tr} \int \frac{d^4k}{(2\pi)^4} e^{-ikx} \gamma_5 T^a [e^{-\not{D}^\dagger \not{D}/M^2} + e^{-\not{D}\not{D}^\dagger/M^2}] e^{ikx} .
\end{aligned}
$$

The calculation of (4.15) becomes transparent if one writes

$$\not{D} = \gamma^\mu(\partial_\mu - iL_\mu)L + \gamma^\mu(\partial_\mu - iR_\mu)R \qquad (4.16)$$

with

$$
\begin{aligned}
L_\mu &= V_\mu + A_\mu \\
R_\mu &= V_\mu - A_\mu \qquad (4.17)\\
L &= (1 - \gamma_5)/2 , \quad R = (1 + \gamma_5)/2 .
\end{aligned}
$$

One can then confirm

$$
\begin{aligned}
\not{D}^\dagger \not{D} &= \not{D}(L)^2 L + \not{D}(R)^2 R \\
\not{D}\not{D}^\dagger &= \not{D}(L)^2 R + \not{D}(R)^2 L \qquad (4.18)
\end{aligned}
$$

and

$$
\begin{aligned}
&e^{-\not{D}^\dagger \not{D}/M^2} \pm e^{-\not{D}\not{D}^\dagger/M^2} \\
&= (L \pm R)e^{-\not{D}(L)^2/M^2} + (R \pm L)e^{-\not{D}(R)^2/M^2} . \qquad (4.19)
\end{aligned}
$$

The evaluation of (4.15) is then essentially reduced to that of the chiral U(1) anomaly in (3.19), and we obtain

$$(-i)\frac{1}{32\pi^2} \mathrm{Tr}T^a \varepsilon^{\mu\nu\alpha\beta}[F_{\mu\nu}(R)F_{\alpha\beta}(R) + F_{\mu\nu}(L)F_{\alpha\beta}(L)] . \qquad (4.20)$$

The calculation of the "covariant" anomaly (4.20) in arbitrary 2n-dimensions proceeds just as in (3.22).

A salient feature of the present prescription is that the Jacobian for the vector transformation (4.11) contains the anomaly

$$(-i)\frac{1}{32\pi^2}\,\mathrm{Tr}T^a\,\varepsilon^{\mu\nu\alpha\beta}[F_{\mu\nu}(R)F_{\alpha\beta}(R) - F_{\mu\nu}(L)F_{\alpha\beta}(L)]\ . \tag{4.21}$$

In particular, if one sets $R_\mu = 0$ and $L_\mu = W_\mu$ in (4.21) one obtains

$$i\,\frac{1}{32\pi^2}\,\mathrm{Tr}T^a\,\varepsilon^{\mu\nu\alpha\beta}F_{\mu\nu}(W)F_{\alpha\beta}(W) \tag{4.22}$$

which leads to the baryon (and lepton) number non-conservation (if one sets $T^a = 1$) in the Weinberg-Salam theory in the presence of instantons.[9]

In summary, the non-Abelian anomalies can be characterized by two different forms of anomalies corresponding to the two different[12] specifications of det$\not{D}$ in (4.6) and (4.13) (and to the two different definitions[13] of composite current operators involved.) The anomaly cancellation condition, for example, becomes identical for those two different forms of the anomaly. The integrable form of the non-Abelian anomaly (4.10) allows the integration of the anomalous identities in the form of the Wess-Zumino term and has interesting applications.[14] The covariant anomaly, on the other hand, has interesting applications in the presence of both of the local gauge and global chiral symmetries.[12,15]

We also note that the "covariant" form of the anomaly in d = 2n dimensions is obtained from the U(1)-type anomaly such as (3.22) in d = 2n + 2 dimensions by the replacement

$$F \rightarrow F + \omega(x) \tag{4.23}$$

and picking up the term linear in the transformation parameter $\omega(x) = \omega^a(x)T^a$, which is the zero-form. Starting from (3.22) in d = 2n + 2 dimensions, one obtains

$$\frac{i^{n+1}}{(2\pi)^{n+1}}\frac{1}{(n+1)!}\,\mathrm{Tr}(F+\omega)^{n+1} \rightarrow \left(\frac{i}{2\pi}\right)\frac{i^n}{2\pi^n}\,\mathrm{Tr}\omega F^n \tag{4.24}$$

which gives the generalization of (4.20) with $L_\mu = 0$, for example. A similar property for the consistent anomaly is well-known.[16] Those properties show that the anomaly cancellation in the level of d = 2n + 2 U(1)-type anomaly ensures the non-Abelian anomaly cancellation in d = 2n.

As an application of (4.24), we note that the term

$$\mathrm{TrF}^2\mathrm{TrF}^4 \tag{4.25}$$

appears in the anomaly consideration of the "zero-slope" limit of the superstring theory.[17] If one makes the replacement (4.23), one obtains

$$2\mathrm{Tr}\omega\mathrm{F}\mathrm{TrF}^4 + 4\mathrm{TrF}^2\mathrm{Tr}\omega\mathrm{F}^3 \tag{4.26}$$

which gives the well-known 1 to 2 ratio of the coefficient. It can be confirmed that the leading term in (4.26) can be cancelled by a suitable local counter term as in Ref.(17). As for the terms with higher powers in the gauge potential, the cancellation becomes more systematic if one first converts the covariant anomaly (4.26) to the consistent form of the anomaly.[13] [We emphasize that in the level of super-string diagrams, the both forms of the anomaly give rise to the equally well cancellation scheme of the anomaly. In the zero-slope limit effective theory, however, a part of the anomaly in the string loop diagram is included as effective couplings.[17] Those effective couplings are naturally constructed to satisfy the Bose symmetry. As a result, the consistent anomaly for the fermion loop, which satisfies the Bose symmetry, gives a more systematic anomaly cancellation, as was illustrated in Ref.(17)].

5. Gravitational Path Integral Measure

The path integral for gravitational interactions has been discussed by various authors in the past.[18] The prescription I am going to utilize is based on the simple requirements: The Jacobian associated with the general coordinate transformations should vanish for each path integral variable separately.[19] In the path integral with gauge fixing, the general coordinate transformations are restricted to the (operator) BRS transformation.[20] We therefore impose the condition that the Jacobian for the BRS transformation should vanish. [Of course, this prescription cannot be implemented for chiral fermions in certain space-time dimensions, and thus leading to the genuine general coordinate anomaly.[21] We comment on this point later].

I would like to illustrate the basic idea on the basis of 4-dimensional Einstein gravity. The Faddeev-Popov effective Lagrangian for the gauge condition

$$\partial_\nu(\sqrt{g}g^{\mu\nu}) = 0 \tag{5.1}$$

is given by

$$\int \mathcal{L}_{eff}dx = \int[-\tfrac{1}{2}\sqrt{g}R + iB_\mu(x)\partial_\nu(\sqrt{g}g^{\mu\nu}) + \xi_\mu M^\mu{}_\nu\eta^\nu]dx \tag{5.2}$$

where $B_\mu(x)$ is the Lagrangian multiplier to implement (5.1), and $M^\mu{}_\nu$ is the corresponding Faddeev-Popov determinant with ξ_μ and η^μ the Faddeev-Popov anti-ghost and ghost, respectively. This effective action is invariant under the BRS transformation [λ is an element of the Grassmann algebra],

$$\begin{aligned}
\delta g_{\mu\nu}(x) &= i\lambda[\eta^\rho\partial_\rho g_{\mu\nu} + (\partial_\mu\eta^\rho)g_{\rho\nu} + (\partial_\nu\eta^\rho)g_{\mu\rho}] \\
\delta g(x) \equiv \delta\det g_{\mu\nu}(x) &= i\lambda[\eta^\rho\partial_\rho g(x) + 2(\partial_\rho\eta^\rho)g(x)] \\
\delta\eta^\mu(x) &= i\lambda\eta^\rho\partial_\rho\eta^\mu(x) \\
\delta\xi_\mu(x) &= \lambda B_\mu(x) \\
\delta B_\mu(x) &= 0 \\
\delta h_a{}^\mu(x) &= i\lambda[\eta^\rho\partial_\rho h_a{}^\mu - (\partial_\rho\eta^\mu)h_a{}^\rho] \\
\delta S(x) &= i\lambda\eta^\rho\partial_\rho S(x)
\end{aligned} \tag{5.3}$$

where the transformation properties of the vielbein $h_a{}^\mu$ and the world scalar S are also shown for the later use.

We first note that the weight 1/2 world scalar $\tilde{S}(x) \equiv g(x)^{1/4}S(x)$ gives an invariant measure.

$$\delta\tilde{S}(x) = \delta[g(x)^{\frac{1}{4}}S(x)] = i\lambda[\eta^\rho\partial_\rho\tilde{S}(x) + \tfrac{1}{2}(\partial_\rho\eta^\rho)\tilde{S}(x)] \tag{5.4}$$

then

$$\begin{aligned}
\prod_x \mathcal{D}\tilde{S}'(x) &= \exp\{i\lambda \mathrm{Tr}[\eta^\rho\partial_\rho\delta(x-y) + \tfrac{1}{2}(\partial_\rho\eta^\rho)\delta(x-y)]\}\prod_x\mathcal{D}\tilde{S}(x) \\
&= \prod_x\mathcal{D}\tilde{S}(x)
\end{aligned} \tag{5.5}$$

The vanishing Jacobian is in fact confirmed as

$$\mathrm{Tr}[\eta^\rho\partial_\rho\delta(x-y) + \frac{1}{2}(\partial_\rho\eta^\rho)\delta(x-y)]$$
$$\equiv \sum_n\int dxdy\phi_n(x)[\eta^\rho\partial_\rho\delta(x-y) + \frac{1}{2}(\partial_\rho\eta^\rho)\delta(x-y)]\phi_n(y)$$
$$= \int dx\partial_\rho[\sum_n \phi_n(x)\phi_n(x)\eta^\rho(x)] = 0 \tag{5.6}$$

for <u>any</u> choice of the basis set $\{\phi_n(x)\}$.

Applying the criterion (5.6), we fix the BRS invariant measure (in n-dimensions)[19]

$$d\mu = \prod_x \mathcal{D}[g^{(\frac{n-4}{4n})}g_{\alpha\beta}(x)]\mathcal{D}[g^{(\frac{n+2}{4n})}\eta^\mu(x)]\mathcal{D}\xi_\mu(x)\mathcal{D}B_\mu(x)\mathcal{D}[g^{\frac{1}{4}}S(x)]$$
$$\equiv \prod_x \mathcal{D}\tilde{g}_{\alpha\beta}(x)\mathcal{D}\tilde{\eta}^\mu(x)\mathcal{D}\xi_\mu(x)\mathcal{D}B_\mu(x)\mathcal{D}\tilde{S}(x) \tag{5.7}$$

The important point is that <u>each</u> integration variable is fixed. This situation differs from the previously known prescriptions.

<u>Comment on Conformal (Weyl) Anomaly</u>

The above path integral prescription is convenient to understand the conformal (Weyl) anomaly in a transparent manner. We illustrate this point for the electromagnetic field in the curved space-time. The action (in 4-dimensions)

$$-\frac{1}{4}\int dx\sqrt{g}g^{\mu\nu}(x)g^{\alpha\beta}(x)F_{\mu\alpha}(A)F_{\nu\beta}(A) \tag{5.8}$$

is invariant under the Weyl transformation

$$g_{\mu\nu}(x) \to e^{-2\alpha(x)}g_{\mu\nu}(x)$$
$$g^{\mu\nu}(x) \to e^{2\alpha(x)}g^{\mu\nu}(x) \tag{5.9}$$
$$A_\mu(x) \to A_\mu(x)$$

where $\alpha(x)$ is a localized infinitesimal parameter.

The gauge potential A_μ is scalar under the Weyl transformation. We now define $A_a(x)$ by using the vierbein as

$$A_a(x) \equiv h_a{}^\mu(x)A_\mu(x) \to e^{\alpha(x)}A_a(x) \tag{5.10}$$

namely, A_a carries the canonical weight. The weight 1/2 variable $\tilde{A}_a(x) \equiv g^{1/4}A_a(x)$ then transforms as

$$\tilde{A}_a(x) \to e^{-\alpha(x)}\tilde{A}_a(x) \ . \tag{5.11}$$

The general coordinate invariant measure

$$\prod_x \mathcal{D}\tilde{A}_a(x) \tag{5.12}$$

thus gives rise to a non-vanishing Jacobian under (5.11). A careful estimate[23] of this Jacobian gives rise to the well-known conformal (Weyl) anomaly.[22,24] Note that the coefficient of the anomaly factor under (5.11) is fixed by the weight factor for the path integral variable.

If the naive measure

$$\prod_x \mathcal{D}A_\mu(x) \tag{5.13}$$

should define the invariant measure, there would appear no conformal anomaly, since $A_\mu(x)$ is a conformal scalar.

6. Bosonic String as Quantum Gravity in d = 2

The string theory received a lot of attention recently in view of the rising prospect for the unified theory with gravity.[17,25] Here I would like to discuss the first quantization of the bosonic string theory from the view point of the quantum gravity in d = 2 and the conformal (Weyl) anomaly.[26,27] We start with the string Lagrangian[25]

$$\mathcal{L} = -\frac{1}{2}\sqrt{g}g^{\mu\nu}\partial_\mu X^a(x)\partial_\nu X^a(x) \tag{6.1}$$

with the string variable $X^a(x)$, $a = 1 \sim D$, and x^μ the two-dimensional parameter. The Einstein Lagrangian $\sqrt{g}R$ is a total divergence in d = 2, and one can regard (6.1) as a total Lagrangian for the gravitational variable $g^{\mu\nu}$ and the "World scalar" variable $X^a(x)$.

To quantize (6.1), we take the conformally Euclidean gauge

$$g_{\mu\nu}(x) = \rho(x)\delta_{\mu\nu} \tag{6.2}$$

or equivalently ($\tilde{g}_{\mu\nu}(x) \equiv g_{\mu\nu}(x)/g^{1/4}$)

$$\begin{aligned} &\tilde{g}_{12}(x) = \tilde{g}_{21}(x) = 0 \\ &\frac{1}{2}[\tilde{g}_{11}(x) - \tilde{g}_{22}(x)] = 0 \ . \end{aligned} \tag{6.3}$$

The Faddeev-Popov effective Lagrangian is then given by

$$\mathcal{L}_{eff} = -\frac{1}{2}\sqrt{g}g^{\mu\nu}\partial_\mu X^a\partial_\nu X^a + iB_1(x)\tilde{g}_{12}(x) + iB_2(x)\frac{1}{2}[\tilde{g}_{11}(x) - \tilde{g}_{22}(x)]$$
$$+ \text{Faddeev-Popov term} \qquad (6.4)$$

and the path integral measure in n = 2 dimensions in (5.7)

$$d\mu = \prod_x \mathcal{D}[g_{\mu\nu}(x)/g^{1/4}]\mathcal{D}[g^{\frac{1}{2}}\eta^\mu(x)]\mathcal{D}\xi_\mu(x)\mathcal{D}B_\mu(x)\mathcal{D}[g^{1/4}X^a(x)]$$
$$\equiv \prod_x \mathcal{D}\tilde{g}_{\mu\nu}(x)\mathcal{D}\tilde{\eta}^\mu(x)\mathcal{D}\xi_\mu(x)\mathcal{D}B_\mu(x)\mathcal{D}\tilde{X}^a(x) \quad . \qquad (6.5)$$

The partition function is given by[27)]

$$Z = \int d\mu \exp[\int \mathcal{L}_{eff}dx]$$
$$= \int \prod_x \mathcal{D}\sqrt{\rho(x)}\mathcal{D}\tilde{X}^a(x)\mathcal{D}\tilde{\eta}^\mu(x)\mathcal{D}\xi_\mu(x)\exp\{\int[-\frac{1}{2}\partial_\mu(\frac{\tilde{X}^a}{\sqrt{\rho}})\partial_\mu(\frac{\tilde{X}^a}{\sqrt{\rho}}) + \xi\sqrt{\rho}\not{\partial}\frac{1}{\rho}\tilde{\eta}]dx\}$$
$$(6.6)$$

after integration over the multiplier fields $B_1(x)$ and $B_2(x)$. In (6.5)

$$\tilde{X}^a(x) \equiv \sqrt{\rho(x)}X^a(x)$$

$$\xi(x) = \begin{pmatrix} \xi_1(x) \\ \xi_2(x) \end{pmatrix}, \quad \tilde{\eta}(x) = \begin{pmatrix} \tilde{\eta}^1(x) \\ \tilde{\eta}^2(x) \end{pmatrix}, \quad \tilde{\eta}^\mu(x) \equiv \rho(x)\eta^\mu(x) \qquad (6.7)$$

$$\not{\partial} \equiv \sigma^1\partial_1 + \sigma^3\partial_2 \; ; \quad \sigma_k \text{ Pauli-metrices} \; .$$

We emphasize that the conformal freedom $\rho(x)$ explicitly appears in (6.6) even after the conformal gauge (6.2).

The $\rho(x)$ dependence in (6.6) is the "pure gauge" form, and it may be removed by

$$\tilde{X}^a(x) \to \sqrt{\rho}\tilde{X}^{a\prime}$$
$$\xi(x) \to \frac{1}{\sqrt{\rho}}\xi(x)' \qquad (6.8)$$
$$\tilde{\eta}(x) \to \rho\tilde{\eta}'(x)$$

In reality, one picks up the anomalous Jacobian factor for the transformation (6.8) and thus the well-known Liouville action.[26)] In general, the pure gauge-type dependence combined with the anomaly gives rise to the so-called Wess-Zumino term.[14)]

The conformal Jacobian factor for the transformation

$$\tilde{X}^a(x) \to e^{\varphi(x)\delta t}\, \tilde{X}^a(x)$$

$$\xi(x) \to e^{-\varphi(x)\delta t}\, \xi(x) \quad ; \qquad \rho(x) \equiv e^{2\varphi(x)}$$

$$\tilde{\eta}(x) \to e^{2\varphi(x)\delta t}\, \tilde{\eta}(x) \tag{6.9}$$

is given by

$$G(\varphi, t) = -\left(\frac{26-D}{24\pi}\right)\int[-2\partial^2 \varphi(x)(1-t) + 2\mu^2 e^{2\varphi(x)(1-t)}]dx \tag{6.10}$$

if one replaces $\rho(x)$ in (6.6) by

$$\rho_t(x) = \exp[2\varphi(x)(1-t)] \ . \tag{6.11}$$

See Ref.(27) for further details of the calculation of the Jacobian factor. Then the accumulated Jacobian factor arising from the repeated application of the transformation (6.9) is given by

$$\Gamma_{WZ}(\varphi) = \int_0^1 dt \int \varphi(x) G(\varphi, t) dx$$

$$= -\left(\frac{26-D}{24\pi}\right)\int[\tfrac{1}{2}(\partial_\mu \varphi)^2 + \tfrac{1}{2}\mu^2(e^{2\varphi} - 1)]dx \tag{6.12}$$

which is the well-known Liouville action.[26∿28)]

The partition function (6.6) then becomes

$$Z = \int \Pi \mathcal{D}\sqrt{\rho(x)}\,\mathcal{D}\tilde{X}^a \mathcal{D}\xi \mathcal{D}\tilde{\eta}\ \exp\{\int[-\tfrac{1}{2}\partial_\mu \tilde{X}^a \partial_\mu \tilde{X}^a + \xi \not{\partial} \tilde{\eta}]dx + \Gamma_{WZ}(\varphi)\} \tag{6.13}$$

For D = 26, one may set $\Gamma_{WZ} \to 0$, and the conformal (Weyl) symmetry in the original Lagrangian (6.1)

$$g^{\mu\nu}(x) \to e^{2\alpha(x)} g^{\mu\nu}(x)$$
$$X^a(x) \to X^a(x) \tag{6.14}$$

is recovered. One may then apply the conformal gauge fixing

$$\rho(x) = 1 \quad \text{or} \quad \varphi(x) = 0 \tag{6.15}$$

and

$$Z(D=26) = \int \pi \mathcal{D} X^a \mathcal{D} \xi \mathcal{D} \eta \exp\{\int_0^T [-\frac{1}{2}\partial_\mu X^a \partial_\mu X^a + \xi \not\partial \eta] d\sigma d\tau\}$$

$$= \text{constant} \times \frac{\det \not\partial}{[\det \partial^2]^{D/2}} = \text{constant} \times [\det \partial^2]^{(\frac{2-D}{2})} . \tag{6.16}$$

Namely, only the D-2 = 24 physical degrees contribute to the Casimir effect[29)]

$$Z(D=26) \underset{T\to\infty}{\longrightarrow} \exp[-E_0 T] \tag{6.17}$$

with the ground state "energy"

$$E_0 = -\frac{D-2}{24} \times (1 \text{ or } 2) \tag{6.18}$$

where the last factor 1 or 2 corresponds to the open ($0 \leq \sigma \leq \pi$) or closed ($0 \leq \sigma \leq 2\pi$) string, respectively.[29)]

Virasoro Condition and Wheeler-DeWitt Equation

The Virasoro condition[30)] in the string theory corresponds to the Wheeler-DeWitt equation[31)] in the conventional quantum gravity. It is therefore interesting to see how it appears in the present path integral formalism. We start with the equations of motion for the gauge variables $\tilde{g}_{12}(x)$ and $(\tilde{g}_{11}(x) - \tilde{g}_{22}(x))/2$ in (6.4);

$$\begin{aligned} &\langle T^{12}(x) + iB_1(x)\rangle = 0 \\ &\langle T^{11}(x) - T^{22}(x) + iB_2(x)\rangle = 0 \end{aligned} \tag{6.19}$$

where we defined

$$T^{\mu\nu}(x) \equiv \frac{\delta}{\delta \tilde{g}^{\mu\nu}(x)} [S_{\text{string}} + S_{\text{Faddeev-Popov}}] . \tag{6.20}$$

The BRS invariance of the action (6.4) and the path integral measure (6.5) then give rise to, for example,

$$\langle \xi_1(x)\rangle = \langle \xi_1(x)\rangle + \lambda \langle B_1(x)\rangle \tag{6.21}$$

under the transformation (5.3). We thus obtain

$$\langle B_1(x)\rangle = \langle B_2(x)\rangle = 0 \tag{6.22}$$

namely,[27)]

$$\begin{aligned} &\langle T^{12}(x)\rangle = 0 \\ &\langle T^{11}(x) - T^{22}(x)\rangle = 0 \ . \end{aligned} \tag{6.23}$$

The second relation of (6.23) corresponds to the Wheeler-DeWitt equation (i.e., the equation of motion associated with $g^{oo}(x)$).

For D = 26 and in Minkowski metric, (6.23) becomes

$$\begin{aligned} &\langle \frac{\partial X^a}{\partial \tau}\frac{\partial X^a}{\partial \sigma} + \text{Faddeev-Popov}\rangle = 0 \\ &\langle \frac{\partial X^a}{\partial \tau}\frac{\partial X^a}{\partial \tau} + \frac{\partial X^a}{\partial \sigma}\frac{\partial X^a}{\partial \sigma} + \text{Faddeev-Popov}\rangle = 0 \ . \end{aligned} \tag{6.24}$$

A notable feature of (6.24) is that the Virasoro condition (and the Wheeler-DeWitt equation in general) contains the Faddeev-Popov ghost fields. One can re-derive the ground state energy (6.18) by a careful treatment[32)] of the second relation of (6.24).

7. Comment on Gravitational Anomalies

A detailed treatment of gravitational anomalies in general is found in the original work of Alvarez-Gaumé and Witten.[21)] We here comment on some of the aspects of the gravitational anomalies from a view point of Ward-Takahashi identities.[33)]

We first recall that the local Lorentz and general coordinate anomalies are summarized as[33)]

$$\sqrt{g}\langle T_A^{\mu\nu}(x)\rangle = A^{\mu\nu}(x) \tag{7.1}$$

$$\sqrt{g}D_\nu\langle T_S^{\mu\nu}(x)\rangle = D_\nu A^{\nu\mu}(x) + g^{\mu\nu}A_\nu(x) \tag{7.2}$$

where $T_A^{\mu\nu}$ and $T_S^{\mu\nu}$ stand for the anti-symmetric and symmetric parts of the energy-momentum tensor, respectively. $A^{\mu\nu}(x)$ and $A_\nu(x)$ corresponds to the Jacobian factors associated with the local Lorentz transformation and the (covariantized) general coordinate transformation defined by

$$\delta_{cov}(\xi^\mu) = \delta_{GC}(\xi^\mu) + \delta_{LL}(\xi^\mu A_{\mu mn}) \tag{7.3}$$

respectively. In (7.3), $\delta_{GC}(\xi^\mu)$ stands for the general coordinate transformation with a parameter ξ^μ, and $\delta_{LL}(\xi^\mu A_{\mu mn})$ the local Lorentz transformation with the parameter $\xi^\mu A_{\mu mn}$; $A_{\mu mn}$ is the spin connection.

We note that the local Lorentz anomaly $A^{\mu\nu}(x)$ in (7.1) can be freely modified by a suitable local counter-term,[13] but the right-hand side of (7.2) remains invariant under the local counter term; the variations in $A^{\nu\mu}$ and A_ν caused by the local counter term cancel in (7.2). [The form of the gravitational anomaly appearing in the analysis of Green and Schwarz,[17] for example, corresponds to the use of a local counter term which sets $A_\nu = 0$. The anomaly (7.1) in this case becomes equivalent to the anomaly in (7.2) which has an invariant meaning under the local counter term. Besides, the "consitent" form of the anomaly is utilized in Ref.(17)].

The above property is important to understand the results of the <u>explicit</u> evaluation of local Lorentz[34] and general coordinate anomalies.[21] We illustrate it by considering the anomalies for J = 3/2 field in d = 2 dimensions. If one takes $\psi_\mu(x)$ as a basic path integral variable, one obtains

$$\begin{aligned}\sqrt{g}<T_A^{\mu\nu}(x)> &= -\frac{i}{4\pi}\frac{\sqrt{g}R}{24}\,\varepsilon^{\mu\nu}\\ \sqrt{g}D_\mu<T_S^{\mu\nu}(x)> &= -\frac{i}{4\pi}\,\partial_\mu\left(\frac{23}{24}\sqrt{g}R\,\varepsilon^{\mu\nu}\right)\end{aligned} \tag{7.4}$$

whereas one obtains

$$\begin{aligned}\sqrt{g}<T_A^{\mu\nu}(x)> &= -\frac{i}{4\pi}\frac{25}{24}\sqrt{g}R\,\varepsilon^{\mu\nu}\\ \sqrt{g}D_\mu<T_S^{\mu\nu}(x)> &= -\frac{i}{4\pi}\,\partial_\mu\left(\frac{23}{24}\sqrt{g}R\,\varepsilon^{\mu\nu}\right)\end{aligned} \tag{7.5}$$

if one uses the variable $\psi_a(x)$ defined by

$$\psi_a(x) \equiv h_a{}^\mu(x)\psi_\mu(x) \tag{7.6}$$

as a basic path integral variable.[33] The extra vielbein field $h_a{}^\mu$ in (7.6) <u>effectively</u> corresponds to adding a particular form of the local counter term discussed in, for example, Ref.(13). One can see that the anomaly for $T_A^{\mu\nu}$ changes under (7.6), but the anomaly for $T_S^{\mu\nu}$

remains invariant. The anomaly cancellation condition should, of course, be discussed on the basis of the invariant anomaly for $T_S^{\mu\nu}$.

References

1) S. L. Adler, Lectures in Elementary Particles and Quantum Theory, edited by S. Deser et al. (MIT, Cambridge, Mass., 1970).
R. Jackiw, Lectures on Current Algebra and its Applications, edited by S. Treiman et al. (Princeton Univ. Press, Princeton, N.J., 1972), and references therein.
2) K. Fujikawa, Phys. Rev. Lett. 42 (1979) 1195; 44 (1980) 1733; Phys. Rev. D21 (1980) 2848; 22 (1980) 1499 (E).
3) J. Schwinger, Phys. Rev. 82 (1951) 914; 91 (1953) 713, 728.
4) C. S. Lam, Nuovo Cimento 38 (1965) 1755.
5) K. Symanzik, Zeits. Naturf. 9 (1954) 809.
6) F. Berezin, The Method of Second Quantization (Academic press, New York, 1966).
7) B. Davison, Proc. R. Soc. London, A225 (1954) 252.
W. Burton and A. DeBorde, Nuovo Cim. 2 (1955) 197.
S. Coleman, in Proceedings of the International School of Subnuclear Physics, Erice, 1977, edited by A. Zichichi (Plenum, N.Y., 1979).
8) M. Atiyah and I. M. Singer, Ann. Math. 87 (1968) 484.
M. Atiyah, R. Bott and V. Potodi, Invent. Math. 19 (1973) 279.
See also
R. Jackiw and C. Rebbi, Phys. Rev. D16 (1977) 1052.
A. Schwartz, Phys. Lett. 67B (1977) 172.
J. Kiskis, Phys. Rev. D15 (1977) 2329.
L. Brown et al., Phys. Rev. D16 (1977) 417.
N. Nielsen and B. Schroer, Nucl. Phys. B127 (1977) 314.
M. Ansourian, Phys. Lett. 70B (1977) 301.
9) G. 't Hooft, Phys. Rev. Lett. 37 (1976) 8.
R. Jackiw and C. Rebbi, Phys. Rev. Lett. 37 (1976) 172.
C. Callan, R. Dashen, and D. Gross, Phys. Lett. 63B (1976) 334.
10) W. Bardeen, Phys. Rev. 184 (1969) 1848.
11) A. P. Balachandran et al., Phys. Rev. D25 (1982) 2718.
M. B. Einhorn and D. R. T. Jones, Phys. Rev. D29 (1984) 331.
S-K. Hu, B-U. Young and D. W. Mckay, Phys. Rev. D30 (1984) 836.

A. Andrianov and L. Bonora, Nucl. Phys. B233 (1984) 232.
R. E. Gamboa-Saravi et al., Phys. Lett. 138B (1984) 145.
L. Alvarez-Gaumé and P. Ginsparg, Nucl. Phys. B243 (1984) 449.
K. Ishikawa, H. Itoyama, B. Sakita and M. Yamawaki (unpublised).
A. Manohar and G. Moore, Nucl. Phys. B243 (1984) 421.
See also
R. Roskies and F. A. Schaposnik, Phys. Rev. D23 (1981) 558.
A. Dhar and S. Wadia, Phys. Rev. Lett. 52 (1984) 959.
R. E. Gamboa-Saravi, F. A. Schaposnik and J. E. Solomin, Phys. Rev. D30 (1984) 1353.
D. Gonzales and A. N. Redlich, Phys. Lett. 147B (1984) 150.

12) K. Fujikawa, Phys. Rev. D29 (1984) 285; D31 (1985) 341.
See also
P. H. Frampton and T. W. Kephart, Phys. Rev. Lett. 50 (1983) 1343, 1347.

13) W. Bardeen and B. Zumino, Nucl. Phys. B244 (1984) 421.
L. Alvarez-Gaumé and P. Ginsparg, Harvard report HUTP-84/A016.

14) J. Wess and B. Zumino, Phys. Lett. 37B (1971) 95.

15) R. Jackiw, MIT report, CTP #1230 (1984), and CTP #1251 (1985) and references therein.

16) R. Stora, Cargese lectures (1983).
B. Zumino, Y. S. Wu and A. Zee, Nucl. Phys. B239 (1984) 477.
L. Alvarez-Gaumé and P. Ginsparg, Nucl. Phys. B234 (1984) 449.
T. Sumitani, J. Phys. A17 (1984) L811.

17) M. B. Green and J. H. Schwarz, Phys. Lett 149B (1984) 117.

18) See, for example, E. S. Fradkin and G. A. Vilkovisky, Phys. Rev. D8 (1973) 4241, and references therein.

19) K. Fujikawa, Nucl. Phys. B226 (1983) 437.
K. Fujikawa and O. Yasuda, Nucl. Phys. B245 (1984) 436.
See also
B. S. DeWitt, in General Relativity, edited by S. Hawking and W. Israel (Cambridge Univ. press, 1979).
M. K. Fung, P. van Nieuwenhuizen, and D. R. T. Jones, Phys. Rev. D22 (1980) 2995.
K. S. Stelle and P. C. West, Nucl. Phys. B140 (1978) 285.
V. de Alfaro, S. Fubini and G. Furlan, Nuovo Cim. 74B (1983) 365.

20) C. Becchi, A. Rouet and R. Stora, Ann. of Phys. 98 (1976) 28.

21) L. Alvarez-Gaumé and E. Witten, Nucl. Phys. B234 (1984) 269.

22) S. W. Hawking, Comm. Math. Phys. 55 (1977) 133.

23) K. Fujikawa, Phys. Rev. D23 (1981) 2262.
R. Endo, Prog. Theor. Phys. 71 (1984) 1366.

24) D. Capper and M. J. Duff, Nuovo Cim. 23A (1974) 173.
M. J. Duff, S. Deser and C. Isham, Nucl. Phys. B111 (1976) 45.
J. Dowker and R. Critchley, Phys. Rev. D13 (1976) 3234; D16 (1977) 3390.
L. Brown and J. Cassidy, Phys. Rev. D15 (1977) 2810.
H. Tsao, Phys. Lett. 68B (1977) 79.
T. Yoneya, Phys. Rev. D17 (1978) 2567.

25) The detailed accounts of the string theory and related references are found in the notes by J. Schwarz, M. Green, D. Gross, L. Brink, J. Gervais, P. Frampton, P. Goddard, C. Rebbi, P. Candelas, S. Wadia and J. Pati in these Proceedings.
As for the earlier path integral approach to the strings, see
D. B. Fairlie and H. B. Nielsen, Nucl. Phys. B20 (1970) 637.
C. S. Hsue, B. Sakita, and M. A. Virasoro, Phys. Rev. D2 (1970) 2857.
M. Kaku and K. Kikkawa, Phys. Rev. D10 (1974) 1110, 1823.

26) A. Polyakov, Phys. Lett. 103B (1981) 207 and 211.
The role of the conformal (Weyl) freedom in the explicit construction of the scattering amphitude was discussed by S. Weinberg, Phys. Lett. 156B (1985) 305.
H. Aoyama, A. Dhar and M. A. Namazie, SLAC-PUB-3623 (1985).
D. Friedan and S. H. Shenker, Chicago report, EFI 85-09.

27) K. Fujikawa, Phys. Rev. D25 (1982) 2584.

28) E. Fradkin and A. Tseytlin, Phys. Lett. 106B (1981) 63.
B. Durhuus, H. B. Nielsen, P. Olesen, and J. L. Petersen, Nucl. Phys. B196 (1982) 498.
P. Di Vecchia, B. Durhuus, P. Olesen, J. L. Peterson, Nucl. Phys. B198 (1982) 157; B207 (1982) 77.
O. Alvarez, Nucl. Phys. B216 (1983) 125.
E. Martinec, Phys. Rev. D28 (1983) 2604.
D. Friedan, Proceedings of the 1982 Les Houches Summer School, J. B. Zuber and R. Stora eds. (North Holland, 1984).

29) L. Brink and H. B. Nielsen, Phys. Lett. 45B (1973) 332.
M. Lüscher, K. Symanzik, and P. Weisz, Nucl. Phys. B173 (1980) 365.
See also J. Polchinski, Texas report, UTTG-13-85.
30) M. Virasoro, Phys. Rev. D1 (1970) 2933.
31) J. A. Wheeler, in Battelle Rencontres, edited by C. DeWitt and J. A. Wheeler (Benjamin, New York, 1968).
B. S. DeWitt, Phys. Rev. 160 (1967) 1113.
32) K. Fujikawa, (to be published).
33) K. Fujikawa, M. Tomiya and O. Yasuda, Z. Phys. C28 (1985) 289.
34) L. N. Chang and H. T. Nieh, Phys. Rev. Lett. 53 (1984) 21.
See also
L. Alvarez-Gaumé, S. Della-Pietra and G. Moore, Harvard report A028 (1984).
O. Alvarez, I. M. Singer and B. Zumino, LBL report 17672 (1984).
F. Langouche, T. Schucker, R. Stora, CERN-TH 3898 (1984).
T. Sumitani, Tokyo report, Komaba 12 (1984).
S. Watamura, Kyoto report, RIFP-565 (1984).
L. Bonora, P. Pasti, and M. Tonin, Padova report-12 and 23 (1984).
35) As for chiral U(1)-type gravitational anomalies, see
T. Kimura, Prog. Theor. Phys. 42 (1969) 1191.
P. Delbourgo and A. Salam, Phys. Lett. 40B (1972) 381.
T. Eguchi and P. G. O. Freund, Phys. Rev. Lett. 37 (1976) 1251.
N. K. Nielsen, G. T. Grisaru, H. Romer and P. van Nieuwenhuizen, Nucl. Phys. B140 (1978) 477.
S. M. Christensen and M. J. Duff, Phys. Lett. 76B (1978) 571.
M. J. Perry, Nucl. Phys. B143 (1978) 114.
R. Critchly, Phys. Lett. 78B (1978) 410.
As for path integral approach, see
R. Endo and T. Kimura, Prog. Theor. Phys. 63 (1980) 683.
J. M. Gipson, Phys. Rev. D29 (1984) 2989; Virginia Report VPI-HEP-84/15.
R. Endo and M. Takao, Prog. Theor. Phys. 73 (1985) 803.
T. Matsuki, Louisiana report, DOE/ER/05490-70 (1985).

Note added:

The path integral approach to the chiral anomaly in supersymmetric Yang-Mills theory has been discussed by

K. Konishi and K. Shizuya, Pisa report IFUP-TH 5/85.

H. Itoyama, V. P. Nair and H-C. Ren, Princeton report (March, 1985).

K. Harada and K. Shizuya, Tohoku report TU/85/280.

VERTEX OPERATORS AND ALGEBRAS

Peter Goddard

Department of Applied Mathematics and Theoretical Physics,
University of Cambridge, Silver Street, Cambridge CB3 9EW
U.K.

ABSTRACT

The use of the vertex operators of string theory to obtain algebras and their representations is described. The origin of vertex operators in string theory is reviewed and it is shown how associated to the points of squared length two on any suitable lattice is a Lie algebra and a corresponding Lie group with a definite structure. The compact algebras obtained in the Euclidean case and the Kac-Moody algebras obtained in the non-Euclidean case are described. The Frenkel-Kac construction is derived. The points of squared length one on the lattice lead to fermionic oscillators, and the fermion-boson equivalence. Fermionic constructions of Kac-Moody algebras, including a transcendental construction for E_8, are given.

1. Introduction

In these lectures I shall describe some aspects of the ways that dual vertex operators, originally constructed in the dual string model, are in a natural way related to certain algebras, including both finite dimensional and infinite dimensional Lie algebras (and, in particular, the affine Kac-Moody algebras) as well as Clifford algebras and their affine infinite dimensional generalisations. The central idea of this is the Frenkel-Kac[1] construction (also obtained by Segal)[2]. The approach we shall adopt here is that of ref.[3] (a similar approach was also adopted in ref.[4]) in which the relation to the motion of strings on tori is evident.

Until recently, the use of vertex operators to represent algebras was an interesting link between mathematics and theoretical physics, in which the flow of ideas had been mainly towards mathematics. In the last year or so however the mathematical ideas have been used to construct a model which has a surprising degree of physical plausibility and which has attracted a very great deal of interest[5]. More generally, the importance of the representation theory of affine Kac-Moody algebras, in which vertex operators play a crucial role, and the Virasoro algebra, another product of string theory, in various aspects of theoretical physics has been increasingly realised in the last couple of years.

Even more exotic algebras than Kac-Moody, Virasoro and Clifford algebras are represented by vertex operators. Frenkel, Lepowsky and Meurman[6] have used these operators to construct a representation of the Greiss algebra and thus a concrete realisation of the natural representation space (the so-called "moonshine module") for the Fischer-Greiss Monster group, the largest sporadic finite simple group. To list all the possible connections and applicaticns of these vertex operators would be to risk becoming too discursive (see e.g. ref.[7]) but it seems likely that their applications in both mathematics and physics are capable of much further development.

In these lectures we shall begin in section 2 by reviewing the way vertex operators, and the spaces in which they act, arise in string theory. (For a review, see e.g. ref.[8].) In section 3 we show how, if the momentum of the string is restricted to a lattice, the points of squared length 2 on the lattice give rise to a Lie algebra. This algebra is the algebra of a compact Lie group if the momenta are Euclidean. We discuss examples of this phenomenon, especially the case of cubic lattice $\mathbb{Z}^d$, leading to the group SO(2d). The triality which occurs when d=4 and the extension of the construction to E_8 when d=8 are described. Next, in section 4, non-Euclidean lattices are considered and in particular the affine Kac-Moody algebras obtained and the Frenkel-Kac construction derived.. Points of squared length 1 are considered in section 5 and shown to lead to fermionic oscillators[3], from which can be constructed representations of affine Kac-Moody algebras bilinear in the fermions, equivalent via a fermion-boson equivalence to the vertex operator representation obtained in the previous section. In section 6 more general consideration is given to the construction of representations of Kac-Moody algebras bilinear in fermions. The level 1 representations, occurring in the heterotic string theory, for the groups $\mathrm{Spin}(32)/\mathbb{Z}_2$ and $E_8 \times E_8$ are constructed from fermions, in the former case only using independent fermionic fields but in the latter using interdependent ones, some of which are transcendental functions of the others[9]. The construction for E_8 generalises to produce the "magic square" of Frendenthal[10].

2. Strings and Vertices

To obtain algebras such as (affine) Kac-Moody algebras from the vertex operators describing the interactions of dual strings it is necessary to require that the momentum of the strings (or at least certain of their components) lie on a lattice. To have momenta restricted to a discrete lattice corresponds physically to a periodicity in space (or space-time), or, equivalently, to the string moving on a torus (though to get a physically plausible picture requires some further structure to be specified). To set the scene for introducing the vertex operators we review briefly the formalism for an open string moving in d-dimensional Minkowski space. [For more details, see e.g. ref.[8].]

A history of the string is described by a world sheet $x^\mu(\sigma,\tau)$, $0 \leq \sigma \leq \pi$, $-\infty < \tau < \infty$, where μ ranges over d values. The motion of the string is determined by the action

$$S = k \int [(\dot{x}.x')^2 - \dot{x}^2 x'^2]^{1/2} d\sigma\, d\tau \tag{2.1}$$

where

$$\dot{x} = \frac{\partial x}{\partial \tau}, \quad x' = \frac{\partial x}{\partial \sigma} \quad \text{and} \quad k = T_o/c \tag{2.2}$$

is a constant, with T_0 having the dimensions of tension. If the reparametrisation invariance of S is used to impose the orthonormality conditions,

$$\dot{x}.x' = 0\ , \quad \dot{x}^2 + x'^2 = 0\ , \tag{2.3}$$

the equations of motion and boundary conditions take the form

$$\ddot{x} = x'' \tag{2.4}$$

and

$$x' = 0 \quad \text{at} \quad \sigma = 0,\pi. \tag{2.5}$$

These have the general solution

$$\frac{1}{\kappa} x^\mu(\sigma,\tau) = q^\mu + p^\mu\tau + \sum_{n=1}^{\infty} \{q_n^\mu \cos n\tau + \dot{q}_n^\mu \frac{\sin n\tau}{n}\} \cos n\sigma \tag{2.6}$$

where the constant κ has been introduced to absorb the dimensions of length. It is convenient to introduce harmonic oscillator variables

$$\alpha_n^\mu = \frac{1}{2} \{\dot{q}_n^\mu - in\, q_n^\mu\} \tag{2.7}$$

so that

$$\alpha_n^{\mu\dagger} = \alpha_{-n}^\mu\ , \tag{2.8}$$

where, classically, the dagger denotes complex conjugation, to become hermitian conjugation on quantisation. We can then rewrite eq.(2.6) as

$$\frac{1}{\kappa} x^\mu(\sigma,\tau) = q^\mu + p^\mu\tau + \frac{i}{2} \sum_{n\neq 0} \{\frac{\alpha_n^\mu}{n} e^{-in(\tau+\sigma)} + \frac{\alpha_n^\mu}{n} e^{-in(\tau-\sigma)}\} \tag{2.9a}$$

$$\equiv \frac{1}{2} Q^\mu(e^{i(\tau+\sigma)}) + \frac{1}{2} Q^\mu(e^{i(\tau-\sigma)}) \tag{2.9b}$$

for

$$Q^\mu(z) = q^\mu - ip^\mu \log z + i \sum_{n\neq 0} \frac{\alpha_n^\mu}{n} z^{-n}\ . \tag{2.10}$$

If we introduce the momentum density

$$\pi(\sigma) = k\dot{x}(\sigma) \ , \tag{2.11}$$

the canonical quantisation conditions take the form

$$[x^\mu(\sigma),\pi^\nu(\sigma')] = i\hbar\,\delta(\sigma-\sigma')g^{\mu\nu} \tag{2.12}$$

provided that we quantise covariantly. In terms of the harmonic oscillator variables (2.7) this implies

$$[\alpha_m^\mu,\alpha_n^\nu] = m\,g^{\mu\nu}\delta_{m,-n} \ , \tag{2.13}$$

$$[q^\mu,p^\nu] = ig^{\mu\nu} \tag{2.14}$$

provided that

$$\kappa^2 k\pi = \hbar \ . \tag{2.15}$$

The creation and annihilation operators α_n^μ act in a Hilbert space which they generate from certain "vacuum" states $|\gamma>$ which carry only momentum,

$$\alpha_n^\mu|\gamma> = 0 \ , \quad n > 0 \ ; \quad p^\mu|\gamma> = \gamma^\mu|\gamma> \ . \tag{2.16}$$

and can be generated from a true vacuum $|0>$ in the usual way

$$|\gamma> = e^{iq.\gamma}|0> \ . \tag{2.17}$$

The constraints (2.3) correspond, in terms of α_n^μ, to the vanishing of

$$L_n = \frac{1}{2}\sum_{m=-\infty}^{\infty} :\alpha_m.\alpha_{n-m}: \tag{2.18}$$

where the normal ordering has been introduced to ensure that

$$L_o = \frac{1}{2}p^2 + \sum_{n>0}\alpha_{-n}.\alpha_n \tag{2.19}$$

is well-defined. So defined, these operators satisfy the Virasoro algebra

$$[L_m,L_n] = (m-n)L_{m+n} + \frac{d}{12}m(m^2-1)\delta_{m,-n} \ . \tag{2.20}$$

In this covariant treatment of the quantum theory the classical constraints (2.3) are applied quantum mechanically as conditions

$$L_n|\psi> = 0 \ , \quad n > 0 \ , \tag{2.21a}$$

$$L_o|\psi> = \lambda|\psi> \ , \tag{2.21b}$$

where λ is an arbitrary constant which can be regarded as an ambiguity associated with normal ordering. It can be shown that the space of physical states, defined by eq.(2.21), is free of ghosts, i.e. negative norm states, resulting from the Lorentzian metric in eq.(2.13), if and only if[11-13]

$$\underline{\text{either}} \qquad d = 26 \qquad \text{and} \qquad \lambda = 1 \tag{2.22a}$$

$$\underline{\text{or}} \qquad 1 \leq d \leq 25 \qquad \text{and} \qquad \lambda \leq 1 \,. \tag{2.22b}$$

However the spectrum of states for the free open string that we have been discussing is much neater if $d = 26$ and $\lambda = 1$ and it is only in this case that an apparently reasonably consistent theory of interacting strings exists. Even this theory possesses a tachyon, the ground state of the string described by unexcited states $|\gamma\rangle$, of the form of eq.(2.17), satisfying eqs.(2.21) with $\lambda = 1$ and so having

$$\gamma^2 = 2 \,. \tag{2.23}$$

With the metric we are using, such a momentum γ is space-like. To avoid this difficulty one needs to move to the $d = 10$ theory of Neveu, Schwarz and Ramond[14] which also involves fermionic degrees of freedom and has a sector which is tachyon-free and supersymmetric in ten-dimensional Minkowski space, the 'superstring' theory of Green and Schwarz[15].

Living with the tachyon, the interactions of this $d=26$ bosonic string are defined perturbatively, starting from tree diagram Born terms. If we introduce the vertex operator

$$U(\gamma,z) = :\exp\{i\gamma.Q(z)\}: \tag{2.24a}$$

$$\equiv \exp\{i\gamma.Q_<(z)\}\, \exp\{i\gamma.Q_o(z)\}\, \exp\{i\gamma.Q_>(z)\} \tag{2.24b}$$

where

$$Q^\mu_>(z) = \sum_{n>0} \frac{\alpha^\mu_n}{n} z^{-n} \,, \quad Q^\mu_<(z) = \sum_{n<0} \frac{\alpha^\mu_n}{n} z^{-n} \tag{2.25}$$

and

$$Q^\mu_o(z) = q - ip^\mu \log z \tag{2.26a}$$

so that

$$e^{i\gamma.Q_o(z)} = z^{\frac{1}{2}\gamma^2} e^{i\gamma.q} z^{\gamma.p} \tag{2.26b}$$

the Born term amplitude for the interaction of N ground state strings of momenta $\gamma_1,\ldots,\gamma_N$ (with $\gamma_i^2 = 2$, $\Sigma\gamma_i = 0$) is obtained by symmetrising

$$\int \langle 0|\, U_{\gamma_1,\ldots\gamma_N}(z_1,\ldots,z_N)\,|0\rangle \prod_{i=1}^{N} \frac{dz_i}{z_i} \Big/ d\gamma \tag{2.27}$$

where

$$U_{\gamma_1,\ldots,\gamma_N}(z_1,\ldots,z_N) = U(\gamma_1,z_1)U(\gamma_2,z_2)\ldots U(\gamma_N,z_N) \ , \qquad (2.28)$$

$$d\gamma = \frac{dz_a\, dz_b\, dz_c}{(z_a-z_b)(z_b-z_c)(z_c-z_a)} \ , \qquad (2.29)$$

a,b,c being any fixed three of the indices $1,\ldots,N$, and the integral is taken with the variables $z_1,\ldots,z_N$ varying over the real line, subject to the constraints that they maintain their order $z_1 > z_2 > \ldots > z_N$, and the points z_a,z_b,z_c remain fixed (as is implied by dividing by $d\gamma$). Independence of the choice of z_a,z_b,z_c follows from the invariance of the integrand of eq.(2.27) under the group of Möbius transformations

$$z \to (az+b)/(cz+d). \qquad (2.30)$$

The way that one proves this Möbius invariance and, more generally, the fact that the only states that couple as intermediate states in such amplitudes are physical states in the sense of eqs.(2.21) is to exploit the relationship of the vertex (2.24) to the Virasoro algebra (2.18), which has the Möbius algebra as the three-dimensional subalgebra given by $n = 1, 0, -1$,

$$[L_n,U(\gamma,z)] = z^n\{ z\frac{d}{dz} + \frac{\gamma^2}{2} n\} U(\gamma,z). \qquad (2.31)$$

A particular consequence, if $\gamma^2 = 2$, is that then

$$[L_n,U(\gamma,z)] = z\frac{d}{dz}\{ z^n U(\gamma,z)\} \qquad (2.32)$$

It follows that

$$A_\gamma = \frac{1}{2\pi i}\oint \frac{dz}{z} U(\gamma,z) \ , \qquad (2.33)$$

where the integration contour encircles the origin once, is a physical state creation operator,

$$[L_n,A_\gamma] = 0 \ , \qquad (2.34)$$

i.e. it maps physical states into physical states, provided that it is well-defined. It will only be well-defined when acting on states whose momenta have integral inner products with γ so that (2.26b) is single-valued.

Finally in this brief summary let us remark that to evaluate (2.27) it is necessary to normal order the operator product (2.28). This is achieved by repeated use of the identity

$$\exp\{i\beta.Q_>(z)\}\ \exp\{i\gamma.Q_<(\zeta)\}$$

$$= (1-\zeta/z)^{\beta.\gamma}\exp\{i\gamma.Q_<(\zeta)\}\ \exp\{i\beta Q_>(z)\} \quad \text{if} \quad |\zeta| < |z| \tag{2.35}$$

which follows from

$$[Q_>^\mu(z), Q_<^\nu(\zeta)] = -g^{\mu\nu}\log(1-\zeta/z) \qquad \text{if} \quad |\zeta| < |z| \tag{2.36}$$

and

$$e^A e^B = e^{[A,B]}\, e^B e^A\,, \tag{2.37}$$

which holds if $[A,B]$ is a c-number.

At this point we shall rather abruptly leave our discussion of dual model amplitudes because we have enough to motivate the introduction of vertex operators, even though this motivation has necessarily been rather sketchy in order to avoid too big a digression.

3. Algebras and Lattices

Now let us consider the effect of restricting the permitted values to lie on a lattice[3] (i.e. a discrete set of points closed under addition and with no point of accumulation). This is not such an unfamiliar idea. Consider a particle moving on a torus. We can picture the torus as Euclidean space $\mathbb{R}^d$ with points identified if they are separated by a displacement contained in some lattice Γ, i.e. $\mathbb{R}^d/\Gamma$. A momentum wave function for such a particle,

$$e^{i\gamma.x/\hbar}\,, \tag{3.1}$$

will be single valued on the torus if and only if

$$\frac{\gamma.a}{2\pi\hbar} \in \mathbb{Z} \quad \text{for all} \quad a \in \Gamma\,. \tag{3.2}$$

This leads to the idea of the dual lattice

$$\Gamma^* = \{\ \phi : \phi.\gamma \in \mathbb{Z} \quad \text{for all} \quad \gamma \in \Gamma\} \tag{3.3}$$

Clearly, condition (3.2) is equivalent to requiring $\gamma \in 2\pi\hbar\,\Gamma^* = \Lambda$, say.

Let us return to the dual model formalism reviewed in section 2, but with momenta taken from a lattice Λ. For the purposes of exploring the mathematics of the construction, we need not suppose the lattice to be in Minkowski space. We could take a space with an arbitrary signature; it could be Euclidean or there could even be certain directions which were totally null, i.e. the metric tensor $g^{\mu\nu}$ might be singular. (More complicatedly it could be that only the projection of the momenta on a certain subspace lies on a discrete lattice, corresponding to a configuration space of the form $\mathbb{R}^m \times T$ where T is some torus; this is indeed the situation in the applications which are considered to be promising from the point of view of physical relevance.)

Specifically we consider the Hilbert space, $\mathcal{H}$, generated by the operators α_n^μ, $n \in \mathbb{Z}$, $1 \le \mu \le d$, satisfying the commutation relations

$$[\alpha_m^\mu, \alpha_n^\nu] = m\, g^{\mu\nu} \delta_{m,-n} \tag{3.4}$$

and the hermiticity conditions

$$\alpha_m^{\mu\dagger} = \alpha_{-m}^\mu , \tag{3.5}$$

from states $|\gamma\rangle$, $\gamma \in \Lambda \subset \mathbb{R}^d$, obeying

$$p^\mu |\gamma\rangle = \gamma^\mu |\gamma\rangle , \tag{3.6a}$$

$$\alpha_m^\mu |\gamma\rangle = 0 , \quad m > 0 , \tag{3.6b}$$

$$\langle\gamma'|\gamma\rangle = \delta_{\gamma\gamma'} , \tag{3.6c}$$

where $p^\mu \equiv \alpha_0^\mu$. In this context we cannot define the position operator q^μ but we can define $e^{i\gamma.q}$ for $\gamma \in \Lambda$ by

$$e^{i\gamma.q}|\gamma'\rangle = |\gamma+\gamma'\rangle \tag{3.7}$$

and the condition it commutes with α_n^μ, $n \neq 0$. Then the whole of $\mathcal{H}$ is generated from the vacuum vector $|0\rangle$ by α_n^μ and $e^{i\gamma.q}$, $\gamma \in \Lambda$.

We can still introduce vertex operators $U(\gamma,z)$ by equations (2.24), (2.25) and (2.26b). Moreover, $U(\gamma,z)$ will actually be single-valued, as a function of complex z, for all states provided that

$$e^{i\gamma.Q_o(z)}|\gamma'\rangle \equiv z^{\frac{1}{2}\gamma^2+\gamma.\gamma'}|\gamma+\gamma'\rangle ,$$

is for all $\gamma' \in \Lambda$. In our present framework this would seem to require $\gamma.\gamma' \in \mathbb{Z}$ and $\gamma^2 \in 2\mathbb{Z}$, which can be achieved by restricting our attention to lattices for which

$$\gamma.\gamma' \in \mathbb{Z} \quad \text{for all} \quad \gamma,\gamma' \in \Lambda \tag{3.9}$$

and points γ on them which have even squared length. A lattice satisfying (3.9) is said to be <u>integral</u>. If <u>all</u> the points on it have even squared length (i.e. $\gamma^2 \in 2\mathbb{Z}$ for all $\gamma \in \Lambda$) it is said to be <u>even</u>, but this is more than we need. An equivalent way of writing condition

(3.9) is $\Lambda \subset \Lambda^*$.

We shall assume, therefore, that our momenta γ lie on an integral lattice Λ and we shall concentrate on vertex operators with $r \in \Lambda$ with $r^2 = 2$. It is convenient to use Λ_2 to denote end points:

$$\Lambda_2 = \{r \in \Lambda : r^2 = 2\} . \tag{3.10}$$

Perhaps it is not surprising in view of (2.34) that vertex operators $U(\gamma,z)$, $\gamma \in \Lambda_2$, have particularly interesting properties.

We now wish to investigate the algebra satisfied by the

$$A_r = \frac{1}{2\pi i} \oint \frac{dz}{z} U(r,z) \tag{3.11}$$

where $r \in \Lambda_2$. To this end, we use eq.(2.35) to obtain

$$\begin{aligned} U(r,z)U(s,\zeta) &= \exp\{ir.Q_<(z) + isQ_<(\zeta)\} \\ &\quad .(z-\zeta)^{r.s} z\zeta\, e^{i(r+s).q} z^{r.p} \zeta^{s.p} \\ &\quad . \exp\{ir.Q_>(z) + isQ_>(\zeta)\} \text{ for } |\zeta| < |z| \\ &\equiv U_{r,s}(z,\zeta). \end{aligned} \tag{3.12}$$

The right hand side of eq.(3.12) is a regular function of z,ζ except for $\zeta = 0$, $z = 0$. If we calculate the product in the other order we see that

$$U(s,\zeta)\ U(r,z)$$

will be given by the same function of z and ζ apart from a factor of $(-1)^{r.s}$, which could be either +1 or -1. Thus what we can easily calculate is

$$A_r A_s - (-1)^{r.s} A_s A_r = \frac{1}{(2\pi i)^2} \Big\{ \underset{|z|>|\zeta|}{\oint \frac{d\zeta}{\zeta} \oint \frac{dz}{z}} - \underset{|\zeta|>|z|}{\oint \frac{d\zeta}{\zeta} \oint \frac{dz}{z}} \Big\} U_{r,s}(z,\zeta) \tag{3.13}$$

The z contours in these two integrals, illustrated in figure 1 have a difference which is equivalent to a small loop about ζ. Thus

$$A_r A_s - (-1)^{r.s} A_s A_r = \frac{1}{(2\pi i)^2} \oint_0 \frac{d\zeta}{\zeta} \oint_\zeta \frac{dz}{z} U_{r,s}(z,\zeta) \tag{3.14}$$

where the z contour positively encircles ζ once and the ζ contour positively encircles the origin.

In certain circumstances (3.14) is easy to evaluate. Firstly, it

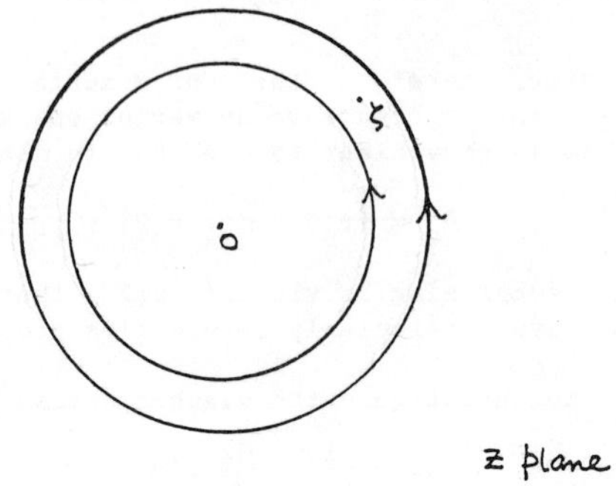

Fig. 1

vanishes if $r.s \geq 0$ because in this case $U_{r,s}(z,\zeta)$ is non-singular at $z = \zeta$. Secondly if $r.s = -1$, there is a simple pole at $z = \zeta$ and so

$$A_r A_s - (-1)^{r.s} A_s A_r = \frac{1}{2\pi i} \oint d\zeta\, e^{i(r+s).q}\, \zeta^{(r+s).p} \exp\{i(r+s).Q_<(\zeta)\} \exp\{i(r+s).Q_<(\zeta)\}$$

$$= \frac{1}{2\pi i} \oint d\zeta\, U(r+s,\zeta) \tag{3.15}$$

$$= A_{r+s}$$

using eqs.(2.24-6) and since $(r+s)^2 = 2$ in this case. Thirdly, if $r.s = -2$ there is a double pole at $z = \zeta$ whose contribution is particularly easy to evaluate in the special case where $r = -s$. In this case it gives

$$A_r A_s - (-1)^{r.s} A_s A_r = \frac{1}{2\pi i} \oint \frac{d\zeta}{\zeta}\, r.p(\zeta) \,, \tag{3.16}$$

where

$$p^\mu(z) = iz \frac{dQ^\mu(z)}{dz} \tag{3.17a}$$

$$= \sum_{n=-\infty}^{\infty} \alpha_n^\mu z^{-n} \,. \tag{3.17b}$$

Thus we have calculated the results,

$$A_r A_s - (-1)^{r.s} A_s A_r = 0 \qquad \text{if} \quad r.s \geq 0 \,, \qquad (3.18a)$$

$$= A_{r+s} \qquad \text{if} \quad r.s = -1 \,, \qquad (3.18b)$$

$$= r.p \qquad \text{if} \quad r = -s \,. \qquad (3.18c)$$

Additionally we have that

$$[p^\mu, A_r] = r^\mu A_r \qquad (3.18d)$$

because A_r carries momenta r. The possibilities listed in eqs.(3.18) are exhaustive if we are dealing with a Euclidean lattice Λ because then, if $r,s \in \Lambda_2$, $|r.s| \leq 2$ and $r.s = -2$ if and only if $r = -s$.

Clearly the factor $(-1)^{r.s}$ is annoying. Without it we would have a Lie algebra. It can be removed by a trick introduced by Frenkel and Kac[1]: we just multiply the vertex operators A_r by certain suitably chosen functions of momenta (which are functions of a discrete variable, since the momenta lie on a lattice). Suppose we denote by Λ_R the sub lattice of Λ generated by the points $r \in \Lambda_2$. For each $u \in \Lambda_R$, we introduce a function of momentum c_u with properties such that if $\tilde{c}_u = e^{iq.u} c_u$

$$\tilde{c}_u \tilde{c}_v = (-1)^{u.v} \tilde{c}_v \tilde{c}_u \,, \qquad (3.19)$$

$$\tilde{c}_u \tilde{c}_{-u} = 1 \,, \qquad (3.20)$$

$$\tilde{c}_u \tilde{c}_v = \varepsilon(u,v)\, \tilde{c}_{u+v} \,, \qquad (3.21)$$

where for each pair $u,v \in \Lambda_R$, $\varepsilon(u,v)$ is either $+1$ or -1. Such a $\tilde{c}_u$ is called in the mathematical literature a <u>cocycle</u>. If we can find such an object it will solve our problems because if we then set

$$E^r = A_r c_r \,, \quad r \in \Lambda_2 \qquad (3.22)$$

we have

$$[E^r, E^s] = 0 \qquad \text{if} \quad r.s \geq 0 \,, \qquad (3.23a)$$

$$= \varepsilon(r,s) E^{r+s} \qquad \text{if} \quad r.s = -1 \,, \qquad (3.23b)$$

$$= r.p \qquad \text{if} \quad r = -s \,, \qquad (3.23c)$$

$$[p^\mu, E^r] = r^\mu E^r \,, \qquad (3.23d)$$

and

$$[p^\mu, p^\nu] = 0 \,. \qquad (3.23e)$$

Further, provided that c_u is hermitian, p^μ, E^r satisfy the hermiticity conditions

$$E^{r\dagger} = E^{-r} \ , \quad p^{\mu\dagger} = p^\mu \ . \qquad (3.24)$$

In the Euclidean case, eqs.(3.23) define a closed finite dimensional Lie algebra g_Λ , which we can regard as the Lie algebra of some compact Lie group G_Λ associated with the integral lattice Λ . The momentum operators p^μ $1 \le \mu \le d$ form a Cartan subalgebra for g_Λ and the E_r step operators so that

$$\text{rank}\ g_\Lambda = d \qquad (3.25a)$$

$$\dim g_\Lambda = d + |\Lambda_2| \ , \qquad (3.25b)$$

using $|\Lambda_2|$ for the number of points of length squared 2 on Λ , and Λ_2 is the set of roots of g_Λ . [We have not completely established that there are no elements of g_Λ commuting with all the p^μ which are not linear combinations of them. For this see ref.[3]. The algebra g_Λ will be non-abelian provided that Λ_2 is non-empty, and it will be semi-simple provided that $\dim\Lambda_2 = \dim\Lambda$.

We should now return to indicate how one might construct the functions c_u . A convenient way for a theoretical physicist to do this is to consider a basis $\{e_i, 1 \le i \le m\}$ for the lattice Λ_R , i.e. the points $u \in \Lambda_R$ of the form $\sum n_i e_i$, $n_i \in \mathbb{Z}$. Then suppose we can introduce γ matrices γ_i such that

$$\gamma_i\gamma_j = (-1)^{e_i . e_j}\gamma_i\gamma_j \qquad (3.26)$$

[This can always be done, e.g. by embedding Λ_R as a sublattice of a cubic lattice in higher dimensions and introducing anti-commuting γ-matrices associated in a similar way to the points of unit length on the cubic lattice.] Then if $u = \sum u_i e_i$, $u_i \in \mathbb{Z}$, we define

$$\gamma^u = \varepsilon_u\, \gamma_1^{u_1}\gamma_2^{u_2} \ldots \gamma_m^{u_m} \qquad (3.27a)$$

where the choice of sign $\varepsilon_u = \pm 1$ is a matter of convention which has to be settled for each point of Λ_R. It does not matter how we do this provided that we arrange that

$$(\gamma^u)^{-1} = \gamma^{-u} \qquad (3.27b)$$

Then, for $u, v \in \Lambda_R$ we have

$$\gamma^u\gamma^v = \varepsilon(u,v)\ \gamma^{u+v} \ , \qquad (3.28)$$

where $\varepsilon(u,v) = \pm 1$, and

$$\gamma^u\gamma^v = (-1)^M\gamma^v\gamma^u \ , \qquad (3.29a)$$

where

$$M = \Sigma\, u_i v_j\, e_i . e_j$$

$$= u.v \tag{3.29b}$$

as $u,v \in \Lambda_2$. Thus we could use γ^u to remove the sign problem in eqs.(3.18). However, this would involve enlarging the space we are considering by tensoring it with a representation space for the γ matrix algebra (3.26). This is unnecessary because we can just abstract what we need from these γ-matrices and incorporate it in functions c_u . Consider first the case of states with momenta $v \in \Lambda_R$ and define

$$\tilde{c}_u |w\rangle = \varepsilon(u,w)\,|u+w\rangle \ , \quad u,w \in \Lambda_R \tag{3.30}$$

But from (3.28) it follows that

$$\varepsilon(u,v) = (-1)^{u.v}\varepsilon(v,u) \ , \tag{3.31a}$$

$$\varepsilon(u,0) = \varepsilon(u,-u) = +1 \ , \tag{3.31b}$$

and

$$\varepsilon(u,v+w)\ \varepsilon(v,w) = \varepsilon(u,v)\ \varepsilon(u+v,w) \ , \tag{3.31c}$$

and hence

$$\tilde{c}_u \tilde{c}_v = (-1)^{u.v} \tilde{c}_v \tilde{c}_u \ , \tag{3.32a}$$

$$\tilde{c}_u \tilde{c}_{-u} = 1 \tag{3.32b}$$

$$\tilde{c}_u \tilde{c}_v = \varepsilon(u,v)\ \tilde{c}_{u+v} \ . \tag{3.32c}$$

To extend the definition of c_u to states $|\lambda\rangle$ for any $\lambda \in \Lambda$ we divide Λ into disjoint cosets $\{\Lambda_R + \lambda_j\}$ and define

$$\tilde{c}_u |v+\lambda_j\rangle = \varepsilon(u,v)\,|u+v+\lambda_j\rangle \ . \tag{3.33}$$

We can generalise what we have said here a little: it is not really necessary for the lattice Λ to be integral. We need only that the operators E^r are well-defined for $r \in \Lambda_2$ and this requires that

$$e^{ir.Q_o(z)} |\lambda\rangle = z^{1+\lambda.r} |\lambda+r\rangle \tag{3.34}$$

be single valued [cf. eq.(3.8)]. This happens if and only if $\lambda.r \in \mathbb{Z}$ whenever $r \in \Lambda_2$ and $\lambda \in \Lambda$. This condition can be equivalently written

$$\Lambda_R \subset \Lambda^* \ , \tag{3.35}$$

which is a weaker condition than that we previously imposed, that the lattice be integral, i.e. $\Lambda \subset \Lambda^*$.

Let us consider some examples of Euclidean lattices. The simplest possibility is to take $\Lambda = \mathbb{Z}^d$, $d > 1$, the hypercubic lattice in d dimensions. There are $2d(d-1)$ points of squared length 2 on this lattice. These are the points for which only two of the coordinates are non-zero and these are each ± 1. Thus Λ_R consists of points of the form

$$(n_1, n_2, \ldots, n_d), \quad n_i \in \mathbb{Z}, \quad \Sigma n_i \in 2\mathbb{Z} \tag{3.36}$$

It is not difficult to check that Λ_2 is the root system of $SO(2d)$ and, of course, the dimensions check

$$d + |\Lambda_2| = d(2d-1) = \dim SO(2d). \tag{3.37}$$

If we exponentiate the generators we have constructed we get a specific group G_Λ with g_Λ as algebra, that is we get a specific global structure. The torus from which we have started is essentially the maximal torus, the maximal abelian subgroup of G_Λ. We can determine this global structure by looking at which eigenvalues of the generators p^μ, $1 \leq \mu \leq d$, occur, and this is given exactly by the original lattice Λ. In other words Λ is the set of <u>weights</u> of G_Λ. The condition that λ be a weight for g_Λ is that

$$2r.\lambda/r^2 \in \mathbb{Z} \quad \text{for all roots } r \tag{3.38}$$

and this is precisely (3.35) because in our construction all the roots of g_Λ inevitably have $r^2 = 2$. In the case $\Lambda = \mathbb{Z}^d$, the points of length 1, which are just those with only one non-zero coordinate and that coordinate being ± 1, correspond to the $2d$ weights of the vector representation of SO(2d). Since these weights generate additively all of $\mathbb{Z}^d$, we have

$$G_{\mathbb{Z}^d} = SO(2d). \tag{3.39}$$

The largest lattice permitted if we are to have g_Λ equal to so(2d) is the dual of (3.36), namely the lattice consisting of the points $(n_1, n_2, \ldots, n_d)$ where

$$\underline{\text{either}}\ n_i \in \mathbb{Z},\ 1 \leq i \leq d, \ \underline{\text{or}}\ n_i + \tfrac{1}{2} \in \mathbb{Z},\ 1 \leq i \leq d. \tag{3.40}$$

This is the lattice of all weights of so(2d). Taking this lattice instead of $\mathbb{Z}^d$, we still have $g_\Lambda = so(2d)$, provided that $d \neq 8$, but we have $G_\Lambda = Spin(2d)$, the simply connected covering group of SO(2d). On the other hand, the smallest possible choice is to take Λ to be just Λ_R, in which case we would have

$$G_\Lambda = SO(2d)/\mathbb{Z}_2, \quad \text{where} \quad \mathbb{Z}_2 = \{\pm 1\} \subset SO(2d).$$

If d is even, there are yet other possible choices for Λ. We could take Λ to be the lattice consisting of Λ_R together with the points

$$(n_1+\tfrac{1}{2}, n_2+\tfrac{1}{2}, \ldots, n_d+\tfrac{1}{2}), \quad n_i \in \mathbb{Z}, \quad \Sigma n_i \in 2\mathbb{Z} \tag{3.41}$$

This lattice is only integral if d is a multiple of 4 and, in that case it is, like $\mathbb{Z}^d$, self-dual, i.e. $\Lambda = \Lambda^*$. But when d is a multiple of 8 it has the additional property, not possessed by $\mathbb{Z}^d$, of being <u>even</u>, i.e. x^2 is even for all $x \in \Lambda$. (Even self-dual Euclidean lattices have $\dim\Lambda$ equal to a multiple of 8.)[16] In general, this choice of Λ produces a group G_Λ, which is Spin(2d) divided by a different $\mathbb{Z}_2$ subgroup, not isomorphic to SO(2d). The exceptions occur when $d = 4$ and 8. If $d = 8$, the lattice has extra points of squared length 2, namely the 128 permitted points each of whose coordinates is $\pm\frac{1}{2}$. The corresponding operators E^r increase from 120 to 248 and in fact $G_\Lambda = E_8$. [Of course, it is open to us not to add these extra points and stick with SO(16)]. If $d = 4$, the resulting group G_Λ is actually isomorphic to SO(8). Actually the condition in (3.41) that Σn_i be even could be replaced by the condition that Σn_i be odd. For any d this would produce an isomorphic G_Λ; for $d = 4$ it produces a third copy of SO(8). These three SO(8) algebras are related by outer isomorphisms associated with the symmetry of the corresponding Dynkin diagram. This is known as the <u>triality</u> property of SO(8).

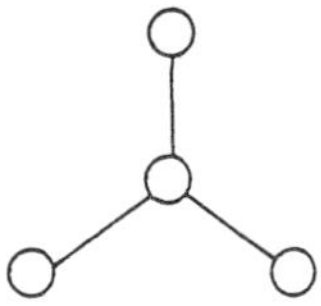

Fig. 2

When $d = 16$ the lattice specified by adding the points of (3.41) to Λ_R, is again even and self dual. The only other possibility for such a lattice, with dimension 16, is the direct sum of two E_8 lattices[16]. These even self-dual lattices produce the groups Spin(32)/$\mathbb{Z}_2$ and $E_8 \times E_8$, respectively, which have excited much interest recently[17,5].

Another exercise is to consider the possibilities in one dimension. There if Λ is to satisfy (3.35) and have points γ with $\gamma^2 = 2$ it must be either $\sqrt{2}\,\mathbb{Z}$ or $(1/\sqrt{2})\mathbb{Z}$. In the former case $G_\Lambda = SO(3)$ in the latter SU(2). In two dimensions taking $\Lambda = \mathbb{Z}^2$ gives $G_\Lambda = SO(4) = SU(2) \times SU(2)/\mathbb{Z}_2$ whilst taking $\Lambda = \sqrt{2}\,\mathbb{Z}^2$ gives $G_\Lambda = SU(2) \times SU(2)$.

In general we see that Λ is the weight lattice of some group with algebra g_Λ, a sublattice of the lattice of all weights of g_Λ, or, equivalently of the weight lattice of the simply connected group with algebra g_Λ. This Lie algebra is inevitably <u>simply laced</u>, that is all the roots have the same squared length, 2. If g_Λ is simple, the only possibilities are that is one of $A_n \cong su(n+1)$, $D_n \cong so(2n)$, E_6, E_7 or E_8; otherwise it is a direct sum of copies of these and u(1) factors. All simply-laced compact Lie groups can be obtained in this way.

Note that eq.(3.8) implies that

$$E^r|\lambda\rangle = 0 \quad \text{if} \quad r.\lambda \geq 0 \tag{3.42}$$

If we choose a set of position roots for g_Λ and consider weight λ dominant with respect to this choice, the "vacuum" states $|\lambda\rangle$ are highest weight states for the corresponding representation. Thus we get all irreducible representations of G_Λ.

As we said earlier it may be that only certain of the components of the momenta lie on a lattice, i.e. the momenta take values in $\mathbf{R}^{m-1,1} \times \Lambda$. We can then perform the construction of g_Λ ignoring the continuous components and it still follows that g_Λ will commute with the Virasoro generators

$$[L_n, E^r] = 0 \;, \quad [L_n, p^\mu] = 0 \;. \tag{3.43}$$

Thus the physical states, satisfying eqs.(2.21) fall into representations of G_Λ ; it is a symmetry of the perturbative theory.

4. Non-Euclidean Lattices and Kac-Moody Algebras

Let us now give a little consideration to what happens if Λ is not Euclidean. Then g_Λ is defined to be the Lie algebra generated by the p^μ, $1 \leq \mu \leq d$; E^r, $r \in \Lambda_2$. Thus it is spanned by elements of the form

$$E' = [E^{r_1}, [E^{r_2}, [\ldots [E^{r_{m-1}}, E^{r_m}] \ldots]]] \;. \tag{4.1}$$

This is a step operator corresponding to the root

$$v = \sum_{i=1}^{m} r_i \;. \tag{4.2}$$

Using dual model techniques we can show that

$$E' = 0 \quad \text{unless} \quad v^2 \leq 2 \;. \tag{4.3}$$

Further it follows from eq.(3.42) that

$$[L_n, E'] = 0 \;, \tag{4.4}$$

that is the whole of g_Λ commutes with the Virasoro algebra. In the Lorentzian case we can think of the elements of g_Λ as being associated with states of a string moving on a Minkowski torus $2\pi\hbar\, \mathbb{R}^d/\Lambda$. Actually an isomorphism can be set up between g_Λ and a subspace of the physical states, which has interesting consequences if Λ is Lorentzian[3,4].

A new feature occurs when we move away from Λ being Euclidean: there is no longer in general just one step operator for non-zero root v. Roots occur with variable multiplicities. To illustrate this let us consider the calculation of

$$[E^r, E^s] \qquad r,s \in \Lambda_2 \tag{4.5}$$

a little further in the non-Euclidean case. This vanishes if $r.s \geq 0$ and the case $r.s = -1$ is as before. The first new feature occurs when $r.s = -2$. This no longer forces $r = -s$ but it does imply that $r+s$ is light-like,

$$(r+s)^2 = 0 \ . \tag{4.6}$$

There is a double pole in eq.(3.14) and evaluation of the appropriate residue gives

$$[A_r, A_s] = \frac{1}{2}(r-s)_\mu A^\mu_{r+s} \tag{4.7}$$

where for light-like vectors $k \in \Lambda_R$ we define

$$A^\mu_k = \frac{1}{2\pi i}\oint \frac{dz}{z} :p^\mu(z)\exp\{ikQ(z)\}: \ . \tag{4.8}$$

This operator satisfies

$$k_\mu A^\mu_k = 0 \tag{4.9}$$

$$[L_n, \varepsilon_\mu A^\mu_k] = 0 \quad \text{if} \quad \varepsilon.k = 0 \ . \tag{4.10}$$

Note that $(r-s).(r+s) = 0$. The operators $\varepsilon.A_k$ describe the coupling of massless states with polarisation vectors ε. They are the original "photon" vertex operators of Del Gindice, Di Vecchia and Fubini[18]. For each $k \in \Lambda_R$ with $k^2 = 0$, there are $d-2$ independent step operators $\varepsilon.A_k$ corresponding to the possible states of a massless vector particle in d dimensions. The root multiplicity of such a light-like vector $k \in \Lambda_R$ is thus $d-2$. For a Lorentzian lattice these root multiplicities grow fast with $-v^2$.

A particularly interesting case to consider is the affine case[3], where we take a d-dimensional Euclidean lattice Λ and enlarge it by the addition of an orthogonal light-like vector k to obtain a new lattice Λ'. Thus

$$k^2 = 0 \ , \quad x.k = 0 \quad \text{for all} \quad x \in \Lambda \tag{4.11}$$

and

$$\Lambda' = \{x+nk : x \in \Lambda,\ n \in \mathbb{Z}\}. \tag{4.12}$$

Now one needs to be a little careful because the metric tensor $g^{\mu\nu}$ for Λ' is singular, in that $\det g = 0$, and so the inverse $g_{\mu\nu}$ does not exist. In consequence we cannot construct the Virasoro algebra (2.18) for this lattice.

For the affine lattice, the points of squared length 2 are

$$\Lambda_2' = \{r+nk : r \in \Lambda_2,\ n \in \mathbb{Z}\} \tag{4.13}$$

and, using the construction of the last section with oscillators α_n^μ $1 \le \mu \le d+1$, $g_{\Lambda'}$ has a basis consisting of

$$E_n^r \equiv E^{r+nk} , \quad r \in \Lambda_2,\ n \in \mathbb{Z}, \tag{4.14}$$

$$A_n^\mu = \oint \frac{dz}{z}\, p^\mu(z)\exp\{inkQ(z)\} , \quad 1 \le \mu \le d . \tag{4.15}$$

These satisfy the commutation relations

$$[A_m^\mu, A_n^\nu] = m\, \delta^{\mu\nu}\, \delta_{m,-n}\, k.p \tag{4.16a}$$

$$[A_m^\mu, E_n^r] = r^\mu\, E_{m+n}^r \tag{4.16b}$$

$$[E_m^r, E_n^s] = 0 , \qquad r.s \ge 0 , \tag{4.16c}$$

$$= \varepsilon(r,s)\, E_{m+n}^{r+s} , \qquad r.s = -1 , \tag{4.16d}$$

$$= r.A_{m+n} + m\, k.p\, \delta_{m,-n} , \qquad r = -s . \tag{4.16d}$$

The oscillators $k.\alpha_n$ commute with the whole of $g_{\Lambda'}$. What we have obtained in this way is an <u>affine Kac-Moody algebra</u>. Associated to any Lie algebra g, which we take for convenience to be finite-dimensional and simple, there is a corresponding affine Kac-Moody algebra, $\hat{g}$. If g takes the form

$$[T^a, T^b] = if_{abc}\, T^c \tag{4.17}$$

using a basis in which the structure constants f_{abc} are totally antisymmetric, $\hat{g}$ has generators T_m^a, $m \in \mathbb{Z}$, satisfying an algebra of the form

$$[T_m^a, T_n^b] = if_{abc}\, T_{m+n}^c + Km\, \delta_{m,-n}\, \delta^{ab} . \tag{4.18}$$

Here K is a central element, i.e.

$$[K,T^a_m] = 0 \tag{4.19}$$

which can be assigned a definite value in any irreducible representation. In fact K is quantised, if we consider highest weight unitary representations, that is representations in which

$$T^{a\dagger}_m = T^a_{-m} \tag{4.20}$$

and the whole space of states is built up from "vacuum states" ψ satisfying

$$T^a_m\psi = 0 \quad \text{if} \quad m > 0 \;. \tag{4.21}$$

In such representations K has to satisfy

$$2K/r^2 \in \mathbb{Z} \tag{4.22}$$

for all roots r of g. (For reviews of Kac-Moody algebras see refs. [19], [20] or [21].) If g is simply laced with the roots normalised so that $r^2 = 2$, this condition just says that K must be integral. The integer K is called the <u>level</u> of the representation. This ties in with what we have found in eqs.(4.16) by calculation. If we rewrite a simply-laced g in a Cartan-Weyl basis, selecting a Cartan subalgebra p^μ, $1 \leq \mu \leq d$, it takes the form of eqs.(3.23). In this basis (4.18) reads

$$[p^\mu_m, p^\nu_n] = Km\ \delta^{\mu\nu}\delta_{m,-n} \tag{4.23a}$$

$$[p^\mu_m, E^r_n] = r^\mu E^r_{m+n} \tag{4.23b}$$

$$[E^r_m, E^s_n] = 0 \;, \qquad r.s \geq 0 \tag{4.23c}$$

$$= \varepsilon(r,s)\ E^{r+s}_{m+n} \;, \qquad r.s = -1 \;, \tag{4.23d}$$

$$= k.p_{m+n} + Km\ \delta_{m,-n} \;, \qquad r.s = -2 \;. \tag{4.23e}$$

which we see is isomorphic to (4.16) with $A^\mu_m \leftrightarrow p^\mu_m$, $k.p \leftrightarrow K$. For the definitions (4.14) and (4.15) of E^r_n and A^μ_n to make sense we need that k.p take only integral values, in line with the general result we have quoted from the representation theory of Kac-Moody algebras.

If we restricted the eigenvalues of p to the lattice Λ', k.p would vanish identically but it is not necessary to do this; p can take values in any lattice $\tilde{\Lambda} \supset \Lambda'$ with the property that $x.y \in \mathbb{Z}$ if $x \in \tilde{\Lambda}$, $y \in \Lambda'$. A convenient construction to take $\tilde{\Lambda}$ to be the

Lorentzian lattice obtained by adjoining to Λ a second light-like vector k' with

$$k.k' = 1\ ,\quad k'^2 = 0 \tag{4.24}$$

Then we have

$$\tilde{\Lambda} = \{x+mk+nk' : x \in \Lambda,\ m,n \in \mathbb{Z}\} \tag{4.25}$$

and

$$\Lambda' = \{x \in \tilde{\Lambda} : k.x = 0\}\ . \tag{4.26}$$

We can now reintroduce the Virasoro operators if we wish and

$$[L_n,\hat{g}] = [k.\alpha_n,\hat{g}] = 0\ . \tag{4.27}$$

Our construction of the Kac-Moody algebra $\hat{g}$ is then very analogous to constructing the physical states of dual string theory in a covariant formalism. There is also an analogue of working directly in the light-cone gauge. This is the original construction of Frenkel and Kac[1]. In this construction one considers all the Laurent coefficients of $U(r,z)$ and abandons properties like (4.27). If we consider again a Euclidean lattice Λ and a lattice $\Lambda_R \subset \Lambda$ generated by points of squared length 2 in Λ, satisfying (3.35) and define

$$E^r_n = \frac{1}{2\pi i}\oint \frac{dz}{z} z^n U(r,z)c_r\ , \tag{4.28}$$

so that

$$U(r,z)c_r = \sum_{n=-\infty}^{\infty} E^n_r\, z^{-n}\ , \tag{4.29}$$

calculations exactly like those of section 3 show that these E^r_n satisfy eqs.(4.23) with α^μ_m replacing p^μ_m and $K=1$. Thus we have obtained a level 1 representation of the affine Kac-Moody algebra $\hat{g}_\Lambda$.

This is not really a very different way of doing things. We could deduce the algebra satisfied by E^r_n, as defined by eq.(4.28), and α^μ_m as follows. Since the $k.\alpha_n$ commute with all the operators involved in the definitions of E^r_n, A^μ_n by eqs.(4.14) and (4.15), we will not affect the algebra (4.16) if we set

$$k.\alpha_n \to 0\ ,\quad n \neq 0\ ,\quad k.p \to 1 \tag{4.30}$$

Doing this yields the definition (4.28) of E^r_n and sends

$$A^\mu_n \to \alpha^\mu_n\ . \tag{4.31}$$

5. Points of Unit Length

So far we have concentrated on the points Λ_2, of squared length 2 on the lattice Λ. We can also consider other points of the lattice, in particular those of unit length

$$\Lambda_1 = \{e \in \Lambda : e^2 = 1\} . \tag{5.1}$$

Vertex operators associated with such points are naturally fermions rather than bosons. Consider the vertex operator of eq.(2.33),

$$A_\gamma = \frac{1}{2\pi i} \oint \frac{dz}{z} U(\gamma, z) \tag{5.2}$$

$\gamma = e, f \in \Lambda_1$. Suppose that Λ is Euclidean. We shall need that Λ_1 generates an integral sublattice of Λ , i.e. $e.f \in \mathbb{Z}$ for all $e, f \in \Lambda_1$, in order to avoid fractional powers of $(z-\zeta)$, which would prevent us easily evaluating integrals involving

$$U(e,z)U(f,\zeta) = \exp\{ie.Q_<(z) + if.Q_<(\zeta)\}$$

$$. (z-\zeta)^{e.f} z^{\frac{1}{2}} \zeta^{\frac{1}{2}} e^{i(e+f).q} z^{ep} \zeta^{f.p}$$

$$. \exp\{ie.Q_>(z) + if.Q_>(\zeta)\} \quad \text{for} \quad |\zeta| < |z|$$

$$\equiv U_{e,f}(z,\zeta) \tag{5.3}$$

since $|e.f| \leq 1$ this means that either $f = \pm e$, or $f.e = 0$. Assuming that we can make $U_{e,f}(z,\zeta)$ single valued as a function of z and ζ , from (3.13) we find

$$A_e^2 = 0 \tag{5.4a}$$

$$[A_e, A_f] = 0 , \qquad e.f = 0 , \tag{5.4b}$$

$$\{A_e, A_{-e}\} = 1 . \tag{5.4c}$$

Before we go further with this discussion we should address the question of single-valuedness. The question turns on whether

$$\exp\{ie.Q_o(z)\}|\lambda\rangle = z^{\frac{1}{2}+e.\lambda}|e+\lambda\rangle \tag{5.5}$$

is single valued, and this requires that $\lambda.e \in \mathbb{Z} + \frac{1}{2}$. To see how this can be arranged, let us make matters more concrete. Since we need $e.f = 0. \pm 1$ for each $e, f \in \Lambda_1$, we are dealing with orthogonal unit vectors. This means that Λ_1 generates a lattice $\mathbb{Z}^d$, which as we saw in section 3 is the weight lattice of SO(2d). The full weight lattice of the algebra so(2d) consists of four cosets with respect to its root lattice Λ_R, given by

$$\Lambda_R \ , \quad \Lambda_v \ , \quad \Lambda_s \ , \quad \Lambda_{\bar{s}} \tag{5.6}$$

where

$$\Lambda_v = \Lambda_R + \lambda_r \ , \quad \lambda_v = (1,0,0,\ldots,0) \ , \tag{5.7a}$$

$$\Lambda_s = \Lambda_R + \lambda_s \ , \quad \lambda_s = (\tfrac{1}{2},\tfrac{1}{2},\tfrac{1}{2},\ldots,\tfrac{1}{2}) \ , \tag{5.7b}$$

$$\Lambda_{\bar{s}} = \Lambda_R + \lambda_{\bar{s}} \ , \quad \lambda_{\bar{s}} = (-\tfrac{1}{2},\tfrac{1}{2},\tfrac{1}{2},\ldots,\tfrac{1}{2}) \ . \tag{5.7c}$$

Taking $\Lambda = \Lambda_R \cup \Lambda_v$ gives $G_\Lambda = SO(d)$, whilst, as we remarked before, taking $\Lambda = \Lambda_R \cup \Lambda_s$ or $\Lambda_R \cup \Lambda_{\bar{s}}$ gives some other group of the form spin(2d)/$\mathbb{Z}_2$.

Let us fix our consideration on

$$\Lambda = \mathbb{Z}^n = \Lambda_R \cup \Lambda_v \tag{5.8}$$

which always has points of unit length. If $e \in \Lambda_1$ and

$$\lambda \in \Lambda_s \cup \Lambda_{\bar{s}} \tag{5.9}$$

$e.\lambda \in \mathbb{Z} + \frac{1}{2}$ and (5.5) is single-valued. Thus eqs.(5.4) hold provided that we restrict all the momenta to lie in the spinor lattices (5.9).

Now consider the relationship between A_e, $e \in \Lambda_1$ and the operators A_r , $r \in \Lambda_2$, which, apart from the factor c_r , represent SO(2d). These operators preserve the condition that the momentum of a state is of the form (5.9). A calculation, which is by now straightforward, shows that

$$A_e A_r - (-1)^{e.r} A_r A_e = 0 \ , \qquad e.r \geq 0 \ , \tag{5.10a}$$

$$A_e A_r + A_r A_e = A_{e+r} \ , \qquad e.r = -1 \ . \tag{5.10b}$$

This exhausts all the possibilities in Euclidean space. It remains to remove the awkward signs again. Looking at the fermion operators alone, in eqs.(5.4), we see that what this needs is equivalent to a Klein transformation. We consider the particular case of $\mathbb{Z}^d$, and introduce the usual γ-matrix algebra

$$\{\gamma^i,\gamma^j\} = 2\delta^{ij} \tag{5.11}$$

We associate[3] to $u = (n_1,n_2,\ldots,n_d)$,

$$\gamma^n = \varepsilon_u \gamma_1^{n_1} \gamma_2^{n_2} \cdots \gamma_d^{n_d} , \tag{5.12a}$$

and as in eq.(3.27a) we choose the signs $\varepsilon_u = \pm 1$ so that

$$(\gamma^u)^{-1} = \gamma^{-u} . \tag{5.12b}$$

Then, again defining $\varepsilon(u,v)$ for $u,v \in \Lambda$ by

$$\gamma^u \gamma^v = \varepsilon(u,v) \gamma^{u+v} , \tag{5.13}$$

we now have

$$\gamma^u \gamma^v = (-1)^M \gamma^v \gamma^u , \tag{5.14}$$

where

$$\begin{aligned} M &= \sum_{i \neq j} u_i v_j \\ &= (\textstyle\sum u_i)(\sum v_j) - u.v \\ &= (\textstyle\sum u_i^2)(\sum v_j^2) - u.v \qquad \text{(mod 2)} \end{aligned}$$

$$\varepsilon(u,v) = (-1)^{u.v + u^2 v^2} \varepsilon(v,u) \tag{5.15}$$

which is just the property we need.

We define

$$\tilde{c}_n |w\rangle = \varepsilon(u, w-\lambda_o) |u+w\rangle , \tag{5.16}$$

where $\lambda_o = \lambda_s$ or $\lambda_{\bar{s}}$ depending on whether $w \in \Lambda_R + \lambda_s$ or $\Lambda_R + \lambda_{\bar{s}}$, and defining $c_u = e^{-q.u} \tilde{c}_u$, we set

$$E^r = A_r c_r \tag{5.17}$$

$$B^e = A_e c_e \tag{5.18}$$

In this way, we obtain a nicely graded algebra:

$$\{B^e, B^f\} = \delta_{e,-f} \tag{5.19a}$$

$$[E^r, B^e] = 0 \qquad r.e \geq 0 \tag{5.19b}$$

$$= \varepsilon(r,e) B^{r+e} \qquad r.e = -1 \tag{5.19c}$$

$$[p^\mu, B^e] = e^\mu B^e \tag{5.19d}$$

and the other commutation relations of eqs.(3.23). This is a Lie superalgebra.

The algebra of eq.(5.19a) is just a Clifford algebra for SO(2d). If e_i, $1 \leq i \leq d$, is a basis for $\mathbb{Z}^d$, setting

$$\gamma^i = B^{e_i} + B^{-e_i} \tag{5.20a}$$

$$\gamma^{d+i} = i(B^{e_i} - B^{-e_i}) \tag{5.20b}$$

for $1 \leq i \leq d$, we obtain

$$\{\gamma^i, \gamma^j\} = 2\delta^{ij} \qquad 1 \leq i,\ j \leq 2d \tag{5.21}$$

and

$$\gamma^{i\dagger} = \gamma^i \quad \text{as} \quad B^{e\dagger} = B^{-e} . \tag{5.22}$$

Now we have obtained the Dirac algebra for $G_{\underset{\sim}{\Lambda}} = SO(2d)$, not the one we used as a method of constructing the cocycle c_n, which is associated with $SO(d)$ and the maximal torus of $SO(2d)$.

Clearly we can now extend this construction to non-Euclidean lattices to obtain infinite dimensional Lie superalgebras. Again, of particular interest is the affine case: we adjoint to Λ a single orthogonal light-like vector k. Applied as in section 4 to $\Lambda = \mathbb{Z}^d$, this yields the affine Kac-Moody algebra $\hat{D}_d$, and we can make the substitution (4.30) to give the Frenkel-Kac construction. We repeat the same process using points of length squared one as well to obtain

$$B^{n+ek} \to d_n^e = \frac{1}{2\pi i}\oint \frac{dz}{z} z^n U(e,z)\, c_e , \quad n \in \mathbb{Z}, \tag{5.23}$$

(where we have used the notation d_n^e because d_n is the transitional notation for the Ramond fermion oscillators in the Neveu-Schwarz-Ramond model[14], and we shall see that that is exactly what these operators are). They satisfy

$$\{d_m^e, d_n^f\} = \delta_{m,-n}\,\delta_{e,-f} , \quad m,n \in \mathbb{Z} \tag{5.24a}$$

$$d_m^{e\dagger} = d_{-m}^e , \tag{5.25b}$$

$$d_m^e|w\rangle = 0 \quad \text{if} \quad m+e.w \geq \frac{1}{2} \tag{5.25c}$$

The correspondence with the Ramond oscillators is made clearer if we change bases as in eqs.(5.20)

$$d_n^i = \frac{1}{\sqrt{2}}(d_n^{e_i} + d_n^{-e_i}) ; \quad d_n^{d+i} = \frac{i}{\sqrt{2}}(d_n^{e_i} - d_n^{-e_i}) \tag{5.26}$$

so that

$$\{d_m^i, d_n^j\} = \delta^{ij}\,\delta_{m,-n} , \; 1 \leq i, j \leq 2d , \tag{5.27a}$$

$$d_m^{i\dagger} = d_{-m}^i . \tag{5.27b}$$

Further it is not difficult to see that the 2^d states $|w\rangle$ such that coordinates of w are each $\pm\frac{1}{2}$ form a subspace invariant under d_o^e, or, equivalently, d_o^i. These are the Ramond vacuum states which transform like a Dirac spinor under SO(2d).

Since we have constructed the Ramond oscillators, it seems sensible to see if we can construct the half-integrally moded Neveu-Schwarz oscillators. Consider again the single-valuedness of the integrand in eq.(5.23), i.e. of

$$\frac{1}{2\pi i}\oint \frac{dz}{z} \exp\{ie.Q_<(z)\}\; e^{ie.q}\; z^{\frac{1}{2}+n+e.p}\; \exp\{ie.Q_>(z)\}. \tag{5.28}$$

Another way to achieve this is to have $n+\frac{1}{2} \in \mathbb{Z}$ and $e.p \in \mathbb{Z}$; that is we consider the operator acting on states with momenta in Λ_R or Λ_v, and then it has a natural expansion in terms of half-odd-integrally moded oscillators. So we define for $e \in \Lambda_1$,

$$b_t^e = \frac{1}{2\pi i}\oint \frac{dz}{z}\, z^t U(e,z) c_e \;, \quad t \in \mathbb{Z}+\tfrac{1}{2} . \tag{5.29}$$

Then,

$$\{b_t^e, b_n^f\} = \delta_{t,-n}\,\delta_{e,-f} \;, \quad t,u \in \mathbb{Z}+\tfrac{1}{2} , \tag{5.30a}$$

$$b_t^{e\dagger} = b_{-t}^{-e} , \tag{5.30b}$$

$$b_t^e|w\rangle = 0 \quad \text{if} \quad t+e.w \geq \tfrac{1}{2} . \tag{5.30c}$$

So, in particular,

$$b_t^e|0\rangle = 0 \quad \text{if} \quad t > 0 . \tag{5.30d}$$

Again we can define components b_t^i, $1 \leq i \leq 2d$, with the usual SO(2d) vector indices, exactly as in eq.(5.26).

So we see that we can consider $U(e,z)$ as acting on a state with any momentum which is a weight of so(2d), but when this weight lies in $\mathbb{Z}^n$, the operator has an expansion in half-odd-integral modes (Neveu-Schwarz oscillations) whilst, when it lies on either of the spinor lattices, it has an expansion in integral modes (Ramond oscillators).

We have obtained anticommuting fields d_n^e, b_t^e out of commuting fields α_n^μ. Can this process be reversed? To answer this question we must consider quantities bilinear in the fermion fields and these products must be normal ordered with respect to the fermion oscillators. This is simplest to do in the Neveu-Schwarz case. We define

$$\times b_t^e b_u^f \times = b_t^e b_u^f \quad \text{if} \quad n > 0 \tag{5.31a}$$

$$= -b_u^f b_t^e \quad \text{if} \quad n < 0 \tag{5.31b}$$

in the usual way. If we write

$$U(e,z)c_e \equiv \sum b_t^e z^{-t} = b^e(z) \tag{5.32}$$

it follows that

$$b^e(z)\, b^f(\zeta) = {}^\times_\times b^e(z) b^f(\zeta) {}^\times_\times + \frac{z^{\frac{1}{2}}\zeta^{\frac{1}{2}}}{z-\zeta}\,\delta_{e,-f} \quad \text{if} \quad |\zeta| < |z| \,. \tag{5.33}$$

Thus

$$b^e(z)b^e(z) = {}^\times_\times b^e(z)b^e(z){}^\times_\times = 0 \,, \tag{5.34a}$$

$${}^\times_\times b^e(z)b^f(z){}^\times_\times = b^e(z)b^f(z)$$

$$= \varepsilon(e,f)U(e+f,z)\, c_{e+f} \equiv \varepsilon(e,f)\, E^{e+f}(z) \quad \text{if} \quad e.f = 0 \tag{5.34b}$$

and

$${}^\times_\times b^e(z)b^{-e}(\zeta){}^\times_\times = b^e(z)b^{-e}(\zeta) - z^{\frac{1}{2}}\zeta^{\frac{1}{2}}/(z-\zeta)$$

$$= \frac{z^{\frac{1}{2}}\zeta^{\frac{1}{2}}}{z-\zeta}\,[\exp\{ie(Q_<(z)-Q_<(\zeta))\}z^{e.p}\zeta^{-e.p}\exp\{ie(Q_>(z)-Q_>(\zeta))\} - 1]$$

$$\to iz\frac{d}{dz}(e.Q) = e.P(z) \quad \text{letting} \quad z \to \zeta \,. \tag{5.34c}$$

So we can express all the elements of $\hat{g}_\Lambda$ and, in particular, all the basic bosonic oscillators α_m^μ as bilinear quantities in the fermion oscillators. This is the boson-fermion equivalence which dates back to the work of Skyrme[22,23]. Fermion fields are expressed as exponential quantities in boson fields and boson fields as bilinear quantities in the fermion fields.

Similar expressions are obtained in the case of the Ramond fields. In this case normal ordering is a little more complicated because of the zero mode; we define

$$ {}_{\times}^{\times} d_m^e \, d_n^f {}_{\times}^{\times} = d_m^e \, d_n^f \qquad \text{if} \quad n > 0 \; , \tag{5.35a} $$

$$ = \frac{1}{2}[d_m^e, d_o^f] \; , \qquad \text{if} \quad n = 0 \; , \tag{5.35b} $$

$$ = - d_n^f \, d_m^e \; , \qquad \text{if} \quad n < 0 \; . \tag{5.35c} $$

It is now true that, writing

$$ U(e,z) \; c_e \equiv \sum d_n^e \, z^{-n} = d^e(z) \; , \tag{5.36} $$

$$ d^e(z) \; d^f(\zeta) = {}_{\times}^{\times} d^e(z) \; d^f(\zeta) {}_{\times}^{\times} + \frac{1}{2} \frac{z+\zeta}{z-\zeta} \, \delta_{e,-f} \; . \tag{5.37} $$

Again it follows that, as before,

$$ {}_{\times}^{\times} d^e(z) \; d^e(z) {}_{\times}^{\times} = 0 \tag{5.38a} $$

$$ {}_{\times}^{\times} d^e(z) \; d^f(z) {}_{\times}^{\times} = \varepsilon(e,f) \; E^{e+f}(z) \; , \qquad \text{if} \quad e.f = 0 \; , \tag{5.38b} $$

$$ {}_{\times}^{\times} d^e(z) \; d^{-e}(z) {}_{\times}^{\times} = e.P(z) \; . \tag{5.38c} $$

Thus we have seen the vertex operators E_n^r associated with points $r \in \mathbb{Z}^d$ with $r^2 = 2$ define, together with α_n^μ , $1 \leq \mu \leq d$, a level 1 representation of the Kac-Moody algebra associated with $so(2d)$. Since

$$ E_n^r|\lambda\rangle = 0 \quad \text{if} \quad n+r.\lambda \geq 0 \; , \tag{5.39a} $$

and

$$ \alpha_n^\mu|\lambda\rangle = 0 \quad \text{if} \quad n > 0 \; , \tag{5.39b} $$

we see that $|\lambda\rangle$ will be a vacuum state if $\lambda = 0$, λ_v, λ_s or $\lambda_{\bar{s}}$ or one of the states corresponding to a value of λ on the weight lattice of $so(2d)$ which is a rotation of one of these points. In this way, from 0, λ_v, λ_s and $\lambda_{\bar{s}}$, one obtains the weights of the scalar, vector, and two distinct spinor representations of $so(2d)$, respectively.

The vacuum states, as defined by eq.(4.21) always fall into representations of the algebras $\{T_o^a\} \cong g_\Lambda$, whose affine version $\hat{g}_\Lambda$ we are discussing. For an irreducible representation of $\hat{g}_\Lambda$ the vacuum representation will be irreducible and the representation of $\hat{g}_\Lambda$ is characterised by the level K and this vacuum representation. There are only a finite number of possibilities for the vacuum representation for a given value of K . For $K = 1$ and $g_\Lambda = so(2d)$ the four possibilities we have obtained are all that there are.

We have seen that elements E_n^r, α_n^μ of the Kac-Moody algebra can be represented by bilinears in fermions, Neveu-Schwarz fermions on states with momenta in Λ_R or Λ_v and Ramond fermions on states with momenta in Λ_s or $\Lambda_{\bar{s}}$. We denote these spaces by $\mathfrak{H}^b$ and $\mathfrak{H}^d$, respectively. Thus the representation of this Kac-Moody algebra by bilinears in $b^i(z)$ splits into the $K=1$ scalar and vector representations, which possess highest weight states $|0\rangle$ and $b^i_{-\frac{1}{2}}|0\rangle$ respectively, whilst the representation in terms of bilinears in $d^i(z)$ splits up into the two $K=1$ spinor representations with highest weight states $|\lambda\rangle$ where each coordinate of λ is $\pm\frac{1}{2}$. These form an irreducible representation for the Dirac algebra d^i_o.

Let us consider the relationship of these operators to the Virasoro operators given by (2.18) or, equivalently,

$$\sum_{n=-\infty}^{\infty} L_n z^{-n} = \frac{1}{2} :P(z)^2:$$

which satisfy the algebra (2.20). It is straightforward to use eq.(2.31) to show

$$[L_n, E^r_m] = -m\, E^r_{m+n} \tag{5.41a}$$

$$[L_n, b^e_t] = -t\, b^e_{n+t} \tag{5.41b}$$

$$[L_n, d^e_m] = -m\, d^e_{m+n} \tag{5.41c}$$

Now there are other operators obeying (5.41b) and (5.41c) namely, those given by

$$\sum_{n=-\infty}^{\infty} L^b_n z^{-n} = \frac{1}{2} z \,{}^\times_\times b'(z) b(z) {}^\times_\times , \tag{5.42}$$

and

$$\sum_{n=-\infty}^{\infty} L^d_n z^{-n} = \frac{1}{2} z \,{}^\times_\times d'(z) d(z) {}^\times_\times + \frac{d}{8} , \tag{5.43}$$

respectively; that is

$$[L^b_n, b^e_t] = -t\, b^e_{n+t} , \tag{5.44a}$$

$$[L^d_n, d^e_m] = -m\, d^e_{n+m} . \tag{5.44b}$$

Since the space $\mathfrak{H}^b$ is generated from $|0\rangle$ by the operators b^e_t, $t \in \mathbb{Z}+\frac{1}{2}$, $e \in \Lambda_1$, and since L_n and L^b_n clearly agree on $|0\rangle$ for $n \geq 0$ it follows that

$$L_n = L^b_n \quad \text{on} \quad \mathfrak{H}^b \tag{5.45a}$$

and similarly

$$L_n = L_n^d \quad \text{on} \quad \mathfrak{H}^d . \tag{5.45b}$$

Incidentally we can exploit this fermion equivalence to calculate the trace of L_0 over each of $\mathfrak{H}^b, \mathfrak{H}^d$ in two different ways. In the one dimension case this yields the identities

$$\sum_{n\in\mathbb{Z}} z^{\frac{1}{2}n^2} \prod_{m=1}^{\infty} (1-z^m)^{-1} = \prod_{t=\frac{1}{2}}^{\infty} (1+z^t)^2 \tag{5.46a}$$

and

$$\sum_{t\in\mathbb{Z}+\frac{1}{2}} z^{\frac{1}{2}t^2} \prod_{m=1}^{\infty} (1-z^m)^{-1} = 2z^{1/8} \prod_{n=1}^{\infty} (1+z^n)^2 \tag{5.46b}$$

which are both equivalent to the Jacobi triple product formula. In higher dimensions, powers of this identity are obtained.

6. Fermionic Constructions of $\hat{E}_8$ and other Algebras

In the last section we saw how the level 1 representations of the affine Kac-Moody algebra associated with so(2d) could be constructed first using vertex operators and then, equivalently by virtue of the boson-fermion equivalence, using bilinears in fermions. This fermionic construction generalises in a way familiar from current algebra. Suppose we consider a representation of a Lie algebra g, specified as in eq.(4.17) in terms of N dimensional real antisymmetric matrices M^a_{ij}, $1 \leq i,j \leq N$, so that

$$[M^a, M^b] = f_{abc} M^c , \tag{6.1}$$

normalised so that

$$\mathrm{tr}(M^a M^b) = -\kappa\, \delta^{ab} . \tag{6.2}$$

Let us assume for convenience that g is simple. Then it has roots of at most two lengths. Let ψ be a root with the longer of these two lengths. Then κ/ψ^2 is called the <u>Dynkin index</u> of the representation. It is independent of the way the basis $\{T^a\}$ of g has been normalised and it has to be an integer. In the cases produced by the construction of section 3 all roots have squared length 2 and so for such algebras g

$$\kappa \in 2\mathbb{Z} . \tag{6.3}$$

We can use this representation to construct a representation of the associated Kac-Moody algebra $\hat{g}$ if we introduce an N component fermion field $H^i(z)$ which is either periodic (Ramond case) or anti-periodic (Neveu-Schwarz case) on the unit circle,

$$H^i(z) = \sum_{n \in \mathbb{Z}} d^i_n z^{-n} \quad [R] , \tag{6.4a}$$

$$H^i(z) = \sum_{t \in \mathbb{Z}+\frac{1}{2}} b^i_t z^{-t} \quad [NS] , \tag{6.4b}$$

where

$$\{d^i_m, d^j_n\} = \delta^{ij} \delta_{m,-n} , \quad m,n \in \mathbb{Z} , \tag{6.5a}$$

or

$$\{ b^i_t, b^j_n\} = \delta^{ij} \delta_{t,-n} , \quad t,u \in \mathbb{Z}+\tfrac{1}{2} , \tag{6.5b}$$

and the space is built up from vacuum states ψ_o satisfying

$$d^i_m \psi_o = 0 , \ m > 0 , \quad \text{or} \quad b^i_t \psi_o = 0 , \quad t > 0 . \tag{6.6}$$

In the Neveu-Schwarz case we can take the vacuum ψ_o to be non-degenerate, $\psi_o \equiv |0\rangle$, whilst in the Ramond case the vacuum states form a representation of the Clifford algebra d^i_o and so can be taken to be the unique $2^{N/2}$ or $2^{(N-1)/2}$ dimensional irreducible representation of this algebra (depending on whether N is even or odd). We define

$$T^a(z) \equiv \sum_{n \in \mathbb{Z}} T^a_n z^{-n} = \frac{i}{2} H^i(z) M^a_{ij} H^j(z) \tag{6.7}$$

(where it is not necessary to normal order explicitly because of the antisymmetry of M^a). Then calculation shows that

$$[T^a_m, T^b_n] = i f_{abc} T^c_{m+n} + \frac{\kappa}{2} m \, \delta^{ab} \delta_{m,-n} . \tag{6.8}$$

(See e.g. ref.[24].) This is a representation of $\hat{g}_\Lambda$ of level κ/ψ^2, which in the simply-laced cases we have been considering is $\kappa/2$, an integer as it should be.

From such representations, and their products we can form all the representations of $\hat{g}$ if g is one of the classical algebras so(n), su(n) or sp(n). [For more details see ref.20]. In particular the level 1 representations of so(n) are obtained by taking M to be the n dimensional vector representation as in the last section. The corresponding vacuum representations of so(n) are the scalar, the vector and the spinor. The level 1 representations of su(n) are obtained from the inclusion $SU(n) \subset SO(2n)$ and of sp(n) from $Sp(n) \subset SU(2n) \subset SO(4n)$. This construction gives a fermionic description of the level 1 representations of the Kac-Moody algebra corresponding to so(32). In the heterotic string theory, such representations occur as a spectrum generating algebra and the corresponding weights lie on the even self-dual lattice $\Lambda_R \cup \Lambda_s$, the weight lattice of the $\mathrm{Spin}(32)/\mathbb{Z}_2$ group not isomorphic to SO(32). To get this global structure from the fermionic description we see from the last section that we must take the even b-number sector of the Neveu-Schwarz space together with the even chirality part of the Ramond space.

If we consider the other possible group for the heterotic string, $E_8 \times E_8$, we are not able to produce such an easy fermionic description of the level 1 representation. Let us consider E_8. This time there is just one level one representation, and in this the vacuum representation is just a scalar, so that all states are built up from a vacuum $|0\rangle$, annihilated by T^a_n, $n \geq 0$. The smallest representation of E_8 is the adjoint with dimension 248 and Dynkin index 30 so that the fermionic construction described in this section gives a level 30 representation of $\hat{E}_8$. However the bosonic construction of section 4 applied to the root lattice of E_8 does give a level 1 representation. To do this we use all the 240 points of squared length 2 on this even self-dual lattice, which is $\Lambda_R \cup \Lambda_s$, where Λ_R, Λ_s are defined by (5.7) with $d = 8$.

Clearly part of this can be described by fermions because we can give a fermionic description of the $K=1$ representation corresponding to $SO(16) \subset E_8$. To relate this to the bosonic construction we use the points of unit length on the weight lattice of so(16). The key to enlarging this to E_8 is to consider how this description relates to an $SO(8) \times SO(8) \subset SO(16)$ and then to exploit the triality property of the SO(8) subgroups[9].

If we use $\Lambda_R^{(4)}$, $\Lambda_v^{(4)}$, $\Lambda_s^{(4)}$, $\Lambda_{\bar{s}}^{(4)}$ for the lattices associated with the first SO(8) factor and $\Lambda_R^{(4)'}$, $\Lambda_v^{(4)'}$, $\Lambda_s^{(4)'}$, $\Lambda_{\bar{s}}^{(4)'}$ for the second SO(8) factor, we have the root lattice of SO(16)

$$\Lambda_R^{(8)} = [\Lambda_R^{(4)} \oplus \Lambda_R^{(4)'}] \cup [\Lambda_v^{(4)} \oplus \Lambda_v^{(4)'}] \tag{6.10}$$

and the vector weights of SO(16) lie on

$$\Lambda_v^{(8)} = [\Lambda_R^{(4)} \oplus \Lambda_v^{(4)'}] \cup [\Lambda_v^{(4)} \oplus \Lambda_R^{(4)'}] . \tag{6.11}$$

The sixteen points of unit length on $\Lambda_v^{(8)}$ consist of eight points $\{\pm e_i\}$ of unit length or $\Lambda_v^{(4)}$ and eight points $\{\pm e_i'\}$ of unit length on $\Lambda_v^{(4)'}$. The fermionic representation of the generators of SO(16) then takes the form

$$b^{e_i}(z) b^{e_j}(z), \quad b^{e_i'}(z) b^{e_j'}(z), \quad b^{e_i}(z) b^{e_j'}(z) \tag{6.12a}$$

when acting on states with momentum in $\Lambda_R^{(8)}$ and

$$d^{e_i}(z) d^{e_j}(z), \quad d^{e_i'}(z) d^{e_j'}(z), \quad d^{e_i}(z) d^{e_j'}(z) \tag{6.12b}$$

when acting on states with momenta in

$$\Lambda_s^{(8)} = [\Lambda_s^{(4)} \oplus \Lambda_s^{(4)'}] \cup [\Lambda_{\bar{s}}^{(4)} \oplus \Lambda_{\bar{s}}^{(4)'}] \qquad (6.13)$$

In this way we get 28 generators which transform as the (adjoint, 1) representation of $SO(8) \times SO(8)$, 64 generators transforming as the (v,v) representation and 28 generators transforming as the (1,adjoint).

The remaining generators of E_8 correspond to points like

$$(\tfrac{1}{2},\tfrac{1}{2},\tfrac{1}{2},\tfrac{1}{2},\tfrac{1}{2},\tfrac{1}{2},\tfrac{1}{2},\tfrac{1}{2}) \quad \text{or} \quad (\tfrac{-1}{2},\tfrac{1}{2},\tfrac{1}{2},\tfrac{1}{2},\tfrac{-1}{2},\tfrac{1}{2},\tfrac{1}{2},\tfrac{1}{2}) \qquad (6.14)$$

and so transform like (s,s) or $(\bar{s},\bar{s})$, respectively, with respect to $SO(8) \times SO(8)$. We can thus form these using fermions associated with the unit length points $\{\pm f_i\}$ on $\Lambda_s^{(4)}$, $\{\pm f_i'\}$ on $\Lambda_s^{(4)'}$, $\{\pm g_i\}$ on $\Lambda_{\bar{s}}^{(4)}$ and $\{\pm g_i'\}$ on $\Lambda_{\bar{s}}^{(4)'}$. From these we can form the extra generators

$$d^{f_i}(z)d^{f_j'}(z), \; d^{g_i}(z)d^{g_j'}(z) \qquad (6.15a)$$

acting on $\Lambda_R^{(8)}$ and

$$b^{f_i}(z)b^{f_j'}(z), \; b^{g_i}(z)b^{g_j'}(z) \qquad (6.15b)$$

acting on $\Lambda_s^{(8)}$. These 128 generators added to the 120 we already have form the whole of the E_8 Kac-Moody algebra. We know that they will close on this algebra because the fermion bilinears equate to the bosonic vertex operators which give the Kac-Moody algebra for E_8 by the general construction of section 4. The 56 fields corresponding to the generators of $SO(8) \times SO(8)$ could equally well be formed from $b^{f_i}(z)b^{f_j}(z)$, etc., as $b^{e_i}(z)b^{e_j}(z)$, so we have three ways of constructing them from bilinears in fermions related by triality.

It is convenient to regard $b^{e_i}(z)$, $d^{e_i}(z)$ as part of the same fermion field $B^{e_i}(z)$, equal to $b^{e_i}(z)$ when acting on states with momentum in $\Lambda_R^{(4)} \cup \Lambda_V^{(4)}$ and $d^{e_i}(z)$ when acting on states with momentum in $\Lambda_s^{(4)} \cup \Lambda_{\bar{s}}^{(4)}$. The sixteen fermion fields $B^{e_i}(z)$, $B^{e_j'}(z)$ are independent and can be used to generate the whole space from a finite number of vacuum states. The fields $B^{f_i}(z)$, $B^{f_j'}(z)$ are independent of one another but not of the $B^{e_i}(z)$, $B^{e_j'}(z)$. Since the $B^{f_i}(z)$, $B^{f_j'}(z)$ are exponentials of the bosonic oscillators which are themselves quadratic in the $B^{e_i}(z)$, $B^{e_j'}(z)$, the $B^{f_i}(z)$, $B^{f_j'}(z)$ are transcendental functions of the $B^{e_i}(z)$, $B^{e_j'}(z)$, being exponentials of quadratic forms in them. This is reminiscent of the fermion emission vertex[25], the moments of which form the basic oscillators in the superstring formalism of Green and Schwarz[15] and indeed that vertex can be understood precisely in those terms. Evidently $B^{g_i}(z)$ and $B^{g_j'}(z)$ are dependent on $B^{e_i}(z)$ and $B^{e_j'}(z)$ in an exactly similar way.

Thus we have obtained a construction of the level one representation of E_8 Kac-Moody algebra starting from 16 independent fermions, which will give a construction of the level one representation of the $E_8 \times E_8$ Kac-Moody algebra using 32 independent fermions. The same number of independent fermions is required for $spin(32)/\mathbb{Z}_2$ but the difference between the cases is that the $spin(32)/\mathbb{Z}_2$ case all the generators are bilinear in these free fermions whilst in the $E_8 \times E_8$ case some of them are transcendental.

This construction we have given for E_8 generalises[10] to other situations in which we have a triality property analogous to that for SO(8) and leads to the "magic square" of Frendenthal. Suppose we have a lattice Λ with the following triality properties:

(a) $\Lambda^{(o)} \subset \Lambda$ is an even integral lattice;

(b) $\Lambda/\Lambda^{(o)} = \{\Lambda^{(o)}, \Lambda^{(1)}, \Lambda^{(2)}, \Lambda^{(3)}\} \cong \mathbb{Z}_2 \times \mathbb{Z}_2$ as an abelian group;

(c) $\Lambda^{(o)} \cup \Lambda^{(i)}$ is an integral lattice, $1 \leq i \leq 3$;

(d) if $\lambda^i \in \Lambda^{(i)}$, $1 \leq i \leq 3$, then $\lambda_i . \lambda_j \in \mathbb{Z} + \frac{1}{2}$ if $i \neq j$;

(e) $\Lambda^{(i)} \cong \Lambda^{(j)}$, $1 \leq i, j \leq 3$.

We can give three examples of such a system, the first being the one familiar from SO(8):

(i)
$$\begin{aligned}
\Lambda^{(o)} &= (n_1, n_2, n_3, n_4), \textstyle\sum n_j \in 2\mathbb{Z}, \ n_j \in \mathbb{Z} \\
\Lambda^{(1)} &= (n_1, n_2, n_3, n_4), \textstyle\sum n_j \in 2\mathbb{Z}+1, \ n_j \in \mathbb{Z} \\
\Lambda^{(2)} &= (n_1, n_2, n_3, n_4), \textstyle\sum n_j \in 2\mathbb{Z}, \ n_j \in \mathbb{Z}+\frac{1}{2} \\
\Lambda^{(3)} &= (n_1, n_2, n_3, n_4), \textstyle\sum n_j \in 2\mathbb{Z}+1, \ n_j \in \mathbb{Z}+\frac{1}{2}
\end{aligned}$$

(ii)
$$\begin{aligned}
\Lambda^{(o)} &= \sqrt{2}(n_1, n_2, n_3), \quad n_j \in \mathbb{Z} \ ; \\
\Lambda^{(i)} &= \sqrt{2}(n_1, n_2, n_3), \quad n_i \in \mathbb{Z}, \ n_j \in \mathbb{Z}+\frac{1}{2}, \ j \neq i \ .
\end{aligned}$$

(iii)
$$\begin{aligned}
\Lambda^{(o)} &= (n_1, \sqrt{3}n_2), \quad n_i \in \mathbb{Z}, \ n_1+n_2 \in 2\mathbb{Z}, \\
\Lambda^{(1)} &= (n_1, \sqrt{3}n_2), \quad n_i \in \mathbb{Z}, \ n_1+n_2 \in 2\mathbb{Z}+1, \\
\Lambda^{(2)} &= (n_1, \sqrt{3}n_2), \quad n_i \in \mathbb{Z}+\frac{1}{2}, \ n_1+n_2 \in 2\mathbb{Z}, \\
\Lambda^{(3)} &= (n_1, \sqrt{3}n_2), \quad n_i \in \mathbb{Z}+\frac{1}{2}, \ n_1+n_2 \in 2\mathbb{Z}+1 \ .
\end{aligned}$$

In these three cases $g_{\Lambda^{(o)}}$ is $so(8)$, $su(2) \oplus su(2) \oplus su(2)$ and $u(1) \oplus u(1)$ respectively, but we can construct more interesting algebras if we append to the points of squared length 2 on $\Lambda^{(o)}$ those of squared length 1 on each of the $\Lambda^{(i)}$. These form the root system for a Lie algebra, L_n, which is not in general simply laced. To see that this follows from properties (a) to (d) we have to check the following, where α_a denote points of squared length 2 in $\Lambda^{(o)}$ and β_a points of squared length 1 in the $\Lambda^{(i)}$:

$$2\alpha_a.\alpha_b/\alpha_a^2 = \alpha_a.\alpha_b \in \mathbb{Z} \quad \text{by (a);}$$

$$2\alpha_a.\beta_b/\alpha_a^2 = \alpha_a.\beta_b \in \mathbb{Z} \quad \text{by (c);}$$

$$2\beta_a.\beta_b/\beta_a^2 = 2\beta_a.\beta_b \in \mathbb{Z} \quad \text{by (c) or (d) .}$$

The algebras we obtain in this way in the three cases are F_4, sp(3) and su(3), respectively.

Now we can write down a bosonic representation for $\hat{g}_{\Lambda}(o)$ using the Frenkel-Kac construction and using momenta taking their values in Λ. Since, in each of our examples, every $\alpha^{\mu} \in \Lambda_2^{(o)}$ can be expressed as $e_i + f_i$ where $e_i, f_i \in \Lambda_1^{(i)}$, the set of points of unit length on $\Lambda^{(i)}$ we can express this bosonic space as a fermionic space in three different ways, depending on the choice of i :

$$E^{e_i+f_i}(z) = B^{e_i}(z)B^{f_i}(z) \; , \tag{6.16}$$

$$e_i.P(z) = {}^{\times}_{\times}B^{e_i}(z)B^{-e_i}(z)^{\times}_{\times} \tag{6.17}$$

where $B^{e_i}(z)$ acts as a Neveu-Schwarz field $b^{e_i}(z)$ on states with momentum in $\Lambda^{(o)}$ or $\Lambda^{(i)}$ and as a Ramond field on states with momentum in $\Lambda^{(j)}$, $j \neq i$.

Now suppose we have two examples of such triality structures, not necessarily isomorphic, $\Lambda = \bigcup_{a=0}^{3} \Lambda^{(a)}$ and $\Lambda' = \bigcup_{a=0}^{3} \Lambda^{(a)\prime}$. If we consider the lattice

$$\Gamma = \sum_{a=0}^{3} \Lambda^{(a)} \oplus \Lambda^{(a)\prime} \tag{6.18}$$

then Γ_2 is the union of $\Lambda_2^{(o)}$, $\Lambda_2^{(o)\prime}$, and $\Lambda_1^{(i)} \times \Lambda_1^{(i)\prime}$, i = 1,2,3 and g_Γ for the various choices of h_Λ, $h_{\Lambda'}$ is shown in the table:

h_Λ \ $h_{\Lambda'}$	A_2	C_3	F_4
A_2	$A_2 \oplus A_2$	A_5	E_6
C_3	A_5	D_6	E_7
F_4	E_6	E_7	E_8

which is essentially the Frendenthal magic square.

The algebra $\hat{g}_\Lambda$ may be constructed in exactly the same way as we have constructed E_8, adding to the generators associated with $g_\Lambda(o) \times g_\Lambda(o)$,

$$B^{e_i}(z)\ B^{e'_i}(z) \quad i = 1,2,3 \tag{6.19}$$

$e_i \in \Lambda_1^{(i)}$, $e'_i \in \Lambda_1^{(i)'}$.

In these lectures we have not been able to cover all the aspects of the representation theory of Kac-Moody algebras which we should have liked to discuss. For further discussion from similar viewpoints see refs. 3,20,21,26,27. In particular we have not had time to develop the relationship between the representations of the Virasoro and Kac-Moody algebras [5,24,28-30] and the relevance of this to the fermion-boson equivalence between σ-models and free fermion theories[31].

Acknowledgements

I wish to thank the organizers of the ICTP Summer Workshop on High Energy Physics and Cosmology for the opportunity to present these lectures and Professor Abdus Salam for the hospitality of the ICTP where they were written. I am very grateful to Adrian Kent, Werner Nahm, Adam Schwimmer and particularly David Olive for many discussions about infinite dimensional algebras and their applications.

References

[1] I.B. Frenkel and V.G. Kac, Inv. Math. 62 (1980) 23.
[2] G. Segal, Comm. Math. Phys. 80 (1981) 301.
[3] P. Goddard and D. Olive, Algebras, Lattices and Strings, in Ref.7, p.51.
[4] I.B. Frenkel, American Mathematical Society Lectures in Applied Mathematics 21 (1985) 325.
[5] D.J. Gross, J.A. Harvey, E. Martinec and R. Rohm, Phys. Rev. Lett. 54 (1985) 502; Princeton preprint 1985.
[6] I.B. Frenkel, J. Lepowsky and A. Meurman, A Moonshine Module for the Monster, in Ref.7, p.231.
[7] J. Lepowsky, S. Mandelstam and I.M. Singer (eds.), Vertex Operators in Mathematics and Physics (Springer-Verlag, 1984).
[8] J. Scherk, Rev. Mod. Phys. 47 (1975) 123.
[9] P. Goddard, D. Olive and A. Schwimmer, Imperial College Preprint TP84-85/23, to be published in Physics Letters.
[10] P. Goddard, W. Nahm, D. Olive and A. Schwimmer, to appear.
[11] R.C. Brower, Phys. Rev. D6 (1972) 1655.
[12] P. Goddard and C.B. Thorn, Phys. Lett. 40B (1972) 235.
[13] C.B. Thorn, A Proof of the No-Ghost Theorem using the Kac Determinant, in Ref.7, p.411.
[14] A. Neveu and J.H. Schwarz, Nucl. Phys. B31 (1971) 86; P. Ramond, Phys. Rev. D3 (1971) 2415.
[15] J.H. Schwarz, Phys. Rep. 89 (1982) 223; M.B. Green, Surveys in High Energy Physics 3 (1983) 127.
[16] J.P. Serre, A Course in Arithmetic (Springer-Verlag, 1973).
[17] M.B. Green and J.H. Schwarz, Phys. Lett. 149B (1984) 117.
[18] E. Del Gindice, P. Di Vecchia and S. Fubini, Am. Phys. 70 (1972) 378.
[19] V.G. Kac, Infinite dimensional Lie algebras (Birkhauser, 1983).
[20] P. Goddard, Kac-Moody Algebras: Representations and Applications, DAMTP preprint 85/7.
[21] D. Olive, Kac-Moody Algebras: an introduction for physicists, Imperial College preprint TP84-85/14.
[22] T.H.R. Skyrme, Proc. Roy. Soc. A247 (1958) 260; A252 (1959) 236; A260 (1961) 127; A262 (1961) 237.
[23] R.F. Streater and I.F. Wilde, Nucl. Phys. B24 (1970) 561; S. Coleman, Phys. Rev. D11 (1975) 2088; S. Mandelstam, Phys. Rev. D11 (1975) 3026.
[24] P. Goddard and D. Olive, Nucl. Phys. B257 (1985) 226.
[25] C.B. Thorn, Phys. Rev. D4 (1971) 1112; E. Corrigan and D.I. Olive, Nuovo Cimento, 11A (1972) 749; E. Corrigan and P. Goddard, Nucl. Phys. B68 (1974) 189.
[26] A. Kent, Unitary Representations of the Virasoro Algebra, DAMTP preprint 85/8.
[27] P. Goddard, Critical Exponents, Infinite Dimensional Lie Algebras and Symmetric Spaces, DAMTP preprint 85/14.
[28] D. Friedan, Z. Qiu and S. Shenker, Conformal Invariance, Unitarity and Two Dimensional Critical Exponents in Ref.7; Phys. Rev. Lett. 52 (1984) 1575.

[29] P. Goddard, A. Kent and D. Olive, Phys. Lett. 152 (1985) 88; DAMTP preprint, to appear.

[30] P. Goddard, W. Nahm and D. Olive, Symmetric Spaces and Sugawara's Energy Momentum Tensor in Two Dimensions, Imperial College preprint TP/84-85/25, to appear in Physics Letters.

[31] E. Witten, Commun. Math. Phys. 92 (1984) 455.

SUPERSTRING PHENOMENOLOGY

P. Candelas

Center for Theoretical Physics
University of Texas, Austin TX 78712

Gary T. Horowitz

Department of Physics, University of California
Santa Barbara, California 93106

Andrew Strominger

The Institute for Advanced Study
Princeton, N.J. 08540

Edward Witten

Joseph Henry Laboratories, Princeton University
Princeton, N.J. 08544

ABSTRACT

We discuss some recent work on Kaluza-Klein compactifications of superstring theories and the resulting low energy phenomenology. The string theory approach to compactification is emphasized as opposed to the effective field theory approach. An attempt is made to keep the discussion non-technical.

A number of remarkable developments have recently taken place[1,2] in the theory of superstrings.[3] These new developments inspire optimism that superstrings provide both the first mathematically consistent quantum theory of gravity and the first truly unified theory of all forces and matter. The ultimate test of any physical theory, however, is not mathematical consistency, but confrontation with experiment. Historically, this confrontation has been crucial in the development as well as the testing of new theories. This has made the development of a quantum theory of gravity exceedingly difficult, since quantum gravitational effects are presumably only directly observable at 10^{19} GeV, and are thus beyond the reach of present day experiment.

A remarkable feature of superstring theories is that they should in principle lead to many low energy predictions. An attempt at exploring these predictions was made in ref. [4] and will be reviewed here. The basic reason predictions are possible is that it is very difficult to construct a consistent quantum theory of gravity, and those theories that have been found to survive the many consistency requirements have almost no freedom to adjust coupling constants,[5] alter the gauge group[1] or add or subtract particles.[*3] Thus, if the theory is correct, it must predict all the masses and couplings of the observed elementary particles!

Superstring theories appear to be consistent only if the dimension of spacetime is ten and the gauge group, if any, is $E_8 \times E_8$ or SO(32). On the one hand these constraints are desirable in that they are precisely the type of restrictions one would hope to arise in a truly unified theory. On the other hand, they are at first sight disappointing since we certainly do not observe ten dimensions or SO(32) or $E_8 \times E_8$ gauge symmetries. Clearly, if these theories are to have some connection with nature we must find a way to reduce both the dimension of spacetime and the gauge group. The idea we are going to discuss is based on the modern incarnation[6] of the old Kaluza-Klein program that some of the spacelike dimensions are compactified and small enough to not be directly observable.

One can discuss compactifications either in terms of a low energy effective field theory determined from the string, or directly in terms of the string theory. Here we discuss only the latter approach. The reader is referred to reference [4] for an effective field theory approach which leads to equivalent conclusions. We will try to review the main results in a non-technical manner and ignore certain generalizations. For further details see reference [4].

As discussed by Green and Schwarz,[3] the superstring theory is a field theory with an infinite number of fields. Among these many fields is a massless spin two field. In view of general arguments,[7]

---

*However, some loss of predictive power could occur if the full quantum theory turns out to have degenerate vacua.

this strongly suggests that superstrings must in a suitable approximation give rise to general relativity. In general relativistic field theories the geometry of spacetime is not fixed, and one can expand around curved backgrounds. One is thereby led to attempt an expansion of the superstring theory around backgrounds other than flat ten dimensional Minkowski space. Phenomenologically interesting backgrounds are of the form $M^4 \times K$, where K is some small six dimensional space and M^4 is four dimensional Minkowski space.

It is not at all obvious that there exists a consistent expansion around such backgrounds. Indeed, a considerable effort was required to demonstrate the consistency of the superstring theory expanded around flat M^{10}. On a curved background, we begin with a generalization of the superstring action:

$$S = -\frac{1}{4\pi\alpha'}\int d\sigma d\tau\, \sqrt{-q}\,(q^{\alpha\beta}g_{\mu\nu}\partial_\alpha X^\mu \partial_\beta X^\nu + \text{superpartners})$$
$$\alpha,\beta = 0,1$$
$$\mu,\nu = 0,1,\cdots,9 \qquad (1)$$

where we simply replace the flat metric $\eta_{\mu\nu}$ on M^{10} with the curved metric $g_{\mu\nu}(X)$ on $M^4 \times K$. $X^\mu(\sigma,\tau)$ is the location of the string world sheet in spacetime and $q_{\alpha\beta}$ is the world sheet metric. S contains no kinetic term for $q_{\alpha\beta}$, so one does not expect to treat $q_{\alpha\beta}$ as a dynamical variable of the theory. Instead, one attempts to solve for $q_{\alpha\beta}$ (up to gauge freedom) via its equation of motion

$$\frac{\delta S}{\delta q_{\alpha\beta}} = 0 \qquad (2)$$

The theory described by equation (1) can be viewed as a two dimensional supersymmetric non-linear σ-model with dynamical fields $X^\mu(\sigma,\tau)$ (plus fermions) and internal space $M^4 \times K$. Equation (2) requires the stress energy tensor $T_{\alpha\beta}$ of this two dimensional field theory to vanish. Classically, this can always be achieved. However, quantum mechanically, (2) becomes an operator equation. The trace of this equation takes the form:

$$T^\alpha_{\ \alpha} = aR + \beta_{\mu\nu}\partial_\alpha X^\mu \partial_\beta X^\nu q^{\alpha\beta} + \text{superpartner} \qquad (3)$$

where R is the scalar curvature of the metric $q_{\alpha\beta}$, a is a coefficient depending on the dimension of spacetime and $\beta_{\mu\nu}$ is the β-"function" associated with the spacetime metric. If this trace does not vanish, then the equation of motion (2) for $q_{\alpha\beta}$ cannot be imposed and additional dynamics must be introduced for $q_{\alpha\beta}$.[8] Whether or not this can be done in a consistent fashion remains an open question. Even if it can, however, the resulting "Polyakov" string theory probably does not contain massless particles and is therefore not relevant as a theory of the fundamental interactions.

Fortunately, there are special cases in which $T^{\alpha}{}_{\alpha} = 0$. This condition is related to a local scale invariance of the two dimensional σ model. Classically, S is invariant under the transformations:

$$\begin{aligned} X^{\mu} &\to X^{\mu} \\ q_{\alpha\beta} &\to \Omega^2 q_{\alpha\beta} \end{aligned} \tag{4}$$

which implies $T^{\alpha}{}_{\alpha} = 0$. However in the quantum theory, ultraviolet regulators usually destroy this scale invariance. For finite theories, however, this is not the case. Supersymmetric non-linear σ models on a flat two dimensional world sheet are known to be finite (assuming the existence of appropriate regulators) for Ricci flat internal spaces and N = 2 or N = 4 supersymmetry, and are believed to be finite for Ricci flat spaces and N = 1 supersymmetry.[9] Scale invariance on a curved world sheet requires in addition that the dimension of spacetime be the critical dimension 10. Thus Ricci flatness of the internal space $M^4 \times K$ is, in the present framework, essential for consistent superstring compactifications.[10]

Of particular interest is the heterotic superstring,[2] because it probably has the best phenomenology (and because it's hard to believe that nature declined such a good opportunity to base itself on exceptional groups). Recall that in this theory the fields are divided into left moving and right moving modes on the world sheet. In the fermionic representation there is a right moving anticommuting ten dimensional Majorana-Weyl spinor S, and 32 left moving anticommuting two dimensional Majorana-Weyl spinors ψ^m $m = 1,\cdots,32$ which are spacetime scalars. For the SO(32) superstring, ψ^m is in the fundamental representation of the group, while for $E_8 \times E_8$ the situation is more complicated.[2] The presence of gauge interactions in this theory suggests that one should be able to expand about a background configuration that has a non-zero Yang-Mills field as well as non-zero curvature. In this case, the light cone action for the heterotic superstring becomes[2,11]:

$$S^H_{LC} = -\frac{1}{4\pi\alpha'}\int d\sigma d\tau(\partial_\alpha X^i \partial^\alpha X^j g_{ij} + i\bar{S}\gamma^- \mathcal{D}_+ S$$

$$+ \psi^m \mathcal{D}_- \psi^m + F^A{}_{ij}\bar{S}\gamma^-\gamma^i\gamma^j S T^A{}_{mn}\psi^m\psi^n) \qquad (5)$$

where i,j = 1,···,8 run only over the transverse space $R^2 \times K$ and S obeys the light cone condition $\gamma^+ S = 0$. $F^A{}_{ij}$ is the SO(32) or $E_8 \times E_8$ background Yang-Mills field, and $T^A{}_{mn}$ the appropriate generators. $\mathcal{D}_\pm$ is a covariant derivative with respect to both the background gauge and spin connections. It acts on the left movers as $\mathcal{D}_-\psi = \partial_-\psi + (\partial_- X^j)A_j\psi$ (with $\partial_\pm = \frac{\partial}{\partial\tau} \pm \frac{\partial}{\partial\sigma}$) and on the right movers as $\mathcal{D}_+ S = \partial_+ S + (\partial_+ X^j)\omega_j S$. It is related to the spacetime gauge covariant derivative $\mathcal{D}_j$ by $\mathcal{D}_\pm = (\partial_\pm X_j)\mathcal{D}_j$.

The heterotic superstring corresponds to an N = ½ supersymmetric non-linear σ model, as the supersymmetry acts on the right movers only. In general this is insufficient to insure that the σ model trace anomaly vanishes, even on a Ricci flat background. However, if you set the background gauge field A_i equal to the spin connection ω_i then a remarkable thing happens. Since ω_i must take values in a subgroup of SO(8) rather than SO(32) or $E_8 \times E_8$, many of the ψ's decouple from the gauge field and just satisfy a free Dirac equation. Their contribution to the trace anomaly will be cancelled by ghosts just as in the expansion about flat space without background gauge fields. The remaining ψ's now enter symmetrically with the components of the spinor S, and one recovers a left right symmetry in this sector of the theory.

An N = 1 (at least) world sheet supersymmetry then results* and the trace anomaly will vanish if the geometry is Ricci flat.

Another consistency requirement is the absence of anomalies in the four dimensional Lorentz algebra. In particular, there are potential anomalies in the commutator $[M^{i-}, M^{j-}]$ associated with the value of the two point function of T_{++} in the σ model. In the critical dimension, for background geometries $M^4 \times K$, this anomaly has been shown not to arise in lowest non-trivial order when K is Ricci-flat,[11] and presumably does not arise at higher orders either.

So far we have found that if one looks for a string state that can be described just by a background geometry, and assumes the relevant conditions are obeyed for all values of the sigma model coupling, then consistency forces the background geometry to be Ricci flat, and the background gauge connection A to equal the background spin connection ω. Have we found all the consistency requirements? Hopefully the answer is no, because these constraints do not uniquely determine a background configuration. Further consistency requirements might be useful in narrowing the range of potential vacuum configurations for the superstring theory.

We have argued that (modulo the above comments) one can construct a consistent first quantized string theory on Ricci flat manifolds with $A = \omega$, but it remains to be shown that such backgrounds provide solutions to the full superstring theory. What is meant by this statement is that the vacuum expectation values of all tadpoles vanish. In a scalar field theory with potential $V(\phi) = \phi$, for example, setting the background scalar field to zero is not the semiclassical approximation to a solution of the theory because the tadpole $\langle 0|\phi|0\rangle$ does not vanish. The superstring theory is a field theory with an infinite number of fields, so one similarly requires that $\langle 0|\Phi|0\rangle = 0$, where Φ is any operator that creates a tadpole.

*Actually, the N = 1 world sheet supersymmetry will not be obvious in the form of the action (5) unless the geometry is not only Ricci flat but Kähler. Otherwise the supersymmetry is apparent in the Ramond Neveu-Schwarz formulation of this theory. See Reference [11] for details.

There is a simple argument which shows that to all orders in the σ-model, but tree level in the string theory, all the tadpoles vanish for the background configurations we have discussed. Consider, for example, the operator $\Phi_D = q^{\alpha\beta} g_{ij} \partial_\alpha X^i \partial_\beta X^j$, which creates a scalar dilaton field. For backgrounds with no trace anomaly, this operator scales as $\Phi_D \to \Omega^{-2} \Phi_D$ under the two dimensional scale transformations (4). However, since the world sheet in tree level string calculations can (after stereographic projection) be taken to be a plane which is invariant under these transformations, the expectation value $\langle 0|\Phi_D|0\rangle$ must vanish. Similar arguments can be made for the other tadpoles.

Having shown that Ricci flat manifolds with $A = \omega$ provide solutions of the superstring theory at tree level, one still needs to check their classical stability before they can be called vacuum solutions.[12] A general argument for perturbative stability can be made in the special case where the resulting theory has N = 1 spacetime supersymmetry (as distinct from world sheet supersymmetry). This may also be the most phenomenologically interesting case, since without spacetime supersymmetry at the compactification scale there is likely to be a serious hierarchy problem. With N = 1 spacetime supersymmetry, the Hamiltonian for perturbations around the supersymmetric background can be written $H = Q^+ Q$, where Q is the supercharge operator for spacetime supersymmetry. In light cone gauge the absence of ghosts is manifest, so H acts on a positive norm Hilbert space and its expectation value in any state is therefore non-negative. This ensures perturbative vacuum stability.

What is the condition for N = 1 spacetime supersymmetry? In flat space, the action (5) is invariant under the supersymmetry transformations[2,3]

$$\delta_\varepsilon X^i = (p^+)^{-\frac{1}{2}} \bar{\varepsilon} \gamma^i S$$

$$\delta_\varepsilon S = i(p^+)^{-\frac{1}{2}} \gamma_- \gamma_\mu (\partial_- X^\mu) \varepsilon \tag{6}$$

where $\mu = 0,\cdots,9$ and ε is a right moving ten dimensional Majorana-Weyl spinor.

The commutator of two supersymmetry transformations is a

translation:

$$[\delta_1,\delta_2]X^i = \xi^-\partial_- X^i + a^i$$

$$[\delta_1,\delta_2]\,S = \xi^-\partial_- S$$

where

$$\xi^- = -\frac{2i}{p^+}\bar{\varepsilon}^{(1)}\gamma_-\varepsilon^{(2)}$$

$$a^i = -2i\bar{\varepsilon}^{(1)}\gamma_i\varepsilon^{(2)} \qquad (7)$$

A curved background admits spacetime supersymmetry if there exists supersymmetries parametrized by $\varepsilon^{(1)}$ and $\varepsilon^{(2)}$ such that a^i is any spatial translation of the transverse $R^2 \times K$. (Four dimensional Lorentz invariance then insures the full four dimensional Poincaré supersymmetry.)

Unlike S, ε need not obey the light cone condition $\gamma^+S = 0$. There are correspondingly two types of supersymmetry transformations characterized by $\gamma^+\varepsilon_+ = 0$ and $\gamma^-\varepsilon_- = 0$ of positive and negative "transverse chirality." If $\varepsilon^{(1)}$ and $\varepsilon^{(2)}$ have the same transverse chirality then a^i can be easily seen to vanish. Both types of world sheet supersymmetries are thus necessary to ensure spacetime supersymmetry. The transformation laws for ε_+ are particularly simple:

$$\delta X^i = 0$$
$$\delta S = -2i\sqrt{p^+}\,\varepsilon_+ \qquad (8)$$

The action (5) will be invariant under this transformation if ε_+ is a covariantly constant spinor on the transverse space.* This follows since the first and third term in (5) are trivially invariant, and the second term gives a total derivative when $\mathcal{D}_+\varepsilon_+ = 0$. The last term also vanishes because when $A = \omega$, the gauge and metric curvatures are also equal and $F^A{}_{ij}\gamma^i\gamma^j\varepsilon_+ = 0$ is just the integrability condition for $\mathcal{D}_j\varepsilon_+ = 0$. Under these conditions there are always ε_- supersymmetries as well since these are just the usual supersymmetries of the non-linear σ model.

*More precisely, ε_+ is the restriction to the world sheet of a covariantly constant spinor on $\mathbb{R}^2 \times K$.

Thus we have learned that N = 1 (or greater) spacetime supersymmetry follows from the existence of a covariantly constant spinor on the internal manifold K. This restriction can be understood in terms of a group known as the holonomy group of K. If a spinor (or vector) is parallel transported around a closed loop in K, it will undergo some O(6) rotation. The group of all rotations that can be generated in this way is the holonomy group. For example, the holonomy group of flat space is trivial while the holonomy group of the standard metric connection on the six sphere is O(6). The existence of a covariantly constant spinor says that the holonomy group of K cannot be the full O(6) since there is at least one spinor that is not rotated under parallel transport. In fact it implies that the holonomy group is at most an SU(3) subgroup of O(6). (We are interested in the case where it is precisely SU(3), otherwise the manifold K must have zero Euler number which we will soon see leads to severe problems with phenomenology.) A manifold with U(3) (or smaller) holonomy is known as a complex Kähler manifold.

The requirement that K be a three complex dimensional Ricci flat Kähler manifold is quite a restrictive one. To better understand the nature of this restriction, we consider the requirements on K individually. Recall that a six dimensional real manifold can be viewed as patches of $\mathbb{R}^6$ which are "glued together" at the edges by identifying points of one patch with those of another in a smooth (i.e. C^∞) manner. Similarly, a three dimensional complex manifold can be viewed as patches of $\mathbb{C}^3$ glued together in a holomorphic i.e. complex analytic manner. Since holomorphic functions are always smooth, it is clear that every complex three manifold can be viewed as a real six manifold. However, the converse is not always the case. For example, although it is not obvious, the familiar manifolds S^6 and $S^2 \times S^4$ cannot be viewed as complex manifolds, although $S^3 \times S^3$ and $S^1 \times S^5$ can.

To understand the Kähler condition we must discuss the metric. The analog of a positive definite metric for a real manifold is a hermitian metric one can define a unique (torsion free, metric compatible) covariant derivative. Now consider a vector V such that for any function f depending only on the complex coordinates $\bar{z}^i$ and not on z^i we have

$$V^i \nabla_i f = 0 \tag{9}$$

Such a vector is called holomorphic. In most cases, if one starts with a holomorphic vector and parallel transports it along a curve, then the vector will not remain holomorphic. However, there are special metrics for which this difficulty does not occur, namely those with U(3) holonomy. These special metrics are called Kähler. Not every complex manifold admits a Kähler metric. For example, no metric on $S^3 \times S^3$ or $S^1 \times S^5$ can be Kähler, although $S^2 \times S^2 \times S^2$ does admit Kähler metrics. One can view a Kähler manifold as the nicest type of complex manifold in that the metric structure and the complex structure are compatible in the above sense.

Having restricted ourselves from arbitrary Riemannian six manifolds to complex manifolds to Kähler manifolds, there remains one final restriction -- our space must be Ricci flat. In general there is a topological obstruction to finding a Ricci flat metric on a Kähler manifold. This is known as the first Chern class. For example, in two dimensions, the first Chern class is related to the Euler number. In this case it is well known that the only compact two manifold which admits a Ricci flat metric is the one with zero Euler number i.e. the torus. Calabi conjectured that in higher dimensions the first Chern class was the only obstruction and that if it vanishes a Ricci flat metric always exists.[13] Calabi's conjecture was proved twenty years later by Yau.[14] This result is of great significance to us for the following reason. Whereas Ricci flat Kähler metrics are in general quite complicated, Kähler manifolds with vanishing first Chern class can be constructed quite easily -- as we now indicate. We call such manifolds Calbi-Yau manifolds (if their Euler number is non-zero) although it should be noted that the theorems proved by these mathematicians are really more general than we consider here.

Perhaps the simplest example of a Calabi-Yau manifold is the following. Start with $\mathbb{C}^5$ with coordinates $(z_1, \cdots, z_5)$ and consider the subspace satisfying

$$\sum_{i=1}^{5} z_i^5 = 0 \tag{10}$$

This is a four dimensional complex manifold. Notice that if (z_i) is a solution so is $\lambda(z_i)$. Now identify each non-zero solution (z_i) with $\lambda(z_i)$ for all complex $\lambda \neq 0$. Let Y_0 denote the resulting three complex dimensional manifold. One can show that Y_0 has anishing first Chern class and hence admits a Ricci flat Kähler metric. Although this space is simple to construct, its topology is surprisingly complicated. For example, it turns out that the Euler number χ of Y_0 is -200. Therefore every vector field on this space must vanish in at least 200 points!*

Many other examples of Ricci flat Kähler spaces can be constructed.[15,16] However, it is important to note that although there are an infinite number of complex three manifolds that admit Kähler metrics, it is likely that only a finite number of these admit Ricci flat Kähler metrics.[15]

Having found a vacuum configuration of the superstring theory, we can now extract the low energy effective field theory. Recall that the gauge connection was assumed to acquire an expectation value equal to that of the spin connection, which is valued in SU(3) (because K has SU(3) holonomy). For the heterotic superstring with an $E_8 \times E_8$ gauge field this breaks the gauge group down to $E_8 \times E_6$. The theory contains fermions (gluinos) in the adjoint representation. The adjoint (248) of E_8 decomposes as $(1,78) \oplus (3,27) \oplus (\overline{3},\overline{27}) \oplus (8,1)$ under $SU(3) \times E_6$.

Let n_{27}^L and n_{27}^R be the number of left and right handed massless fermion multiplets in four dimensions transforming as a 27 under E_6, and let $N = n_{27}^L - n_{27}^R$. Since the difference between left and right is just a matter of definition, the number of generations is $|N|$. Because the fundamental theory contains ten dimensional Weyl fermions massless chiral fermions on M^4 are associated, in the usual way, with chiral zero modes of the Dirac operator on K. Fermions transforming in the 27 of E_6 also transform in the 3 of SU(3). Hence $N = n_{27}^L - n_{27}^R$

---

*Strictly speaking, this statement is only true for vector fields whose zeros are non-degenerate.

is just the index of the (gauge coupled) Dirac operator on K acting on spinors in the 3 representation of SU(3). Now since the SU(3) Yang-Mills potential is equal to the spin connection, this index can depend only on the geometry of K. But the index is a topological invariant. Hence, it can depend only on the topology of K. In fact it turns out that N is simply one half the Euler number χ of K. So the number of generations is $|N| = \frac{1}{2} |\chi(K)|$.

This is a solution of the long-standing chiral fermion problem in Kaluza-Klein theories.[17] To our knowledge there is no previous case of a supersymmetric Kaluza-Klein compactification that produced a chiral spectrum of fermions. In the present case, not only are chiral fermions naturally obtained, but they are in the correct representation (27) of a realistic grand unified group E_6.[18] It is remarkable that superstring theory, which was developed to solve the problem of quantum gravity, also solves the chiral fermion problem.

Returning now to the manifold Y_0, we see that this particular example would predict 100 families, which is considerably more than we need and in serious conflict with cosmology (among other things). Fortunately, there is a general procedure for constructing a new Calabi-Yau manifold from an old one such that the absolute value of the Euler number is reduced. This procedure is applicable whenever there exists a group of discrete symmetries that leave no point fixed (except, of course, for the identity which leaves every point fixed). A simple example of this is provided by S^2, which has a Z_2 symmetry that maps each point to the antipodal point. If we identify points related by this Z_2 we obtain a new manifold $(S^2/Z_2) = RP^2$. Roughly speaking RP^2 is "half as big" as S^2, and has half the Euler number. This can be easily seen by using the formula for the Euler number expressed as an integral of the curvature. In general, if the discrete group has n elements the new Euler number will be 1/n times the old one.

For the manifold Y_0, consider the group generated by

$$A: (z_1,z_2,z_3,z_4,z_5) = (z_5,z_1,z_2,z_3,z_4)$$
$$B: (z_1,z_2,z_3,z_4,z_5) = (\alpha z_1,\alpha^2 z_2,\alpha^3 z_3,\alpha^4 z_4,z_5) \qquad (11)$$

where $\alpha \equiv e^{2\pi i/5}$. It is clear that A and B each take solutions of (10) into solutions and hence are symemtries of Y_0. It is also clear that $A^5 = B^5$ = identity so that the group of symmetries generated by A and B is $Z_5 \times Z_5$. Somewhat less clear but not hard to verify is that no element of this group has fixed points except for the identity. It is this last requirement which rules out more general symmetries such as non-cyclic permutations and multiplying the z_i with powers of α other than those given by powers of B. By identifying points related by this symmetry group, we obtain a new Ricci flat Kähler manifold $Y = Y_0/Z_5 \times Z_5$ with Euler number

$$\chi(Y) = \chi(Y_0)/25 = -8 \tag{12}$$

(Unlike RP^2, Y is still orientable.) So the superstring theory formulated on Y predicts four generations. Other Calabi-Yau manifolds have been found that predict one, two, three or four generations.[15,16]

So far we have seen that the superstring theory can produce a reasonable number of generations of quarks and leptons in the required representation of a realistic grand unified group. However there are at least two more ingredients that are essential for a realistic E_6 theory: mechanisms for breaking E_6 down to a baryon number conserving group at the GUT scale and for Weinberg-Salam symemtry breaking.[18] The Higgs fields responsible for the latter can be found in the superpartners of the quarks and leptons. The former occurs in a fashion unique to Kaluza-Klein theories, as we now explain.

Quotient spaces such as $Y_0/Z_5 \times Z_5$ or S^2/Z_2 are not simply connected. Consider for example the line on the two sphere extending from the north to south pole. On S^2/Z_2 this line becomes a non-contractible loop. However if we go twice around this loop on S^2/Z_2, it can be obtained from a closed loop on S^2 which is contractible (i.e. $\Pi_1(S^2/Z_2) \doteq Z_2$). Similarly Y has loops which are contractible after transversing five times. Now consider the expectation value:

$$U = \langle \exp(i \oint_\Gamma A_m \, dx^m) \rangle \tag{13}$$

where A_m lies in an Abelian subgroup of E_6 and Γ is a non-contractible loop. U is then an element of E_6. In general U can acquire an expectation value even if the E_6 field strength vanishes.[19] However if we wrap five times around Γ Stokes theorem can then be used to relate the path ordered exponential to the integral of the curvature over a two surface spanning the loop. Since the E_6 field strength vanishes for vacuum configurations, this implies $U^5 = 1$.

An expectation value for U breaks the E_6 gauge symmetry. Fermion zero modes that are not neutral under U acquire masses, as do gauge fields that do not commute with U. Thus the effects of U resemble in this respect those of a Higgs field in the adjoint representation. This is precisely what is required for GUT symmetry breaking, and allows one to break E_6 down to, for example, $SU(3) \times SU(2) \times U(1)^3$.

Having obtained these initial successes, one should now proceed to compute particle lifetimes, masses and coupling constants. These quantities turn out to depend on more refined topological properties of the internal manifold K.[16,20] For the Y manifold discussed in this paper there is a serious problem with rapid proton decay.[5,21] There might be a Calabi-Yau manifold with appropriate topological properties for which such difficulties are absent. Another problem is supersymmetry breaking.[12] At present no phenomenologically viable mechanism has been found. Finally, assuming a Calabi-Yau manifold is found that has all the right properties, one would like to understand why superstring theory prefers this one over any other. Or, indeed, why is four spacetime dimensions preferred at all?

These are serious problems and certainly worthy of attention. However, it is an unprecedented and exciting state of affairs that we are judging a potential unification of quantum mechanics and gravity on the basis of low energy phenomenology.

Acknowledgments

We are grateful to E. Martinec for useful conversations. This work was supported in part by NSF Grants PHY80-19754, PHY81-07384, and PHY82-05717.

REFERENCES

1. Green, M. B. and Schwarz, H. J., Phys. Lett. 149B, 117 (1984).
2. Gross, D. J., Harvey, J. A., Martinec, E., and Rohm, R., Phys. Rev. Lett. 54, 502 (1985), and Princeton preprints.
3. Green, M. B. and Schwarz, J. H., Nucl. Phys. B181, 502 (1981); B198, 252 (1982); B198, 441 (1982); Phys. Lett. 109B, 444 (1982); Green, M. B., Schwarz, J. H., and Brink, L., Nucl. Phys. B198, 474 (1982).
4. Candelas, P., Horowitz, G. T., Strominger, A., and Witten, E., to appear in Nucl. Phys. B. (1985).
5. Witten, E., Phys. Lett. 149B, 351 (1984).
6. Scherk, J. and Schwarz, J. H., Phys. Lett. 57B, 463 (1975); Cremmer, E. and Scherk, J., Nucl. Phys. B103, 399 (1976).
7. Feynman, R. P., Proc. Chapel Hill Conference (1957); Weinberg, S., Phys. Rev. D138, 988 (1965); Deser, S., Gen. Relativity and Grav. 1, 9 (1970).
8. Polyakov, A. M., Phys. Lett. 103B, 207, 211 (1981).
9. Freedman, D. Z. and Townsend, P. K., Nucl. Phys. B177, 443 (1981); Alvarez-Gaumé, L. and Freedman, D. Z., Phys. Rev. D22, 846 (1980); Commun. Math. Phys. 80, 282 (1981); Morozov, A. Ya., Perelmov, A. M., and Shifman, M. A., ITEP preprint; Hull, C. M., unpublished; Alvarez-Gaumé, L. and Ginsparg, P., in Proceedings of the Argonne Symposium on Anomalies, Geometry and Topology.
10. This point was independently made by Friedan, D. and Shanker, S., unpublished talk given at the Aspen Summer Institute (1984). See also Witten, E., Comm. Math. Phys. 92, 455 (1984) and Lovelace, C., Phys. Lett. 135B, 75 (1984).
11. Callan, C., Friedan, D., Martinec, E., and Perry, M., Princeton preprint (1985).
12. Whether or not they remain solutions at the quantum level is of course an important and non trivial question that we do not address here. See Dine, M., Rohm, R., Seiberg, N., and Witten, E., to appear in Phys. Lett. B (1985); Dine, M. and Seiberg, N., IAS preprint (1985), and Kaplunovsky, V., to appear.

13. Calabi, E., in Algebraic Geometry and Topology: A Symposium in Honor of S. Lefschetz (Princeton University Press, 1957), p. 78.

14. Yau, S.-T., Proc. Natl. Acad. Sci. 74, 1798 (1977).

15. Yau, S.-T., in Proceedings of the Argonne Symposium on Anomalies, Geometry and Topology.

16. Strominger, A. and Witten, E., to appear in Comm. Math. Phys.

17. Witten, E., Nucl. Phys. B186, 412 (1981). Work on this problem can be found in Chapline, G. and Slansky, R., Nucl. Phys. B209, 461 (1982); Wetterich, C., Nucl. Phys. B223, 109 (1983); Randjbar-Daemi, S., Salam, A., and Strathdee, S., Nucl. Phys. B214, 491 (1983); Witten, E., to appear in the Proceedings of the 1983 Shelter Island II Conference (MIT Press, 1985); Olive, D. and West, P., Nucl. Phys. B217, 248 (1983); Chapline, G. and Grossman, B., Phys. Lett. 125B, 109 (1984); Frampton, P. H. and Yamamoto, K., Phys. Rev. Lett. 125B, 109 (1984); Weinberg, S., Phys. Lett. 138B, 47 (1984); Frampton, P. H. and Kephardt, T. W., Phys. Rev. Lett. 53, 867 (1984); Koh, I. G. and Nishino, H., Trieste preprint ICTP/84/129.

18. Gürsey, F., Ramond, P., and Sikivie, P., Phys. Lett. 60B, 177 (1976); Gürsey, F. and Sikivie, P., Phys. Rev. Lett. 36, 775 (1976); Ramond, P., Nucl. Phys. B110, 214 (1976).

19. Hosotani, Y., Phys. Lett. 129B, 193 (1983).

20. Witten, E., to appear in Nucl. Phys. B.

21. Dine, M., Kaplunovsky, V., Nappi, C., Mangano, M., and Seiberg, N., Princeton preprint (1985).

LOW ENERGY SUPERSTRING THEORY †

Burt A. Ovrut*
The Rockefeller University
Department of Physics
New York, New York 10021

LECTURE 1: E_6 SYMMETRY BREAKING

ABSTRACT

We derive two methods for determining the symmetry breaking of E_6 in the low energy superstring theory, and classify all breaking patterns. A method for calculating the effective vacuum expectation values is presented. We show that there are theories with naturally light SU_2^W Higgs doublets, and classify all theories in which this occurs.

1. Introduction

The theory of superstrings[1], which evolved from the string theories of the early 1970's[2], has recently undergone a great revival of interest, spurred by the work of Green and Schwarz[3]. An anomaly free, d=10 superstring theory is possible with gauge groups O(32)[3] or $E_8 \times E_8$[4]. The subsequent compactification of the ten dimensional space to $M_4 \times K$, with M_4 being Minkowski space and K a compact six-dimensional manifold, places further restrictions on the theory. In particular, the requirement that the compactification leave an unbroken N=1 local supersymmetry in d=4 implies that K has SU_3 holonomy[5]. The existence of spaces with SU_3 holonomy was conjectured by Calabi[6] and proved by Yau[7]. On such spaces one is naturally led to a four dimensional gauge theory with reduced gauge group $E_6 \times E_8$[5]. Below the Planck scale the supergauge multiplet gives rise to an adjoint 496 of gluons and gluinos corresponding to E_8, and an adjoint 78 of gluons and gluinos associated with E_6.

† Invited lectures at the 1985 Summer Workshop in High Energy Physics and Cosmology, ICTP, Trieste, Italy.

* On leave of absence from the University of Pennsylvania, Department of Physics, Philadelphia, Pennsylvania 19104.

Work supported in part by Department of Energy Grant Number DE-AC02-81ER40033B.

In addition there are n^L 27's and n^R $\overline{27}$'s of E_6, each containing left handed fermions and their scalar superpartners. The number of generations, $N = n^L - n^R$, is restricted on topological grounds. For simply connected Calabi-Yau spaces K_0, N turned out to be hopelessly large, $N \geq 36$. This led the authors of Ref. 5 to consider multiply connected spaces $K = K_0/H$, where H is a discrete group which acts freely on K_0. For a specific choice of K_0, and $H = Z_5 \times Z_5$, the number of generations is reduced to N = 4. There is an additional benefit to having K be a multiply connected manifold. Define the Wilson loop

$$U = P\, e^{-i\int_\gamma T^a A^a_m\, dx^m} \qquad (1)$$

where A^a_m is the vacuum state E_6 gauge field, T^a, are the group generators and γ is a contour in K. Then, as pointed out in Ref. 5, one can have $U \neq 1$ even though the vacuum state gauge field strength $F_{mn}{}^a$ vanishes everywhere.[8] The reason for this is that, because of the "holes" in K, $A_m{}^a$ cannot necessarily be globally gauged to zero, even when $F_{mn}{}^a$ vanishes globally. Therefore, as long as contour γ is non-contractible, U is not necessarily unity. It follows from Eqn. (1) that U is an element of E_6. For a given vacuum configuration, $A_m{}^a$, there can be many inequivalent U matrices. Loosely speaking, there is one U for every "hole" in manifold K. If we define $\mathcal{H} = \{U\}$, then, with respect to matrix multiplication, $\mathcal{H}$ is a discrete subgroup of E_6. As an abstract group $\mathcal{H} \subseteq H$. For example, if $A_m{}^a$ vanishes everywhere then $\mathcal{H} = \{1\}$. For non-trivial $A_m{}^a$, however, one can find that $\mathcal{H} = H$. Let $\mathcal{H}(=H)$ and $\mathcal{H}'(=H)$ correspond to vacua $A_m{}^a$ and $A_m{}^{a'}$ respectively. Then $\mathcal{H}$ and $\mathcal{H}'$ may be two inequivalent embeddings of H into E_6. The possible existence of non-trivial discrete group $\mathcal{H}$ has important ramifications. If $A_m{}^a$ is a fixed vacuum state and $\mathcal{H}$ the associated discrete subgroup, then denote by $\mathcal{G}$ the subgroup of E_6 that commutes with $\mathcal{H}$. Then, as discussed in Ref. 5, at energies below the Planck scale, E_6 will be spontaneously broken to $\mathcal{G}$.

This result is of fundamental importance in model building. For practical purposes the symmetry breaking is due to effective vacuum expectation values (VEV's)

$$\int_\gamma T^a A^a_m \, dx^m \qquad (2)$$

in the adjoint 78 representation of E_6. In this talk we derive methods for calculating patterns of E_6 symmetry breaking. For concreteness, we focus our discussion on the multiply connected manifold $K = K_0/H$ with $H = Z_5 \times Z_5$ and four generations. Many of our results, however, are valid for other manifolds and discrete groups.

2. E_6 Symmetry Breaking

We want to study the symmetry breaking patterns of E_6 on multiply connected Calabi-Yau manifolds $K = K_0/H$. In general, H can be any abelian or non-abelian discrete group that acts freely on K_0. The problem of E_6 breaking is simplified, however, if we restrict H to be an abelian group. For concreteness, we focus on the group $H = Z_5 \times Z_5$, (and manifold $K_0/(Z_5 \times Z_5)$ with four generations) although our discussion is valid for any abelian group. Let $A_m{}^a$ be a fixed vacuum configuration, $\mathcal{W}$ (= $\{U\}$) the discrete group of Wilson loops associated with it, and $\mathcal{G}$ the subgroup of E_6 that commutes with $\mathcal{W}$. Also, denote the $SU_3{}^c \times SU_2{}^w \times U_1$ gauge group of the standard model by $\tilde{\mathcal{G}}$. The following observations will be helpful in deriving E_6 symmetry breaking patterns. First, the two Wilson loop generators of $Z_5 \times Z_5$ commute and, hence, the associated effective VEV's can be simultaneously diagonalized. These VEV's can be extended to form a basis for the E_6 Cartan subalgebra. All the elements of this basis commute and, therefore, $\mathcal{G}$ must have the same rank as E_6, namely six. Second, $\mathcal{G}$ must be at least as large as $\tilde{\mathcal{G}}$ in order to successfully describe known phenomenology. Therefore, we must find the embeddings of $\tilde{\mathcal{G}}$ in E_6. This problem has already been studied extensively, in particular by Slansky[9], who has listed all the possible symmetry breaking patterns that preserve $\tilde{\mathcal{H}}$ and given the vector boson masses in terms of the symmetry breaking direction in weight space. Combined with our first observation it follows that $\mathcal{H}$ must be at least $SU_3^C \times SU_2^W \times U_1 \times U_1 \times U_1$. Third, note that if U is a Wilson loop in Z_5 then $U^5 = 1$. This follows from the fact that a non-contractable path repeated five times is contractable. Finally, note that if $\mathcal{H} = \{1\}$, E_6 remains unbroken. We now discuss our first method of determining E_6 breaking patterns.

A) Decomposition into maximal subgroups

E_6 has three maximal subgroups with rank 6: $SU_3^C \times SU_3^W \times SU_3$, $SU_2 \times SU_6$, and $SU_{10} \times U_1$. Let $\mathcal{H} = Z_5 \times Z_5$ and consider $SU_3^C \times SU_3^W \times SU_3$. If $U \in \mathcal{H}$ a possible form for U is

$$\begin{pmatrix} 1 & & \\ & 1 & \\ & & 1 \end{pmatrix} \times \begin{pmatrix} \alpha^j & & \\ & \alpha^j & \\ & & \alpha^{-2j} \end{pmatrix} \times \begin{pmatrix} \beta^k & & \\ & \beta^k & \\ & & \beta^{-2k} \end{pmatrix} \qquad (3)$$

where j, k are integers, $\alpha^5 = \beta^5 = 1$, and, hence, $U^5 = 1$. Letting j and k vary from 0 to 4, we generate 25 distinct U's, which form a representation of the group $Z_5 \times Z_5$. The group $\mathcal{H}$ can be read off from the form of Eqn.(3). It is

$$\mathcal{H} = SU_3^C \times SU_2^W \times U_1 \times SU_2 \times U_1 \qquad (4)$$

A second possible form for U is

$$\begin{pmatrix} 1 & & \\ & 1 & \\ & & 1 \end{pmatrix} \times \begin{pmatrix} \alpha^j & & \\ & \alpha^j & \\ & & \alpha^{-2j} \end{pmatrix} \times \begin{pmatrix} \beta^k & & \\ & \beta^{4k} & \\ & & 1 \end{pmatrix} \qquad (5)$$

Again j, k are integers, $\alpha^5 = \beta^5 = 1$, and letting j and k vary from 0 to 4, we generate 25 distinct U's which form a representatoin of $Z_5 \times Z_5$. The group $\mathcal{G}$ can be read off from the form of Eqn. (5). It is

$$\mathcal{G} = SU_3^{\,C} \times SU_2^{\,W} \times U_1 \times U_1 \times U_1 \tag{6}$$

Eqs.(3) and (5) are the only possible embeddings of $\mathcal{H} = Z_5 \times Z_5$ into $SU_3^C \times SU_3^W \times SU_3$ that preserve $\tilde{\mathcal{G}}$. The embeddings of $\mathcal{H} = Z_5$ into $SU_3^C \times SU_3^W \times SU_3$ can be worked out in the same manner. The results are that $\mathcal{G}$ can be $SU_3^C \times SU_2^W \times SU_3^C \times SU_2$, and $SU_6 \times U_1$. The remaining breaking patterns can be found by embedding $Z_5 \times Z_5$ and Z_5 into maximal subgroups $SU_2 \times SU_6$ and $SO_{10} \times U_1$. It is clearly preferable to have a more general method, particularly one that allows a simple determination of the associated effective VEV's Such a method is most easily found by using the Dynkin formalism[10].

B) Method of Weyl weights

We can write the most general Wilson loop U as $\exp\{i\sum\lambda^i H^i\}$, where the H^i are the six generators of the Cartan subalgebra of E_6 and the λ^i are six real parameters. Arrange the λ^i into a vector $\lambda \equiv (a,b,c,d,e,f)$. Let α be a root vector of E_6. The mass of the vector boson corresponding to α is proportional to the inner product (λ, α). Since the vector bosons corresponding to $\tilde{\mathcal{G}}$ must be massless, $\lambda = (-c,c,a,b,c,0)$. We can then write U as a 27-dimensional diagonal matrix that depends on the three real parameters a,b, and c. The diagonal elements are given in Table 1, together with the transformation properties under $SU_3^C \times SU_2^W$, SO_{10}, and SU_5 of the elements of the 27-plet on which they act. Throughout this paper we label elements of the 27-plets that transform as an A under SO_{10} and a B under SU_5 by [A,B].

To find the group $\mathcal{G}$, we simply embed the discrete group in the above form for U. For $H = Z_5 \times Z_5$, we could let $e^{ia} = \alpha$, $e^{ib} = \beta$, and $e^{ic} = \alpha^j \beta^k$, where $\alpha^5 = \beta^5 = 1$ and j and k are integers.

$\mathcal{H} = Z_5 \times Z_5$ corresponds to two of the parameters a,b, and c being independent, while for $\mathcal{H} = Z_5$ there is only one independent parameter. Furthermore, by using Table 20 in Slansky's report[9], the relevant part of which is reproduced here as Table 2, we can find which $\mathcal{H}$ corresponds to a given $\mathcal{G}$.

As an example, let us find all the embeddings of $\mathcal{H} = Z_5 \times Z_5$ that break E_6 to $SU_4 \times SU_2{}^W \times U_1 \times U_1$ (Pati-Salam).[11] From Table 1 we see that if the (e,ν) and (u,d) are to lie in a 4 of SU_4 we must have $b=-c$. (We have used the fact that the (e,ν) corresponds to the weak doublet in the $[16,\overline{5}]$ and the (u,d) to the doublet in the $[16,10]$.) We can then either let a be independent or fix a = 0,c,2c,3c, or 4c. (Because of the Z_5 symmetries $5c \equiv 0$, $6c \equiv c$,etc.)
For a independent we consult Table 2 and find that (λ,a) is zero only for $\alpha = \pm(000001)$, $\pm(0100\text{-}10)$, $\pm(0\text{-}10011)$, $\pm(10001\text{-}1)$, $\pm(100100)$, $\pm(\text{-}1101\text{-}1\text{-}1)$, $\pm(\text{-}10010\text{-}1)$, and the six roots (000000). So we have 20 massless vector bosons and these roots span a representation of $SU_4 \times SU_2{}^W \times U_1 \times U_1$. Similarly for a=0 we have 46 massless vector bosons and the gauge symmetry is $SO_{10} \times U_1$; for a = c we have $SU_4 \times SU_2{}^W \times U_1 \times U_1$; for a = 2c, $SU_5 \times SU_2 \times U_1$; for a = 3c, $SU_5 \times SU_2 \times U_1$; and for a = 4c, $SU_4 \times SU_2{}^W \times U_1 \times U_1$. We break to the Pati-Salam group in three cases: i) a independent, $\mathcal{H} = Z_5 \times Z_5$, $e^{ic} = \alpha^j$, $e^{ib} = \alpha^{-j}$, $e^{ia} = \beta^k$; ii) a = c, $\mathcal{H} = Z_5$, $e^{ic} = \alpha^j$, $e^{ib} = \alpha^{-j}$ $e^{ia} = \alpha^j$; iii) a = -c, $\mathcal{H} = Z_5$, $e^{ic} = \alpha^j$, $e^{ib} = \alpha^{-j}$, $e^{ia} = \alpha^{-j}$.

By examining all the values for a,b, and c allowed by the discrete symmetry, one can exhaust all the symmetry breaking induced by $\mathcal{H}$. In particular we recover the embeddings found by the first method. In addition, we can find for which symmetries the effective VEV's are zero in some directions. We use this method in the next section to generate naturally light Higgs doublets.

3. Naturally Light Higgs Doublet Problem

We would like to have light Higgs doublets to set the electroweak scale and give masses through Yakawa couplings to the ordinary fermions. In particular, it is necessary that at least one of the light doublets be in the $[10,\overline{5}]$ or $[10,5]$ representations under $[SO_{10}, SU_5]$. At the same time, supersymmetric E_6 theories contain extra color triplets that can mediate nucleon decay via dimension 4 or 5 baryon number violating, $\mathcal{H}$ invariant, operators. These triplets must be given very large masses. It is very difficult, even with fine tuning, to keep the doublet light while making the triplet heavy. Fortunately, the same mechanism that breaks the gauge symmetry gives us a natural method for splitting the doublets from the triplets. Although the method does not depend on which discrete symmetry we use, for simplicity we assume that $H = Z_5 \times Z_5$.

The nontrivial gauge fields that give rise to Wilson loops different from unity can also lead to a mass term through the coupling of the four-space part of the chiral superfields to the gauge fields in the compactified dimensions.[12] In other words, one of the 27's can couple to the $\overline{27}$ through a term involving an effective 78 VEV ($27 \times \overline{27} \times 78$ contains a singlet). These fields thus acquire a mass of order the inverse radius of the compactified dimensions, presumably the Planck mass, while the other four 27's remain massless. (Note that we cannot have a $27 \times \overline{27}$ bare mass term since, until E_6 is broken, the chiral superfields are all massless zero modes.) Let us call the 27 and $\overline{27}$ that pair off χ and $\overline{\chi}$, and the remaining 27's ψ. We then expect all the components of χ and $\overline{\chi}$ to gain huge masses and disappear from the spectrum. As we now show, however, it is possible for some of these components to remain naturally massless. This occurs when the diagonal entries of the U's that multiply these components are unity. That is, the associated effective VEV's are zero in these directions.

As an example, let us use Table 1 to find which components of χ and $\overline{\chi}$ are left massless for the breaking to $SU_4 \times SU_2^W \times U_1 \times U_1$

given in Section 2: i) for $b = -c$, a independent, none of the U^{diag} is one, the effective VEV then has no zeros, and all the components of χ and $\overline{\chi}$ are massive; ii) for $b = -c$, $a = c$, $U^{diag} = 1$ for the color triplet in the $[16,\overline{5}]$ and the singlet in the $[16,10]$, and hence those components of χ and $\overline{\chi}$ remain massless; iii) for $b = -c$, $a = -c$ the color triplet weak singlet in the $[16,10]$ and the $[16,1]$ remain massless.

Using Tables 1 and 2, we can find for which values of the parameters a,b, and c we obtain light doublets and what the resulting gauge symmetries are. The light doublets in χ can be used as Higgs fields to break $SU_2^W \times U_1$ and to give masses to quarks and leptons. The corresponding light doublets in $\overline{\chi}$ cannot couple to ordinary matter, and hence we ignore them. We list below all the cases in which at least one weak doublet in χ is light while the color triplets are all heavy:

i) $b = a+c$, a and c arbitrary
massless: doublet in $[10,5]$
gauge symmetry: $SU_3^C \times SU_2^W \times U_1 \times U_1 \times U_1$

ii) $b = 4c$, $a = 3c$
massless: doublet in $[10,5]$
gauge symmetry: $SU_5 \times SU_2 \times U_1$

iii) $b = 0$, $a = 4c$
massless: doublets in $[10,5]$ and $[16,\overline{5}]$, the singlet $[16,1]$
gauge symmetry: $SU_3^C \times SU_2^W \times U_1 \times U_1$

iv) $a = 2c$, b arbitrary
massless: doublet in $[10,\overline{5}]$
gauge symmetry: $SU_3^C \times SU_2^W \times U_1 \times U_1$

v) $a = 2c$, $b = 0$
massless: doublets in $[10,\overline{5}]$ and $[16,\overline{5}]$, singlet in $[16,10]$
gauge symmetry: $SU_3^C \times SU_2^W \times U_1 \times U_1$

vi) $a = 2c$, $b = 3c$
massless: doublets in $[10,5]$ and $10,\overline{5}]$, the singlet $[1,1]$
gauge symmetry: $SU_3^C \times SU_2^W \times SU_2 \times U_1 \times U_1$

vii) $a = 2c$, $b = 4c$
massless: doublet in $[10,\bar{5}]$
gauge symmetry: $SU_5 \times SU_2 \times U_1$

viii) $b = 0$, a and c arbitrary
massless: doublet in $[16,\bar{5}]$
gauge symmetry: $SU_3{}^c \times SU_2{}^w \times U_1 \times U_1 \times U_1$

ix) $b = 0$, $a = 3c$
massless: doublet in $[16,\bar{5}]$
gauge symmetry: $SU_5 \times SU_2 \times U_1$

The embeddings of $\mathcal{H}$ that give rise to these values for a,b, and c can be readily found. For example, in case i) let $e^{ia} = \alpha^j$, $e^{ic} = \alpha^k$, and $e^{ib} = \alpha^{j+k}$, $\alpha^5 = 1$. This corresponds to $\mathcal{H} = Z_5 \times Z_5$. In case ii) let $e^{ic} = a^j$, $e^{ia} = \alpha^{3j}$, $e^{ib} = \alpha^{4j}$, which corresponds to $\mathcal{H} = Z_5$.

It is worth re-emphasizing that our light Higgs doublets were not obtained by fine tuning. Setting, for example, $b = 3c$, $a = 2c$ as in case vi) is a choice of parameters (optimistically a minimum for the vacuum configuration), but not a fine tuning, since e^{ia}, e^{ib}, and e^{ic} are restricted by the discrete symmetry to be fifth roots of unity. These Higgs doublets are totally massless until supersymmetry is broken spontaneously.

Acknowledgements

The results discussed in this talk were derived in collaboration with J. Breit and G. Segre. Similar results were also obtained independently by E. Witten and A. Sen.

Table 1. Diagonal elements of the Wilson loops U

U^{diag}	$SU_2^W \times SU_3^c$	SO_{10}	SU_5
$\exp\{i(b-3c)\}$	(1,1)	1	1
$\exp\{i(c+a-b)\}$	(2,1)	10	5
$\exp\{2ic\}$	(1,3)		
$\exp\{i(2c-a)\}$	(2,1)	10	$\overline{5}$
$\exp\{i(c-b)\}$	$(1,\overline{3})$		
$\exp\{\{-i(a+c)\}$	(1,1)	16	1
$\exp\{ib\}$	(2,1)	16	$\overline{5}$
$\exp\{i(a-c)\}$	$(1,\overline{3})$		
$\exp\{i(a-b-2c)\}$	(1,1)	16	10
$\exp\{i(b-a)\}$	$(1,\overline{3})$		
$\exp\{-ic\}$	(2,3)		

Table 2. Nonzero E_6 Roots

Root α	(λ, α)	Root α	(λ, α)
(000001)	0	(0100-10)	0
(0-10011)	0	(10001-1)	0
(-11001-1)	3c	(-210000)	3c
(0-111-1-1)	a+b-2c	(00-12-10)	2b-a-c
(0-12-10-1)	2a-b-c	(1-11-110)	a-b-c
(101-10-1)	a-b-c	(1-11-11-1)	a-b-c
(0-11-101)	a-b-c	(001-1-10)	a-b-c
(0-11-100)	a-b-c	(00100-1)	a
(0-1101-1)	a	(00100-2)	a
(-1010-10)	a	(-1-11000)	a
(-1010-1-1)	a	(-100100)	b+c
(-1101-1-1)	b+c	(-10010-1)	b+c
(010-110)	2c-b	(000-120)	2c-b
(010-11-1)	2c-b	(-110-101)	2c-b
(-100-111)	2c-b	(-110-100)	2c-b
(-101-110)	a-b+2c	(-111-10-1)	a-b+2c
(-101-11-1)	a-b+2c	(-11-1011)	3c-a
(-12-1000)	3c-a	(-11-1010)	3c-a

LECTURE 2: THE ONE-LOOP EFFECTIVE LAGRANGIAN

ABSTRACT

Using the low energy effective Lagrangian of the superstring, we discuss the breaking of supersymmetry and the internal gauge group. We calculate the quadratically divergent part of the one-loop potential energy and examine the stability of the vacuum.

Of paramount importance is the form of the four-dimensional, low energy N=1 supergravity allowed by the superstring. This problem has been approached from two points of view:

a) A truncation of the d=10, modified Chapline-Manton N=1 supergravity theory[13] to four spacetime dimensions.[14]

b) A new, d=4, "off-shell" irreducible, N=1 supergravity multiplet[15] which incorporates the symmetries and particle spectrum of the superstring[16].

Although these theories do not fully reflect the complexity of compactification on Calabi-Yau manifolds, they provide a reasonable starting point. After elimination of all auxiliary fields the Lagrangian can be written in the form of a minimal supergravity theory[17]. Remarkably, however, the Kahler potential of the theory is fixed by the superstring. The Lagrangian contains in addition to the graviton and gravitino the following fields:

1) A vector superfield transforming as a 248 of E_8 (or one of its subgroups, if E_8 is broken).

2) Vector fields $A_\mu{}^a$ and their fermionic superpartners λ^a that arise from the decomposition of the 78 vector superfield of E_6 under $\mathscr{G}$, the low energy gauge group.

3) Two complex scalar gauge singlet fields S and T and their fermionic partners χ_S and χ_T.

4) Complex scalars C^A and their fermionic partners $\chi_C{}^A$ that arise from the decomposition of the 27 chiral superfields of E_6 under $\mathscr{G}$. They include quarks, leptons, and Higgs bosons.

The Kahler potential and matter superpotential are given by[14,16]

$$K = M^2 \left[- \ln\left(\frac{S+\bar{S}}{M}\right) - 3 \ln \left\{ \left(\frac{T+\bar{T}}{M}\right) - 2 \frac{\bar{C}_A C^A}{M^2} \right\}\right] \qquad (1)$$

$$W_C = d_{ABC}\, C^A C^B C^C \qquad (2)$$

where $M = M_{Planck}/\sqrt{8\pi}$ and d_{ABC} are proportional to the invariant tensors of $\mathcal{G}$. We have suppressed family indices for the matter fields in K and W_C. It was shown in Ref. 14 that the kinetic energy term for the $\mathcal{G} \times E_8$ supergauge fields is of the form

$$\delta_{ab}\ \mathrm{Re}\ \frac{S}{M} W^{a\alpha} W^b_\alpha \qquad (3)$$

where $W^{a\alpha}$ is the superfield covariant curl of the gauge vector supermultiplet. The E_8 sector of the theory can become strongly interacting for energies below mass scale, Λ_{cond}. For such energies it is possible, by using (3), to integrate out the E_8 supergauge fields and obtain an effective superpotential for S. It is found to be[18,19]

$$W_S = M^3 b\, e^{-3S/2b_o M} \qquad (4)$$

where b is an arbitrary constant, and b_o is the coefficient of the E_8 gauge coupling beta founction (or the beta function of some subgroup of E_8, if E_8 is spontaneously broken during compactification). Typically, Λ_{cond} is much larger than 10^{10} GeV. In this paper we want to discuss energy scales of $O(10^{10}$ GeV) or smaller. Therefore, we omit all E_8 supergauge fields from the effective Lagrangian, and add W_S to the superpotential. In addition, Dine et al.[18] point out that the antisymmetric tensor field strength can acquire a v.e.v. This leads to an additional term in the superpotential of the form $M^3 a$, where a is a constant. The effective superpotential is then

$$W = M^3(a + b\, e^{-3S/2b_o M}) + d_{ABC}\, C^A C^B C^C \qquad (5)$$

Given K and W one can calculate the effective low energy Lagrangian consistent with the superstring. More generally, one could take

$$W = W_S(S) + d_{ABC}\, C^A C^B C^C \qquad (6)$$

where W_S is not restricted to be of the form in (5). We assume that

W_S is such that the associated potential energy has a minimum for finite S, and that W_S is non-vanishing at this minimum (e.g. $a \neq 0$ in Eqn.(5)). Supersymmetry breaking is then introduced through the W_S sector of the theory. In this lecture we examine how the supersymmetry breaking is communicated to the matter fields C^A and whether the vacuum is stable when radiative corrections are included. First, we calculate the tree-level potential energy. If we define

$$\hat{S} = S + \bar{S}\ , \quad \hat{T} = T + \bar{T}$$

$$\hat{Q} = \hat{T} - 2\,\frac{\bar{C}_A C^A}{M^2} \tag{7}$$

then the Kahler metric is given by

$$g = \left(\begin{array}{c|cc} \frac{M^2}{\hat{S}^2} & 0 & 0 \\ \hline 0 & \frac{3M^2}{\hat{Q}^2} & \frac{-6MC^B}{\hat{Q}^2} \\ 0 & \frac{-6M\bar{C}_A}{\hat{Q}^2} & \frac{6M}{\hat{Q}}\left(\delta_A{}^B + \frac{2\bar{C}_A C^B}{M\hat{Q}}\right) \end{array}\right) \tag{8}$$

The inverse Kahler metric is

$$g^{-1} = \left(\begin{array}{c|cc} \frac{\hat{S}^2}{M^2} & 0 & 0 \\ \hline 0 & \frac{\hat{Q}\hat{T}}{3M^2} & \frac{\hat{Q}C^B}{3M^2} \\ 0 & \frac{\hat{Q}\bar{C}_A}{3M^2} & \frac{\hat{Q}}{6M}\,\delta_A{}^B \end{array}\right) \tag{9}$$

Substituting these results into the bosonic part of the interaction Lagrangian given by Cremmer et al.[17]

$$\begin{aligned} \mathcal{L}_{B,INT}/e \; &= -M^4\, e^G \{G'_i\, (G''^{-1})^i{}_{j^*} G'^{j^*} - 3\} \\ &\quad - \frac{M^4}{2}\, g^2\, \mathrm{Re}\, f^{-1}_{ab}\, (G'_i (T^a)^i_j\, z^j)(G'_k (T^b)^k_\ell\, z^\ell) \end{aligned} \tag{10}$$

where, from (3),

$$f_{ab} = \delta_{ab} \frac{S}{M} \tag{11}$$

and T^a are the Lie algebra generators of $\mathcal{G}$, we find that the tree level potential energy is given by

$$V/e = \frac{M^4}{\hat{S}\hat{Q}^3} \left\{ \frac{\hat{S}^2}{M^2} |D_S W|^2 + \frac{\hat{Q}}{6M} \left| \frac{\partial W}{\partial C^A} \right|^2 \right\} + \frac{18 g^2 M^3}{\hat{Q}^2} \mathrm{Re} \frac{1}{S} (\bar{C}_A (T^a)^A{}_B C^B)^2 \tag{12}$$

Note that V is non-negative. This is due to an exact cancellation of the -3 part of the first term in Eqn. (10). This cancellation is easily traced to two properties of the effective theory induced from the superstring:

1) the $-3 \ln (\hat{Q}/M)$ term in the Kahler potential K, and
2) the T-independence of the superpotential W.

The tree-level vacuum state is determined by minimizing V and is given by

$$\langle D_S W \rangle = 0$$
$$\langle C^A \rangle = 0 \tag{13}$$

The v.e.v. $\langle T \rangle$ is undetermined at tree level. Note that V = 0 at the minimum. The gravitino mass is given by

$$m_{3/2} = \frac{1}{(\langle \hat{S} \rangle \langle \hat{T} \rangle^3)^{1/2}} \langle W_S \rangle \tag{14}$$

Since W is non-vanishing at the minimum, it follows that $m_{3/2}$ is non-zero and that supersymmetry is spontaneously broken. The value of $m_{3/2}$ is not fixed at tree level since $\langle T \rangle$ is undetermined. We conclude that the low energy effective theory has, at tree-level, a "stable" vacuum state ($\langle S \rangle$, $\langle T \rangle$, and $\langle C^A \rangle$ are finite) with a "naturally" vanishing cosmological constant (no fine tuning of parameters). Furthermore, this vacuum spontaneously breaks supersymmetry but does not set the scale of this breaking ($m_{3/2}$ is non-zero but arbitrary). It is important to note that if any one of the terms $D_S W$, $\partial W/\partial C^A$, or $\bar{C}_A (T^a)^A{}_B C^B$ did not vanish then $\langle \hat{S} \rangle \langle \hat{Q} \rangle \to \infty$, at least one of $\langle S \rangle$, $\langle T \rangle$, or $\langle C^A \rangle$ would be infinite, and the vacuum "unstable".

Having found the vacuum state, we expand S, T, and C^A around this vacuum ($z^i = \langle z^i\rangle + z'^i$) and calculate the tree level Lagrangian up to operators of dimension four. The Lagrangian can then be written as

$$\mathcal{L} = \mathcal{L}_{KE(1)} + \mathcal{L}_{KE(2)} + \mathcal{L}_{SUSY} + \mathcal{L}_{SB} + \mathcal{L}_{H} \tag{15}$$

where

$$\mathcal{L}_{KE(1)} = -D^{\mu}\bar{C}_A D_{\mu}C^A - \bar{\chi}_C{}^A{}_L \not{D}\, \chi_{\bar{C}AR} - \frac{1}{4} F^a_{\mu\nu}F^{a\mu\nu} - \bar{\lambda}^a_L \not{D}\, \lambda^a_R \tag{16}$$

$$\mathcal{L}_{KE(2)} = -\partial^{\mu}\bar{S}\partial_{\mu} S - \partial^{\mu}\bar{T}\,\partial_{\mu}T - \bar{\chi}_{SL}\not{\partial}\,\chi_{\bar{S}R} \tag{17}$$

$$\mathcal{L}_{SUSY} = -\left(\left|\frac{\partial W_C}{\partial C^A}\right|^2 + \frac{g^2}{2}(\bar{C}_A(T^a)^A{}_B C^B)^2\right) - \left(\frac{1}{2}\frac{\partial^2 W_C}{\partial C^B\partial C^A}\chi_C{}^A{}_L\,\chi_C{}^B{}_L + i\sqrt{2}\, g(T^a)^A{}_B C^B\bar{\lambda}^a{}_R\,\chi_{\bar{C}AR} + \text{h.c.}\right) \tag{18}$$

$$\begin{aligned}\mathcal{L}_{SB} = &-\Big(\left(\frac{m_{3/2}}{M}\right)\left[(1+\delta^2)|S|^2 - \delta(S^2 + \bar{S}^2)\right]\bar{C}_A C^A \\ &-\left(\frac{m_{3/2}}{M}\right)\left[(S - \delta\bar{S})\, W_C + \text{h.c.}\right]\Big) \\ &-\left(\left(\frac{m_{3/2}}{M}\right)\left[\frac{1}{4}(S - \delta\bar{S})\,\bar{\lambda}^a{}_L\,\lambda^a{}_L + \text{h.c.}\right]\right)\end{aligned} \tag{19}$$

and

$$\delta = -\frac{\langle\hat{S}\rangle^2}{\langle W\rangle}\left\langle\frac{\partial^2 W}{\partial S^2}\right\rangle \tag{20}$$

$\mathcal{L}_H$ is rather complicated. The physically interesting part of $\mathcal{L}_H$ is

$$\mathcal{L}_H = -m^2_{3/2}\left[(1+\delta^2)|S|^2 - \delta(S^2 + \bar{S}^2)\right] + m_{3/2}\left[\frac{\delta}{2}\bar{\chi}_{SL}\,\chi_{SL} + \text{h.c.}\right] + \ldots \tag{21}$$

The section of the Lagrangian involving normal matter only, $\mathcal{L}_{KE(1)} + \mathcal{L}_{SUSY}$, can be shown to be globally supersymmetric. That is,

$$\mathcal{L}_{KE(1)} + \mathcal{L}_{SUSY} = \frac{1}{4} \left[\int d^2\theta \, W^{a\alpha} W^{a}{}_{\alpha} + h.c. \right] + \int d^4\theta \, C^{\dagger}{}_{A} \, e^{2g(T^a V^a)^A{}_B} C^B + \left[\int d^2\theta \, W_C(C^A) + h.c. \right] \tag{22}$$

where C^A is the chiral superfield associated with $(C^A, \chi_C{}^A)$, V^a is the vector superfield associated with $(A_\mu{}^a, \lambda^a)$ and gauge group $\mathcal{G}$, and $W^{a\alpha}$ is the superfield covariant curl of V^a. The potential energy for this part of the Lagrangian is

$$V = \left| \frac{\partial W_C}{\partial C^A} \right|^2 + \frac{g^2}{2} \left(\bar{C}_A (T^a)^A{}_B C^B \right)^2 \tag{23}$$

which is non-negative and quartic in the fields. All mass terms and cubic couplings vanish. Hence, $\mathcal{G}$ is not spontaneously broken at tree-level. One might hope that radiative corrections would spontaneously break $\mathcal{G}$ through a Coleman-Weinberg type mechanism.[20] However, since $\mathcal{L}_{KE(1)} + \mathcal{L}_{SUSY}$ is globally supersymmetric, all radiative corrections to V from this part of the Lagrangian vanish. The Coleman-Weinberg mechanism might be induced by supersymmetry breaking terms in the Lagrangian. The explicit supersymmetry breaking terms involving normal matter are contained in $\mathcal{L}_{SB}$. They differ from the supersymmetry breaking terms in flat Kahler potential theories[21] in two fundamental ways.

1) The terms in $\mathcal{L}_{SB}$ all couple normal matter to "hidden" sector field S.

2) The terms in $\mathcal{L}_{SB}$ are "hard" operators of dimension four.

Note from $\mathcal{L}_H$ that S has mass of $O(m_{3/2})$. It follows that S is not necessarily heavy with respect to the electroweak scale. Therefore S should not be decoupled from the effective Lagrangian. Unfortunately, the inclusion of S in the effective Lagrangian leads to another problem. Since the operators in $\mathcal{L}_{SB}$ are "hard", they induce quadratically and logarithmically divergent terms in the radiatively corrected potential energy for which there are no counterterms. It follows that the divergent terms must be cut off at a scale Λ, and the result added to the effective potential. Assuming $\Lambda \gg m_{3/2}$, we expect the quadratic terms to dominate over the logarithmic one.

We now calculate, to the one-loop level, the quadratically divergent contributions to the C,T field sector of the potential energy which are proportional to $m_{3/2}$ (there are quadratically divergent terms proportional to V but these are uninteresting). First, we must determine the part of the Lagrangian relevant to this specific calculation. The necessary terms may be obtained from the Lagrangian of Cremmer et al.[17] They are (note we are not limiting ourselves to terms with dimension $d \leq 4$).

$$\mathcal{L}_{KIN}/e = -\left(\frac{M}{\hat{S}}\right)^2 \partial_\mu \bar{S}\, \partial^\mu S$$
$$- \Big\{\left(\frac{M}{\hat{S}}\right)^2 \bar{\chi}_{SL} \not{\partial} \chi_{SR} + 3\left(\frac{M}{\hat{Q}}\right)^2 \bar{\chi}_{TL} \not{\partial} \chi_{TR} - \frac{6MC^A}{\hat{Q}^2} \chi_{TL} \not{\partial} \chi_{\bar{C}AR}$$
$$- \frac{6M\bar{C}_A}{\hat{Q}^2} \bar{\chi}_{C\ L}^{\ A} \not{\partial} \chi_{TR} + \frac{12\bar{C}_B C^A}{\hat{Q}^2} \bar{\chi}_{C\ L}^{\ B} \not{\partial} \chi_{\bar{C}AR} + \text{h.c.}\Big\} \tag{24}$$

and

$$\mathcal{L}_{INT}/e = \frac{-M^2}{\hat{S}\,\hat{Q}^3} \left| W - \hat{S}\frac{\partial W}{\partial S} \right|^2 + \frac{M^2}{(\hat{S}\hat{Q}^3)^{1/2}} \Big\{ -\frac{\partial^2 W}{\partial S^2} \bar{\chi}_{SL} \chi_{SL}$$
$$- \frac{6W}{\hat{Q}^2} \bar{\chi}_{TL} \chi_{TL} + 12\frac{\bar{C}_A W}{M\hat{Q}^2} \left[\bar{\chi}_{TL} \chi_{C\ L}^{\ A} + \bar{\chi}_{C\ L}^{\ A} \chi_{TL}\right] \tag{25}$$
$$- 24\bar{C}_A \bar{C}_B \frac{W}{M^2 \hat{Q}^2} \bar{\chi}_{C\ L}^{\ A} \chi_{C\ L}^{\ B} + \text{h.c.}\Big\}$$

Now let $S = \langle S \rangle + S'$ and expand up to terms quadratic in S' (since we are interested in S' loops).

Define

$$\hat{m}_{3/2} = \frac{1}{(\langle\hat{S}\rangle\ \hat{Q}^3)^{1/2}} \quad (\langle W_S\rangle + W_C) \tag{26}$$

The gravitino mass term is given by

$$\hat{m}_{3/2}[\bar{\psi}_{\mu R}\sigma^{\mu\nu}\ \psi_{\nu R} - \bar{\psi}_R \cdot \gamma\ \eta_L] + \text{h.c.} \tag{27}$$

where η_L is the Goldstone fermion defined by

$$\eta_L = \sqrt{3}\ \chi_{TL} \tag{28}$$

Combining the $\bar{\chi}_T \chi_T$ mass term with Eqn. (27) we find

$$\hat{m}_{3/2}[\bar{\psi}_{\mu R}\sigma^{\mu\nu}\ \psi_{\nu R} - \psi_R \cdot \gamma\ \eta_L - \frac{2}{3}\ \bar{\eta}_L \eta_L] + \text{h.c.} \tag{29}$$

The remaining part of the Lagrangian can be written as

$$\mathcal{L} = \mathcal{L}_{KE} + \mathcal{L}_{SUSY} + \mathcal{L}_{SB} \tag{30}$$

where

$$\mathcal{L}_{KE} = -\ \partial_\mu \bar{S}\ \partial^\mu S - \bar{\chi}_{SL} \not{\partial} \chi_{SR} \tag{31}$$

$$\mathcal{L}_{SUSY} = -\ \hat{m}_{3/2}^2\ \ \delta^2 |S|^2 + \frac{\hat{m}_{3/2}}{2}\ \delta[\bar{\chi}_{SL}\,\chi_{SL} + \text{h.c.}] \tag{32}$$

$$\mathcal{L}_{SB} = -\ |\hat{m}_{3/2}|^2\ |S|^2 \tag{33}$$

and δ is defined in Eqn. (20) The first two terms, $\mathcal{L}_{KE} + \mathcal{L}_{SUSY}$, are globally supersymmetric. That is

$$\mathcal{L}_{KE} + \mathcal{L}_{SUSY} = \int d^4\theta\ S^\dagger S + [\int d^2\theta (\frac{-\hat{m}_{3/2}}{2}\ \delta S^2) + \text{h.c.}] \tag{34}$$

where S is the chiral superfield associated with (S, χ_S). It follows that the interactions in $\mathcal{L}_{SUSY}$ do not contribute to quadratic divergences. The quadratically divergent contribution to the C,T potential energy due to $\mathcal{L}_{SB}$ is easily computed. It is found to be

$$+\frac{\hat{m}^2_{3/2}\Lambda^2}{(4\pi)^2} \tag{35}$$

One can choose a gauge in which the $\psi_R \cdot \gamma \eta_L$ term in Eqn.(29) vanishes. In this gauge, the quadratically divergent contribution to the C,T potential energy due to η_L is

$$-\frac{4\,\hat{m}^2_{3/2}\Lambda^2}{(4\pi)^2} \tag{36}$$

Finally, we must add the gravitino contribution,which, following Barbieri and Ceccotti[22], is given by

$$+\frac{2\,\hat{m}^2_{3/2}\,\Lambda^2}{(4\pi)^2} \tag{37}$$

Adding these three terms together, we conclude that the quadratically divergent contribution to the C,T sector of the potential energy, at the one-loop level, is

$$V_{1\text{-loop},\Lambda^2} = -\frac{\hat{m}^2_{3/2}\Lambda^2}{(4\pi)^2} \tag{38}$$

where $\hat{m}_{3/2}$ is a function of C^A and T defined in Eqn. (26).

Also we note that not only does the gaugino mass vanish to tree level, but there is also no quadratically divergent contribution.

Let us now examine the stability of the vacuum. Adding together Eqns. (23) and (38), we have

$$V = \left|\frac{\partial W}{\partial C^A}\right|^2 + \frac{g^2}{2}(\bar{C}_A (T^a)^A{}_B C^B)^2 - \frac{\hat{m}_{3/2}{}^2 \Lambda^2}{(4\pi)^2} \tag{39}$$

which is minimized for $\langle C\rangle = 0$, and $\langle T\rangle = 0$. This implies that $m_{3/2} \to \infty$, and that the vacuum is unstable. This instability arises because, at tree level, we had a flat potential for T, but the radiative corrections introduce a potential $-1/(\hat{T})^3$ for which there is no counterterm. We conclude that using the tree level Lagrangian to calculate one-loop corrections to the action is (unlike flat Kahler metric theories) too naive. A more ambitious approach, using the full effect of the superstring at the one-loop level, is apparently necessary and is currently under study.

ACKNOWLEDGEMENTS

All work discussed in this talk was done in collaboration with J. Breit and G. Segre. Topics of a similar nature have been discussed by M. Mangano, and by P. Binetruy and M. Gaillard. Models of the type discussed in this talk were first introduced by J. Ellis, A. Lahanas, D. Nanopoulos, and K. Tamvakis in Phys. Lett. 134B(1984) 429, within the context of N=1 supergravity theories.

References

1. J.H. Schwarz, Phys. Rep. 89 (1982) 223; M.B. Green "Surveys in High Energy Physics" 3 (1983) 127 are good reviews of the subject.

2. P. Ramond, Phys. Rev. D2 (1971) 2415; A. Neveu and J.H. Schwarz, Nucl. Phys. B31 (1971) 86, Phys. Rev. D4 (1971) 1109; L. Brink, D.I. Olive, C. Rebbi and J. Scherk, Phys. Lett. 45B (1973) 198 describe early models with fermions and bosons.

3. M. Green and J. Schwarz, Phys. Lett. 149B (1984) 117 for N=1 supersymmetry with open or closed strings. L. Alvarez-Gaume and E. Witten, Nucl. Phys. B234 (1983) 269 for N=2 supersymmetry with closed strings.

4. For recent discussions see also E. Witten, Phys. Lett. 149B (1984) 351; D.J. Gross, J. Harvey, E. Martinec and R. Rohm, Phys. Rev. Lett. 54 (1985) 46.

5. We are following here the discussion of P. Candelas, G.T. Horwitz, A. Strominger and E. Witten, "Vacuum Configuration for Superstrings", NSF-TTP-84-170 preprint (1984).

6. E. Calabi in Algebraic Geometry and Topology: A Symposium in Memory of S. Lefschetz (Princeton University Press, 1957), p. 58.

7. S.T. Yau, Proc. Nat. Acad. Sci. 74 (1977) 1798.

8. Similar mechanisms had been proposed for gauge symmetry breaking in Kaluza-Klein theories by Y. Hosotani, Phys. Lett. 129B (1984) 193.

9. R. Slansky, Phys. Rep. 79 (1981) 1.

10. E.B. Dynkin, Am. Math. Soc. Trans. Ser. 2 (1957), 111; 6 (1957) 245.

11. J.C. Pati and A. Salam, Phys. Rev. D10 (1974) 275.

12. Y. Hosotani, Phys. Lett. 126B (1983) 309; D.J. Gross, R. Pisarski and L. Yaffe, Rev. Mod. Phys. 53 (1981) 43.

13) A.H. Chemseddine, Nucl. Phys. B185 (1981) 403. E. Bergshoff, M. De Roo, B. De Wit and P. Van Nieuwenhuizen, Nucl. Phys. B195 (192) 97.
G.F. Chapline and N.S. Manton, Phys. Lett. 120B (1983) 105.
14) E. Witten "Dimensional Reduction of Superstring Models" Princeton preprint 1985.
15) G. Girardi, R. Grimm, M. Muller and J. Wess, Z. Phys. C26 (1984) 427; Phys. Lett. 147B (1984) 81.
16) W. Lang, J. Louis and B.A. Ovrut, Pennsylvania preprint UPR-0280T (1985) and Karlsruhe preprint KA Thep 85-2 (1985).
17) E. Cremmer, B. Julia, J. Scherk, S. Ferrara, L. Girardello and P. van Nieuwenhuizen, Nucl. Phys. B147 (1979) 105.
18) M. Dine, R. Rohm, N. Seiberg and E. Witten, "Gluino Condensation in Superstring Models", I.A.S. preprint 1985.
19) Gluino condensation in the hidden sector is also discussed in J.P. Derendinger, L.E. Ibanez and H.P. Nilles, CERN preprint TH 4123/85 (1985).
20) S. Coleman and E. Weinberg, Phys. Rev. D7 (1973) 1888.
21) L. Hall, J. Lykken, and S. Weinberg, Phys. Rev. D27 (1983) 2369; H. Nilles, M. Srednicki, and D. Wyler,Phys. Lett. 120B (1983) 346.
22) R. Barbieri and S. Cecotti, Z. Phys. C17 (1983) 183.

STRING THEORY AND CONFORMAL INVARIANCE: A REVIEW OF SELECTED TOPICS

Spenta R. Wadia

Tata Institute of Fundamental Research
Homi Bhabha Road, Colaba, Bombay 400 005
INDIA

ABSTRACT

We motivate the principle of conformal invariance in string theory, within the framework of Polyakov's formulation of string quantum mechanics. The relevant formalism of conformal invariant field theory is introduced emphasising an algebraic view point. These ideas are illustrated with strings moving on $R^d \times G$, where G is a compact Lie group.

INTRODUCTION:

Recently there has been a tremendous revival of interest in the string model as a possible candidate for a unified theory of matter and interaction[1]. The basic idea is that the elementary particles with masses much less than the planck mass are viewed as the massless modes of a single entity: the string. If we restrict ourselves to closed strings, then all the interactions between these particles emerge from a single vertex in which a string splits into two strings. The string model not only promises to be a viable theory of all forces below planck mass but can lead to a prediction of phenomena beyond planck mass, like early universe cosmology and black holes.

Unified field theories like supergravity/super Yang-Mills theories are local 'phenomenological lagrangians' which emerge from string theory in the long wavelength limit. Their local symmetries possibly reflect a higher symmetry of the string theory. A formulation of the string field theory which is built on this higher gauge invariance is an outstanding problem.

In these notes we shall discuss some selected aspects of string theory, which may have some bearing on the previous question.

STRINGS AND 2-DIM, QUANTUM FIELD THEORY:

To motivate our considerations let us begin with the string field theory.[2] The closed string field is defined on the space of closed loops. It is local in the sence that it creates a excitation at the Curve C. We de note it by $\Psi(C)$. The lagrangian for the string field which incorporates the string splitting vertex can be schemetically written as

$$\mathcal{L} \sim \Psi \Delta \Psi + \Psi^3 \tag{1}$$

We note that the space of curves on which the string field is defined is determined as a solution to the equations of motion.

We do not know how to write down the kinetic energy operator in (1) which is independent of the vacuum geometry. So let us adopt a self consistent approach. Firstly in perturbation theory when one assumes the vacuum geometry to be flat minkowski space-time, a simple choice for Δ is the Laplacian in loop space plus a potential term.

$$\Delta = \int_0^{\pi} d\sigma\, \eta^{ab} \left(\frac{-\delta^2}{\delta X^a(\sigma)\,\delta X^b(\sigma)} + \frac{\partial X_a}{\partial \sigma} \frac{\partial X_b}{\partial \sigma} \right), \; \eta_{ab} \text{ is the minkowski metric} \tag{2}$$

The string interaction, which is described by the gravitation emission vertex, modifies the minkowski geometry and the effective metric becomes $G^{ab} = \eta^{ab} + h^{ab}$.[3] Here h^{ab} is the symmetric traceless graviton field. The laplacian (2) is replaced by its covariant generalization

$$\Delta = -\int_0^{\pi} d\sigma \frac{1}{\sqrt{G}} \frac{\delta}{\delta X^a(\sigma)} \left(G^{ab} \sqrt{G} \frac{\delta}{\delta X^b(\sigma)} \right) + \int_0^{\pi} d\sigma\, G^{ab} \frac{\partial X_a}{\partial \sigma} \frac{\partial X_b}{\partial \sigma} \tag{3}$$

The perturbative spectrum of the closed bosonic string also includes massless particles corresponding to the anti-symmetric tensor field A_{ab} and the dilaton. The effective interaction of A_{ab} is described by the substitution

$$\frac{\delta}{\delta X^a(\sigma)} \rightarrow \frac{\delta}{\delta X^a(\sigma)} - i\, A_{ab}(X(\sigma))\frac{\partial X^b}{\partial \sigma}(\sigma) \equiv \mathcal{D}_a \tag{4}$$

into (3). We shall discuss the dilaton subsequently.

In this way we can construct an effective coupling of the string to various fields which condense in the vacuum as a result of the interaction. Now we can study the properties of the operator (3), (4). Of course in general G^{ab} is not near the identity and defines a manifold M. It is best to write a representation for the string propagator i.e. Δ^{-1}, in terms of a sum over random surfaces propagating between 2 fixed string configurations which could also be chosen to be points as a limiting case.

By analogy with point field theory where the field propagator can be represented as a sum over random paths,[4,5,6,7] we write

$$\langle c|\Delta^{-1}|c'\rangle = \langle \Psi(c)\Psi(c')\rangle = \int_c^{c'} \mathcal{D}X(\xi)\,\mathcal{D}g_{\mu\nu}(\xi)\, e^{-A(x;g)}$$

$$A(x;g) = \frac{1}{4\lambda^2}\int_{\sigma_2}\sqrt{g}\, g^{\mu\nu}\partial_\mu X_a \partial_\nu X_b\, G^{ab}(x) + i\int_{\Sigma_2} dx_a \wedge dx_b\, A^{ab}(x) + \mu^2\int_{\sigma_2} d^2\xi\sqrt{g} \tag{5}$$

(5) is a 2-dim. field theory in parameter space of (ξ_1, ξ_2) which parametrize the world sheet σ_2. $X_a(\sigma)$ are the string co-ordinates valued in the manifold M, with metric $G^{ab}(X)$. Σ_2 is the image of σ_2 in M. We have introduced in (5) the 2 dimensional metric $g_{\mu\nu}(\sigma)$ in order that the action A is reparametrization invariant. This world sheet symmetry is forced upon us because (5) represents a string green's function which is only a function of the 'end points' C and C'. Another symmetry of the string action is Weyl invariance: $g_{\mu\nu}(\xi) \rightarrow e^{\alpha(\xi)}\, g_{\mu\nu}(\xi)$. The Weyl transformation scales physical

distances on the world sheet.

We emphasize that here we have not proved the representation (5). More precisely we have not constructed the precise form of the string operator defined by (5). This will be presented elsewhere. For the present purpose we simply accept (5) as a definition of Δ and study the corresponding 2-dim. field theory.

The anti-symmetric tensor field $A^{ab}(X)$ in (5) may admit singularities. In case C and C' in (5) are pinched to points we are dealing with vacuum diagrams which involve closed surfaces. In this case (5) can be written in terms of the field strength F = d A. Employing Stokes theorem

$$\int_{\Sigma_2} dX_a \wedge dX_b\, A^{ab}(X) = \int_{\Sigma_3} dX_a \wedge dX_b \wedge dX_c\, F^{abc}\,, \quad \partial\Sigma_3 = \Sigma_2 \tag{6}$$

This is possible only if every image Σ_2 of the string world sheet σ_2 is a boundary of some 3 surface Σ_3 in M. This requires the second Betti number of M to vanish: $\beta_2(M) = 0$. (6) is a generalization of the Wess-Zumino term to string theory.[6] Here since in general σ_2 is homeomorphic to a sphere with handles, the usual homotopy requirement $\pi_2(M) = 0$ is replaced by $\beta_2(M) = 0$. We further require that F, though closed, is not exact i.e. the 3rd Betti number is non-zero: $\beta_3(M) \neq 0$. Furthermore F is an integral harmonic form. These are analouges of monopoles.

We mention that the couplings of the metric G^{ab} in (5) is general co-ordinate invariant reflecting this properly of the laplacian (3). Similarly the coupling of A^{ab} is invariant under $A \to A + d\Lambda$, reflecting the gauge invariance of covariant derivative $\mathcal{D}_a$ in (4). The coupling of the dilaton field to the string is quite subtle.[7] In the path integral representation it corresponds to the term $\int d^2\xi \sqrt{g}\, R^{(2)} \Phi$. $R^{(2)}$ is the 2-dim. curvature scalar. Note that the vacuum expectation value of Φ multiplies the Euler invariant $\chi = \int d^2\xi \sqrt{g}\, R^{(2)} = 2 - 2H$, where H is the number of handles. Hence $e^{\langle\Phi\rangle}$ is the coupling constant of the string field theory and the path integral in (5) is represented as

$$\langle \Psi(C)\, \Psi(C') \rangle = \sum_{H=0}^{\infty} \left(\frac{1}{N}\right)^{H} Z_H \tag{7}$$

where $N = \exp(2\langle\Phi\rangle)$ and Z_H denotes the path integral with σ_2 having H handles.

REPARAMETRIZATION AND WEYL INVARIANCE:

We have already mentioned that the 2 dim. field theory (5) must be invariant under reparametrizations of σ_2. It is possible to define (5) by a cutoff procedure which respects this symmetry. In general the quantum field theory with cutoff violates Weyl invariance since this symmetry corresponds to scaling of physical distances on the world sheet. However a consistent string theory requires the quantum theory to maintain this symmetry. The combined requirement of reparametrization and Weyl invariance can serve to restrict the 'condensates' G^{ab}, A^{ab} etc.

In what follows we first study some general principles related to the study of strings coupled to non-trivial background fields and then illustrate these ideas by a concrete example.

In order being over the techniques of conformal invariant field theory we need to use the reparametrization invariance of (5) to bring the metric locally to the conformal gauge: $g_{\mu\nu} = e^{\phi}\,\delta_{\mu\nu}$. Physical distances on the world sheet are given in this gauge by $(ds)^2 = \int e^{\phi}\, dz\, d\bar{z}$, where $Z = \xi_1 + i\,\xi_2$ and $\bar{Z} = \xi_1 - i\,\xi_2$. This distance is invariant under the residual analytic reparametrizations $z \to \omega(z)$, $\bar{z} \to \bar{\omega}(\bar{z})$ and $\phi \to \phi - \ln|\omega'(z)|^2$. We call these conformal transformations. Further $(dS)^2 \to e^{\alpha}\,(dS)^2$, under the Weyl transformation that is a constant.

An important notion in the conformal gauge is that of conformal tensors, which transform in a particularly simple way under conformal transformations. The components of these tensors are labeled by Z and $\bar{Z}$ and the transformation law when $z \rightarrow \omega(z)$ is

$$T_{\underbrace{zz\cdots z}_{n}\,\underbrace{\bar{z}\cdots\bar{z}}_{m}} \rightarrow \left(\frac{\partial\omega}{\partial z}\right)^{n}\left(\frac{\partial\bar{\omega}}{\partial\bar{z}}\right)^{m} T_{z\cdots z\,\bar{z}\cdots\bar{z}} \tag{8}$$

n + m is the conformal dimension of the tensor and (n-m) is the conformal spin. In particular the components of the metric are $g_{Z\bar{Z}} = g_{11} + g_{12}$, $g_{ZZ} = g_{11} - g_{22} + 2ig_{12}$ and $g_{\bar{Z}\bar{Z}} = \bar{g}_{ZZ}$. Raising and lowering of indices is simply multiplication by $g^{Z\bar{Z}}$ or $g_{Z\bar{Z}}$. Further covariant differentiation is particularly simple

$$\begin{aligned} \nabla_{\bar{z}} T_{z\cdots} &= \partial_{\bar{z}} T_{z\cdots} \\ \nabla_{z} T_{z\cdots} &= (g_{z\bar{z}})^{n}\, \partial_{z}\left[(g^{z\bar{z}})^{n} T_{z\cdots} \right] \end{aligned} \tag{9}$$

Now it is easy to calculate the Faddeev-Popov determinant in the conformal gauge. To do this let us parametrize the metric near the point $g^{\circ}_{\mu\nu} = \delta_{\mu\nu} e^{\phi_{0}}$ on the gauge slice by infinitisimal reparametrizations $Z \rightarrow Z + v^{Z}$. Writing $g = g_{0} + \delta g$

$$\begin{aligned} \delta g_{z\bar{z}} &= (\nabla^{\circ}_{z} v^{z} + \nabla^{\circ}_{\bar{z}} v^{\bar{z}})\, g^{\circ}_{z\bar{z}} + \delta\phi\, e^{\phi_0} \\ \delta g_{zz} &= 2\nabla^{\circ}_{z} v_{z} \,, \quad \delta g_{\bar{z}\bar{z}} = 2\nabla^{\circ}_{\bar{z}} v_{\bar{z}} \end{aligned} \tag{10}$$

since the volume in the space of metrics is

$$\prod_{\xi} \delta g_{zz}\, \delta g_{z\bar{z}}\, \delta g_{\bar{z}\bar{z}} = \prod_{\xi} \delta\phi\, \delta v^{z}\, \delta v^{\bar{z}}\, \Delta_{FP} \tag{11}$$

The Faddeev-Popov determinant works out to be $\Delta_{FP} = \det \nabla^{Z} \det \nabla^{\bar{Z}}$.[5] We have not worried about the global aspects of gauge fixing associated with the zero modes of the Faddeev-Popov operators. Introducing ghost fields b_{ZZ} and C^{Z}, the Faddeev-Popov determinant can be written in terms of local action

$$\Delta_{FP} = \int e^{\int d^2\xi \sqrt{g}\,(b_{zz} \nabla^{z} C^{z} + c.c.)} \prod_{\xi} db_{zz}\, dC^{z} \tag{12}$$

In the conformal gauge the ghosts are free fields and their action works out to be

$$A(b,c;g) = \int d^2\xi \left(b_{zz}\, \partial_{\bar{z}} C^z + c.c.\right) \tag{13}$$

Hence in the conformal gauge the classical action (5) plus the ghost action are

$$A(x;g) + A(b,c;g) = \frac{1}{4\lambda^2}\int_{\sigma_2} d^2\xi\, \partial_\mu X_a \partial_\nu X_b G^{ab} + i\int_{\Sigma_2} dX_a \wedge dX_b A^{ab}(x) + \int d^2\xi \left(b_{zz}\partial_{\bar{z}} C^z + c.c.\right) \tag{14}$$

(14) describes a non-linear σ -model plus a ghost action. The path integral (7), henceforth denoted by Z becomes

$$Z = \int \mathcal{D}b_{zz}\mathcal{D}C^z\mathcal{D}X_a\; e^{-[A(x;g)+A(b,c;g)]} \tag{15}$$

CONFORMAL ANOMALY AS A JACOBIAN OF LOCAL RESCALINGS, WARD IDENTITIES AND OPERATOR PRODUCT EXPANSIONS

Let us denote a string of operator porducts of the string theory by $X(Z) \equiv X(X_a(Z))$. (For example X can be the operator product of the currents $J^\alpha = e^\alpha_a\, \partial_z X^a$, e^α_a is a vierbien such that $e^\alpha_a\, e^\alpha_b = G_{ab}$: $X = J^\alpha(z,\bar{z})\, J^\beta(w,\bar{w}) \ldots\ldots\ldots$). Now consider the greens function

$$\langle X\rangle_g = \int \frac{\mathcal{D}X_a}{Z}\; X\; e^{-[A(x;g)+A(b,c;g)]} \tag{16}$$

Under infinitisimal conformal transformations $Z \to Z + v^Z$ and $\bar{Z} \to \bar{Z} + v^{\bar{Z}}$, the fields transform as

$$\begin{aligned} \delta X_a(z,\bar{z}) &= v^z \partial_z X_a + v^{\bar{z}} \partial_{\bar{z}} X_a \\ \delta b_{zz} &= v^z \partial_z b_{zz} + 2\, b_{zz}\, \partial_z v^z \\ \delta C^z &= v^z \partial_z C^z - 1\, C^z \partial_z v^z \end{aligned} \tag{17}$$

$$\delta A = \int d^2z \left(v^z \partial_{\bar{z}} T_{zz} + c.c.\right) \tag{18}$$

where T_{ZZ} is the traceless part of the stess tensor given by

$$T_{zz} = \frac{1}{4\lambda^2} G^{ab} \partial_z X_a \partial_z X_b + 2 b_{zz} \partial_{\bar{z}} C^z + \partial_z b_{zz} C^z \tag{19}$$

The measure is not necessarily invariant and transforms by a non-trivial jacobian which in the conformal gauge is the local expression $J \simeq (1 + \int d^2 z \sqrt{g} [v^z b_z + v^{\bar{z}} b_{\bar{z}}])$. Then invariance of the functional integral under this change of variables leads to the ward identity

$$\langle \delta X(z') \rangle = \langle \int d^2 z \sqrt{g} \{ v^z (b_z + \nabla^z T_{zz}) X(z') + c.c. \} \rangle \tag{20}$$

In the derivation (20), it is important to note that we held the world sheet metric $g_{\mu\nu}$ fixed. Hence the conformal transformation indeed changes distances on the world sheet and hence is not the conformal reparametrizations discussed before, but local Weyl rescalings[6].

Let us derive some consequences of this Ward identity. If we put X = 1, we get the equation $b_Z = \nabla^Z T_{ZZ}$. Now a local dim 1 tensor $\nabla^Z T_{ZZ}$ can only be proportional to $\nabla_Z R$, where R is the 2 dim. scalar curvature on the world sheet. Hence we get the important equation $b_Z = \frac{-\lambda}{24} \nabla_Z R$, where λ is an undetermined constant. In fact it is possible to rewrite b_Z such that

$$\nabla^z T_{zz} + \frac{\lambda}{24} \nabla_z R = g^{z\bar{z}} \partial_{\bar{z}} \left[T_{zz} + \frac{\lambda}{24} (\partial_z \phi \partial_z \phi - 2 \partial_z \partial_z \phi) \right]$$

The Ward identity now reads

$$\langle \delta X(z') \rangle = \int d^2 z \{ v^z \partial_{\bar{z}} \langle T^0_{zz} X(z') \rangle + z \to \bar{z} \} \tag{21}$$

where $T^0_{ZZ} = T_{ZZ} + \frac{\lambda}{24} (\partial_z \phi \partial_z \phi - 2 \partial_z \partial_z \phi)$. The second term is the stress tensor of the Liouwille field theory[4,5]. λ is called the 'Conformal' anomaly coefficient. Actually it is the Weyl anomaly.

Now as long as Z and Z' are not close by one get the conservation law $\partial_{\bar{z}} \langle T^0_{zz} X(z') \rangle = 0$. When X = 1, this says that T^0 is analytic. In case Z and Z' are close to each other the operator product $T^0_{ZZ} X(Z')$ can have singularities at Z = Z'. In this case (21) becomes

$$\langle \delta X(z') \rangle = \langle \oint_C dz\, v^z T^0_{zz}\, X(z') \rangle \tag{22}$$

The contour C in (22) is around the singular points of the operator product. In particular for $T^0_{ZZ} \equiv T^0(Z)$ itself we have

$$\langle \cdots \delta T^0(z') \cdots \rangle = \langle \cdots \oint_C dz\, T^0(z) T^0(z') \cdots \rangle \tag{23}$$

Since T^0 is a conformal tensor of rank 2, its conformal transformation is given by

$$\delta T^0 = v(z) T^{0\prime}(z) + 2 T^0(z) v'(z) + \frac{C}{2} v'''(z) \tag{24}$$

In (24) we have added an unknown C-number term. (24) and (23) imply the operator product expansion[8]

$$T^0(z) T^0(z') \sim \frac{C/2}{(z-z')^4} + \frac{2T^0(z')}{(z-z')^2} + \frac{T^{0\prime}(z')}{(z-z')} + \text{reg. terms.} \tag{25}$$

It is an important theorem that $C = -\lambda$: The central charge C is (upto a sign) the same as the coefficient of the Liouville stess tensor. We shall not go into the proof here[6].

VIRASORO ALGEBRA

Equation (25) is one of the fundamental equations of string theory. It is equivalent to the familiar Virasoro algebra. Let us derive this. The Virasoro generators L_n are defined by $L_n = \oint dz\, z^{n+1} T^0(z)$. Then consider the commutator in a green function

$$\langle [L_n, T^0(z)] \rangle = \langle \oint_{C_z} dw\, w^{n+1} T^0(w) T^0(z) - \oint_{C_0} dw\, w^{n+1} T^0(z) T^0(w) \rangle \tag{26}$$

C_Z is chosen to circle the origin and Z and C_0 encircles the origin but not Z. Then by deforming the contour we get

$$\langle [L_n, T^\circ(z)] \rangle = \oint_C d\omega\, \omega^{n+1} \langle T^\circ(\omega) T^\circ(z) \rangle \tag{27}$$

Substituting from the operator product expansion (25) we get using Canchy's theorem

$$[L_n, T^\circ(z)] = z^{n+1} \frac{\partial}{\partial z} T^\circ(z) + 2(n+1) z^n T^\circ(z) + \frac{c}{12} n(n^2-1) z^{n-2} \tag{28}$$

from which the Virasoro algebra immediately follows

$$[L_n, L_m] = (n-m) L_{n+m} + \frac{c}{12} n(n^2-1) \delta_{n+m,0} \tag{29}$$

From (19) we see that the central charge C of the Virasoro algebra is a sum of contributions from matter fields and the ghost. The matter field contribution is a characteristic of the system, and in general difficult to compute. The ghost part is easily calculable since we know from the action that $b_{ZZ} C^W \sim (Z-W)^{-1}$. Then[4,5]

$$T_{ghost}(z) T_{ghost}(\omega) \sim \frac{-26/2}{(z-\omega)^4} + \cdots \tag{30}$$

Hence the ghost contribution to the central charge is $C = -26 < 0$.

If we denote by C_i the contribution to the central charge coming from the ith matter field, then $C = \sum_i C_i - 26$. Since matter fields by definition belong to unitary representation of the Virasoro algebra, $C_i \geqslant 0$ and further we require C 0.

WEYL INVARIANCE

Uptil now we have studied some consequences of reparametrization invariance and in particular found that maintaining this symmetry necessarily leads to an anomaly for Weyl invariance. As we had mentioned the central charge of the Virasoro algebra is in fact the coefficient of the Liouville stress tensor. Therefore Weyl invariance is ensured if $C = \sum_i C_i - 26 = 0$. This means that when $C = 0$, physics on the string world sheet is truly scale invariant and the trace of stress tensor $T_{Z\bar{Z}} = 0$. Around the perturbative vacuum $G^{ab} = \eta^{ab}$, where the matter central charge is the number of space-time dimensions, this fact implies the closure of the lorentz algebra for d = 26. It is important to mention that C = 0 is also a consequence of the nilpotency of the BRS charge corresponding to the matter + ghost system (14).

STRINGS ON A GROUP MANIFOLD

To illustrate some of the general concepts we study a particularly useful example where the manifold on which a string lives is $M = R_d x G$. R_d is d-dimensional minkowskispace and G is a compact manifold. Let us denote the d-coordinates of R_d by $X_i(Z,\bar{Z})$ and those of the compact manifold G by $\theta_j(Z,\bar{Z})$. The metric on G can be written in terms of vierbiens: $G_{ab} = \sum_\alpha e^\alpha_a e^\alpha_b$. Then the traceless part of the stress tensor $T(Z,\bar{Z})$ has the Sugawara form:

$$T(z,\bar{z}) = \frac{1}{K_1} \alpha^i_z \alpha^i_z + \frac{1}{K_2} J^\alpha_z J^\alpha_z \tag{31}$$

where $\alpha^i_z = \partial_z x^i$ and $J^\alpha_z = e^\alpha_a \partial_z \theta^a$ are local currents.

K_1 and K_2 are normalizations of the stess-tensor which may not be identical to those following from the classical Noether's theorem. In fact K_2 receives quantum corrections. For the case of a general manifold if one knows the current algebra of J_Z , then one can completely characterize the system. Unfortunately we have not made progress on this general problem and hence will hence forth restrict G to be a compact Lie group[6].

Let us denote the group elements by $U(z,\bar{z}) = e^{i t_a \theta^a}$. The anti-hermetian generators satisfy the lie algebra $[t_a, t_b] = f^{abc} t_c$ and are normalized as $\mathrm{tr}\, t_a t_b = -2\delta_{ab}$. In this case the action (14) becomes

$$A = \frac{1}{2}\int_{\sigma_2} d^2\xi\, \partial_\mu x_i \partial_\mu x_i + \frac{1}{4\lambda^2}\int_{\sigma_2} d^2\xi\, \mathrm{tr}\, \partial_\mu U \partial_\mu U^{-1} \tag{32}$$

$$+ \frac{ik}{N_G}\int_{\Sigma_3} d^3\xi\, \mathrm{tr}\, \partial_\mu U U^{-1} \partial_\nu U U^{-1} \partial_\lambda U U^{-1} \epsilon_{\mu\nu\lambda} \quad + \text{ghost terms}$$

(32) is the Wess-Zumino type model. k is an integer and N_G is a normalization factor depending on the group.

A very instructive one loop calculation valid for small λ around a background field $U_0(Z,\bar{Z})$ gives rise to the following effective action

$$S_{eff}(\phi) = \frac{(26-d-\dim M)}{48\pi}\int d^2\xi\left(\tfrac{1}{2}\partial_\mu\phi\partial_\mu\phi + \mu^2 e^{\phi}\right) + \frac{C_V}{8\pi}(1-\alpha^2)\int d^2\xi\, \mathrm{tr}\,\partial_\mu U_0\partial_\mu U_0^{-1}\ln\left(\frac{\Lambda^2}{\mu^2}e^{\phi}\right) \qquad (33)$$

Λ is a cutoff and μ is a fixed scale. C_V is the casimir of the adjoint representation: $f_{adc}f_{bdc} = C_V\delta_{ab}$ and $\alpha^2 = \left(\frac{6\lambda^2 k}{N_G}\right)^2$. From (33) we can easily calculate the trace of the stress tensor

$$T_{z\bar{z}} = \frac{\delta S_{eff}}{\delta\phi} = \frac{1}{48\pi}(26-d-\dim M)\left(\sqrt{g}R^{(2)}+\mu^2\right) + \frac{C_V}{8\pi}(\alpha^2-1)\,\mathrm{tr}\,\partial_\mu U_0\partial_\mu U_0^{-1} \qquad (34)$$

Reparametrization and Weyl invariance imply $T_{Z\bar{Z}} = 0$, i.e. $\alpha^2 = 1$ and $26 = d + \dim M$. From this simple calculation we see that these requirements constrain the parameters of this model.

We now proceed to demonstrate that at $\alpha^2=1$, the model indeed forms a representation of the Virasoro algebra. The crucial observation in this direction is due to Knizhnik and Zamolodchikov,[9,10] who pointed out that at $\alpha^2=1$ (32) has a local symmetry: $U(Z,\bar{Z}) \longrightarrow \Omega(Z)U(Z,\bar{Z})\tilde{\Omega}(\bar{Z})$ and $X_i(Z,\bar{Z}) \rightarrow X_i(Z,\bar{Z}) + f_i(Z) + g_i(\bar{Z})$. It is easy to see that invariance leads to conservation laws for the currents $J = \frac{12k}{N_G}\partial_z U U^{-1}$, $\bar{J} = \frac{12k}{N_G}U^{-1}\partial_{\bar{z}}U$, $\alpha_z^i = \partial_z X^i$, $\alpha_{\bar{z}}^i = \partial_{\bar{z}}X^i$:

$$\partial_{\bar{z}}J = \partial_z\bar{J} = \partial_{\bar{z}}\alpha_z^i = \partial_z\alpha_{\bar{z}}^i = 0 .$$

The local transformation laws lead to the operator product expansions of these currents. To see this we simply note that infinitisimal gauge transformations for $J(Z)$ and α_Z^i are

$$\delta_\omega J^b(z) = f^{abc}\omega^a(z)J^c(z) + \frac{12k}{N_G}\partial_z\omega^b(z) \qquad (35)$$

$$\delta_\epsilon \alpha^i(z) = i\,\partial_z\epsilon^i(z) \qquad (36)$$

Now since

$$\delta_\omega J^b(z) = \oint dz'\, \omega^a(z') J^a(z') J^b(z) \tag{37}$$

$$\delta_\epsilon \alpha^i(z) = \oint dz'\, \epsilon^j(z')\alpha^j(z')\alpha^i(z) \tag{38}$$

it is straight forward to see that the following operator product expansions (OPE) are valid

$$J^a(z)J^b(\omega) \sim \frac{12k/N_G}{(z-\omega)^2} + f^{abc}\frac{J^c(\omega)}{(z-\omega)} + \cdots \tag{39}$$

$$\alpha^i(z)\alpha^j(\omega) \sim \frac{i}{(z-\omega)^2} \tag{40}$$

Similar OPE are valid for $\bar{J}$ and $\alpha^i_{\bar{Z}}$.

Another important ingredient is the conformal transformation properties of these conserved dimension 1 currents which leads to the OPE

$$T(z)J^a(\omega) \sim \frac{J^a(\omega)}{(z-\omega)^2} + \frac{J^{a\prime}(\omega)}{(z-\omega)} + \cdots \tag{41}$$

$$T(z)\alpha^i(\omega) \sim \frac{\alpha^i(\omega)}{(z-\omega)^2} + \frac{\alpha'^i(\omega)}{(z-\omega)} + \cdots \tag{42}$$

The OPE (40) and (42) is the Heisenberg system. (39) and (41) is the Kac-Moody system. It is a simple matter to check (using Wick's theorem) that the definition (31) of T(Z) is consistent with (41) and (42) only if

$$K_1 = \frac{1}{2} \quad , \quad K_2 = -\left(C_V + \frac{48\pi k}{N_G}\right) \tag{43}$$

Note that C_V appearance of C_V is a purely quantum effect. After this it is straightforward to check that the Virasoro algebra (25),(29) is satisfied by this system only if the central charge is

$$\sum_i C_i = d + \frac{k \dim G}{k + C_v\left(\frac{N_G}{48\pi}\right)} \tag{44}$$

and Weyl invariance requires this to cancell the ghost contribution of -26. Hence we have the basic formula

$$d + \frac{k \dim G}{k + C_v\left(\frac{N_G}{48\pi}\right)} = 26 \tag{45}$$

It turns out that $N_G = 48$ for SU(n) and $N_G = 24$ for orthogonal and excptional groups.

Take for example the group $E_8 \times E_8$. Then $C_v = 60$ and (45) becomes

$$d + 2\left(\frac{248\,k}{k+30}\right) = 26 \tag{46}$$

The extra factor of 2 is there since there are 2 copies of E_8. If we choose k = 1, then (46) becomes

$$d + 2(\text{rank of } E_8) = 26$$

and hence d = 26 - 16 = 10! This is the choice for the heterotic string.

Before ending this section we mention that formula (45) reflects a local property of the string world sheet. It is independent of the global properties of the world sheet like the appearance of handles. Hence higher loops in the string field theory do not renormalize this true level result.

SPECTRUM:

We briefly sketch the perturbative spectrum of the above model[6]. We restrict ourselves to a world sheet without handles and consider writing the currents $J^a(Z)$, $\bar{J}^a(\bar{Z})$, α_Z^i, $\alpha_{\bar{Z}}^i$ in terms of oscillators, which are coefficients of the laurent expansions:

$$J^a(z) = \sum_{-\infty}^{+\infty} \frac{J^a_n}{z^n} \quad , \quad \bar{J}^a(\bar{z}) = \sum_{-\infty}^{+\infty} \frac{\bar{J}^a_n}{\bar{z}^n}$$

$$\alpha^i_z = \sum_{-\infty}^{+\infty} \frac{\alpha^i_n}{z^n} \quad , \quad \alpha^i_{\bar{z}} = \sum_{-\infty}^{+\infty} \frac{\bar{\alpha}^i_n}{\bar{z}^n} \tag{47}$$

Now the OPE (39) and (40) imply the Kac-Moody and Heisenberg algebras for the oscillators

$$[J^a_n, J^b_m] = f^{abc} J^c_{n+m} + n \frac{24\pi k}{N_G} \delta_{n+m,0} \tag{48}$$

$$[\alpha^i_n, \alpha^j_m] = -n \delta^{ij} \delta_{n+m,0} \tag{49}$$

As was noted earlier the Virasoro operators are defined by the laurent expansion of the stress tensor T(Z)

$$T(z) = \sum_{-\infty}^{+\infty} \frac{L_n}{z^{n+2}} \tag{50}$$

Then the OPE (41), (42) imply the equations

$$[L_n, J^a_m] = -m J^a_{n+m} \quad , \quad [L_n, \alpha^i_m] = -m \alpha^i_{n+m} \tag{51}$$

Further the Sugawara form (31) implies that

$$L_n = \frac{1}{K_1} : \sum \alpha^i_m \alpha^i_{n-m} : + \frac{1}{K_2} : \sum J^a_m J^a_{n-m} :$$

$$\bar{L}_n = \frac{1}{K_1} : \sum \bar{\alpha}^i_m \bar{\alpha}^i_{n-m} : + \frac{1}{K_2} : \sum \bar{J}^a_m \bar{J}^a_{n-m} : \tag{52}$$

we have indicated in (52) that L_0 and $\bar{L}_0$ are normal ordered i.e.: $:J^a_n J^a_{-n}: = J^a_{-n} J^a_n$, $n > 0$. Formulae (43) and (45) can also be derived using the oscillator representation.

The 'vacuum' is defined by

$$J_n^a|0\rangle = \bar{J}_n^a|0\rangle = \alpha_n^i|0\rangle = \bar{\alpha}_n^i|0\rangle = 0\,,\; n>0 \tag{53}$$

and all physical states of the string satisfy the Virasoro conditions

$$L_n|\Psi\rangle = 0\,,\; n>0 \tag{54}$$

$$(L_0 + \bar{L}_0 - 2)|\Psi\rangle = 0 \tag{55}$$

$$L_0|\Psi\rangle = \bar{L}_0|\Psi\rangle \tag{56}$$

These conditions are a consequence of BRS invariance[13]. (56) says that the string state has zero conformal spin. Equation (55) is by definition the 'equation of motion'. It contains the zero mode oscillators α_0^i and $\bar{\alpha}_0^i$ which can be identified with the momentum of the string: $\alpha_0^i = \bar{\alpha}_0^i = p^i$. Then (55) implies.

$$\frac{1}{4}(\text{Mass})^2 = -\frac{P_i^2}{4} = \frac{1}{K_1}\sum_{n\neq 0}\left(\alpha_{-n}^i\alpha_n^i + \bar{\alpha}_{-n}^i\bar{\alpha}_n^i\right) + \frac{1}{K_2}\sum_n\left(J_{-n}^a J_n^a + \bar{J}_{-n}^a\bar{J}_n^a\right) - 2 \tag{57}$$

From (57) we see that the vacuum is a tachyon:

$$(\text{Mass})^2|0\rangle = -2|0\rangle$$

and the massless levels are given by

$\alpha_{-1}^i\bar{\alpha}_{-1}^j|0\rangle$	gravitational multiplet
$J_{-1}^a\bar{\alpha}_{-1}^i|0\rangle$	vector boson, G x 1
$\alpha_{-1}^i\bar{J}_{-1}^a|0\rangle$	Vector boson, 1 x G
$J_{-1}^a\bar{J}_{-1}^a|0\rangle$	charged scalar, G x G

The traceless symmetric, antisymmetric and the trace part of the gravitational multiplet are identified with the graviton, anti-symmetric tensor and dilaton, massless particle states. There are 2 types of charged vector bosons since the gauge group is G x G. There are also charged scalars. The Virasoro conditions $L_n|\Psi\rangle = 0$, $n > 0$ serve to tell us that the longitudinal modes of the gravitons and vector particles are unphysical modes that do not satisfy the gauge conditions $L_n|\Psi\rangle = 0$, $n > 0$. For details of this we refer to standard review articles.

We remark that the current algebra method of introducing gauge quantum numbers in string theory is equivalent to the method of Frenkel and Kac. In fact the currents we mentioned act as vertex operators.

It would be very useful to generalize this algebraic method we have presented to more complicated manifolds. Some how it should be possible to code the local structure of the string manifold in the OPE of an appropriate current algebra. Progress on this problem will have implications for the type of issues we began these notes with, viz. the problem of writing string theories independent of particular choice of background geometries which in turn must be determined as a consequence of the equations of motion.

CONCLUDING REMARKS

There is no question that the principle of conformal invariance has deep consequences for string theory. We have presented certain aspects of this program: motivating how this principle arises, establishing the framework of conformal invariant field theory within Polyakov's formulation of string quantum mechanics and illustrating these methods by the case of a string moving on a group manifold. It would be worthwhile taking back these insights to string field theory.

This program has been carried out in other directions by several authors. Notably the equation $T_{z\bar{z}} = 0$, has been computed upto 2 loops and the equations of motion for the background fields G_{ab}, A_{ab}, Φ and the Yang-Mills fields A^a_μ have been written down to this order, in 10-spacetime dimensions[14].

We want to make a comment on one consequence of these equations. One of the very remarkable solutions to these set of equations is the embedding of the spin connection of space time in the gauge group. The spin connection being determined by Einsteins equations $R_{ij} = 0$. This equation has a black-hole as its solution. Now consider string propagation in presence of the black hole. The string path integral (5) must now be interpreted as a thermodynamic partion function at a temperature proportional to planck mass. Now it is well known that at such temperatures the free energy diverges. There is a possible phase transition. We remark that this phase transition is well known, however what we find remarkable is that the string itself has produced a configuration which gives us an inkling that one may need new degrees of freedom for distances shorter than Planck length[15].

ACKNOWLEDGEMENT

I wish to thank Professor Abdus Salam and the organisers of the Workshop Professors Pati, Iengo, Shafi and Furlan for the opportunity to present these lectures. I also wish to thank ICTP for its hospitality.

REFERENCES

1) See for example articles in Proceeding of Symposium on "Anomalies, Geometry and Topology", Ed. Bardeen, W.A. and White, A.R., World Scientific Publishing Co.
2) Kaku, M. and Kikkawa, K., Physics Rev. D10 1110, 1823 (1974).
3) Das, S.R. and Wadia, S., Fermi Lab/TIFR preprint 1985; Sen, A. Fermi Lab preprint; Fradkin, E.S. and Tseytlin, A.A., Lebedev Inst. Report (1985).
4) Polyakov, A.M., Phys. Lett. 103B, 207 (1981).
5) Friedan, D., Les Houches (1982); Alvarez, O., Nucl. Phys. B216, 125 (1983); Fujikawa, K., Phys. Rev. D25, 2584 (1982).
6) Jain, S., Shankar, R., and Wadia, S., Phys. Rev. D32 (1985).
7) Fradkin, E.S. and Tseytlin, A.A., Lebedev Inst. preprint N261(1984)
8) See for example Belavin, A., Polyakov, A. and Zamolodchikov, A., Nucl. Phys. B241, 333 (1980).
9) Knizhnik, V.G. and Zamolodchikov, A.B., Nucl. Phys. B247, 83 (1984).
10) Witten, E., Com. Math. Phys. 92, 455 (1984).
11) This result was derived in ref. 6. It was independently derived by P. Goddard and D. Olive, Nucl. Phys. B257, 226 (1985); Nemeschansky, D. and Yankielowicz, S., Phys. Rev. Lett. 54, 620 (1985).
12) Gross, D.J., Harvey, J.A., Martinec, E. and Rohm, R., Phys. Rev. Lett. 54, 502 (1985).
13) See Fujikawa, K. and Gervais, J.L., in this proceedings.
14) Sen, A., Fermi Lab preprint; Callan, C., Friedan, D., Martinec, E. and Perry, M., Princeton preprint.
15) These thoughts have also been expressed by several other workers including Green, M., Goddard, P., Gross, D. and Nahm, W.

LOWERING THE CRITICAL DIMENSION FOR HETEROTIC STRINGS

E. Sezgin
International Centre for Theoretical Physics, Trieste
ITALY

Currently the most promising and the most extensively studied string theory is the $d = 10$ heterotic superstring theory [1] with (1,0) worldsheet supersymmetry [*] and $E_8 \times E_8$ Yang-Mills symmetry. In order to make contact with the real world, clearly it is desirable to lower the critical dimension from ten down to four. One approach to this problem is spontaneous compactification on $M_4 \times K$ where M_4 is the four dimensional Minkowski spacetime and K is a six dimensional compact manifold [3]. However, this is not so easy in view of the fact that a covariant interacting field theory for the heterotic string is yet to be constructed. Nevertheless, one can couple the first quantized heterotic string to a curved background consisting of the massless sector of the heterotic string (i.e. $d = 10$ supergravity plus Yang-Mills [4] and by demanding the consistency of the coupling obtain information on the nature of the allowed background. Such a theory is expected to be described by a two dimensional σ-model whose target manifold is the $d = 10$ curved superspace augmented with additional 32 fermionic coordinates needed for the coupling of Yang-Mills[5),6)]. The theory presumably would have the $d = 2$ general coordinate, scale, Lorentz and Siegel [7] invariances as well as the $d = 10$ general coordinate, Lorentz, supersymmetry, tensor gauge and Yang-Mills gauge invariances. Consistency of the string coupling would then require the absence of anomalies in all these local symmetries. In particular, the absence of an anomaly in the local scale invariance amounts to the vanishing of all the β-functions. Furthermore, the anomalies in the local $d = 10$ symmetries will manifest themselves as the $d = 2$ "σ-model anomalies" [2),8)]. Showing that all the anomalies vanish to all orders in the σ-model perturbation theory would establish the

---

*) The $1 + 1$ dimensional supersymmetry algebra of type (p,q) is defined [2] to be that generated by p right-handed Majorana-Weyl supercharges and q left-handed ones.

consistency of the theory at the string tree-level. To show the consistency at the string 1-loop level one has to show that the worldsheet global gravitational anomaly vanishes [9] (i.e. the modular invariance holds).

The problem as stated above, to our knowledge, has not been tackled so far. However, various aspects of it, in the case of the Neveu-Schwarz-Ramond representation (as opposed to the Green-Schwarz representation) of the heterotic string in purely bosonic backgrounds have been studied by several authors [2],[3],[10] with many interesting results. By far the most extensively studied consistent heterotic string vacuum configuration is $M_4 \times$ Calabi-Yau manifold (i.e. a 3-complex dimensional Ricci flat Kahler manifold of SU(3) holonomy) where the SU(3) spin connection is identified with the SU(3) Yang-Mills field embedded in one of the E_8's [3].

An alternative approach to the problem of lowering the critical dimension of the heterotic string has been considered very recently by Narain [11]. Instead of first compactifying the 26 dimensional left moving sector to 10 dimensions and then the 10 dimensions of the left and right moving sector to $d = 4$, the author of Ref.(11) directly compactifies the (26-d) and (10-d) dimensions of the left and right moving sectors into tori, thereby going down to d dimensions in one step. Remarkably, it turns out that [11] if the lattice used in this construction is an even self-dual Lorentzian lattice with signature of (26-d) plus signs and (10-d) minus signs, then the d-dimensional heterotic string theory is consistent and one loop finite! Moreover, the massless sector of these theories in $d = 6$ and 4 turn out to correspond to the $N = 2$ and 4 supergravity coupled to rank (26-d) super Yang-Mills [11]. Clearly this is an important new development in the subject which is likely to trigger new investigations.

Another approach to the problem of lowering the critical dimension of the heterotic string, the one which we shall consider here, is the following. We consider a (1,0) heterotic σ-model with Wess-Zumino term [12] coupled to conformal supergravity in two dimensions [13] where the target manifold of the σ-model is $M_d \times G$, i.e. a d-dimensional Minkowski spacetime times an arbitrary group manifold. In addition to the (1,0) conformal supergravity fields, and the left and right moving $M_d \times G$ coordinates, the model also contains the left moving fermionic partners of these coordinates and an arbitrary number of right moving fermionic coordinates. The model is solvable provided that the coefficient of the Wess-Zumino term is related to that of the kinetic term. In this case, to find a consistent vacuum configuration for this heterotic string, we quantize the theory in the light cone gauge and establish the super Virasoro symmetry as well as the M_d global Lorentz symmetry. This yields two equations involving d, dim G, the second Casimir eigenvalue in the adjoint representation, the coefficient of the Wess-Zumino term and the number of the right moving fermionic coordinates. These two equations must be solved simultaneously, and their solution correspond to a d-dimensional heterotic string theory consistent at the string tree level and to all orders in the σ-model perturbation theory. In fact, we do find such solutions as we discuss below, and they include $d = 4$ and 6.

Until the $d = 2$ global gravitational anomalies are shown to be absent, and a reasonable spectrum is shown to exist, both of which we do not attempt to do here and deserve further investigation, the following can be regarded as an instructive exercise in the search for a consistent heterotic string theory in $d = 4$ (or 6). Before we begin with the derivation of the critical dimension formula, let us remark that, the analogous investigation was carried out for the bosonic string in Ref.(14) and for the vectorlike (1,1) supersymmetric spinning string in Ref.(15).

The general Lagrangian for the (1,0) heterotic σ-model with Wess-Zumino term coupled to $d = 2$ conformal supergravity is given by [13] (Convention: $\eta_{ab} = (-+)$, $\varepsilon_{01} = 1$),*)

$$
\begin{aligned}
(2\pi\alpha')e^{-1}\mathcal{L} = & -\frac{1}{2} g^{\mu\nu}\partial_\mu\phi^i\partial_\nu\phi^j g_{ij} - \frac{1}{2}\lambda\varepsilon^{\mu\nu}\partial_\mu\phi^i\partial_\nu\phi^j b_{ij} \\
& -\frac{i}{2}\bar{\chi}_i\gamma^\mu(\partial_\mu\chi^i + \Gamma^i_{jk}\partial_\mu\phi^j\chi^k) - i\bar{\psi}_\mu\gamma^\nu\gamma^\mu\chi_i\partial_\nu\phi^i \\
& +\frac{1}{3}\lambda T_{ijk}\bar{\psi}_\mu\chi^i\bar{\chi}^j\gamma^\mu\chi^k
\end{aligned}
$$

---

*) Note the misprint in the coefficient of the $T\psi\chi^3$ term in Ref.(13).

$$-\frac{i}{2}\bar{\psi}^A e_a^\mu \gamma^a \left(\partial_\mu \psi^B + A_i{}^B{}_C(\phi)\partial_\mu \phi^i\right)\psi^C G_{AB}(\phi)$$

$$-\frac{1}{8}\hat{F}_{ij\,AB}\,\bar{\chi}^i \gamma^\mu \chi^j\, \bar{\psi}^A \gamma_\mu \psi^B \tag{1}$$

where α' is the string tension, λ is an arbitrary constant, (e^a_μ, ψ_μ) are the fields of the off-shell $d = 2$ (1,0) conformal supergravity, ϕ^i are the coordinates of an arbitrary Riemannian manifold M, $\chi^i = -\gamma_5\chi^i$ are their fermionic partners, and $\psi^A = +\gamma_5\psi^A$ $(A = 1,..\ell)$ are the fermions with no bosonic partners. Furthermore, g_{ij} is the metric on M, and Γ^i_{jk} is the torsionful connection on M given by

$$\Gamma^i_{jk} = \left\{ {i \atop j\,k} \right\} - \lambda T_{jk}{}^i \tag{2}$$

The torsion T_{ijk} is totally antisymmetric and is defined in terms of the torsion potential (tordion) $b_{ij}(\phi) = -b_{ji}(\phi)$ as

$$T_{ijk} = \frac{1}{2}\left(\partial_i b_{jk} + \partial_k b_{ij} + \partial_j b_{ki}\right) \tag{3}$$

The field $A_i{}^B{}_C(\phi)$ is the Yang-Mills connection of $SO(\ell)$ (or a subgroup thereof), and $G_{AB}(\phi) = G_{BA}(\phi)$ is an arbitrary metric function [2]. The field $\hat{A}_i{}^B{}_C(\phi)$ is defined by [2]

$$\hat{A}_i{}^B{}_C(\phi) = A_i{}^B{}_C + \frac{1}{2}G^{BA}G_{AC,i} \tag{4}$$

Note that $\hat{F}_{ijAB}$ is the curvature of this connection.

The action of the Lagrangian (1) is invariant under the $d = 2$ local supersymmetry (ε), special supersymmetry (η), dilatation (Λ_D) and Lorentz transformations (Λ_M) as follows [13]

$$\delta e^a_\mu = 2i\bar{\varepsilon}\gamma^a\psi_\mu - \Lambda_D e^a_\mu + \Lambda_M \varepsilon^{ab} e^b_\mu$$

$$\delta\psi_\mu = \left(\partial_\mu + \frac{1}{2}\omega_\mu(e,\psi)\right)\varepsilon + \gamma_\mu\eta - \frac{1}{2}\Lambda_D\psi_\mu - \frac{1}{2}\Lambda_M\psi_\mu$$

$$\delta\phi^i = -i\bar{\varepsilon}\chi^i$$

$$\delta\chi^i = -\gamma^\mu(\partial\!\!\!/_\mu\phi^i + i\bar{\psi}\!\!\!/_\mu\chi^i)\varepsilon + \tfrac{1}{2}\Lambda_D\chi^i + \tfrac{1}{2}\Lambda_M\chi^i$$

$$\delta\psi^A = i\bar{\varepsilon}\chi^i\,\hat{A}_i{}^A{}_B\psi^B + \tfrac{1}{2}\Lambda_D\psi^A - \tfrac{1}{2}\Lambda_M\psi^A \qquad (5)$$

where $\omega_\mu(e,\psi) = \varepsilon^{\rho\sigma}(-e_\mu^a\partial_\rho e_{\sigma a} + i\bar{\psi}_\rho\gamma_\mu\psi_\sigma)$. Finally, we note that all the spinors are Majorana-Weyl.

We now specialize to $M_d \times G$ with the following coordinates

$$\phi^i = \begin{pmatrix} x^\alpha \\ y^I \end{pmatrix} , \qquad \begin{matrix} \alpha = 0,1,\cdots,d-1 \\ I = 1,\cdots,d_G \end{matrix} \qquad (6)$$

Furthermore, in order that the path integral is well defined we must set

$$\lambda = \frac{k\alpha'}{4a^2} , \qquad k \in \mathbb{Z} \quad \text{for all } G \qquad (7)$$

where a is the characteristic size of the group manifold. For a detailed discussion of this point we refer the reader to Ref.(16). Since we do not know how to solve the model exactly in the presence of the Yang-Mills filed $A_i{}^B{}_C(\phi)$ and the metric $G_{AB}(\phi)$, we shall furthermore set

$$A_i{}^B{}_C = 0 , \qquad G_{AB} = \delta_{AB} \qquad (8)$$

Recalling that on a group manifold

$$\left\{ {I \atop JK} \right\} = -L_K{}^a L^I{}_{a,J} + \frac{1}{2a} f_{JK}{}^I$$
$$T_{IJK} = a^{-1} f_{IJK} \qquad (9)$$

from (2) and (7) we see that

$$\Gamma^I_{JK} = -L_K{}^a L^I{}_{a,J} + \frac{1}{2a}\left(1 - \frac{k\alpha'}{2a^2}\right) f_{JK}{}^I \qquad (10)$$

where f_{IJK} are the structure constants of the Lie algebra of G, and $L_I^a(\phi)$ are the left invariant basis elements on G defined by

$$g^{-1}\partial_\mu g = a^{-1} L_I{}^a T_a \partial_\mu y^I \tag{11}$$

Here, g is a group element and T^a are the antihermitian generators of the Lie algebra of G. From (1) and (10) it is now clear that if we set (k > 0 without loss of generality)

$$\frac{k\alpha'}{2a^2} = 1 \tag{12}$$

and define

$$\chi^a = \chi^I L_I{}^a \tag{13}$$

then the χ-kinetic term becomes simply $\bar{\chi}^a \not\partial \chi^a$. Next, let us fix the superconformal gauge;

$$e_\mu{}^a = f(\sigma,\tau)\,\delta_\mu^a \quad , \quad \psi_\mu = \gamma_\mu \lambda(\sigma,\tau) \tag{14}$$

and define the left and right moving currents

$$\begin{aligned} J^{\mu a}_{(R)} &= \tfrac{1}{2}(\eta^{\mu\nu} + \varepsilon^{\mu\nu})\,\partial_\nu y^I\, L_I{}^a \\ J^{\mu a}_{(L)} &= \tfrac{1}{2}(\eta^{\mu\nu} - \varepsilon^{\mu\nu})\,\partial_\nu y^I\, R_I{}^a \end{aligned} \tag{15}$$

where $R_I^a(\phi)$ is the right invariant basis element on G defined by

$$g\,\partial_\mu g^{-1} = -\,a^{-1} R_I{}^a T_a \partial_\mu y^I \tag{16}$$

From (6)-(10) and (13)-(15) it now follows that the Lagrangian (1) yields the following field equations

$$\begin{aligned} &\not\partial \lambda^\alpha = 0 \quad , \quad \not\partial \chi^a = 0 \quad , \quad \not\partial \psi^A = 0 \\ &\partial_\mu J^{\mu a}_{(R)} = 0 \quad , \quad \partial_\mu J^{\mu a}_{(L)} = 0 \end{aligned} \tag{17a}$$

and constraints

$$\delta g^{\mu\nu}: \quad T_{\mu\nu} = 0 \quad , \quad \delta\psi_\mu : \quad S^\mu = 0 \tag{17b}$$

where the energy momentum tensor $T_{\mu\nu}$ and the supercurrent S^μ are given by

$$\begin{aligned} T_{\mu\nu} = & \left(x^\alpha{}_{,\mu}\, x^\beta{}_{,\nu} - \tfrac{1}{2}\eta_{\mu\nu}\, x^\alpha{}_{,\lambda}\, x^{\beta,\lambda} \right) \eta_{\alpha\beta} \\ & + \left(y^I{}_{,\mu}\, y^J{}_{,\nu} - \tfrac{1}{2}\eta_{\mu\nu}\, y^I{}_{,\lambda}\, y^{J,\lambda} \right) g_{IJ} \\ & - \tfrac{i}{2}\bar\lambda\gamma_\nu\partial_\mu\lambda - \tfrac{i}{2}\bar\chi^a\gamma_\nu\partial_\mu\chi^a - \tfrac{i}{2}\bar\psi^A\gamma_\nu\partial_\mu\psi^A \end{aligned}$$

$$\begin{aligned} S^\mu = & \left(\eta^{\mu\nu} + \varepsilon^{\mu\nu} \right) \Big(\partial_\nu x^\alpha\, \lambda^\alpha + J^a_{\nu(L)}\, \chi^a \\ & - \tfrac{i}{12} f_{abc}\, \bar\chi^a\gamma_\nu\chi^b\, \chi^c \Big) \end{aligned} \tag{18}$$

The advantage of having started with the (1,0) σ-model coupled to conformal supergravity (as opposed to having started with just globally supersymmetric (1,0) σ-model) is now evident: It has given us the equation $S^\mu = 0$, i.e. the fermionic analog of the Virasoro constraint $T_{\mu\nu} = 0$.

The solutions to the free field equations (17a) are $(2\alpha' = 1)$

$$\begin{aligned} & x^\alpha = q^\alpha + \tfrac{1}{2} p^\alpha (u+v) + \tfrac{i}{2}\sum_{n\neq 0}\frac{1}{n}\left(\alpha^\alpha_n e^{-2inu} + \tilde\alpha^\alpha_n e^{-2inv} \right) \\ & J^a_{+(R)} = \sqrt{2}\sum_{n=-\infty}^{\infty} \beta^a_n e^{-2inu} \quad , \quad J^a_{-(L)} = \sqrt{2}\sum_{n=-\infty}^{\infty} \tilde\beta^a_n e^{-2inv} \\ & \lambda^\alpha = \sum_{n=-\infty}^{\infty} d^\alpha_n e^{-2inu} \\ & \chi^a = \sum_{n=-\infty}^{\infty} S^a_n e^{-2inu} \quad , \quad \psi^A = \sum_{n=-\infty}^{\infty} \tilde S^A_n e^{-2inv} \end{aligned} \tag{19}$$

where $u = \tau + \sigma$, $v = \tau - \sigma$, and by λ, χ and ψ^A we mean the single nonvanishing components of these spinors. Moreover, for x and J $n \in \mathbb{Z}$, while for λ, χ and ψ there are two possibilities:

$$n \in \mathbb{Z} \qquad \text{periodic b.c. (Ramond fermions)}$$
$$n \in \mathbb{Z} + \frac{1}{2} \qquad \text{antiperiodic b.c. (Neveu-Schwarz bosons)} \tag{20}$$

In order to solve the constraints, (17b), it is convenient to work in the light cone gauge *)

$$x^+ = p^+ \tau \quad , \quad \lambda^+ = 0 \tag{21}$$

From (17b),(19) and (21) we see that $T_{++} = 0$ allows us to solve for α_n^-. Defining $L_n = p^+ \alpha_n^- / 2$, and normal ordering the operators one finds

$$L_n = b\,\delta_{n,0} + \frac{1}{2}\sum_{m=-\infty}^{\infty}\Big(:\alpha^i_{n-m}\alpha^i_m: + \gamma :\beta^a_{n-m}\beta^a_m: + \left(m - \frac{n}{2}\right) :d^i_{n-m} d^i_m: + \left(m - \frac{n}{2}\right) :S^a_{n-m} S^a_m: \Big) \tag{22}$$

where b and γ are constants which take the values 0 and 1 respectively, if we did not normal order the operators. However, due to normal ordering, demanding [17] that L_n's obey the Virasoro algebra with a central extension (regardless of what the value of the central extension is) one finds that

$$b = \varepsilon(d-2) + \varepsilon' d_G \quad , \quad \gamma = \left(1 + \frac{C_A}{2k}\right)^{-1} \tag{23}$$

where d_G is the dimension of G, $C_A = f_{abc} f_{abc} / d_G$ and $(\varepsilon, \varepsilon')$ are (0)1/16 for (anti)periodic boundary condition for (λ, χ) respectively.

---

*) Convention: On the worldsheet $\xi^\pm = (\tau \pm \sigma)\sqrt{2}$, on M_d $x^\pm = (x^0 \pm x^{d-1})/\sqrt{2}$, and from here on $i = 1,\dots,d-1$.

We recall that in computing the commutator algebra of L_n, we use the standard commutation relations for p^α, q^α and all the oscillators [18]. In particular, the group oscillators obey [19] the Kac-Moody algebra

$$[\beta^a_m , \beta^b_n] = \frac{-i}{\sqrt{k}} f^{abc} \beta^c_{m+n} + n\,\delta_{m+n,0}\,\delta^{ab} \tag{24}$$

Turning to the constraints (17b), we see that $T_{--} = 0$ allows us to solve for $\tilde{\alpha}^-_n$. Defining $\tilde{L}_n = p^+\tilde{\alpha}^-_n/2$ and normal ordering one finds

$$\tilde{L}_n = \tilde{b}\,\delta_{n,0} + \frac{1}{2}\sum_{m=-\infty}^{\infty}\Big(:\tilde{\alpha}^i_{n-m}\tilde{\alpha}^i_m: + \tilde{\gamma} :\tilde{\beta}^a_{n-m}\tilde{\beta}^a_m: + (m-\tfrac{n}{2}) :\tilde{S}^A_{n-m}\tilde{S}^A_m: \Big) \tag{25}$$

As before, $\tilde{b}$ and $\tilde{\gamma}$ are fixed by demanding the closure of the $\tilde{L}_n$ algebra to be

$$\tilde{b} = \varepsilon''\ell \quad , \qquad \tilde{\gamma} = \left(1 + \frac{c_A}{2k}\right)^{-1} \tag{26}$$

where, ℓ is the number of ψ-fermions (i.e. $A = 1,\ldots,\ell$) and ε'' is (0)1/16 for (anti)periodic boundary condition on ψ^A.

Finally, in (17b) we see that $S^- = 0$ is satisfied automatically while $S^+ = 0$ allows us to solve for d^-_n. Defining, $F_n = p^+d^-_n/2$ and normal ordering one finds

$$F_n = \sum_{m=-\infty}^{\infty}\Big(:d^i_{n-m}\alpha^i_m: + h :\beta^a_{n-m}S^a_m: - \frac{ih}{6\sqrt{k}} f^{abc}\sum_{\ell=-\infty}^{\infty} :S^a_{n-\ell-m}S^b_m S^c_\ell: \Big) \tag{27}$$

where, h = 1. in the absence of normal ordering, but in the presence of normal ordering, it is fixed (from the requirement that $[F_n, F_m]$ closes) to be

$$h = \left(1 + \frac{c_A}{2k}\right)^{-1/2} \tag{28}$$

Finally, we establish the quantum conformal invariance of the theory, by finding that the operators L_n, $\tilde{L}_n$ and F_n given in (22), (25) and (27) respectively, obey the following super Virasoro algebra

$$[L_n, L_m] = (n-m) L_{n+m} + \frac{c}{12} n(n^2-1) \delta_{m+n,0}$$

$$\{F_n, F_m\} = 2 L_{n+m} + \frac{1}{3} c \left(n^2 - \frac{1}{4}\right) \delta_{m+n,0}$$

$$[L_n, F_m] = \left(\frac{1}{2}n - m\right) F_{n+m}$$

$$[\tilde{L}_n, \tilde{L}_m] = (n-m) \tilde{L}_{n+m} + \frac{\tilde{c}}{12} n(n^2-1) \delta_{m+n,0} \tag{29}$$

where the central extensions are given by

$$c = (d-2) + \frac{d_G}{1 + \frac{C_A}{2k}} + \frac{1}{2}(d-2) + \frac{1}{2} d_G \tag{30a}$$

$$\tilde{c} = (d-2) + \frac{d_G}{1 + \frac{C_A}{2k}} + \frac{1}{2} \ell \tag{30b}$$

In (30a) the contributions of α^i, β^a, d^i and S^a, and in (30b) the contributions of $\tilde{\alpha}^i$, $\tilde{\beta}^a$ and $\tilde{S}^A$ oscillators are denoted separately.

Since we worked in the light cone gauge we must now check the Lorentz invariance in M_d. The only relevant commutator is $[M^{i-}, M^{j-}]=0$. Using (29) one finds that this commutator holds provided that [20]

$$c = 12 \quad , \quad \tilde{c} = 24 \tag{31}$$

$$M^2 = 8\left(L_0 - \frac{1}{2}\right) - p^i p^i \tag{32}$$

$$L_0 - \frac{1}{2} = \tilde{L}_0 - 1 \tag{33}$$

Note that the subtractions in L_0 in (32) and (33) are due to the normal ordering ambiguities [20)] which arise in the computation of $[M^{i-},M^{j-}]$.

Finally, from (30) and (31) one deduces that critical dimension d must satisfy both of the following two equations

$$d = 10 - \frac{2}{3}\frac{d_G}{1 + \frac{c_A}{2k}} - \frac{1}{3} d_G \tag{34}$$

and

$$d = 26 - \frac{d_G}{1 + \frac{c_A}{2k}} - \frac{1}{2}\ell \tag{35}$$

where, we recall that, d_G is the dimension of the group manifold, $k > 0$ is the integer in front of the Wess-Zumino term, ℓ is the number of right moving fermionic coordinates ψ^A, and c_A is the second Casimir eigenvalue in the adjoint representation given by

G	SU(n)	SO(2n+1)	Sp(n)	SO(2n)	G_2	F_4	E_6	E_7	E_8
c_A	2n	4n-2	2n+2	4n-4	8	18	24	36	60

The solutions of (34) for simple Lie groups were already given in Ref.(15). It is clear that for all those solutions (35) will also be satisfied such that ℓ is fixed. The result is the following

d	8	8	6	5	4	4	3	2
G	SO(3)	SU(2)	SU(3)	SO(5)	SO(5)	SU(3)	SU(4)	SO(5)
k	1	2	1	1	2	5	1	7
ℓ	33	33	36	37	36	34	40	34

(36)

For semisimple Lie groups we give the following two examples:

d	6	4
G	SU(2) × SU(2)	SU(2) × SU(2) × SU(2)
k	1	1
ℓ	34	35

(37)

Note that $d_G = (10-d)$ is never a solution of (34) unless d_G is (10-d) dimensional torus. It is interesting that the perturbative $d = 2$ Lorentz anomaly [21] which is proportional to $(\ell-d-d_G-22)$ vanishes for all the cases listed in (36) and (37). (The gravitino contributes 22 units, and a spinor with the same chirality as the gravitino contributes 1 unit).

It is worth mentioning that, in the case of $d = 6,3$ listed in (36), $\ell/2 + \text{rank}\ G = (26-d)$. Considering the bose-Fermi equivalences in two dimensions [22] this may prompt the conjecture that in these cases the theory is equivalent to a kind discussed by Narain [11] where $(26-d)$ dimensions are compactified into tori. In that case, in $d = 6$ for example, a gauge group of the form $SU(3) \times H$ where rank $H = \ell/2 = 18$. (e.g. $E_8 \times E_8 \times SU(3)$ or $E_8 \times E_8 \times SU(2) \times U(1)$, may in principle emerge *)). Unfortunately however, our mass formula differs from that of Ref.(11) essentially due to the fact that the $\varepsilon' d_G$ term which appears in L_0 (see (23)) has a value which differs from that of the tori case where it would have read $\varepsilon'(26-d)$. (Note that $d_G = (26-d)$ is never a solution of (34), and moreover the space-time fermions, i.e. λ^α, must obey periodic boundary conditions). In fact it is due to this difference in the mass formula that in our case massless fermions in M_d do not even seem to arise [15),16)]. Nevertheless, as it has been pointed out in Ref.(16), if one compactifies the $(d-4)$ dimensions into a tori such that the internal momentum lies on a weight lattice of an appropriate representation of a rank $(d-4)$ group, then massless fermion states can arise in 4 dimensions. The reason for why this can work is that the tachyonic mass which arises in d dimensions can be utilized in creating massless fermionic solitons [1]. In the context of 10 to 4 compactification of the NSR model, this phenomenon has been shown in Ref.(23), and in the context of 6 to 4 compactification in Ref.(16).

It is a pleasure to thank Professor Abdus Salam for discussions which led to the idea of lowering the critical dimension for heterotic strings. I also wish to thank Professors J. Strathdee, E. Bergshoeff and H. Sarmadi for helpful discussions.

---

*) Note that these two groups are large enough to contain the anomaly free $N = 2$, $d = 6$ supergravity model of Ref.(24) with $E_6 \times E_7 \times U(1)$ symmetry.

References

1) D.J. Gross, J. Harvey, E.J. Martinec and R. Rohm, Phys. Rev. Lett. 54, 502 (1985), Nucl. Phys. B256, 253 (1985).
2) C.M. Hull and E. Witten, "Supersymmetric σ-models and the heterotic string", Princeton preprint (1985).
3) P. Candelas, G.T. Horowitz, A. Strominger and E. Witten, Nucl. Phys. B258, 46 (1985).
4) A. M. Chamseddine, Nucl. Phys. B185, 403 (1981); E. Bergshoeff, M. de Roo, B. de Wit and P. van Nieuwenhuizen, Nucl. Phys. B195, 97 (1982); G.F. Chapline and N.S. Manton, Phys. Lett. 120B, 105 (1983).
5) E. Witten, "Twister-like transform in ten dimensions", Princeton preprint (1985).
6) E. Bergshoeff, E. Sezgin and P.K. Townsend, "Superstring actions in $d = 3,4,6,10$ curved superspace", Trieste preprint (1985).
7) W. Siegel, Phys. Lett. 128B, 397 (1983).
8) G. Moore and P. Nelson, Phys. Lett. 53, 1519 (1984); J. Bagger, D. Nemeschansky and S. Yankielowicz, in "Symposium on Anomalies and Topology", Ed. A. White (World Scientific, 1985) and references therein.
9) E. Witten, in "Symposium on Geometry, Anomalies and Topology", Ed. A. White (World Scientific, 1985).
10) C.G. Callan, E. Martinec, M.J. Perry and D. Friedan, "Strings in background fields", Princeton preprint (1985); A. Sen, "The heterotic string in arbitrary background field", Fermilab preprint (1985); E.S. Fradkin and A.A. Tseytlin, "Effective field theory from quantized strings", Lebedev preprint (1984); I. Bars, D. Nemeschansky and S. Yankielowicz, "Compactified superstrings and torsion", SLAC preprint (1985).
11) K.S. Narain, "New heterotic string theories in uncompactified dimensions < 10", Rutherford preprint (1985).
12) S.J. Gates Jr., C.M. Hull and M. Rocek, Nucl. Phys. B248, 157 (1984); T. Curtwright and C. Zachos, Phys. Rev. Lett. 53, 1799 (1984); E. Braaten, T. Curtwright and C. Zachos, "Geometry, topology and supersymmetry in nonlinear models", Florida preprint (1985); P. Howe and G. Sierra, Phys. Lett. 148B, 451 (1984).

13) E. Bergshoeff, E. Sezgin and H. Nishino, "Heterotic σ-models and conformal supergravity in two dimensions", Trieste preprint (1985).
14) D. Nemeschansky and S. Yankielowicz, Phys. Rev. Lett. 54, 620 (1985); S. Jain, R. Shankar and S. Wadia, "Conformal invariance and string theories in compact space bosons", Tata Institute, Bombay preprint (1985).
15) E. Bergshoeff, S. Randjbar-Daemi, Abdus Salam, H. Sarmadi and E. Sezgin, "Locally supersymmetric σ-model with Wess-Zumino term in two dimensions and critical dimensions for strings", Trieste preprint (1985).
16) E. Bergshoeff, H. Sarmadi and E. Sezgin, to appear in the Proceedings of the Cambridge Workshop on Supersymmetry and its Applications (1985).
17) P. Goddard, A. Kent and D. Olive, "Unitary representations of the Virasoro and super Virasoro algebras", Cambridge preprint (1985).
18) J.H. Schwarz, Phys. Rep. 89, 223 (1982).
19) E. Witten, Comm. Math. Phys. 92, 455 (1984).
20) P. Goddard, J. Goldstone, C. Rebbi and C.B. Thorn, Nucl. Phys. B56, 109 (1973).
21) L. Alvarez-Gaume and E. Witten, Nucl. Phys. B234, 269 (1983).
22) See for example W. Nahm's contribution in this proceedings.
23) L. Castellani, "Non-abelian gauge fields from $10 \rightarrow 4$ compactification of closed superstrings", Torino preprint (1985).
24) S. Randjbar-Daemi, Abdus Salam, E. Sezgin and J. Strathdee, Phys. Lett. 151B, 351 (1984).

CONFORMALLY INVARIANT QUANTUM FIELD THEORIES IN 2 DIMENSIONS

Werner Nahm
Physikalisches Institut
der Universität Bonn
Nussallee 12, 5300 Bonn 1
FRG

Conformally invariant quantum field theories have been studied for a long time, both from the axiomatic point of view and in the context of specific examples, like the massless Thirring model. Recently new examples have been discovered, with applications ranging from statistical physics[1]) to the Rubakov effect[2]).

These theories are among the few quantum field theories which can be treated in a mathematically rigorous way. In order to do this, we shall introduce a technical simplification, restricting the system to a finite space interval of length L. This procedure has the additional advantage that it facilitates applications to string theory. For definiteness we shall use periodic or anti-periodic boundary conditions, but other conditions would not make too much of a difference.

Quantum field theories should be described by defining a Hilbert space $\mathcal{H}$ and the action of local field operators on $\mathcal{H}$. One also would like to describe the set of all local fields of the theory. There are many of them, as from each local field one can construct others by taking derivatives and products. Thus one should consider an infinite dimensional vector space $\mathcal{F}$ of local fields and the action of various operations on it. Later this space $\mathcal{F}$ will be discussed in more detail.

The canonical local fields to start with are given by the energy-momentum density. If the action of these fields on $\mathcal{H}$ is known, one also knows the time evolution, as the Hamiltonian H can be obtained by integration.

Conformally invariant QFTs form a rather restricted set. Often constructions which look very different yield isomorphic theories. A good first test for isomorphisms is obtained from the number of field degrees of freedom. In 2 dimensions one finds

$$\mathrm{tr}\,\theta(E - H) \sim \exp\left(\frac{2\pi}{3} cEL\right)^{1/2} , \tag{1}$$

for a space interval of length L, independent of the boundary conditions. For a free Bose field with N components one has c=N, thus c is a reasonable measure for the field degrees of freedom. N free Majorana fermions yield c=N/2, but in general c need not be integral or half integral.

We shall need names for the various models which will be discussed. For N Majorana spinors we'll use $M(\psi, N)$. A Dirac spinor will be regarded as made out of two Majorana spinors. For the number c we just write $c(\psi, N) = N/2$ etc.

$M(\varphi, T^N)$ will denote a scalar field which has the same Lagrangian as a free one, but takes values $\varphi(x,t)$ on an N-dimensional torus T^N. Different metrics on T^N yield different systems, but always $c(\varphi, T^N) = N$. In two dimensions, scalar fields are to be measured in units of $\hbar^{1/2}$, and the same applies to lengths on T^N.

M(G,k) will denote the conformally invariant chiral model, with variable $g(x,t) \in G$, where G is a compact Lie group. The action has to include a Wess-Zumino term to achieve conformal invariance. Otherwise it is unique up to an integral multiplier, which we denote by k. One finds

$$c(G, k) = \frac{k \dim G}{k + \tilde{h}(G)} , \tag{2}$$

where $\tilde{h}(G)$ is the dual Coxeter number of G, given by the following table:

G	SU(N)	SO(N), N≥5	Sp(2N)	E_6	E_7	E_8	F_4	G_2
$\tilde{h}$	N	N - 2	N + 1	12	18	30	9	4

For the simply laced Lie groups SU(N), SO(2N), E_6, E_7, E_8 this number is equal to the ordinary Coxeter number h(G), for which

$$h(G) = \frac{\dim (G)}{\mathrm{rank}(G)} - 1. \tag{3}$$

In this case

$$c(G,1) = \mathrm{rank}(G). \tag{4}$$

In fact there is an isomorphism[3,4)]

$$M(G,1) = M(\varphi, T_{max}(G)) \tag{5}$$

where $T_{max}(G)$ is a maximal torus of G, or in other words a Cartan subgroup. Nothing like this can be true for Lie groups which are not simply laced, as the degrees of freedom are different.

For abelian G one has $\tilde{h}(G) = 0$ and there is no Wess-Zumino term. Still the normalization of the kinetic term has to be fixed in a similar way, if one has applications to string theory in mind and wants to avoid global conformal anomalies. A useful definition is

$$M(U(1),\ k) = M(\varphi,\ S^1(k)), \tag{6}$$

where $S^1(k)$ is the circle with circumference $(\pi k\hbar)^{1/2}$.

Another interesting isomorphism is

$$M(SO(N),1) = M(\psi, N),\ N \geq 4 \text{ or } N = 2. \tag{7}$$

For N=3 one has instead

$$M(SO(3),2) = M(\psi,3) \tag{8}$$

For N=2, eq.(7) states the Skyrme-Mandelstam bosonization of a Dirac fermion, for greater N one obtains Witten's non-abelian generalization[5)]. Eq.(8) is a special case of the general isomorphism[6)]

$$M(G,\tilde{h}(G)) = M(\psi, \dim(G)) \tag{9}$$

The equality of the constants c can be checked easily.

There is an even more general isomorphism of this kind, which applies to all symmetric spaces G'/G. Let L(G) be the Lie algebra of G and put

$$L(G') = L(G) \oplus K. \tag{10}$$

Let

$$G = \bigotimes_i G_i \tag{11}$$

with simple or abelian G_i. Each G_i is represented on K. Let k_i be the Dynkin index of this representation. This is defined as follows. For simple G an invariant form on L(G) is given by $tr(T^a T^b)$, where T^a, $T^b \in L(G)$ and the trace is carried out in some representation. All these forms are proportional, and the proportionality coefficient is the Dynkin index of the representation. For the adjoint representation it is normalized to $\tilde{h}(G)$. For U(1) we shall use k=dim(K)/2. Note that for reducible representations the Dynkin index is additive. One finds

$$\bigotimes_i M(G_i, k_i) = M(\psi, \dim(K)). \tag{12}$$

Eq.(9) arises for the symmetric space (G x G)/G, eq.(7) for SO(N+1)/SO(N). Another example is M(SO(16),16) = M(ψ,128), which comes from the symmetric space E_8/SO(16).

Now consider the model $\hat{M}(\psi, N, G)$ which arises from M(ψ,N), when some subgroup G of SO(N) is gauged. Again we have factors G_i of G and Dynkin indices k_i given by their representations on the fermion fields. The model $\hat{M}(\psi, N, G)$ is not conformally invariant, but in general it will have a conformally invariant massless sector, which does not interact with the massive states. Let this sector be denoted by M(ψ,N,G). One way to obtain it is by letting the coupling constant go to infinity. One finds

$$c(\psi,N,G) = N/2 - \sum_i \frac{k_i \dim(G_i)}{k_i + \tilde{h}(G_i)} \tag{13}$$

Using the subgroups U(1) and SU(2) given by

$$U(1) \times SU(N) \subset SO(2N) \tag{14a}$$

$$SU(2) \times Sp(2N) \subset SO(4N) \tag{14b}$$

one finds

$$M(\psi,2N,U(1)) = M(SU(N),1) \tag{15}$$

$$M(\psi,4N,SU(2)) = M(Sp(2N),1). \tag{16}$$

Now use the subgroup

$$Sp(2) \times Sp(2N-2) \subset Sp(2N) \tag{17}$$

as additional gauge group. Then[7)]

$$c(N) = c(\psi,4N,SU(2) \times Sp(2N-2)) = 1 - \frac{6}{(N+1)(N+2)} . \tag{18}$$

This series of values c(N), N ≧ 2 is of particular interest, as it corresponds to field theories coming from statistical mechanics. It is well known that statistical systems at second order phase transitions can be regarded as field theories. In fact, they are determined uniquely by the condition c=c(N), and we shall denote them by M(N). Here M(2) can be obtained from the Ising model, M(3) from the tricritical point of the Ising model, and M(4) from the 3-states Potts model. One has

$$M(2) = M(\psi,1) \tag{19}$$

which is the isomorphism underlying Onsager's solution of the 2 dimensional Ising model. For M(3) and M(4) one has simpler constructions than those coming from eq.(18), namely

$$M(3) = M(SO(7),1,G_2), \tag{20}$$

$$M(4) = M(G_2,1,SU(3)). \tag{21}$$

Here $M(G',k,G)$ is obtained from $M(G',k)$ in the same way as $M(\psi,N,G)$ from $M(\psi,N)$.

One may ask in which cases $M(\psi,N,G)$ and $M(G',k,G)$ are trivial. This is rather easy to check from the number c of field degrees of freedom, which must vanish. For $M(\psi,N,G)$ this happens exactly, when G together with its N dimensional representation on the fermions forms a symmetric space, a statement which may be regarded as the inverse of the isomorphism (12). Because of eq.(7) the analogous statement applies to $M(SO(N),1,G)$. The general condition has not yet been worked out.

Let us now see, how such isomorphism are derived. For conformally invariant QFTs the energy momentum tensor is traceless, as $T^{\mu\nu}\delta g_{\mu\nu}$ must vanish for a change of the metric $g_{\mu\nu}$ by a scale factor. Then in 2 dimensions

$$\partial_\mu T^{\mu\nu} = 0, \quad T^{\mu\nu} = T^{\nu\mu} \tag{22}$$

implies in Minkowski space

$$(\partial_0 \pm \partial_1)(T^{00} \pm T^{10}) = 0 \tag{23}$$

and in euclidean space

$$(\partial_0 \pm i\partial_1)(T^{00} \pm iT^{10}) = 0. \tag{24}$$

$T^{00} + T^{01}$ is the right moving energy density, $T^{00} - T^{01}$ the left moving one. The corresponding momentum densities agree with the energy densities up to a sign. Denote the Fourier components of $T^{00} + T^{01}$ by L_n, L'_n. When suitably normalized, they form Virasoro algebras, namely

$$[L_n, L_m] = (m-n)L_{n+m} + \frac{c}{12}m(m^2-1)\delta_{m+n} \tag{25}$$

for L_n and analogously for L'_n. For ordinary field theories one usually expects the same number c in both cases, but not necessarily for strings. Note that

$$[L_n, L'_m] = 0. \tag{26}$$

By definition one has

$$L_n^+ = L_{-n}, \; L_n'^+ = L'_{-n}. \tag{27}$$

The eigenvalues of L_o, L'_o must be non-negative. They yield a grading of the Hilbert space $\mathcal{H}$ and therefore unitary representations of the Virasoro algebras on $\mathcal{H}$ which have a highest (or lowest) weight. These representations are well known. The irreducible one are characterized by the eigenvalues of L_o, L'_o on their unique lowest eigenstate $|\psi_0\rangle$

$$L_o|\psi_o\rangle = h\,|\psi_o\rangle\,, \quad L'_o|\psi_o\rangle = h'\,|\psi_o\rangle. \tag{28}$$

It turns out that for $c \geqq 1$ all values $h \geqq 0$ are possible, whereas for $c \geqq 1$ only the $c=c(N)$ yield unitary representations. In this case, h is restricted to the finite set

$$h_{p,q} = \frac{((N+2)p-(N+1)q)^2-1}{4(N+1)(N+2)}\,, \quad 1 \leqq q \leqq p \leqq N\,, \tag{29}$$

where q, p are integers. For $c=c(N)$ the Hilbert space decomposes into a finite direct sum of such representation spaces $\mathcal{H}(h,h')$, whereas for $c \geqq 1$ in general an infinite number of irreducible representations occur.

Let us consider M(2). Here one has

$$h_{11} = 0, \; h_{21} = \frac{1}{2}\,, \; h_{22} = \frac{1}{16}\,. \tag{30}$$

The relevant $\mathcal{H}(h,h')$ can be described as Fock spaces for a single Majorana fermion. Both periodic and anti-periodic boundary conditions have to be taken into account. For anti-periodic boundary conditions there are no zero modes and the Fock space decomposes according to even and odd fermion number, both for right and left movers. Together they yield

$$\mathcal{H}_a = \mathcal{H}(0,0) \oplus \mathcal{H}(\tfrac{1}{2},0) \oplus \mathcal{H}(0,\tfrac{1}{2}) \oplus \mathcal{H}(\tfrac{1}{2},\tfrac{1}{2}). \tag{31}$$

For periodic boundary conditions, the Fock space yields two copies of $\mathcal{H}(1/16,\ 1/16)$.

By definition, the local fields obtained from the energy-momentum tensor only act within a given $\mathcal{H}(h,h')$. Other local fields intertwine between different representations. The Majorana fermion e.g. yields two local fields ψ_L and ψ_R, which intertwine between the summands of $\mathcal{H}_a$ on one hand and the two copies of $\mathcal{H}(1/16,\ 1/16)$ on the other hand. One may expect another local field intertwining between $\mathcal{H}(0,0)$ and $\mathcal{H}(1/16,\ 1/16)$, and indeed this is the object which is most commonly observed by experimentalists working on 2 dimensional statistical systems. From the field theoretical point of view this object is peculiar, as it changes the spin structure on S^1, replacing periodic by anti-periodic boundary conditions and vice versa.

The Virasoro algebra has a finite SU(1,1) subalgebra given by L_{-1}, L_o, L_1. Its representations are trivial or infinite dimensional and can be interpreted as having fixed particle numbers. For a local field $\varphi(x)$ the states $\varphi(x)|0\rangle$ with $x_1 \in S^1$ and $|0\rangle$ = vacuum, form a representation of this algebra. If its ground state has $(h,h')=(h_\varphi, h_\varphi')$, one obtains at small distance x-x'

$$\langle 0|\varphi(x)\,\varphi(x')|0\rangle \sim (z-z')^{-2h_\varphi}\,(\bar{z}-\bar{z}')^{-2h'_\varphi} \tag{32}$$

where in euclidean space $z = x_o+ix_i$. Obviously the behaviour under scale transformations $z \to \lambda z$, $\lambda > 0$ is given by

$$d_\varphi = h_\varphi + h'_\varphi \tag{33}$$

which is called the dimension of φ. Euclidean rotations change z and $\bar{z}$ by opposite phases. Thus

$$s_\varphi = h_\varphi - h'_\varphi \tag{34}$$

is the (conformal) spin.

The strange field s(x) of the Ising model has $d_s=1/8$, $s_s=0$, which yields equal time correlation functions

$$\langle 0|s(x_1)\ \ s(y_1)|0\rangle \sim (x_1-y_1)^{-1/4}. \tag{35}$$

The exponent 1/4 is a well known critical exponent of statistical physics. In ordinary Minowski space it seems to be impossible to accomodate such dimensions, which is related to the fact that fermions in string theories come in groups of eight.

For $c \geq 1$ we have mentioned that $h \geq 0$ can be an arbitrary nonnegative number. Already for c=1 one can easily construct corresponding operators. Consider the model $M(\varphi,S^1(r))$, where r need not be integer. The Hilbert space $\mathcal{H}$ has a countable number of representations of the Virasoro algebras, labelled by the winding number of φ and the excitations of the zero mode φ = const. The vertex operators[8] change the winding number and the zero mode excitation by one unit. For them $(h,h') = (0,r^2/2)$ or $(r^2/2,0)$.

Many of the systems we considered have continuous symmetries, in fact all of them except the M(N) which have to small Hilbert spaces to admit them. These symmetries yield currents, with (h,h')=(0,1) and (1,0). The Fourier components in non-abelian currents form special Kac-Moody algebras[9], namely extensions of loop algebras by c-numbers. If one wants unitary representations, the central term must be an integer multiple of the minimal one which is compatible with unitarity. The representation theory of these algebras is well understood, and one can decompose $\mathcal{H}$ into such unitary representations. These representations are bigger than the Virasoro algebra representations, and most of the models mentioned above just decompose into a finite number of such representations. The exceptions are the systems with abelian symmetries only, e.g. $M(\varphi,S^1(r))$. Only for integer r one can define an abelian analog of a Kac-Moody loop group[4]. Topologically, this group decomposes into infinitely many components. The component of unity is given by exponentiation of the abelian currents, whereas group elements in the other components have the properties of baryon creation and annihilation operators. Of special interest are $M(\varphi,S^1(1)) = M(\varphi,2)$, where these operators are just the fermion fields, and $M(\varphi,S^1(2)) = M(SU(2),1)$, where they are SU(2) currents. For general integer k they have the properties of k-fold products of fermions. Even for these models with abelian symmetries the number of representations relevant at grade n only increases atmost polynomially with n, such that the number c of field degrees of freedom is still the one given by the abelian currents. In fact the statements concerning c for most of the models mentioned above just come from the representation theory of the Kac-Moody algebras[9].

For a given central extension, a Kac-Moody algebra only has a finite number of different representations, which together form the Hilbert space of the theory. Very little is known about local field operators which intertwine between the representations. One such operator is given by the quantum analog of $g(x,t)$ for $M(G,k)$. For any irreducible representation ρ of G of dimension $d(\rho)$ one can define a local field $\rho(g)$ with $d(\rho) \times d(\rho)$ components. Its transformation properties are given by

$$h_{\rho(g)} = h'_{\rho(g)} = \frac{\kappa \ \dim G/d(\rho)}{\kappa + \tilde{h}(G)} \tag{36}$$

where κ is the Dynkin index of the representation [10].

When isomorphic models are constructed in different ways, for one way some local fields may be easy to obtain, which look rather unexpected for the other construction. For statistical mechanics the field $s(x)$ coming from the Ising model is a very natural object, whereas for a long time it was overlooked in field theory. The isomorphisms stated above are only valid, if one does not restrict some system by forgetting of such possibilities. $M(\psi,1)$ restricted to anti-periodic boundaries only is obviously not isomorphic to $M(2)$.

Moreover, it is not always obvious, how big the final Hilbert space should be. For $M(E_8,1)$ the Kac-Moody algebra only has a single representation, and it is natural to regard it as the whole Hilbert space. $M(SO(8) \times SO(8),1)$ has more sectors, and local fields which are fermions. Out of these fields one may construct the E_8 currents[8]. In fact $M(E_8,1)$ is a subsystem of $M(SO(8) \times SO(8),1)$. Now $M(E_8,1)$ or rather the rigth moving part of it is an essential ingredient in the heterotic string. If its fermionic description makes physical sense, it should be considered as a subsystem of a bigger one which does not have the complete E_8 symmetry.

Now let us consider systems like $M(\psi,N,G)$ where some symmetry is gauged. In the gauge $A_r=0$ the gauge interaction is ultralocal in time and a Hamiltonian description is easy. Integrating out A_o one finds an interaction term

$$H_{int} = :\int dx_1 dy_1 \; J(x)G(x,y)J(y): \,, \tag{37}$$

where G(x,y) is a Green's function, which only depends on the space components. To see which states remain massless, one just has to check which states are annihilated by H_{int}. Obviously, these are the ones for which

$$J_n|\psi\rangle = 0 \quad \text{for} \quad n \leqq 0 \,, \tag{38}$$

i.e. the ground states of the irreducible unitary representations of the gauge group Kac-Moody algebra, into which $\mathcal{H}$ decomposes. As has been mentioned, there is only a finite number of isomorphism classes for these algebras, as the central extension is fixed. Thus we may write

$$\mathcal{H} = \bigoplus_i (\mathcal{H}_i \otimes \mathcal{H}_i^{\prime}) \tag{39}$$

where i labels the isomorphism classes, and the $\mathcal{H}_i$ are the representation spaces of them. If $\mathcal{H}_o$ is the unique isomorphism space for which the ground state is a singlet under G, we obtain for the conformally invariant sector of the gauged model

$$\mathcal{H}(G) = \mathcal{H}_o^{\prime} \,. \tag{40}$$

If one takes a tensor product

$$\mathcal{H} = \mathcal{H}_1 \otimes \mathcal{H}_2 \tag{41}$$

then the number of degrees of freedom is additive

$$c = c_1 + c_2 . \tag{42}$$

Thus the gauging of some group reduces c by the corresponding amount, which is the result used above.

Finally let us try to describe the space $\mathcal{F}$ of all local fields, graded by their dimensions and spins, i.e. by (h,h'). Note first that local fields with dimension zero cannot depend on x, i.e. they are trivial. For a free Majorana fermion one can construct local fields of the form

$$(\partial_+^{n_1}\psi_R)(\partial_+^{n_1}\psi_R)\;.....\;(\partial_-^{m_1}\psi_L)\;..... \tag{43}$$

with

$$(h,h') = \left(\sum_i (n_1 + \tfrac{1}{2}),\; \sum_i (m_i + \tfrac{1}{2})\right). \tag{44}$$

Thus the space of these local fields has the structure of a Fock space, which gives hope for an understanding of the space of all local fields. In fact this is possible along the following lines[10)]:
Consider the operator product expansion of $T^{00} + T^{01}$ with a given field $\varphi(x)$ and write the local fields which appear in this expansion as $\mathcal{L}_n\varphi(x)$,

$$(T^{00} + T^{01})(z)\varphi(x') \sim \sum_n (z - z')^{-2+n}\,\mathcal{L}_n\varphi(x'). \tag{45}$$

This defines a family of operations $\mathcal{L}_n$ on $\mathcal{F}$. The Lie algebra generated by these operators on $\mathcal{F}$ turns out to be a Virasoro algebra, isomorphic to the one acting on $\mathcal{H}$. Note, moreover, that this algebra also is defined, if the space is not compactified to a circle. One also can define $\mathcal{L}_n'$ in the same way, which commute with the $\mathcal{L}_n$, as expected.

The same procedure can be applied using currents instead of the energy-momentum tensor, which yields the action of a Kac-Moody algebra on $\mathcal{F}$. Moreover, $\mathcal{F}$ can be given the structure of a Hilbert space, by defining scalar products to vanish for fields with different (h,h'), whereas for fields with the same (h,h') a scalar product is given by the coefficient of the scalar term in the operator product expansion,

$$\varphi^+(x)\,\varphi'(x') \sim \langle\varphi|\varphi'\rangle\,(z-z')^{-d_\varphi - d_{\varphi'}}\,(\bar{z}-\bar{z}')^{-d'_\varphi - d'_{\varphi'}} + \ldots \tag{46}$$

Now, almost all of the preceding discussion for $\mathcal{H}$ can be carried over to the space $\mathcal{F}$. An exception is given by the group transformations of the loop group of U(1) which are not connected to the identity, but perhaps this also can be generalized in some way.

When $\mathcal{H}$ decomposes into a finite number of representations of the Virasoro algebra of a Kac-Moody algebra, the same is true for $\mathcal{F}$. The ground states of the irreducible representations are now called primary fields. Theories with a finite number of primary fields are rather well understood. The other cases need additional work. Little is known e.g. about the structure of $\mathcal{H}$ and $\mathcal{F}$ for supersymmetric sigma models on arbitrary Ricci flat spaces. One might hope to use new Lie algebras with even bigger representations than for the Virasoro algebra and the Kac-Moody algebras of loop type. However, it is known that all new graded Lie algebras which might appear are rather big, in fact the number of generators must increase exponentially with the grade. Another promising line of research is the study of integrable non-conformal models in two dimensions, where many results have been obtained by physicists, though little is known about the overall mathematical structure.

References

1) A.A. Belavin, A.M. Polyakov, A.B. Zamolodchikov, Nucl. Phys. B241 (1984) 333;
D. Friedan, Z. Qiu, S. Shenker, Phys. Rev. Lett. 52 (1984) 1575.

2) N.S. Craigie, W. Nahm, Phys. Lett. 147B (1984) 127, 152B (1985) 203.

3) I.B. Frenkel, V.G. Kac, Invent. Math. 62 (1980) 2366.

4) G. Segal, Comm. Math. Phys. 81 (1981) 301.

5) E. Witten, Comm. Math. Phys. 92 (1984) 455.

6) P. Goddard, W. Nahm, D. Olive, Imperial College preprint TP/84-85/25.

7) P. Goddard, D. Olive, Nucl. Phys. B257 (1985) 226.

8) P. Goddard's contribution to this school.

9) V.G. Kac, Infinite dimensional Lie algebras, Birkhäuser 1983.

10) V.G. Knizhnik, A.B. Zamolodchikov, Nucl. Phys. B247 (1984) 83.

SUPERSTRINGS AND PREONS

Jogesh C. Pati

Department of Physics and Astronomy
University of Maryland
College Park, MD 20742, U.S.A.

1. GRAND UNIFICATION AND BEYOND

The grand unification hypothesis[1,2] provides answers to a few fundamental questions such as (i) why there exist quarks and leptons; (ii) why there exist weak, electromagnetic and strong interactions with vastly varying strengths and (iii) why is electric charge quantized? It also provides a compelling reason for the non-conservations of baryon and lepton numbers which we need to account for the generation of baryon-excess in the early universe. Furthermore, models of grand unification with a single-stage hierarchy lead to a prediction[3] for $\sin^2\theta_W$ in accord with experiments. For these reasons, while we await the results of the searches for proton-decay,[F1] it seems more than likely to me that grand unification is somehow relevant at least effectively at a basic level.

From its inception it has been clear, however, that grand unification can not provide a complete theoretical framework by itself. This is because it leaves a few basic problems unresolved: in particular, (i) unification of gravity with the other forces, (ii) the problem of the fermion-generation puzzle and of the fermion mass-hierarchy and (iii) the gauge-hierarchy problem. It also does not account for the proliferation of the fundamental fields including the scalars. Furthermore, it does not help pick a particular

---

[F1] The results of the IMB group seem to exclude the minimal SU(5)-model for which proton lifetime is expected to be less than about 2×10^{31} yrs. On the other hand, certain elegant models of grand unification based on symmetries like SO(10) and SU(16), which unify members of one family (in contrast to SU(5)), permit proton lifetimes $\sim 10^{31}$-10^{34} yrs for $\sin^2\theta_W \cong 0.22$-$0.23$, by allowing for a two-stage hierarchy. This is why proton-decay experiments probing into the regions of longer lifetimes ($> 10^{32}$ yrs) and searching for alternative sets of proton-decay modes are essential to test the concept of grand unification.

gauge symmetry. One, therefore, is led to believe that there must exist fundamentally NEW PHYSICS beyond that of grand unification which would provide a resolution of these problems. Hopefully, this new physics would somehow still retain all the successes of grand unification.

2. WHY SUPERSTRINGS AND PREONS?

The discovery[4] of a general mechanism of anomaly cancellation for ten dimensional superstring theories, together with the observation that such a cancellation occurs only provided that the gauge symmetry in $d = 10$ is either SO(32) or[5] $E_8 \times E_8$, generates the exciting new prospect that one may finally have a consistent and unified quantum theory of all interactions including gravity. Such an anomaly free superstring theory is essentially unique in its gauge and field-content and it has no free parameters at the basic level except the Planck-length.

In spite of many attractive features, however, these theories including the heterotic $E_8 \times E_8$ theory[5] appear so far to be beset with serious difficulties in describing the real world as long as one identifies the fundamental four dimensional fields with the quarks and the leptons. Although one obtains chiral fermions belonging to a desirable representation[6] (27) of E_6, one seems to face either the problem of rapid proton-decay,[7,8] or the problem of inconsistency of renormalization group-analysis[9] or both. Other possible difficulties pertain to the questions of the origin of supersymmetry-breaking and that of the lower mass-scales like m_W. Even if these difficulties are somehow resolved, the greatest stumbling block, it seems to me, would be the problem of the fermion mass-hierarchy and fermion-mixings.[F2]

The purpose of these lectures is to suggest[10] that the advantages of the superstring theories as regards (a) uniqueness, (b)

[F2] The only handle one has to address to this problem is topology [see Ref. 7] and one naturally wonders whether topology alone can account for the intricate multistep mass-hierarchy of quarks and leptons spanning over more than five orders of magnitudes, and their mixing angles, as well.

parameterlessness and (c) good quantum gravity may be retained and yet the difficulties listed above may be circumvented if the fundamental four dimensional fields are identified with preons[11] rather than with quarks and leptons.[F3] Such an identification would also enhance the prospect of a resolution of the fermion mass-hierarchy-problem because of new dynamics and new symmetries, which naturally arise within preonic theories, especially those with supersymmetry.

3. GENERAL CHARACTERISTICS OF A VIABLE PREON MODEL

While *a priori* there can be many alternative preon models, a few general characteristics seem to be rather crucial for realizing a viable and an economical preon model. They are the following:

(i) There must exist effectively at least two preonic scales[14,15] -- i.e. a light scale $\Lambda_H \sim 1$ TeV as well as a superheavy scale $\Lambda_M \sim 10^{14}$-10^{15} GeV -- associated with a "hypercolor" and a "metacolor" gauge force respectively.

(ii) There must also exist supersymmetry preferably in its local form at the underlying preonic level.[16]

That there must exist the light scale $\Lambda_H \sim 1$ TeV follows by demanding that $SU(2)_L \times U(1)$-breaking and quark-lepton masses arise simultaneously through preonic condensates rather than through the VEV of an elementary Higgs field. These considerations in fact lead one to the conclusion (see Appendix I) that the inverse size of the heaviest τ and/or a fourth τ' family can not exceed one to few TeV:[14]

$$(1/r_0)_{\tau,\tau'} \lesssim 1 \text{ To few TeV} . \qquad (1)$$

The need for the superheavy scale (Λ_M) arises on rather general grounds by demanding consistency of the model with cosmological issues such as the generation of baryon-excess.[15,17] Furthermore,

----

[F3]Remarks of a similar nature pertaining to the identification of fields in higher dimensional supergravity theories were made in Ref. 12. For some speculative remarks, which are similar in spirit, see also Ref. 13.

rare processes like $K^o - \bar{K}^o$ and $K_L \to \bar{\mu}e$ require that the inverse sizes of the e and the μ-families should exceed at least (30 to 100) TeV.[14,18] In Appendix II, I discuss the relevant arguments and also show how under certain circumstances the e and the μ-families can have much bigger sizes ~ 1 TeV^{-1} without conflicting with the limits from $K_L \to \bar{\mu}e$ and K^o-$\bar{K}^o$ processes.

The need for supersymmetry is strongly suggested because it opens the door for evading[19] certain no-go theorems pertaining to (a) chiral symmetry[20] breaking and (b) vectorial symmetry-preservation[21] which apply to *ordinary* QCD-type theories, but which need to be evaded for *preonic* theories. Local supersymmetry also opens the door for evading the no-go theorem[22] regarding the formation of massless composite gauge particles, which are needed within certain preon models. Furthermore, a supersymmetric theory provides a natural basis for a fermion-boson or the so-called flavon-chromon type preon-models[23] which have many attractive features. For example, (a) they help satisfy the 't Hooft's anomaly-matching condition in a very simple manner[24] without needing proliferation, and (b) they naturally conserve quantum numbers like B and L at a basic level, which are violated spontaneously. Thus they do not have an intrinsic proton-decay problem unlike some other models.

4. A MINIMAL MODEL WITH FAMILY-REPLICATION

A very economical model incorporating these two features (i) and (ii) -- with N=1 <u>local</u> supersymmetry -- has been proposed recently.[15] I discuss at the end how such a model can be obtained for example from the heterotic superstring theory. In Appendix III, I exhibit a few variants of the model.

The model, which we refer to as the "minimal" SUSY model, introduces just four left plus four right-handed chiral superfields, i.e. $\Phi_+^{a,i} = (f_L|C_I)^{a,i}$ and $\Phi_-^{a,i} = (f_R|C_{II})^{a,i}$, with a=1 to 4, each transforming as a fundamental representation "i" of the metacolor gauge symmetry G_M, having a scale-parameter $\Lambda_M \sim 10^{14\pm1}$ GeV. It is assumed (a) that supersymmetry breaks dynamically through the formation of the metacolor gaugino-condensate:[25] $\langle\lambda_M,\lambda_M\rangle \sim \Lambda_M^3$; thereby the gauginos become superheavy and <u>decoupled</u>. This induces soft SUSY breaking mass-terms of order $(\Lambda_M^3/M_{p\ell}^2) \sim 300$ GeV. At this stage, the eight spin-1/2 components

$$f^{a,i}_{L,R} = (u,d|c,s)^i_{L,R} \equiv (f^e|f^\mu)^i_{L,R} \tag{2a}$$

and the eight spin-0 components

$$C^i \equiv (C_I|C_{II})^i = (r,y,b,\ell|r',y',b',\ell')^i \tag{2b}$$

of the superfields define independent commuting flavor $SU(4)_L \times SU(4)_R$ and SU(8)-color symmetries respectively.[26] The symmetry SU(8)-color contains $SU(4)^C$ and $SU(4)^{HC}$ acting on the unprimed and the primed chromons -- i.e. the sets $\{C_I\}$ and $\{C_{II}\}$ -- respectively. A word regarding notation: the fields (u,d,c,s) denote preons carrying the respective flavors but no color. Thus, they are distinct from quarks carrying the same flavors, which, in this paper, will be denoted by (q_u,q_d,q_c,q_s).

It is assumed that the approximate (exact within "maximal" SUSY models) global flavor-color symmetry $G = SU(4)_L \times SU(4)_R \times SU(8)^C$ breaks dynamically at the metacolor scale Λ_M into the anomaly-free subgroup $G_0 \equiv SU(2)_L \times U(1)_Y \times SU(3)^C \times SU(4)^{HC}$, which emerges as an effective low energy gauge symmetry, for $E < \Lambda_M$, through the formation of composite gauge bosons of sizes $\sim \Lambda_M^{-1}$. The descent of G into G_0 may take place through the formation of metacolor singlet preonic condensates of the type shown below:

$$\text{(i)} \quad \sigma^i_{L,R} \sim (15_{L,R},1^C)^{i=1,2} \sim \begin{array}{l} (\bar{f}_{L,R}\gamma_\mu\vec{\lambda}f_{L,R})(\bar{f}\gamma_\mu f) \\ \text{and/or } (\bar{f}^a_L f^b_R)(\bar{f}^c_R f^d_L) \end{array}$$

$$\text{(ii)} \quad \zeta_{\alpha\beta} \sim (1_{L,R},63^C) \sim \text{e.g. } (\bar{f}_L\gamma_\mu f_L + \bar{f}_R\gamma_\mu f_R)(C^\star_\alpha \overleftrightarrow{\partial}_\mu C_\beta)$$

$$\text{(iii)} \quad \Delta^j_R \sim (1_L.3_R,\overline{10})^j \sim (f_R f_R C^\star_I C^\star_I)^j \,. \tag{3}$$

The transformation-properties of $\sigma^i_{L,R}$ and ζ are with respect to $SU(4)_{L,R} \times SU(8)^C$, while that of Δ_R is with respect to the familiar subgroup $SU(2)_L \times SU(2)_R \times SU(4)^C$, where $SU(2)_{L,R}$ act on the doublets $(u,d)_{L,R}^i$ and $(c,s)_{L,R}^i$ and $SU(4)^C$ on the quartet (r,y,b,ℓ). The matrices $\tilde{\lambda}$ and $\mathbb{1}$ denote the familiar λ-matrices and the unit matrix respectively for $SU(4)$. The index j signifies three different family-index combinations -- i.e. Δ_R^{ee}, $\Delta_R^{e\mu}$ and $\Delta_R^{\mu\mu}$ -- corresponding to the entries $f_R^e f_R^e$, $f_R^e f_R^\mu$ and $f_R^\mu f_R^\mu$ in (2). Two distinct σ_L's ($i = 1$ and 2) are needed with a pattern of VEV's as shown below to break[27] $SU(4)_L$ to the GIM subgroup $SU(2)_L^{e+\mu}$:

$$\sigma_L^1 = \begin{pmatrix} 1 & & & \\ & 1 & & \\ & & -1 & \\ & & & -1 \end{pmatrix} v_1, \quad \sigma_L^2 = \begin{pmatrix} & 1 & & \\ 1 & & & \\ & & & 1 \\ & & 1 & \end{pmatrix} v_2. \tag{4}$$

Likewise for $SU(4)_R$. The action of the condensates shown in (2) is shown below:

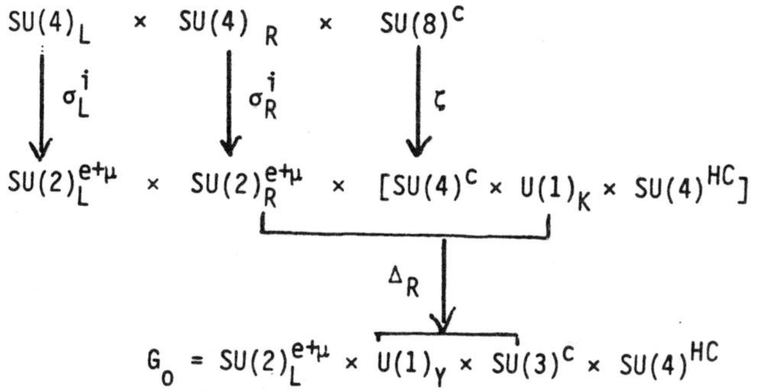

$$\begin{array}{ccccc}
SU(4)_L & \times & SU(4)_R & \times & SU(8)^C \\
\downarrow \sigma_L^i & & \downarrow \sigma_R^i & & \downarrow \zeta \\
SU(2)_L^{e+\mu} & \times & SU(2)_R^{e+\mu} & \times & [SU(4)^C \times U(1)_K \times SU(4)^{HC}] \\
& & & & \downarrow \Delta_R \\
G_0 & = & SU(2)_L^{e+\mu} & \times & U(1)_Y \times SU(3)^C \times SU(4)^{HC}
\end{array} \tag{5}$$

All four condensates are governed by one and the same scale parameter Δ_M. The symmetry $U(1)_K$ corresponds to the generator Y_K belonging to $SU(8)^C$ which is $(+1|-1)$ in the space of $(C_I|C_{II})$. Note that Δ_R breaks one linear combination of flavon and chromon numbers; it also breaks $SU(2)_R$ and $B-L \subset SU(4)^C$. In fact only the component of Δ_R^{ee} having the composition $u_R u_R \ell^* \ell^*$, and likewise the components $u_R c_R \ell^* \ell^*$ and $c_R c_R \ell^* \ell^*$ of $\Delta_R^{e\mu}$ and $\Delta_R^{\mu\mu}$ respectively, are electrically neutral. Thus only these components having the quantum numbers of dineutrino $(\nu_R \nu_R)$ acquire VEV. They break lepton number L associated with the ℓ-chromon and give a heavy Majorana mass to the composite ν_R's. In the context of a Higgs-mechanism for left-right symmetric theories[28] with elementary Higgs-fields, it has been shown that Δ_R can acquire a large VEV ($\gg m_{W_L}$) while $\langle \Delta_L \rangle$ remains essentially zero. Following this result as a guidance, one assumes that an analogous breaking pattern holds for the case of dynamical symmetry breaking as well.

As mentioned above, Δ_R breaks lepton number L but not baryon number B associated with (r,y,b) chromons. We assume that additional metacolor-singlet condensates form which break B and other surviving global quantum numbers -- e.g. the condensate $\Sigma \sim f_R^a f_R^b f_R^c f_R^d C_I^* C_I^* C_I^* C_I^*$ breaks B and L satisfying $\Delta B = \Delta L = -1$. These, together with Δ_R's, help generate baryon-excess[17] and induce proton-decay. The superheaviness of the metacolor scale $\gtrsim 10^{14}$ GeV turns out to be crucial for the generation of baryon-excess. Note furthermore that the spontaneous breaking of the global quantum numbers of the model like flavon and chromon-numbers will generate Goldstone bosons which will, however, be weakly coupled and thus "invisible" owing to the metacolor scale Λ_M being superheavy. These are some of the reasons -- based on cosmological considerations -- why the metacolor scale is chosen to be superheavy in the first place.

One can also derive a value for the metacolor scale from independent considerations based on renormalization group-equations. Since $SU(3)^C$ and $SU(4)^{HC}$ are unified within $SU(8)^C$ at a momentum scale $\sim \Lambda_M$, corresponding to the scale of the condensates ζ and Δ_R which break $SU(8)^C$, one obtains:

$$(\Lambda_M/\Lambda_H)^{b_4} = (\Lambda_M/\Lambda_{QCD})^{b_3} \qquad (6)$$

The exponents b_3 and b_4 determine the β-functions for $SU(3)^C$ and $SU(4)^{HC}$ respectively. With only gauge boson-contributions, we have $(b_3/b_4) = 3/4$, and thus $\Lambda_M \approx (\Lambda_H^4/\Lambda_{QCD}^3)$. Substituting $\Lambda_{QCD} \approx$ (100 to 200) MeV and $\Lambda_H \approx 1$ TeV (deduced from the scale of $SU(2)_L \times U(1)$-breaking, see later), we thus obtain:

$$\Lambda_M \sim 10^{15}\text{-}10^{14} \text{ GeV} . \qquad (7)$$

It is remarkable that this scale, derived on the basis of SU(8)-renormalization group-analysis, happens to agree with the constraints arising from cosmology mentioned above and that it also nearly coincides with the typical grand unification-scale.[F4]

Quantum Metacolor Dynamics and Massless Composites: To determine the spectrum of massless composites made by the metacolor force, we make the following dynamical assumptions:

(i) Chiral $SU(2)_L$ and $U(1)_Y$ gauge symmetries are not broken dynamically at the metacolor scale. In turn, these symmetries protect the masses of all the flavons and also of the spin-1/2 composite fermions -- like quarks and leptons -- if they form, although the full global chiral symmetry $SU(4)_L \times SU(4)_R$ breaks partially as shown in (4).

(ii) The spin-0 chromons C_I and C_{II} and also the lowest configuration metacolor singlet spin-1/2 and spin-0 composites, i.e. fC^*V and CC^* remain massless in the scale of Λ_M (V denotes metacolor gluon).[F5]

(iii) There is a saturation of binding at the level of the lowest configuration metacolor singlet composites like fC^*V and CC^* (see below).

The assumptions (i) and (ii) leading to masslessness of the spin-1/2 composites require that we satisfy 't Hooft's anomaly

---

[F4] Indirectly, we infer that the hypercolor gauge symmetry would have to be SU(4) for a two-scale model. If it were any "smaller", its scale-parameter could not be much higher than that of $SU(3)^c$, assuming that it is still unified with $SU(3)^c$. It can not be "bigger" than SU(4), either, otherwise Λ_M would be too low. For example, if $G_{HC} = SU(5)$, we would have $\Lambda_M \cong (\Lambda_H/\Lambda_{QCD})^{3/2}\,\Lambda_H \cong 100$ TeV. This conclusion can, of course, alter if one permits a third scale $\sim 10^5$ GeV, lying intermediate between 10^{15} and 10^3 GeV. See Appendix II for the sketch of a variant model incorporating such an intermediate scale.

[F5] The identifications of massless q_L and q_R with two-body composites -- i.e. $q_L = f_L C^*$ and $q_R = f_R C^*$ pose the following dilemma. Since the vertex involving the transition $q_L \to f_L C^*$ should be of the form $[\bar{f}_L \gamma_\mu (p_f - p_q)_\mu q_L]/\Lambda_M$, the residue of the amplitude for $f_L C^* \to f_L C^*$ scattering at the composite q_L-pole would be damped by $[m_f - m_q)/\Lambda_M]^2 \to 0$. No such dilemma arises if we identify $q_{L,R}$ with metacolor singlet $(f_{L,R}C^*V)$ or $(f_{L,R}C^*VV)$-composites in appropriate gauge invariant forms where V represents the metagluon.

matching condition. Since the vectorial symmetries $U(1)_f$ and $U(1)_C$ are broken dynamically at Λ_M, the only relevant symmetry for anomaly-matching might have been the full global flavor symmetry $SU(4)_L \times SU(4)_R$. The corresponding anomalies of the preons and of the fC^*V-composites would match if the metacolor gauge symmetry $G_M = SU(N_M)$ with N_M = number of chromons = 8 (for the present model). On the other hand, since $SU(4)_L \times SU(4)_R$ breaks dynamically to $SU(2)_L \times SU(2)_R$ at Λ_M (see (4)), the anomaly of the preons match trivially that of the composites since both vanish without any restriction on G_M.

Anomaly-matching is, however, only a necessary condition, but not sufficient, to yield massless composite fermions. One may naturally ask: What does protect the $SU(2)_L$ symmetry which in turn forces the flavons and the composite quarks and leptons to remain massless? If quantum metacolor dynamics (QMD) was exactly like QCD, one would have expected the $\langle \bar{f}^a f^a \rangle$ -condensate to form, like the $\langle \bar{q}q \rangle$ -condensate, and this would have broken even $SU(2)_L$-symmetry. It is good to note, however, that none of the standard proofs showing that chiral symmetry must break[20] in QCD and furthermore that vectorial global symmetries including "baryon number" and "isospin" can not break dynamically in vector-like QCD type theories,[21] apply to supersymmetric QCD-type theories.[19] One can in fact argue that for locally supersymmetric QCD theories a solution permitting preservation of chiral symmetry but a breakdown of supersymmetry through the formation of the gaugino-condensate is at least a consistent one. One is thus led to conjecture that the underlying reason for the protection of some chiral symmetry like $SU(2)_L$ and the simultaneous breaking of vectorial symmetries like B, L and $SU(4)_{L+R}$ including "isospin" through condensates like σ^i and Δ_R is local supersymmetry. Of course, this conjecture still awaits a proof.

To sustain our assumption (iii) of masslessness of the spin-0 chromons and the spin-0 CC^*-composites, we need to invoke once again supersymmetry and chiral symmetry. If chiral symmetry protects the masses of the fermions, supersymmetry would protect the masses of the bosonic partners. In our case, N=1 local SUSY breaks through the formation of the gaugino-condensate $\langle \vec{\lambda} \cdot \vec{\lambda} \rangle \sim \Lambda_M^3$. It has been shown[25] that in this case the mass m_C of the chromons is protected by the Planck mass: $m_C \sim (\Lambda_M^3/M_{P\ell}^2) \sim$ few hundred GeV for $\Lambda_M \sim 10^{13.5}$ GeV, as desired. With these to provide some rationale for sustaining our dynamical assumptions (i) and (ii), we turn to the spectrum of the massless composites.

Metacolor Binding (size $\sim \Lambda_M^{-1}$): Following the saturation-assumption, we list the set of lowest configuration metacolor singlet composites of the (f,c,V)-system which form under the influence of the metacolor force. For simplicity of writing, we shall consistently suppress the gluonic component V in the composites fC^*V [See F5]. The composites are classified under $SU(4)^C \times SU(4)^{HC}$; their chiral flavor-transformation-properties are not exhibited, but should be apparent:

$$\phi^a_{L,R} = (f^a_{L,R}\, C^*_I)_{4^*_c,1_H} \equiv \phi^{(o)e,\mu}_{L,R}$$

$$\xi^a_{L,R} = (f^a_{L,R} C^*_{II})_{1_c,4^*_H} \longrightarrow \text{Hyperfermions}$$

$$\mathcal{D}_0 = (C_I C^*_{II})_{4_c,4^*_H} \longrightarrow \text{Colored hyperbosons}$$

$$\mathcal{E} = (C_{II}C^*_{II})_{1_c,(1+15)_H}\ ;\ \mathcal{H} = (C_I C^*_I)_{(15+1)_c,1_H}\ . \tag{8}$$

Consistent with our assumptions (i) and (ii)[F6] these composites are massless in the scale of Λ_M. They all have sizes $\sim \Lambda_M^{-1} \sim (10^{14}$ GeV$)^{-1}$. The composites ϕ carrying flavor and color yield two quark-

[F6] Note that we have not listed the spin-0 composites $\phi = \bar{f}f$ in (8). This is because in the limit of supersymmetry such composites correspond to the auxiliary component of the composite superfield $\Phi_+\Phi^*_-$ and thus should effectively have infinite mass. In the presence of SUSY-breaking at a scale M_S, we expect ϕ to have a mass of order $\Lambda_M^2/M_S \sim 10^{15}$-$10^{17}$ GeV at Λ_M. However, the effective mass of ϕ can be much lower at Λ_H due to renormalization through Yukawa coupling. See remarks later on the possible role of ϕ .

lepton-famililes which we identify with the fermions of the "bare" e and μ-families.

Hypercolor Binding (size $\sim \Lambda_H^{-1} \sim 1\ \mathrm{TeV}^{-1}$): The hyperfermions ξ and the colored hyperbosons $\mathcal{D}_0^*$ which are almost point-like composites with sizes $\sim \Lambda_M^{-1}$, bind through the hypercolor force to yield two new families of much bigger sizes $\sim \Lambda_H^{-1} \sim (1\ \mathrm{TeV})^{-1} >>> \Lambda_M^{-1}$ having precisely the same quantum numbers as the ϕ-composites:

$$\chi^a_{L,R} = (\xi^a_{L,R}\mathcal{D}_0^*)_{4^*_C,1_H} \equiv \chi^{(o)\tau,\tau'}_{L,R} \ . \tag{9}$$

These new composites are naturally identified with the bare $\tau^{(o)}$ and a fourth $\tau'^{(o)}$ -families, which are replicase of the bare $e^{(o)}$ and $\mu^{(o)}$-families respectively.

Note that the second-stage fermionic composites $\chi = (\xi\mathcal{D}^*) = (fC^*_{II})(C^*_I C_{II})$ have precisely the same flavor and color-attributes as the first-stage composites ϕ in their core. Yet they can not be regarded as ordinary radial, orbital or quantum pair-excitations of ϕ because their sizes are very much bigger than those of ϕ. For all probes of momenta $<< \Lambda_M$, the composites χ will appear as two rather than four-body composites with constituents ξ and $\mathcal{D}^*$, which are distinct from those of ϕ (see Fig. 1).

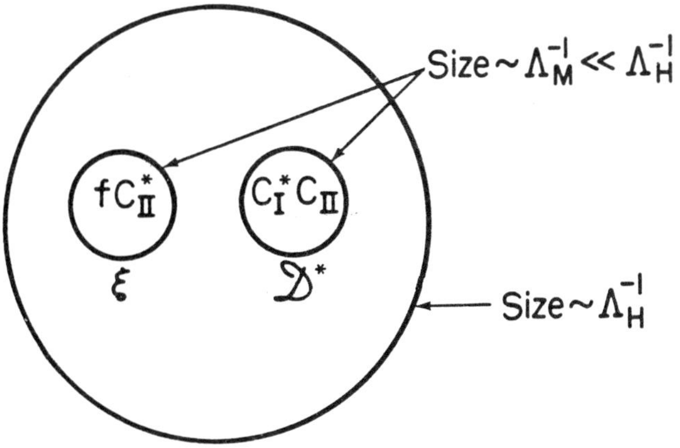

Fig. 1 τ and τ' families as composites of composites.

By utilizing the existence of the two scales, the model has thus generated a mechanism for two-fold replication of quark-lepton families, which in turn predicts four rather than three families.

The replication-idea appears to be a particularly desirable feature, because it leads to preon-models which are (a) economical, (b) viable and (c) thoroughly testable (see discussions later). Furthermore, it greatly simplifies the structure of the fermion mass-matrix by reducing the 2N-family problem to essentially that of an N-family problem (see discussions below).

At a deeper level, if one can understand why there must be four left-handed plus four right-handed chiral superfields, one would understand why there is the muon accompanying the electron. This would answer Rabi's famous question: "Who ordered that?" Simultaneously, owing to the replication-mechanism, one can explain why there must be the τ and predict a τ'. I believe that the apparent arbitrariness in the choice of the number and the representation-content of the preonic superfields will ultimately be removed by appealing to higher dimensional theories, in particular the superstring theories. I return to this question later.

5. FERMION MASSES

By assumption, the hypercolor force makes hyperfermion-condensates $\langle \xi_L^a \, \xi_R^b \rangle \equiv A_{ab} \Lambda_\xi^3$, which gives masses to the quarks and the leptons and also break $SU(2)_L \times U(1)_Y$ dynamically. Here, A_{ab} denotes a 4×4-matrix in the chiral flavor space with entries of order unity or zero, whose structure needs to be determined non-perturbatively and $\Lambda_\xi \sim \Lambda_H \sim 1$ TeV. Unlike the case of QCD, however, A_{ab} need not be a unit matrix owing to the presence of supersymmetry at scales $\gtrsim \Lambda_M$, which can trigger the breakdown of even vectorial symmetry at the metacolor-scale and in turn at the hypercolor-scale. We return to this question later.

The fermion masses and mixings depend additionally on the strengths of the effective four fermion-transitions:

(i) $\phi_L^a + \xi_R^b \to \phi_R^b + \xi_L^a$,

(ii) $\phi_L^a + \xi_R^b \to \chi_R^b + \xi_L^a$ and

$$(iii)\ \chi_L^a + \xi_R^b \to \chi_R^b + \xi_L^a\ . \tag{10}$$

Ordinarily the strengths of the first two processes would be strongly damped by Λ_M^{-2} and thus the masses of the e and the μ-families would only be of order $(\Lambda_H^3/\Lambda_M^2) \sim 10^{-16}$ MeV. But the situation alters drastically due to the presence of the spin-0 $\mathcal{D}_0$-particles with masses $\sim (\Lambda_M^3/M_{p\ell}^2) \sim 300$ GeV. These can bind with the hypergluons V_μ^H to yield light spin-1 composites $\mathcal{D}_\mu$ with the same internal quantum numbers as $\mathcal{D}_0$ and with masses $\sim \Lambda_H$. Now, $\mathcal{D}_\mu$ -exchange would contribute to all three processes listed above; this would be the only relevant contribution to the first two processes, while for the third there can be additional contribution of comparable magnitude. Thus, we have an 8×8 mass-matrix of the following form:

$$M_{8\times 8} \simeq \begin{array}{c} \quad\ \psi(e^o,\mu^o) \qquad\quad \chi(\tau^o,\tau'^o) \\ \left[\begin{array}{c|c} \eta^2 \cdot m_{ab} & \eta \cdot m_{ab} \\ \hline \eta \cdot m_{ab} & (1+X)m_{ab} \end{array}\right] \end{array} . \tag{11}$$

Here, $m_{ab} \equiv g^2(\Lambda_\xi^3/m^2)A_{ab}$ and $\eta \equiv (h/g)$, where g and h are the dimensionless effective coupling constants for the vertices $\chi \to \xi + \mathcal{D}_\mu^*$ and $\psi \to \xi + \mathcal{D}_\mu^*$ respectively. Since these two vertices conserve all strong interaction quantum numbers and are governed by dimensionless coupling constants, one may naively expect g and h to be comparable and thus η to be of order unity.[F7] The quantity X denotes contributions from mechanisms other than the one involving $\mathcal{D}_\mu$-exchange. Thus $X \sim \mathcal{O}(1)$ and conservatively $1/10 \le X \le 10$.

The simplification brought about by the family-replication idea

---

[F7]Strictly speaking, one must examine whether the fact that ψ is a very small-size composite compared to χ could lead to a strong dynamical suppression of h relative to g (e.g. $(h/g) \sim (\Lambda_H/\Lambda_M)^2$). In the presence of such a strong suppression, the mechanism for the generation of masses of the e and the μ-families would have to be altered drastically without, however, giving up the replication-idea. See Appendix II for some discussions on this point.

is obvious. The same 4 × 4-matrix m_{ab} appears in each of the four blocks in (11). As a result, the 8 × 8-matrix M, which normally would require, e.g. 20 parameters, if it is charge-conserving, real and symmetric, requires at best 8 -- i.e. η, X and six more to describe m_{ab}. Even these eight parameters are in principle calculable in terms of the only two parameters[F8] of the model -- i.e. Λ_M and Λ_H.

Regardless of the structure of the condensate matrix m_{ab}, we can diagonalize (11) in the ϕ-χ space. This yields (for $\eta \lesssim 1/3$):

$$m^i_{e,\mu} = (\eta^2\omega)m^i_{\tau,\tau'}, \quad (i = U,D) \ . \tag{12}$$

Here $\omega \approx X/(1+X)^2$ and the superscript i = U,D corresponds to the up and down-flavors in each family. Since $\omega_{max} = 1/4$ and $\omega \approx 1/10$ for $X \approx 1/10$ or 10, we see that for a value of η in its natural range -- e.g. $1/5 \lesssim \eta \lesssim 1/10$, which is certainly within reason, we get

$$m^i_{e,\mu} \approx (10^{-2} \text{ to } 10^{-3})m^i_{\tau,\tau'} \ . \tag{13}$$

In other words, without choosing η to be too small and for a large range of reasonable values for X, we obtain a hierarchy-ratio of order 10^{-2} to 10^{-3}. The model thus provides a very simple explanation for the two major steps of the mass-hierarchy, i.e. why $(m^i_e/m^i_\tau) \ll 1$ and simultaneously why $(m^i_\mu/m^i_{\tau'}) \ll 1$, in accord with the concept of "naturalness" advocated by Dirac. In so doing, it introduces a novel feature. The physical particles, owing to $\bar{\phi}\chi$-mixing (see (11)), are now mixtures of the small size ($\phi^{(o)}$) and the large-size ($\chi^{(o)}$) composites:

$$\phi_{physical} = \phi^{(o)}\cos\alpha + \chi^{(o)}\sin\alpha$$

$$\chi_{physical} = -\phi^{(o)}\sin\alpha + \chi^{(o)}\cos\alpha \tag{14}$$

where $\delta \equiv \tan\alpha = \eta/(1+X) \sim (1/5 \text{ to } 1/10)/(1+X) \sim 3 \text{ to } 10\%$. Thus

---

[F8]Note that if the hypercolor gauge force is generated through composite gauge bosons, as in the minimal SUSY model, then even Λ_H is calculable in terms of Λ_M.

the physical quarks and leptons of the electron and the muon families are mostly very small size-composites $(\phi^{(o)})$, but they do have a small admixture of the large size-composites $(\chi^{(o)})$. This is what makes the replication-idea testable.

The last two steps of the mass-hierarchy which need to be explained involve

(i) The interfamily $(\tau-\tau')$ or equivalently the $(e-\mu)$ splittings -- i.e. why $m_{t'} > m_t$, $m_{b'} > m_b$, $m_c > m_u$ and $m_s > m_d$, and

(ii) Up-Down or "Isospin" splittings within each family -- i.e. why $(m_t/m_b) \simeq 8\text{-}9$, $(m_c/m_s) \simeq 9$, etc.

Barring certain anomalies with regard to the lightest e-family, on which we comment later, the mass-ratios are typically as follows:

$$\begin{aligned} &\left.\begin{array}{l}\text{Replication}\\ \text{Splitting}\end{array}\right\} \quad (\bar{m}_\tau/\bar{m}_e) \sim (\bar{m}_{\tau'}/\bar{m}_\mu) \sim 400 \\ &\left.\begin{array}{l}\text{Interfamily or}\\ e\text{-}\mu \text{ splitting}\end{array}\right\} \quad (\bar{m}_{\tau'}/\bar{m}_\tau) \sim (\bar{m}_\mu/\bar{m}_e) \sim 10\text{-}20 \\ &\left.\begin{array}{l}\text{"Isospin"}\\ \text{splitting}\end{array}\right\} \quad (m_U/m_D)_{\tau,\mu,e} \sim 5\text{-}10\ . \end{aligned} \tag{15}$$

Here, the bar indicates an average mass of a family, the subscripts U and D represent up and down-members of a given family. We have purposely chosen the masses of the members of the electron-family to be somewhat higher than their observed values to exhibit a pattern. This may be justified aposteriori. We see from (15) a progressively decreasing hierarchy. Yet note that not even the Up-Down splitting in any given family is really small to suggest an identification with a perburtative electroweak effect. In other words, all three steps in (15) suggest the relevance of NEW PHYSICS involving perhaps new dynamics and/or new symmetries.

To explain the last two steps of the hierarchy exhibited in (15), one needs to know the structure of the condensate-matrix A_{ab} for which a first-principle calculation is not yet in hand. In the context of a left-right symmetric gauge theory with elementary Higgs fields (like Δ_L, Δ_R and ϕ), it has been shown recently that the Up-Down splittings in any given family can be understood simply in terms of spontaneous breaking of parity[29] -- i.e. of $SU(2)_R$ -- through the VEV $\langle\Delta_R\rangle$, which breaks vectorial "isospin".[F9] It was in fact shown that such a breaking can lead to a sizable "isospin" breaking in the effective Yukawa couplings of the Up and Down quarks with the Higgs multiplet ϕ , already at the scale of $\langle\Delta_R\rangle \gg m_{W_L}$, which in turn can induce large mass-splittings at the electroweak scale. This idea has its analog in a left-right symmetric preon-theory of the type presented here.

One can in fact envisage that a similar mechanism, having its origin in the non-perturbative breaking of the full vectorial flavor symmetry $SU(4)_f$ through the condensates such as $\sigma^i_{L,R}$ and Δ_R (see (5)), triggers $SU(4)_f$-breaking in the effective Yukawa interactions of the composite scalars with the composite fermions at the metacolor scale.[F10] These in turn generate $SU(4)_f$ breaking at the

---

[F9]Note that the idea of spontaneous breaking of vectorial symmetries like "isospin" and in fact flavor-SU(4) as well as B and L is being entertained for the metacolor preon-dynamics, in the first place, because, as mentioned before, SUSY-QCD opens the door for a breakdown of vectorial symmetries unlike ordinary QCD.

[F10]M. Cvetic and I are examining this possibility. The composite scalar which we have in mind is $\phi^{ab} = \bar{f}^a_L f^b_R$. As mentioned before, such a composite is expected to be born superheavy at the metacolor scale with a mass exceeding Λ_M because it arises only as an auxiliary composite field in the limit of SUSY. However, one can show that renormalization group-equations can reduce the effective mass of ϕ $(\bar{m}_\phi)$ evaluated at the hypercolor-scale dramatically by even 10 to 12 orders of magnitude if the relevant Yukawa couplings $(h_y^2/4\pi)$ of ϕ with the composite fermions (ψ,ξ) are of order 1/10, say. Now these effective Yukawa couplings would clearly break full SU(4)-flavor, including of course SU(2)-isospin, in the presence of the condensates $\sigma^i_{L,R}$ and Δ_R . Such breaking-effects, which are finite, are not damped, e.g. by the heavy metacolor-scale, as observed in Ref. 29 for the case of isospin. Cvetic and I are examining whether this SU(4)-breaking effective Yukawa interaction of ϕ superimposed on the dynamics of the hypercolor gauge-force can generate the desired pattern for the condensate-matrix A_{ab} (see text), consistent with the minimization of the energy.

I should add that radiative effects involving ϕ-exchange can induce sizable SU(4)-breaking effects on the vertices of the type $\xi + \mathcal{D}_\mu \to \psi$, which are relevant for the mass-matrix (see text). In the presence of such symmetry-breaking effects, the η-parameter

hypercolor leading to a charge conserving but $SU(4)_f$-breaking condensate-matrix of the form:[F11]

$$A = \begin{bmatrix} A_d & 0 \\ 0 & A_u \end{bmatrix}, \text{ where } A_u = \begin{bmatrix} 0 & \varepsilon \\ \varepsilon & 1 \end{bmatrix}, \quad A_d = \begin{bmatrix} 0 & \varepsilon' \\ \varepsilon' & 1 \end{bmatrix}\kappa' . \tag{14}$$

Here A_u and A_d are 2 × 2 matrices operating on the spaces (ξ^u, ξ^c) and (ξ^u, ξ^d) respectively. With the very few effective dimensionless parameters[F12] ε, ε' and κ' having values in a natural range (1/3 to

---

F10 cont.
(≡ h/g, see text) for the u-type flavors would be different from that for the d-type flavors. Although this will not alter our explanations of the e-τ and the μ-τ' -mass-hierarchies as long as η_u and $\eta_d \lesssim 1/5$, it turns out that the inequality of η_u and η_d is crucial for generating the weak-mixing elements V_{bc} and V_{bu}, which would otherwise vanish in the model. In the text, I have kept $\eta_u = \eta_d$ for the sake of simplicity.

As an additional remark, it is worth noting that the condensate matrix $\langle\bar{\xi}^a\xi^b\rangle$ together with the Yukawa coupling of ϕ with $\bar{\xi}\xi$ would naturally induce a VEV $\langle\phi^{ab}\rangle \sim h_{ab}\langle\bar{\xi}^a\xi^b\rangle/\bar{m}^2_{\phi_{ab}}$, although $\bar{m}^2_\phi$ is positive. In other words, ϕ_{ab} being almost point-like at the hypercolor-scale, acts like a composite Higgs-field. However, in the preonic model (Ref. 15), the $\langle\bar{\xi}\xi\rangle$ -condensate still provides the driving mechanism for symmetry-breaking. The VEV of $\langle\phi\rangle$ can be relevant for the masses of the light e and μ-families provided the effective $\bar{m}_\phi$ at the hypercolor-scale is $\lesssim$ (100-1000) TeV, which is not unlikely. Now $\bar{m}_\phi$ can not be much lighter than about 100 TeV, either, or else it would induce flavor-changing neutral current-processes like $K_L \to \bar{\mu}e$ with too big an amplitude. The joint effects of the composite ϕ and the condensate $\langle\bar{\xi}\xi\rangle$ are under study.

F11 The zeros in A_u and A_d may in fact correspond to entries of order ε^2 and ε'^2 respectively, with $|\varepsilon|$ and $|\varepsilon'| < 1/3$, say. Such patterns for a two-family mass-matrix was suggested by S. Weinberg (Ref. 30). With the replication-idea, one can now revive the useful ness of such patterns for a four-family mass-matrix.

F12 Note that these few effective parameters are also, in principle, calculable within the model.

1/10), such a pattern would account for τ-τ' -splittings (i.e. $(m_{t'}/m_t) \sim \mathcal{O}(-\varepsilon^2)$ and $(m_{b'}/m_b) \sim \mathcal{O}(-\varepsilon'^2)$) and likewise for e-$\mu$ splittings (i.e. $("m_u"/m_c) \sim \mathcal{O}(-\varepsilon^2)$;$("m_d"/m_s) \sim \mathcal{O}(-\varepsilon'^2)$) as well as for Cabibbo-mixing ($\sin\theta_c \simeq \varepsilon-\varepsilon'$) . With such a pattern for A_{ab}, together with the replication-splitting explained before (eq. (13)), we would be able to understand why "m_u" and "m_d" are so small ($\mathcal{O}(10^{-4})$) compared to the masses of the up and down-members of the heaviest family (τ') . We would attribute this small number of order 10^{-4} to $(\eta^2\omega) \times (\varepsilon^2$ or $\varepsilon'^2)$ being that small without any of the parameters η , ω , ε or ε' being too small.

Because "m_u" and "m_d" are that small (~ (10-50 MeV), however, we would expect induced radiative corrections which we have omitted to be particularly important for them. The quotation-marks on m_u and m_d signify that these values for the masses are derived without induced radiative effects and that they are expected to be modified substantially due to such effects, more so than the masses of the other quarks. The radiative corrections may have signs opposite to those of "m_u" and "m_d" and may help account for the rather puzzling observed values of m_u and m_d satisfying $|m_u| < |m_d|$.[F13] Whether a condensate matrix of the form (16) can be derived by minimizing the energy is under study.[F10].

6. CRUCIAL EXPERIMENTAL TESTS

As we saw, because of the replication-property and the special fermion mass-generation-mechanism of the model, the e- and the μ-families are primarily small-size composites with sizes ~ Λ_M^{-1} , while the τ and the τ' are primarily large size composites with sizes ~ Λ_H^{-1} . But fortunately the physical e and the physical μ have an admixture of order $\delta \equiv \eta/(1+X) \simeq$ 3 to 10% (for $1/5 \lesssim \eta \lesssim 1/10$ and $X \lesssim 5$) of the bare large-size composites τ^0 and τ'^0 respectively (see eq. (14)). This permits a host of observable consequences of compositeness involving collider experiments at

---

[F13] If the radiative corrections to masses are as large as nearly 1 GeV, they may even lead to e <—> μ switching. A. Datta and I have recently considered the consequences of this possibility.

relatively low energies $\sim \Lambda_H \sim 1$ TeV on the one hand and several rare processes on the other. Some of these are listed below:

(i) First we expect that the e and the μ-families would exhibit electroweak charge form factors of the form:

$$(1-\delta^2)(-\Lambda_M^2)/(q^2-\Lambda_M^2) + \delta^2(-\Lambda_H^2)/(q^2-\Lambda_H^2)$$

while the τ and τ'-families would exhibit form factors of the form:

$$\delta^2(-\Lambda_M^2)/(q^2-\Lambda_M^2) + (1-\delta^2)(-\Lambda_H^2)/(q^2-\Lambda_H^2) \ .$$

(ii) We also expect that the scattering amplitudes for family non-diagonal processes like

$$(e^-e^+ \text{ or } q_e\bar{q}_e) \to \tau^-\tau^+ \ , \ q_b\bar{q}_b \text{ or } q_t\bar{q}_t \ ,$$

at momenta $< \Lambda_H$, should possess a nonelectromagnetic component $\sim \delta^2(g_H^2/\Lambda_H^2)$ with $g_H^2/4\pi \sim 1$, due to compositeness. This new component would compete favorably with and even exceed the standard electromagnetic amplitude $\sim e^2/q^2$, and would thus show itself prominently through total cross-section, forward-backward asymmetry and polarization-measurements for $q^2 \sim (1/\delta^2)(e^2/g_H^2)\Lambda_H^2 \sim ((1 \text{ to } 3)\ \Lambda_H)^2 \sim (1 \text{ to } 3 \text{ TeV})^2$, for $\delta \sim 1/10$ to $1/30$ and $(g_H^2/4\pi) \simeq 1$. The signal should be visible, of course, already at $|q| \sim (1/3 \text{ to } 1)\Lambda_H \simeq (300 \text{ to } 1000)$ GeV, where the non-electroweak amplitude would be of order 30% compared to the electroweak amplitude.

Note that the family diagonal-processes like $e^-e^+ \to e^-e^+$ and $q_e\bar{q}_e \to e^-e^+$ and even $e^-e^+ \to \mu^-\mu^+$ and $q_e\bar{q}_e \to \mu^-\mu^+$ will not, however, show compositeness to this extent at $q^2 \sim \Lambda_H^2$, since the corresponding amplitude would be of order $\delta^4(g_H^2/\Lambda_H^2)$ or smaller still (by $\sin^2\theta_{Cabibbo}$). We, therefore, predict that the accelerators with CM energies in the range of (1/3 to 1) TeV should discover (a) compositeness of e, μ and τ, primarily through family, non-diagonal processes[F14] like (e^-e^+ or $q_e\bar{q}_e$) $\rightarrow \tau^-\tau^+$, $q_b\bar{q}_b$ or $q_t\bar{q}_t$ and (b) a breakdown of e-μ-τ universality (i.e. $A(e^-e^+ \rightarrow \mu^-\mu^+) \neq A(e^-e^+ \rightarrow \tau^-\tau^+)$ etc.). It is thus especially important to improve the efficiency for τ-detection.

Contrast these from the phenomenological considerations of Eichten, Lane and Peskin,[31] which presumes the relevance of the TeV-scale for the family-diagonal processes involving only the e and the μ-families. The lower limits on the compositeness-scale extending up to two TeV, recently derived from[32] $e^-e^+ \rightarrow e^-e^+$, $\mu^-\mu^+$, and somewhat lower limits derived from[33] $q_e\bar{q}_e \rightarrow e^-e^+$, etc. are based on this analysis of Ref. 31. Mindful of cosmological considerations and the limits from $K_L \rightarrow \bar{\mu}e$ and K^0-$\bar{K}^0$ processes we have, on the other hand, pointed out[14,15] that there are at least very good reasons to believe that the electron and the muon-families have primarily very small sizes $\lesssim$ (30 to 100 TeV)$^{-1}$,[F15] while the τ and/or a heavier τ'-family must have sizes of order 1 TeV^{-1}. In this case, one would not expect to see any noticeable signal of compositeness at CM energies $\lesssim$ 1 TeV-scale by studying processes of the type studied so far at PETRA and at CERN, which involve the fermions of only the electron and the muon-families. We, therefore, can not overemphasize the importance of studying in particular the family

---

[F14] If there is e-μ switching due to radiative effects (see F13) (e^-e^+ or $q_e\bar{q}_e$) $\rightarrow \tau'^-\tau'^+$ would show large departures, instead.

[F15] I have noted elsewhere (Ref. 14) that it is possible to construct at least semi-viable models in which the compositeness scales of the e and the μ-families can be low (~ 1 TeV) without conflicting with the limits from $K_L \rightarrow \bar{\mu}e$ and K^0-$\bar{K}^0$. This can happen, e.g., if the SU(4)-color constituents of the e and the μ-families are different. Such a model is discussed briefly in Appendix II. The consistency of this type of model with fermion mass-hierarchy, mixings and in particular cosmology has not yet been demonstrated. Yet, because of this type of model, we must still keep an open mind regarding the possibility that the e and μ-families are also primarily large-size composites (size ~ 1 TeV^{-1}). We clearly need experimental guidance on this important matter. A model of the type mentioned above was suggested in Ref. 14 and some time later in the second paper of Ref. 34 as well.

non-diagonal processes involving the $(e-\tau)$ and the $(\mu-\tau)$ - combinations for discovering compositeness at the 1 TeV-scale.

(iii) Since the third family consists primarily of large-size composites, the amplitudes for the four-fermion processes involving the fermions of the third family only such as

$$(q_t\bar{q}) \text{ or } (q_b\bar{q}_b) \rightarrow \tau^-\tau^+ \tag{17}$$

would have a component due to compositeness of order

$$g_H^2(1-\delta^2)/\Lambda_H^2 \ . \tag{18}$$

These processes have the virtue that they are not damped due to the smallness of δ^2 , unlike the processes listed in (ii). The amplitude given by (18) would be comparable to the electroweak amplitude $\sim e^2/q^2$ at rather low q^2, i.e.

$$q^2 \sim (e^2/g_H^2)\Lambda_H^2 \sim \Lambda_H^2/100 \sim (100 \text{ GeV})^2 \tag{19}$$

for $g_H^2/4\pi \simeq 1$ and $\Lambda_H \simeq 1$ TeV. Thus for $m_t \simeq (40 \pm 10)$ GeV, we expect that a study of toponium decays into $\tau^-\tau^+$ in the process

$$e^-e^+ \rightarrow \text{toponium} \rightarrow \tau^-\tau^+ \tag{20}$$

should show compositeness rather prominently in the measurements of total cross sections as well as forward-backward asymmetry.[F16] A study of this process is especially important because, as mentioned before, we have argued on rather general grounds,[14] independently of the replication-idea and the detailed mechanism of fermion mass-generation presented here, that the inverse size of the τ-family must be nearly 1 to few TeV. (See Appendix I). In this sense, the

[F16] H. Stremnitzer and I are examining these consequences and also refining the predictions for $K_L \rightarrow \bar{\mu}e$. We thank George A. Snow for kindly drawing our attention to the process involving the production and the decay of the toponium.

prediction that the processes (17) should show compositeness prominently at $q \sim \Lambda_H/10$ is rather compelling.

(iv) One of the crucial predictions of the model which seems to hold for certain variants of the mass-matrix as well is the occurrence of $q_d + \bar{q}_s \to \bar{\mu} e$ and, therefore, $K_L \to \bar{\mu} e$ and $K_L \to \pi \bar{\mu} e$.

The process $q_b^{(o)} + \bar{q}_{b'}^{(o)} \to \tau^{(o)-} + \tau'^{(o)+}$ involving "bare" composites (without mixing) is induced[F16] by the hypercolor force with an amplitude of order g_H^2/Λ_H^2 . Allowing for e-τ and μ-τ' mixings (see eq. (14)), this process in turn induces the transition $q_d + \bar{q}_s \to e^- + \mu^+$ with an amplitude of order $(g_H^2/\Lambda_H^2)\ \sin^4\alpha$, where $\sin\alpha \simeq 3$ to 10%. With constraints from mass-matrix and mixing angles, this leads to a branching ratio $B(K_L \to \bar{\mu} e) \simeq 10^{-7}$ to 10^{-10}, which should be observable in the on-going and forthcoming experiments at BNL. This is a crucial test of the model.

(v) The model naturally predicts the occurrence of $\tau \to e\gamma$ with a branching ratio which is estimated[15] to be $\simeq (1/4)(10^{-5}$-$10^{-6})$. The decay $\tau' \to \mu\gamma$ would be even more prominent[F17], because $m_{\xi^s} \gg m_{\xi^d}$. But $\mu \to e\gamma$ is strongly suppressed.

(vi) With the simplest mechanism of mass-generation presented here, the model predicts that either B_d^o-$\bar{B}_d^o$ or B_S^o-$\bar{B}_S^o$ mixing must be maximal (i.e. $\Delta m \gg \Gamma$) .[35] Such mixing will naturally reflect itself in a pronounced like sign dilepton-production.

(vii) (g-2)-measurements of the electron and the muon should show noticeable departures from standard expectations with an improvement by a factor of 5 to 10 in the accuracy of measurements in each case.[36]

(viii) There must exist a host of new particles with spins 1/2, 0 and 1 which are composites of hyperfermions $\xi^{u,d,s,c}$, with masses of the order of (100 to 2000) GeV. One expects to see an increase in $R(e^-e^+)$ in the energy range of (500-1000) GeV due to excitations of (a) hyperfermion-pairs, (b) spin-0 chromons (C_I), and (c) their point-like composites carrying color and hypercolor like $\mathcal{D}_0$. Some of these will resemble technicolor-spectroscopy, but

---

[F17]For the case of $e \longleftrightarrow \mu$-switching, (F13) one would expect to see $\tau \to \mu\gamma$-decays, instead of $\tau \to e\gamma$.

there will still be distinct elements due to the chromon-component. Note that the hyperfermions have charges ±1/2, while C_I's have charges (1/6, 1/6, 1/6, -1/2) and the C_{II}'s are neutral.

As a digression, it is worth noting that the model utilizes the idea of dynamical symmetry breaking like the technicolor models,[37] but it is much more economical than the technicolor-models involving extended technicolor. One can also show[38] that with four flavors, the model possesses GIM-like mechanism at the preon-level and does not lead to excessive $|\Delta S| = 2$ flavor changing neutral current-processes like K^0-$\bar{K}^0$ unlike familiar technicolor models. Thus the preon-model will have the richness of technicolor-spectroscopy and even more but without the problems of the standard technicolor models.

(viii) There would have to be a fourth family; the expected masses of t', b' and τ' are in the ranges of (250-400), (40-100) and (25-60) GeV respectively.

Thus the model has several rather intriguing and stringent predictions on the basis of which it can live or die. This is contrary to the prevailing notion that composite models are devoid of rather hard predictions. It is worth stressing that the majority of the predictions listed above are crucially tied as much to the replication idea as to the specific mechanism for the generation of fermion masses and mixings which was presented. The predictions pertaining to the processes (17) and (20) are, however, more general.

7. FROM SUPERSTRINGS TO PREONS

I now discuss how the field content of preonic theories of the type presented here may be obtained from superstring theories. I shall first concentrate on the heterotic superstring theory,[5] based on the gauge symmetry $E_8 \times E_8$ in d = 10. As is well known, it is generally assumed that the ten dimensional theory compactifies into $M^4 \times K$, where M^4 is the four dimensional Minkowski space and K a compact six dimensional Calabi-Yau manifold with SU(3) holonomy.[6-8] The existence of such vacuum solutions at the tree level is known. Such a compactification leaves an unbroken N = 1 localsupersymmetry at the Planck (or compactification) scale. It, furthermore, breaks

$E_8 \times E_8$ into $E_8 \times E_6$ if K is simply connected, and into a lower symmetry such as $E_8 \times SU(3) \times SU(2)_L \times [U(1)]^3$, if K is multiply connected. In other words, the topology of K determines the pattern of the primary symmetry breaking. It also determines the massless zero mode matter superfields in d = 4. These turn out to be in the form of several copies of 27's and $\overline{27}$'s of E_6 (even when E_6 is broken). The number of generations N defined by $n_{27} - n_{\overline{27}}$ is given by half the Euler characteristic of K which turns out to be reasonably low if once again K is multiply connected. Models with 1, 2 and 4 generations are known.

Following Ref. 10, I now show that the precise field-content of (four left + four right)-handed superfields of the minimal model sketched above can be derived from the d = 10, $E_8 \times E_8$ superstring theory provided that the compactification to d = 4 leads to two copies of 27's of E_6 (i.e. $N \equiv n_{27} - n_{\overline{27}} = 2$) and that E_6 breaks at the compactification scale to a subgroup $G_o = SU(4)_M \times \tilde{G}$, where $\tilde{G}$ = either $[U(1)]^3$ or $SU(2) \times [U(1)]^2$, or $[U(1)]^2$ or $SU(2) \times U(1)$. The symmetry $SU(4)_M$ is identified with the metacolor gauge symmetry which generates the preon binding force. The symmetry $\tilde{G}$, on the other hand, breaks completely at Λ_M, dynamically, due to preon-condensates. That a two copy-model can arise for several alternative choices of K has been noted in Ref. 7. A possible breaking of E_6 into $SU(4)_M \times \tilde{G}$ with $\tilde{G} = SU(2) \times [U(1)]^2$ has also been shown to exist in the literature [7,8]. Following similar methods, one can argue that $\tilde{G} = [U(1)]^3$ is also permissible.[10]

We now examine the field content of this configuration and show how it can match the one of the minimal preon-model.

The Field Content:

In the absence of the effective Higgs-field VEV $\langle H^a \rangle$, transforming like an adjoint 78 of E_6, we obtain, through compactification, a set of massless 27's and $\overline{27}$'s of E_6 such that $N = n_{27} - n_{\overline{27}} = 2$. In the presence of $\langle H^a \rangle \neq 0$, owing to the effective coupling of $27 \cdot \overline{27} \cdot H(78)$, each of the $\overline{27}$'s pair off with a 27 and become superheavy $\sim \Lambda_C$, while the excess of 27's over $\overline{27}$'s given by N = 2 remains massless. Each of these 27 is a left chiral superfield. To be concrete choose $\tilde{G} = U(1)_L \times U(1)_R \times U(1)_N$, where $U(1)_{L,R}$ denote the diagonal generators of $SU(2)_{L,R} \subset SO(10) \subset E_6$, while $U(1)_N$ commutes with SO(10). (Note that with a preonic identification for the minimal model, none of the components of E_6, not even $U(1)_{L,R}$, are to be identified with the familiar flavor-color symmetries, which arise only through effective local symmetries at the composite level.) Each 27 transforms under $SU(4)_M \times U(1)_L \times U(1)_R \times U(1)_N$ as follows:

$$
\begin{aligned}
27 &= [16_1 + 10_{-2} + 1_4 \text{ of } SO(10) \times U(1)_N] \\
&= A(4,\pm1,0)_1 + B(4^*,0,\pm1)_1 \\
&\quad + C(6,1,1)_{-2} + D(1,\pm1,\pm1)_{-2} + E(1,0,0)_4 . \qquad (21)
\end{aligned}
$$

The $U(1)_N$-charge is denoted in each case by a subscript. The entire set is complex owing to this $U(1)_N$-charge. The spin-1/2 and spin-0 components of the superfields are labelled as follows:

$$
\begin{aligned}
&\underset{\sim}{A}^{4,i} = \left(\frac{f_L}{C_I}\right)^{\underset{\sim}{4},i} , \quad \underset{\sim}{B}^{4,j} = \left(B^{4^*,j}\right)^{\dagger} = \left(\frac{f_R}{C_{II}}\right)^{\underset{\sim}{4},j} \\
&C^I = \left(\frac{\psi_L^{\underset{\sim}{6}}}{S^6}\right)^I , \quad D^J = \left(\frac{\psi^D}{S^D}\right)^J , \quad E^I = \left(\frac{\psi^E}{S^E}\right)^I . \qquad (22)
\end{aligned}
$$

The labels $\underset{\sim}{4}$ and $\underset{\sim}{6}$ denote $SU(4)_M$-representations, while i,j,I and J run over allowed values of other quantum numbers (i.e. $U(1)_{L,R}$) and also allow for the presence of two copies of 27 of E_6.

The metacolor force becomes strong at the scale $\Lambda_M \sim 10^{14}\text{-}10^{15}$ GeV. Following Refs. [15] and [26], we assume first of all that it makes the metacolor gaugino-condensate $\langle\vec{\lambda}_M \cdot \vec{\lambda}_M\rangle \sim \Lambda_M^3$, which in turn breaks supersymmetry and induces soft SUSY breaking terms of order $\Lambda_M^3/M_{P\ell}^2$. This makes the metacolor gauginos superheavy and thereby decoupled.

Second, we assume that owing to the strong metacolor force a bilinear condensate $\langle\psi^6\psi^6\rangle$ and appropriate combinations of the quartic condensates $\langle f_R f_R C^* C^*\rangle$ form as well. These, first of all, break $\mathcal{G}$, i.e. $U(1)_L \times U(1)_R \times U(1)_N$, completely. Second, the condensate $\langle\psi^6\psi^6\rangle$ makes not only the sextets ξ but also the metacolor singlets M and N superheavy. This comes about, because the

$SU(4)_M \times \tilde{G}$ -symmetry, which is operative between M_{Planck} at Λ_M, leads to a superpotential of the form

$$W = \omega_1 A.B.D + \omega_2 A.A.C + \omega_3 B.B.\xi + \omega_4 D.D.E + \omega_5 C.C.E \ . \tag{23}$$

In the presence of the ω_5-coupling, the condensate $\langle\psi^6\psi^6\rangle$ induces an effective VEV $\langle E\rangle \sim \Lambda_M$. This in turn generates a heavy mass of order Λ_M for E as well as for D (due to the ω_4-coupling), but not for A and B. Note that E couples directly only to C and D but not to A and B.

Thus, subject to our assumption about the formation of the condensates which, a priori, is at least feasible, the sextet C as well as the singlets D and E become superheavy and decouple at Λ_M. The only massless fields which remain are the quartets A and B.

Since $U(1)_L \times U(1)_R \times U(1)_N$ is broken completely, we may now drop the distinctions which arise due to the corresponding charges. Allowing still for two copies of 27, we thus have from (15) altogether four left-handed quartets $\underset{\sim}{A}^{4,i=1 \text{ to } 4}$ plus four left-handed antiquartets $\underset{\sim}{B}^{4*,j=1 \text{ to } 4}$ (or equivalently four right-handed quartets $\underset{\sim}{B}^{4,j=1 \text{ to } 4}$). This is precisely the field content of the minimal preon model presented before. We see, incidentally, that for the $E_8 \times E_8$ superstring theory, the metacolor gauge symmetry G_M must be identified with $SU(4)_M$.

8. SUMMARY AND CONCLUDING REMARKS

To summarize, an identification of the fundamental fields in d = 4 with those of the "minimal" supersymmetric preon-model (Ref. [15]) is indeed possible. This can preserve the good features of the superstring theories while circumventing the shortcomings associated with a quark-lepton identification of these fields. In particular, the possible serious difficulties of rapid proton-decay and inconsistency of renormalization group-analysis clearly disappear with a preonic identification. The derivation of the "minimal" model serves as an illustration. In principle, one may search for alternative preon models and examine their possible derivations from the superstring theories. In particular, it is possible to consider the so-called "maximal" supersymmetric preon models (see Appendix III) in which flavor and color are exact global (or possibly even local) symmetries even in the limit of SUSY, together with hypercolor and metacolor. The derivation of such variants from $E_8 \times E_8$ superstring theories is being examined by Hübsch and myself.

As we saw, the preonic identification permits, of course, spontaneous violations of B, L, C and CP at a temperature scale $\sim \Lambda_M \sim 10^{14}$ GeV, and, thereby, an adequate generation of baryon-excess in the early universe.[17] Because $\Lambda_M \sim 10^{14}$ GeV, we would still expect proton to decay with a lifetime in an observable range $\sim 10^{31}$-10^{34} yrs, as in grand-unification models like SO(10) and SU(16).

In another context, the preonic identification retains the good features of dynamical electroweak symmetry breaking associated with the technicolor idea but avoids the proliferations and difficulties of the standard technicolor scenario. The existence of the metacolor force with a high scale (Λ_M) is a pecularity of preonic theories, which indeed serves multiple desirable purposes including spontaneous breaking of SUSY and the binding of the quarks and the leptons.

One advantage of the preonic idea as illustrated by the model of Ref. [15] and the rather general argument (see Appendix I) regarding the inverse size of the τ-family[14] is that it naturally brings with it, contrary to prevailing notion, a host of intriguing crucial and testable predictions involving rare decays such as $K_L \to \bar{\mu}e$ and reactions such as $q_e\bar{q}_e \to \tau^-\tau^+$ and $q_t\bar{q}_t \to \tau^-\tau^+$, etc. with CM energies $\lesssim$ 300-1000 GeV. Thus, if these ideas are even grossly correct, we will, first of all, know about it from experiments in the near future. Most important, that will prevent the idea of a "grand desert" (10^2 to 10^{14} GeV) from being extended to that of a "super-desert" (10^2 to 10^{19} GeV). (See Table I for a summary).

Finally, in case I have conveyed the impression that superstring theories combined with preonic ideas solve all problems, I must state that this is, of course, still far from the truth. One has at best certain plausible scenarios which I believe show promise. But one is still groping in the dark. The resolution of the problem of the vanishing cosmological constant poses a major challenge. Much work needs to be done on the dynamics of compactification of superstring theories on the one hand and on the dynamics of preonic theories involving the questions of (a) supersymmetry breaking, (b) chiral symmetry - preservation at the metacolor scale and (c) spontaneous breaking of vectorial symmetries on the other hand. A derivation of the non-perturbative solution of the condensate matrix is a major task. One is certain of at least one point. Experiments in the very near future can exclude the basic line of construction of the preon-models presented here, if it happens to be wrong.

ACKNOWLEDGEMENT

It is a pleasure to thank M. Cvetic, O.W. Greenberg, A. Hauser, T. Hübsch, R.N. Mohapatra, H. Nishino, S. Nussinov, M. Peskin, Abdus Salam, G.A. Snow and H. Stremnitzer for many helpful discussions. The research was supported in part by the U.S. National Science Foundation.

APPENDIX I

Dynamical Symmetry Breaking Through Preons and the Inverse Size of the Heaviest Family

In this appendix, I show that within a large class of composite models in which at least the heaviest composite quark receives its mass through the formation of bilinear condensates of constituent-fermionic preons rather than via a Higgs-type mechanism, the inverse size of the heaviest family -- which one could identify with the τ-family -- is bounded from above by nearly 1 to 3 TeV.[14]

To see this, consider the class of preon models in which the preons are bound by an underlying QCD-like force to make composite quarks and leptons. Consider the heaviest composite quark q^H, which may correspond to the top quark, or to a t' quark belonging to a new τ'-family, if it exists.

(I) Assume that the composites $q^H_{L,R}$ consist of -- among their constitutents -- spin-1/2 preons $\xi^H_{L,R}$ which, in the chiral limit, define the flavor-chiral transformation properties of $q^H_{L,R}$, such that the system $(q^H_L + \bar{q}^H_R)$ has the same strong-interaction quantum numbers as $(\xi^H_L + \bar{\xi}^H_R)$.

(II) Assume that at least this heaviest composite quark q^H acquires its mass through a dynamical breaking of chiral symmetry owing to the formation of the bilinear fermionic-preon condensates

SCALE	PHYSICS
$\gtrsim M_{PLANCK}$	D = 10 SUPERSTRING THEORIES
	↓
$\Lambda_{COMPACT} \sim M_{PLANCK}$	D = 4, MASSLESS PREONIC MULTIPLETS COUPLED TO METACOLOR GAUGE FORCE, WITH N = 1 LOCAL SUPERSYMMETRY
	↓
$\Lambda_{METACOLOR} \sim 10^{14}$ GeV	• FORMATION OF MASSLESS COMPOSITES $= \{(q,\ell)^a$ & HYPERMATTER $(\xi^a, \mathcal{D}_0)\}_{a=u,d,c,s}$ + {W,γ, GLUONS & HYPERGLUONS}
	• DYNAMICAL BREAKING OF SYMMETRIES: LOCAL SUSY, B,L,C,CP AND GLOBAL $SU(4)_L \times SU(4)_R \times SU(4)^{COL} \times SU(4)^{HC} \longrightarrow G_0 = SU(2)_L^{e+\mu} \times U(1)_Y \times SU(3)^C \times SU(4)^{HC}$
	• FORMATION OF MASSIVE COMPOSITES: $\phi^{ab} = \bar{f}^a f^b$ WITH YUKAWA INT. $\bar{\xi}\xi\phi$ & $\bar{q}q\phi$ WHICH RESPECT ONLY G_0
	• GENERATION OF BARYON EXCESS
	↓
$\Lambda_{HYPERCOLOR} \sim 1$ TEV	• FORMATION OF COMPOSITES OF COMPOSITES $\{(q,\ell)_{\tau,\tau'} = (\xi\mathcal{D}^*)\}$
	• DYNAMICAL BREAKING OF $SU(2)_L \times U(1)_Y$: GENERATION OF HIERARCHICAL FERMION MASSES THROUGH CONDENSATES $\langle\bar{\xi}^a\xi^b\rangle$ (SEE TEXT).

TABLE I. EVOLVING PHYSICS AS ONE GLIDES FROM THE PLANCK TO THE FERMI SCALE: A SCENARIO

$$\langle \xi_L^H \, \xi_R^H \rangle \equiv \Lambda(\xi^H)^3 \tag{A-1}$$

rather than via an effective Higgs-like mechanism, involving either elementary or composite Higgs bosons of very small size (<<1 TeV^{-1}).

Here we are presuming, of course, that the condensate is formed under a preonic technicolor-like force under which ξ^H is non-neutral. Unlike the familiar technicolor models, however, we are assuming, in the spirit of the preon-model, that ξ^H is the constituent of q^H. For this reason, one does not need to introduce extended technicolor.

By the assumption stated above, since ξ^H is a constituent of q^H, we expect the four-fermion process $q_L^H + \xi_R^H \to q_R^H + \xi_L^H$ or equivalently $q_L^H + \bar{q}_R^H \to \xi_L^H + \bar{\xi}_R^H$, which conserves all quantum numbers of the strong binding force, to be represented by an effective interaction:

$$(\kappa^2/\Lambda_{o,H}^2)(\bar{q}_R^H q_L^H)(\bar{\xi}_L^H \xi_R^H) + \text{h.c.} \tag{A-2}$$

Here κ^2 is an effective strong coupling parameter, i.e. $\kappa^2 \approx 1$ to 10, and $\Lambda_{o,H}$ is the inverse size of the heaviest quark q^H, which apriori we may keep distinct from $\Lambda(\xi^H)$. Subject to the formation of the condensate (A-1), this effective interaction leads to a mass term for q^H:

$$m(q^H) = \kappa^2[\Lambda(\xi^H)^3/\Lambda_{o,H}^2] \; . \tag{A-3}$$

Thus,

$$\Lambda_{o,H} = \kappa[\Lambda(\xi^H)^3/m(q^H)]^{1/2} \; . \tag{A-4}$$

Since the condensate $\langle \bar{\xi}^H \xi^H \rangle$, while breaking flavor-chiral symmetry, breaks the electroweak gauge symmetry $SU(2)_L \times U(1)$ as well, it follows from the empirically observed scale of electroweak breaking that $\Lambda(\xi^H) \leq 1/3$ TeV. The equality or near equality would apply if the condensate $\langle \bar{\xi}^H \xi^H \rangle$ is the only source or the dominant source of $SU(2)_L \times U(1)$-breaking. From the experimental searches for the top quark, it appears that its mass lies above about 30 GeV. Thus, by substituting, conservatively, $m(q^H) \geq m_t > 30$ GeV, $\Lambda(\xi^H) \lesssim 1/3$ TeV and $\kappa^2 \lesssim 10$, we obtain an <u>upper bound</u> on the inverse size of the heaviest quark:

$$\Lambda_{o,H} \lesssim 3.3 \text{ TeV} \qquad [\text{for } m(q^H) \geq m_t > 30 \text{ GeV}]. \qquad \text{(A-5)}$$

If the heaviest quark is a t' belonging to a fourth family with a mass $m_{t'} \approx 100$ to 400 GeV,[F18] say, its inverse size would be bounded above by $\Lambda_{o,H} \lesssim 2$ to 1 TeV. This is the reason for our assertion that the inverse size of at least the heaviest composite quark is bounded above by about 1 to few TeV. Assuming that all members belonging to a given family have the same composite structure, the stated upper bound applies to the inverse size of all members of at least the heaviest family. The heaviest family may correspond to the τ or alternatively to a fourth τ'-family. A few remarks are in order.

(i) In the analysis presented so far, we have set aside the question of the origins of the mass splittings within a family represented, for example, by (m_t-m_b), (m_c-m_s), (m_b-m_τ), (m_s-m_μ), etc., and of the mass splittings between the families represented, e.g., by (m_t-m_c) and (m_b-m_s), etc. and by $(m_{t'}-m_t)$, $(m_{b'}-m_b)$ if a fourth family exists. These may, in general, arise (a) due to a nonperturbatively generated hierarchy in the magnitudes of the condensates involving different preon flavors,[F10] or (b) indirectly through dynamical breaking of vectorial symmetries like isospin or SU(4)-flavor due to a dynamical breaking of space-time symmetries like parity,[29] or (c) due to a hierarchy in the sizes of the composites[38] (e.g. $\Lambda_{oe} = \Lambda_{o\mu} \gg \Lambda_{o\tau}$) , or (d) due to special dynamical mechanisms such as the replication-mechanism[15] (see text), or (e) due to symmetries which may provide extra protection to the mass of one family relative to another.[39,40,41] One can, in general, conceive of a combination of these mechanisms (see e.g. Ref. 15 and 41) to solve the full fermion-mass-hierarchy problem.

The point of this appendix is that the assumption of dynamical chiral-symmetry breaking through fermionic constituent preons pro-

[F18] Much heavier masses for t' or rather ξ^c may cause difficulties with the ρ-parameter.

vides a rather low upper bound (see (A-5)) on the inverse size of the heaviest family regardless of the mechanism of the generation of the fermion-mass hierarchy. For this purpose, we have found it reasonable to focus attention on the mass generation of the heaviest quark as this truly represents the scale of $SU(2)_L \times U(1)$-breaking in the context of DSB through preons.

(ii) While we have stated the result (A-5) as an upper bound on $\Lambda_{o,H}$, it is clear from eq. (A-4) that for a reasonable value of $\kappa^2 \simeq 1$ to 10, and with $m(q^H) \simeq 40$ to 400 GeV and $\Lambda(\xi^H) \simeq 300$ GeV, say, we expect

$$\Lambda_{o,H} \simeq (\tfrac{1}{4} \text{ to } 3) \text{ TeV} . \tag{A-6}$$

In other words. the inverse size of at least the heaviest family is in fact equal to about 1 TeV within a factor of 3, say either way. This result, derived on rather general grounds, gives us confidence that the idea of compositeness of quarks and leptons of at least the heaviest family as well as the associated new binding force, with a scale parameter $\sim \Lambda_{o,H}$, would have to be visible prominently in accelerators of the very near future, with CM energies $\lesssim 1$ TeV, if at all the idea is even grossly correct (see text for experimental consequences).

(iii) Since the inverse size of the heaviest family $\Lambda_{o,H}$ given by (A-6) is so close numerically to the chiral-symmetry-breaking condensate parameter $\Lambda(\xi^H)$ which is about 300 GeV, it is natural to presume that one and the same force binds the heaviest family (τ and/or τ') and also breaks chiral symmetry. In this respect, the quantum preon dynamics relevant for the heaviest family or families (QPDH) behaves very much like QCD. Both break chiral symmetry maximally.[F19] The chief difference between the two is that QCD preserves vectorial flavor symmetry while QPDH does not. As alluded to in the text, such a breaking of vectorial symmetry may even be triggered at a superheavy scale much larger than that of QPDH, owing perhaps to the presence of supersymmetry[19] at and above this superheavy scale and a new force associated with such a scale.

APPENDIX II

The Sizes of the Electron and the Muon Families in the Fermion-Boson or the Flavon-Chromon Models:

Constraints from $K_L \to \bar{\mu}e$ Decay:

I will first discuss the constraints from $K_L \to \bar{\mu}e$-decay on the sizes of the e and the μ-families for the fermion-boson or flavon-chromon preon models. This class of models was proposed in Ref. 23 and is being actively pursued in the literature.

The essential festure of this class of models is that quarks and leptons are composites of two types of preons -- those which carry only flavor (f) and those which carry only color (C^*) including lepton number. Quarks and leptons in a given family are composed of the same flavor attributes ("flavons"), i.e.

$$f(q_d) = f(e^-),\ f(q_u) = f(\nu_e)\ ,$$

but they differ from each other only in respect of their color attributes ("chromons"), i.e.

$$C(q_d) = C(q_u) \neq C(e^-) = C(\nu_e)\ .$$

In the interest of economy, the "minimal" models of this type furthermore assume that the e and the μ-families are made of the same set of four chromons -- three for quark colors $C_q = \{C_r, C_y, C_b\}$ plus one for lepton color C_ℓ -- although they may differ from each other

---

F19 Note that owing to the cubic dependence of the mass on $\Lambda(\xi^H)$ (see (A-3)), the mass of the heaviest quark may easily differ by about an order of magnitude from the scale parameter Λ_{QPDH} of the force which binds the heaviest family. For instance, it is not unreasonable to take $\Lambda(\xi^H) \approx (1/2.5)\Lambda_{QPDH}$ and $\Lambda_{o,H} \approx 2.5\,\Lambda_{QPDH}$ with $\kappa^2 \approx 5$. These yield: $m(q^H) \approx (1/19)\Lambda_{QPDH}$. Thus, with $\Lambda(\xi^H) \approx 300$ GeV, i.e. $\Lambda_{QPDH} \approx 750$ GeV, we can identify q^H with the top quark and have $m(q^H) \approx 40$ GeV, which is quite a bit smaller than Λ_{QPDH} .

with respect to their flavon-contents: i.e.

$$C(q_d) = C(q_s) = C(q_u) = C(q_c) \equiv C_q \text{ and}$$

$$C(e^-) = C(\mu^-) = C(\nu_e) = C(\nu_\mu) \equiv C_\ell \ .$$

In addition, these models assume that one and the same force binds the e and the μ families giving them a common inverse size $\Lambda_{oe} = \Lambda_{o\mu} \equiv \Lambda$. One simple realization of this idea is through a set of four spin-1/2 flavons $(f_{u,d,c,s})_{L,R}$ and four spin-0 chromons $C_{r,y,b,\ell}$ each transforming as a representation $\underset{\sim}{n}$ of a metacolor gauge symmetry G. The compositions of the quarks and leptons of the e and the μ-families which are assumed to be singlets of G are given by:

$$\left.\begin{array}{ll} q_u^{r,y,b} = (f_u)\cdot C^*_{r,y,b} \ , & \nu_e = f_u \cdot C^*_\ell \\ q_d^{r,y,b} = (f_d)\cdot C^*_{r,y,b} \ , & e^- = f_d \cdot C^*_\ell \end{array}\right\}$$

$$\left.\begin{array}{ll} q_c^{r,y,b} = (f_c)\cdot C^*_{r,y,b} \ , & \nu_\mu = f_c \cdot C^*_\ell \\ q_s^{r,y,b} = (f_s)\cdot C^*_{r,y,b} \ , & \mu^- = f_s \cdot C^*_\ell \end{array}\right\} \qquad \text{(A-7)}$$

Note that the compositions of the e and the μ families presented above is identical to those of the "minimal" model presented in the text, barring their mixings with the τ and τ' families. In this class of models the four-fermion process $q_d + \bar{q}_s \to e^- + \mu^+$ and, therefore, the decays $K_L \to \bar{\mu}e$, and $K \to \pi\bar{\mu}e$ would consist of an underlying preonic transition $[f(q_d) + C(q_d)^*] + [f(q_s)^* + C(q_s)] \to [f(q_d) + C(e^-)^*] + [f(q_s)^* + C(\mu^-)]$. For the minimal model, this four-fermion process clearly conserves all quantum numbers defined by the preonic strong interactions and would thus be induced by the preonic force involving exchanges of the type shown in Figures 2a and 2b with an amplitude of order A^2/Λ^2 , where $A^2 \sim 1$ to 10, say.

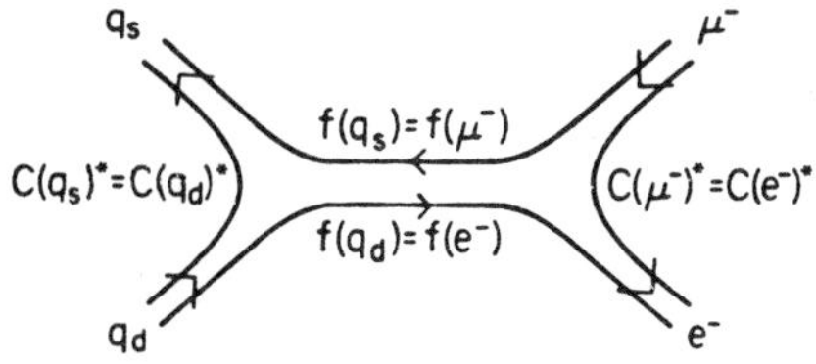

Fig. 2a

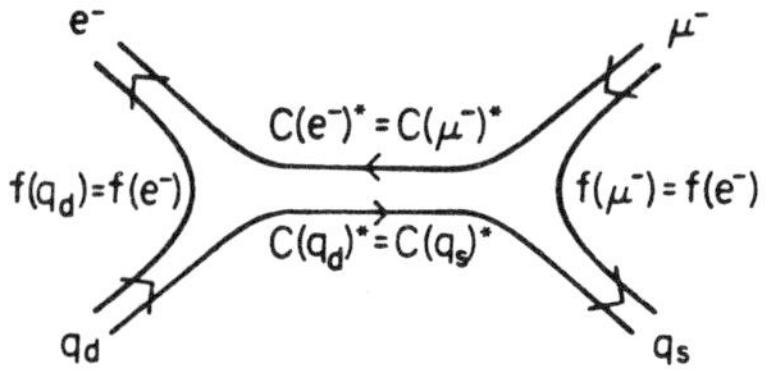

Fig. 2b

From the presently known rough lower limit on the branching ratio of $K_L \to \bar{\mu}e$ which seems to be nearly 10^{-8}, one can thus deduce[18,14,F20] that the inverse size of the e and the μ-families, at least for the minimal fermion-boson models, should exceed about 100 TeV (for $A^2 > 1$):

$$\Lambda_{oe} = \Lambda_{o\mu} \gtrsim 100 \text{ TeV} \quad \text{(For minimal models with e and } \mu \text{ families being made of the same four chromons)} \qquad \text{(A-8)}$$

Constraints from K^0-$\bar{K}^0$ -transitions:

Following Ref. 14, I now present the constraints on the sizes of the e and the μ families for a few alternative preon models on the basis of K^0-$\bar{K}^0$ transition.

(a) The "Most Economical" Pair-Excitation Models:[40] First consider the most economical class of models in which the μ family has pre-

---

[F20]The argument persented here is the exact analog of the one used in deriving the limit of the mass of the leptoquark-gauge boson of SU(4)-color on the basis of $K_L \to \bar{\mu}e$-decay in Ref. 23 and refined in Ref. 42 (compare with Fig. b).

cisely the same preonic quantum numbers (attributes) as the e family, but differs from it, say, by a quantum pair excitation, with the pair being neutral with respect to all conserved quantum numbers. Some recently suggested fermion-boson preon models, based on the quasi-Nambu Goldstone-fermion (QNGF) mechanism,[40] with only two flavors plus four colors in fact belong to this class. In these models, the process

$$q_d + q_d \rightarrow q_s + q_s \tag{A-9}$$

and therefore $K^0 \longleftrightarrow \overline{K}^0$ conserves all strong interaction quantum numbers defined by the preonic force. The process would thus be induced by the preon dynamics through the excitation of relevant pairs with an amplitude of order B^2/Λ^2, where $B^2 \sim$ 1-10, say, even before Cabibbo-mixing. Here Λ denotes the average inverse size of the e and the μ-families. Comparing with the known strength of the real part of the K^0-$\bar{K}^0$ transition, we would need $B^2/\Lambda^2 \lesssim 10^{-12}$ GeV^{-2}, or for $B^2 \gtrsim 1$,

$$\Lambda_{oe} \approx \Lambda_{o\mu} \approx \Lambda \gtrsim 1000 \text{ TeV} \quad \text{(for } \mu \text{ being a quantum-pair excitation of e)} . \tag{A-10}$$

The suggestion[40] that the e and the μ families in these most economical pair-excitation models have rather low inverse sizes $\sim$ 1 TeV is thus inconsistent with the limits from K^0-$\bar{K}^0$ and even $K_L \rightarrow \bar{\mu}e$ transitions.

(b) The "Most Economical" Flavon or Chromon-Excitation Models:[43]
Consider models with just two flavors f = (u,d) plus four colors for which the e and the μ-families have the same net flavor-color quantum numbers, but they differ from each other in respect of, for example, flavon and/or chromon numbers. For instance, suppose $\phi_e = fC^*$ and $\phi_\mu = (fffC)^*$, both ϕ_e and ϕ_μ being neutral with respect to the preon-binding force. We will refer to this model as the flavon-excitation model. A model of this kind has been proposed in Ref. 43. In this case, although $q_d^{(o)} + q_d^{(o)} \rightarrow q_s^{(o)} + q_s^{(o)}$ would not be induced directly by the preon-binding force, we would nevertheless expect, owing to two versus four-body-composite structures,

that the amplitudes for the processes $q_d^{(o)} + q_d^{(o)} \to q_d^{(o)} + q_d^{(o)}$ and $q_s^{(o)} + q_s^{(o)} \to q_s^{(o)} + q_s^{(o)}$ involving bare quarks would differ from each other by terms of order B'/Λ^2, where $B' \sim 1$ to 10 and $\Lambda \sim \Lambda_{oe} \sim \Lambda_{o\mu}$. Following Cabibbo-rotations, these processes would thus lead to $|\Delta S| = 2$-transitions for physical quarks with strengths of order $(B'/\Lambda^2) \sin^2 \theta_C$. From the known strength of K^0-$\bar{K}^0$ transition, we must, therefore have[14] $(B'/\Lambda^2) \sin^2\theta_C \lesssim 10^{-12}$ GeV^{-2}, i.e. for $B' \gtrsim 1$

$$\Lambda_{oe} \sim \Lambda_{o\mu} \sim \Lambda \gtrsim 200 \text{ TeV} \quad \text{(For the "most economical" flavon or chromon-excitation Models)} \quad . \quad \text{(A-10)}$$

The suggestion of an inverse size of about 1 TeV for the e and the μ families for this class of models, is thus inconsistent with the limit from K^0-$\bar{K}^0$ transition.

(c) <u>Preon Models with the e and the μ Families Made of Distinct Flavors</u>: Next consider the class of preon models in which (q_u, q_d, q_c, and q_s) and likewise (ν_e, e^-, ν_μ and μ^-) are made of four distinct flavors $f_{u,d,c,s}$ plus four (or more) colors, as shown in (A-7), with each of the flavons and chromons transforming as representation $\underset{\sim}{n}$ of a metacolor gauge symmetry G. In this model, first of all, the process $q_d^{(o)} + q_d^{(o)} \to q_s^{(o)} + q_s^{(o)}$ will not be induced directly by the preon-binding force.

Furthermore, since the metacolor force is invariant under "rotations" in the (u,d,c,s)-space, and the e and the μ-families have <u>identical preon dynamics</u>, the combined set of processes

(i) $q_d^{(o)} + q_d^{(o)} \to q_d^{(o)} + q_d^{(o)}$

(ii) $q_d^{(o)} + q_s^{(o)} \to q_d^{(o)} + q_s^{(o)}$ and

(iii) $q_s^{(o)} + q_s^{(o)} \to q_s^{(o)} + q_s^{(o)}$,

induced by the preon-binding force, would respect GIM-invariance, i.e. their amplitudes would symbolically have the form $(\bar{q}_d^{(o)} q_d^{(o)} + \bar{q}_s^{(o)} q_s^{(o)})^2$, because the underlying preonic amplitude would have a similar form: $(\bar{f}_d f_d + \bar{f}_s f_s)^2$. So, even after Cabibbo-rotations, the corresponding effective amplitude in terms of the physical fields would still have the GIM-invariant form $(\bar{q}_d q_d + \bar{q}_s q_s)^2$, which has no $|\Delta S| = 2$ piece. Thus the models in which the e and the μ families have identical preon dynamics[F21] but distinct flavors (or more generally some distinct preon-attributes) can permit a GIM-like mechanism at the preon level and, thereby, be safe with respect to $|\Delta S| = 2$ K^0-$\bar{K}^0$ transition even after quark-mass generation and Cabibbo rotation, without requiring the inverse size $\Lambda_{oe} = \Lambda_{o\mu}$ to exceed one to few TeV. In other words, this provides a general mechanism for DSB through preons[38,F22] which is safe with regard to $|\Delta S| = 2$ neutral-current processes, unlike the familiar technicolor models.

It is, of course, still important to ensure that the dynamical generations of quark masses and Cabibbo mixings, which inevitably violate some global flavor symmetries do not produce extra-light pseudogoldstone bosons. Following the works of Ref. 38 and 15, our suggestion for avoiding this problem is this. In a model of the type discussed in the text, which has four basic flavors $f^{a,i}_{L,R} = (u,d,c,s)^i_{L,R}$, the metacolor gauge force, with a scale parameter $\Lambda_M >> 1$ TeV, permits a global-flavor symmetry $G = SU(4)_L \times SU(4)_R \times U(1)^f_{L+R}$. This global symmetry is broken dynamically at the metacolor scale Λ_M to $SU(2)^{e+\mu}_L \times SU(2)^{e+\mu}_R \times U(1)^f_{L+R}$ by condensates like $\sigma_{1,2}$ (see text). With the formation of additional condensates like Δ_R , the full global symmetry of the effective low-energy theory for momenta $< \Lambda_M$ is no more than the effective gauge symmetry $G_0 = SU(2)^{e+\mu}_L \times U(1)_Y \times SU(3)^c \times SU(4)^{HC}$. In particular, the effective Yukawa interactions of the composite hyperfermions $\xi = fC^*_{II}$ and of the composite quarks and leptons $\psi = fC^*_I$ with the composite scalars $\phi = \bar{f}f$ (see F.10 for remarks on the mass of ϕ) , all of

---

[F21]By the same token it follows that if one wishes to consider a model in which the e and the μ families have different preon dynamics, as in the case of the "most economical" flavor or color-excitation models discussed before, one must ensure that such a difference arises only at a scale exceeding about 200 TeV so that $(\sin^2\theta_C/\Lambda^2) < 10^{-12}$ GeV^{-2} (see (A-10)). This would necessitate that both the e and the μ families have sizes exceeding about 200 TeV.

which are formed by the metacolor force, are invariant only under G_0, but they explicitly break[F22] the bigger flavor symmetry $SU(4)_L \times SU(4)_R$. Now the dynamical breaking of the big global symmetry G will produce several massless goldstone bosons,[F23] but these would be "invisible" in that they would couple sufficiently weakly to normal matter if[44] Λ_M exceeds about 10^9 GeV. This is yet another reason why we have argued[15] for at least one superheavy preonic scale.

We must next examine the physics of the effective low energy theory at the hypercolor scale $\Lambda_H \sim 1$ TeV. The hypercolor gauge force operating on the hyperfermions $\xi_{L,R}^{u,d,c,s}$ would, of course, respect again a global flavor $SU(4)_L \times SU(4)_R$ symmetry. The dynamical breaking of this symmetry through hyperfermion condensates, which lead to quark masses and Cabibbo mixings, would generate some massless goldstone bosons and some pseudogoldstone bosons. The former would be absorbed by $W^\pm$ and Z^0 which become massive. The latter would acquire masses of order $(h_{\xi\xi\phi}^2/4\pi)\Lambda_{HC}$, because the corresponding generators of $SU(4)_L \times SU(4)_R$ are broken explicitly by the Yukawa interactions of ϕ. Thus the pseudogoldstone bosons in this picture can be comfortably massive $\sim (\frac{1}{10}$ to $1)$ TeV, for $(h_{\xi\xi\phi}^2/4\pi) \sim (1/10)$ to 1 and $\Lambda_{HC} \sim 1$ TeV. Note that $h_{\xi\xi\phi}$ is expected to be a typical strong interaction coupling constant, since the corresponding coupling is induced by the metacolor force.[F24]

To sum up, we see that to avoid the generation of extra-light pseudogoldstone bosons in a model, which utilizes a dynamical breaking of chiral flavor symmetry through preonic condensates at the

---

[F22]See e.g. Ref. 29 for a demonstration of how isospin breaking is induced in the effective Yukawa couplings due to $\langle\Delta_R\rangle$.

[F23]These goldstone bosons would no doubt be helpful in galaxy formation.

[F24]Note also that unlike the case of elementary-Higgs model, there is no straightforward constraint on even $h_{\psi\phi\phi}$ from the known masses of the e and the μ-families, since these masses now arise through the condensate $\langle\xi\xi\rangle$ (see text) rather than directly through $\langle\phi\rangle$.

scale of 1 TeV, one very likely needs the background of the symmetry breaking effectively "strong" Yukawa interactions. Such interactions arise *naturally* as *effective* interactions between composites without the least bit of proliferations of fundamental parameters and constituents in a preon model with two or more effective scales.[15] One of these scales must be superheavy (> 10^9 GeV) to make massless goldstones essentially invisible and also to permit the generation of baryon-excess.

Returning to the "minimal" fermion-boson or flavon-chromon models which utilize four flavors plus four colors to build the e and the μ-families (so that they are made of identical chromons), one would need their inverse sizes $\Lambda_{oe} = \Lambda_{o\mu}$ to exceed about 100 TeV in order to satisfy the constraint from $K_L \to \bar{\mu}e$ decay (see (A-8)). At the same time, we have shown in Appendix I that the assumption of dynamical mass generation for composite quarks and leptons leads to the conclusion that the inverse sizes of the τ and/or τ'families, must be less than or of order few TeV.

<u>Implications on Model Building</u>: Let me now note certain implications of these two constraints on model building. First observe that the two constraints can not be satisfied within the minimal fermion-boson model if we insist on a universal size for all families: e, μ, τ and any new family τ' to be discovered. These two constraints suggest that, one way or another, one must go beyond the minimal fermion-boson model. There appears to be a few alternative ways of reconciling the two constraints within the framework of the fermion-boson model:

(i) <u>A One-scale model with the e- and the μ-families having different chromons:</u>[14] One may arrange the preon model so as to avoid the constraints of K^0-$\bar{K}^0$ as well as $K_L \to \bar{\mu}e$ and thereby hope to maintain a relatively low universal inverse size of order 1 to few TeV for all three or four families. This could be possible for example by introducing four flavons

$$f_{L,R} = (u,d|c,s)^i_{L,R} \equiv (f_I|f_{II})^i_{L,R} \qquad \text{(A-11)}$$

and eight complex spin-0 chromons[38,15]

$$C = (r,y,b,\ell|r',y',b',\ell) \equiv (C_I|C_{II})^{\underset{\sim}{i}} \quad \text{(A-12)}$$

as in the family-replication model[15] (see text). Assume that each of these flavons and chromons transform as a representation $\underset{\sim}{i}$ of the metacolor gauge symmetry G_M with a scale parameter Λ_M. Assume furthermore that $f_{L,R}$ transform in the usual manner under the familiar flavor gauge symmetry $SU(2)_L \times SU(2)_R$. Unlike the replication model, however, assume that both C_I and C_{II} are quartets of the familiar $SU(4)^{color}$ gauge symmetry. In this case, we can construct four metacolor-singlet composite quark-lepton families, all of inverse size $\sim \Lambda_M$, out of "two-body"[F5] flavon-chromon composites:

$$\phi_1 = f_I C_I^* \, , \, \phi_2 = f_{II} C_I^* \, , \, \phi_3 = f_I C_{II}^* \, , \, \phi_4 = f_{II} C_{II}^* \, . \quad \text{(A-13)}$$

We can identify, these composite families with the e, μ, τ and τ' families in a few alternative ways.

But with the e and μ families being built out of different sets of chromons (and possibly even different sets of flavons), e.g. with $F_e = \phi_1$ and $F_\mu = \phi_3$ or ϕ_4, clearly the straightforward mechanism leading to $K_L \to \bar{\mu}e$ (see Figs. 2a,b) does not apply. One can still induce, due to compositeness, $q_d^{(o)} + \bar{q}_d^{(o)} \to e^{-(o)} + e^{+(o)}$ or $q_d^{(o)} + \bar{q}_d^{(o)} \to \mu^{-(o)} + \mu^{+(o)}$ (for $F_\mu = \phi_3$), with amplitudes of order A^2/Λ_M^2, where $A^2 \sim 1$ to 10. This could induce $q_d + \bar{q}_s \to e^- + \mu^+$ for the physical fermions and, therefore $K_L \to \bar{\mu}e$, only by utilizing Cabibbo rotations in the quark as well as in the lepton-space. Since the leptonic Cabibbo angle can be chosen to be vanishingly small due to the vanishingly small masses of ν_e and ν_μ, however, one need not have a conflict with the observed lower limit on $B(K_L \to \bar{\mu}e)$, although $\Lambda_{oe} = \Lambda_{o\mu} \sim$ few TeV. For reasons mentioned before (see discussions under Model (c), the metacolor force does not induce $\Delta S = 2$ K^o- $\bar{K}^o$ transitions, even after Cabibbo rotations, owing to a GIM-like mechanism in the $(f_I|f_{II})$ as well as $(C_I|C_{II})$-space.

Thus, this class of models can permit all three or four families (e, μ, τ and τ') to have a relatively low inverse size $\sim \Lambda_M \sim$ few to 10 TeV, without conflicting with $K_L \to \bar{\mu}e$ and K^o-$\bar{K}^o$ transitions.

A model of this kind with a single scale parameter $\sim$ few to ten

TeV is, however, likely to suffer from other difficulties involving cosmological issues, as stressed in Ref. 15. First, one must face the problem of the generation of the baryon excess in the early universe, the relevant temperature scale for which far exceeds 10 TeV. Second, in these models, flavon and chromon numbers are good global symmetries of the preonic Lagrangian. Unless they are broken spontaneously, one would expect stable members which are likely to conflict with present energy density. If they are broken spontaneously, as in Ref. 15, there would be massless Goldstone bosons, which would be sufficiently weakly coupled and, therefore, consistent with observation,[44] only provided $\Lambda \gtrsim 10^9$ GeV. It also remains to be seen whether such a one-scale model can successfully generate the mass-hierarchy of all four famillies and their mixings.

(ii) A two-scale model with family-replication: Alternatively, one may arrange the model so as to satisfy both constraints, i.e. $\Lambda_{\sigma\tau} = \Lambda_{\sigma\tau'} \sim 1$ TeV and $\Lambda_{oe} = \Lambda_{o\mu} \gtrsim 100$ TeV. One can start with the same four-flavon-eight-chromon system mentioned above, but assume that only the unprimed chromons $C_I = (r,y,b,\ell)$ transform as a quartet of $SU(4)^{color}$ while the primed chromons $C_{II} = (r',y',b',\ell')$ transform as a quartet of a new hypercolor gauge symmetry $SU(4)^H$ with a scale $\Lambda_H \sim 1$ TeV. In this case, the metacolor gauge force with a scale $\Lambda_M \gtrsim 100$ TeV could be used to bind the e and the μ-families, while the τ and a fourth τ' family could be built as composites of compositees through the hypercolor force. This leads to the replication model which is discussed in detail in the text.

This class of two-scale models[15] with Λ_M being of order 10^{14} GeV and $\Lambda_H \sim 1$ TeV appear to be best suited to meet cosmological constraints. In addition, they considerably lighten the burden of understanding the full fermion mass-hierarchy. At the same time, they offer many testable consequences (see text).

Is the Metacolor Scale Λ_M Superheavy $\sim 10^{14}$ GeV or only Medium Heavy $\sim$ 30 to 100 TeV?

It should be stressed that within the two-scale model,[15] most of our discussions and experimental consequences will be unaltered even if we lower the metacolor scale from 10^{14} to about 30 to 100 TeV. In this case one may perhaps account for the masses of the e and the μ families without involving the $\mathcal{D}_\mu$-exchange; they would be of order $(\Lambda_H/\Lambda_M)^2$ compared to the masses of the τ and τ' families. In other words, one major step in the mass-hierarchy would be explained directly in terms of a hierarchy in the sizes of the composites. But, with $\Lambda_M \sim 30$ to 100 TeV, one would somehow have to invoke a third superheavy scale $\sim 10^{14}$ GeV to account for the generation of baryon excess.

(iii) A hybrid model having family-replication but with $C_e \neq C_\mu$: One can consider a model having 12 complex spin-0 chromons

$$C = \left((r,y,b,\ell)_e \middle| (r,y,b,\ell)_\mu \middle| r',y',b',\ell' \right)^i \equiv (C_e|C_\mu|C_{II})^i \ .$$

Assume that each of the two sets C_e and C_μ transforms as a quartet of familiar $SU(4)^{color}$ while C_{II} transforms as a quartet of $SU(4)^{HC}$. Introduce minimally only two flavons:

$$f = (u,d)^i_{L,R} \ .$$

In this case, the metacolor force (with $\Lambda_M >> 1$ TeV) would generate three sets of small-size composite fermions:

$$\phi_e = fC^*_e \ , \ \phi_\mu = fC^*_\mu \ , \ \xi = fC^*_{II}$$

plus two sets of composite spin-0 colored hyperchromons:

$$\mathcal{D}_e = C_e C^*_{II} \ , \ \mathcal{D}_\mu = C_\mu C^*_{II} \ .$$

We may identify ϕ_e and ϕ_μ with the "bare" e and μ-family fermions. The hypercolor force (with $\Lambda_H \sim 1$ TeV) would make large-size composites $\xi\mathcal{D}^*_e$ and $\xi\mathcal{D}^*_\mu$ which may be identified with the τ and a fourth τ' family. Note that this model is a hybrid of the models presented under (i) and (ii). In this model, $C_e \neq C_\mu$, and $f_e = f_\mu$, but the idea of replication is maintained.

This model would easily meet the constraints of $K_L \to \bar{\mu}e$ and K^0-$\bar{K}^0$ for reasons discussed before. It would lead to fermion masses and mixings, provided we assume the formation of condensates of the type $\langle\xi^a\xi^b\rangle$ as well as $\langle\xi^a\xi^b \mathcal{D}_i^* \mathcal{D}_j\rangle$ where (a,b) run over only (u,d) and (i,j) over (e,μ)-indices. A model of this kind can meet the constraints of cosmology. It is not clear, however, how such a model would emerge naturally from a supersymmetric or supergravity preon model,[F23] without enlarging the flavor sector.

The constraints on the various models are summarized in Table II.

[F23] I may add that a twelve-chromon model permitting a basic SU(12)-gauge symmetry could lead to alternative scenarios including the possibility that it gives rise to $SU(4)^C \times SU(4)_1^{HC} \times SU(4)_2^{HC}$ with two different hypercolor groups, where $\Lambda_{QCD} < \Lambda_{HC1} < \Lambda_{HC2} < \Lambda_{MC}$.

TABLE II

Model	Constraint from $K_L \to \bar{\mu}e$	Constraint from $K^o - \bar{K}^o$	Comments
1) "Minimal" F-B Models with $C_e = C_\mu$ but $f_e \neq f_\mu$ (4 Flavors + 4 Colors).	$\Lambda_{oe} = \Lambda_{o\mu}$ $\gtrsim$ 100 TeV	$\Lambda_{oe} = \Lambda_{o\mu}$ $\gtrsim$ Few TeV	Can implement GIM at preon level. Need new ingredients to build heavier families (See (5) below).
2) The "Most Economical" Pair-Excitation Models (2 Flavors + 4 Colors) "μ" = "e" + $\rho_i \bar{\rho}_i$.	$\Lambda_{oe} \simeq \Lambda_{o\mu}$ $\gtrsim$ 100 TeV	$\Lambda_{oe} \simeq \Lambda_{o\mu}$ $\gtrsim$ 1000 TeV	The Constraints exclude the minimal Quasi-Nambu-Goldstone-Fermion models (see e.g. Refs. 39,40).*
3) The "Most Economical" Flavon or Chromon-Excitation Models (2 Flavors + 4 Colors) "e" = fC*, "μ" = (fffC)*	$\Lambda_{oe} \simeq \Lambda_{o\mu}$ $\gtrsim$ 100 TeV	$\Lambda_{oe} \simeq \Lambda_{o\mu}$ $\gtrsim$ 200 TeV	The Constraints exclude flavon or chromon-Excitation Models (see e.g. Ref. 43).*
4) One-scale Model with $C_e \neq C_\mu$ and $f_e = f_\mu$ or $f_e \neq f_\mu$. (4 Flavors + 8 Colors) (Ref. 14).	$\Lambda_{oe} = \Lambda_{o\mu}$ No strong constraint	$\Lambda_{oe} = \Lambda_{o\mu}$ No strong constraint	(i) Predict four families. (ii) Possible problems with cosmology & mass-hierarchy.
5) A two-scale Model [Ref. 15] with 4 Flavors + 8 Colors leading to $C_e = C_\mu$ but $f_e \neq f_\mu$. The four new chromons generate hypercolor force and lead to replication.	$\Lambda_{oe} = \Lambda_{o\mu}$ $\gtrsim$ 100 TeV	$\Lambda_{oe} = \Lambda_{o\mu}$ No strong constraint	(i) Predict four families. (ii) Consistent with cosmology for $\Lambda_{oe} = \Lambda_{o\mu} \sim \Lambda_M \sim 10^{14}$ GeV. (iii) Promising for mass-hierarchy.
6) A hybrid two-scale Model [Ref. 45] with 12 chromons + minimally two flavons having $C_e \neq C_\mu$ but $f_e = f_\mu$. The model possesses the replication mechanism [see (5) or Ref. 15].	$\Lambda_{oe} = \Lambda_{o\mu}$ No strong constraint	$\Lambda_{oe} = \Lambda_{o\mu}$ No strong constraint	(i) Predict 4 families. (ii) Can be consistent with cosmology (iii) Not clear about supersymmetric origin

*This is because these models need the scale of the preon binding force to be of order 1 TeV to account for the scale of electroweak symmetry breaking which conflicts with the constraints from $K_L \to \bar{\mu}e$ and $K^o - \bar{K}^o$.

APPENDIX III

"Maximal" Supersymmetric Preon Models:

One may introduce the so-called "maximal" supersymmetric preon models in which flavor and color are exact global (or possibly even local) symmetries even in the limit of SUSY, together with hypercolor and metacolor.[45] (This is unlike the "minimal"-SUSY model[15] which is presented in the text in which flavor, color and hypercolor are born as effective symmetries only after the gauginos become superheavy and decoupled and SUSY is broken[16]).

For example, the largest and the most straightforward extension of the "minimal" model would be to introduce twelve positive plus twelve negative chiral superfields $\Phi_{\pm}^{a,i}$, each coupled to a metacolor gauge symmetry G_M as symbolized by the index i which runs over 1 to N corresponding to a representation $\underset{\sim}{N}$ of G_M. The index a runs from 1 to 12 corresponding to chiral (flavor|color|hypercolor) indices = $(u,d,c,s|r,y,b,\ell|r',y',b',\ell')$. The model thus permits at least a global (or local, if we wish, for an anomaly-free subgroup) $SU(12)_L \times SU(12)_R \times U(1)_V$ symmetry even in the limit of SUSY, which contains $[SU(4)_L \times SU(4)_R]_{flavor} \times [SU(4)_L \times SU(4)_R]_{color} \times [SU(4)_L \times SU(4)_R]_{Hypercolor}$. One may assume, as in the "minimal" model, that the metacolor gauge force breaks dynamically this large global (or local) symmetry into $SU(2)_L \times U(1)_Y \times SU(3)^C_{L+R} \times SU(4)^{HC}_{L+R}$, which emerges as an effective low-energy gauge symmetry through composite (or elementary) gauge bosons. The fermions of the flavor superfields are left massless due to $SU(2)_L$, but those of the color and hypercolor superfields become massive (mass $\sim \Lambda_M$) through condensates such as $\langle \bar{\psi}^C_L \psi^C_R \rangle$ and $\langle \bar{\psi}^{HC}_L \psi^{HC}_R \rangle$. The bosonic components of these superfields acquire soft SUSY breaking masses $\sim (\Lambda_M^3/M_{P\ell}^2) \sim 300$ GeV. Thus local and global SUSY are broken. The Goldstone bosons corresponding to spontaneous breaking of $SU(12)_L \times SU(12)_R \times U(1)_V$ are born massless (these are weakly coupled since $\Lambda_M \sim 10^{14}$ GeV), but their fermionic partners are expected to be massive[F25] because SUSY is broken. Let us assume that the bosons of the flavor super-

[F25] This needs further study.

fields become massive ($\sim \Lambda_M$) as well. Thus the fermions of the chiral flavor superfields and only the bosons of the color and hypercolor superfields serve as the massless or light preons.

The model generates composites at the metacolor and the hypercolor scales, as in the "minimal"-SUSY model (see text). However, due to chiral color and hypercolor degrees of freedom, the model would lead to excessive number of quark-lepton families (4 of metacolor size plus 16 of hypercolor size) unless somehow only one linear combination of the bosonic components $\mathcal{G}_L^C$ and $\mathcal{G}_R^C$ and likewise of $\mathcal{G}_L^{HC}$ and $\mathcal{G}_R^{HC}$ remain light and the other combinations become superheavy, in which case there will be just four families as in the "minimal"-SUSY model.

As an intermediate extension, one can consider a model with ten positive plus ten negative chiral superfields $\Phi_{\pm}^{a,\alpha}$ (a = 1 to 10) in which case one would have an $SU(10)_L \times SU(10)_R$ global symmetry (local symmetry permissible for an anomaly-free subgroup), comprising of only two chiral flavor plus four chiral color plus four chiral hypercolor indices. This model would cut down the number of composite families by a factor of two compared to the $SU(12)_L \times SU(12)_R$-model, which still leaves us with too many, i.e. (2 + 8) = 10, families.

As a minimal extension of the "minimal" SUSY model, one can introduce[45] just 8 positive plus 8 negative chiral superfields $\Phi_{\pm}^{a,\alpha}$ (a = 1 to 8) permitting an $SU(8)_L \times SU(8)_R$ global symmetry, with a = 1 to 4 denoting chiral flavor $(u,d,c,s)_{L,R}$ and a = 5 to 8 denoting a chiral primed color (c'). Assume, as in the $SU(12)_L \times SU(12)_R$-model that the fermionic components of the primed color superfields $\Phi_{\pm}^{a,\alpha}$ with a = 5 to 8 become superheavy due to the condensates $\langle \bar{\psi}_L^{c'} \psi_R^{c'} \rangle$; the corresponding bosonic components of $\Phi_{+}^{a,\alpha}$ (a = 5 to 8), which remain light, define the vectorial $SU(4)^{color}$ and those of $\Phi_{-}^{a,\alpha}$ (a = 5 to 8) define the vectorial $SU(4)^{HC}$ symmetry. As before, the fermionic components of the flavor superfields remain massless due to surviving $SU(2)_L$. Subsequent to dynamical breaking of the full global symmetry and the formation of composite gauge bosons due to the metacolor force, the effective low-energy gauge symmetry for energies $< \Lambda_M$ is just $SU(2)_L^{e+\mu} \times U(1)_Y \times SU(3)_{L+R}^C \times$

$SU(4)^{HC}$.

The spectrum of quark-lepton families of this SU(8) × SU(8)-model is identical to that of the "minimal" SUSY model presented in the text. One obtains two families of size ~ Λ_M^{-1} and two of size ~ Λ_H^{-1}. This model thus preserves all the main properties of the fermion mass-matrix and the predictions of the "minimal" SUSY model. Unlike the "minimal" SUSY model, however, the "maximal" extensions[F26] including the SU(8) × SU(8) case allow flavor, color and hypercolor including electric charge to be defined as exact symmetries of the basic lagrangian.

As regards the derivation of these models from (for example) the $E_8 \times E_8$ superstring theory, one would need compact manifolds K leading to six, five and four generations of 27 of E_6 for the SU(12) × SU(12), SU(10) × SU(10) and the SU(8) × SU(8) models respectively. The cases of six and four generations are easy to obtain, while that of five appears to be hard. These questions are being examined by T. Hübsch in detail.

[F26]In exploring the "minimal" or the "maximal" SUSY models, one keeps in mind the question of whether the effective coupling constants of $SU(2)_L$, $SU(3)^C$ and $U(1)_Y$, generated by composite (or elementary) gauge bosons are equal at Λ_M due to constraints of symmetry and dynamics. With these three coupling constants being equal at Λ_M, one would be led to essentially the same predictions for $\sin^2\theta_W$ and the unification scale $M = \Lambda_M$ as those obtained in simple grand unified theories, on the basis of renormalization group equations. As an example, the SU(12) × SU(12)-model, viewed naively, leads to such an equality of coupling constants at Λ_M.

REFERENCES

1) J.C. Pati and Abdus Salam, Phys. Rev. D8 (1973) 1240; Phys. Rev. Lett. 31 (1973) 661; Phys. Rev. D10 (1974) 275.

2) H. Georgi and S.L. Glashow, Phys. Rev. Lett. 32 (1974) 438.

3) H. Georgi, H. Quinn and S. Weinberg, Phys. Rev. Lett. 33 (1974) 451.

4) M. B. Green and J. H. Schwarz, Phys. Lett. 149B (1984) 117; 151B (1985), 21, and references therein.

5) D. J. Gross, J. A. Harvey, E. Martinec and R. Rohm, Phys. Rev. Lett. 54 (1985) 502.

6) P. Candelas, G. T. Horowitz, A. Strominger and E. Witten, Princeton University, Preprint NSF-ITP-84-170 (1984).

7) E. Witten, "Symmetry breaking patterns in superstring theories", Princeton University, Preprint (1985); A. Strominger and E. Witten, Princeton University, preprint (1985).

8) J. D. Breit, B. A. Ovrut and G. Segre, University of Pennsylvania, preprint UPR-0279 T (1985).

9) M. Dine, V. Kaplunovsky, M. Mangano, C. Nappi and N. Seiberg, Princeton University, preprint (1985).

10) T. Hübsch H. Nishino and J. C. Pati, IC/85-66, Phys. Lett. to appear.

11) J. C. Pati and Abdus Salam, Phys. Rev. D10, (1974) 275; For a review and other references see e.g. M. Peskin, Proc. Int. Symp. on Lepton and Photon INter. at High Energies, Bonn (1981) Ed. W. Pfeil, p. 880; L. Lyons, Prog. Part. and Nucl. Phys. 10, (1983) 227; H. Harari, Weizman Institute, preprint (to appear in the Proc. of the 1984 Summer School, St. Andrews).

12) H. Nishino, J.C. Pati and S.J. Gates Jr., Phys. Lett. 154B (1985) 363.

13) V. Silveira and A. Zee, Phys. Lett. 157B (1985) 191.

14) J. C. Pati, Phys. Rev. D Rap. Com. 30, (1984) 1144.

15) J. C. Pati, Phys. Lett. 144B, (1984) 375.

16) J. C. Pati and Abdus Salam, Nuc. Phys. B214 (1983) 109; ibid B234 (1984) 223; R. Barbieri, Phys. Lett. 121B (1983) 43.

17) P. Mohapatra, J. C. Pati and H. Stremnitzer, Md. preprint 85-200.

18) I. Bars, Proc. XVIIth Rencontre de Moriond, Les Arcs, France,

(1982) Ed. J. Tran Thanh Van, Vol. 1, p. 54 (1982).

19) J. C. Pati, H. Sharatchandra and M. Cvetic (in preparation); M. Cvetic, University of Maryland Preprint 85-23.

20) See e.g. G. 't Hooft, Cargese Lectures (1979); D. Weingarten, Phys. Rev. Lett. 51 (1983) 1830; S. Nussinov, Phys. Rev. Lett. 51 2081; E. Witten, Phys. Rev. Lett. 51 (1983) 2351.

21) C. Vafa and E. Witten, Nucl. Phys. B234 (1984) 173.

22) S. Weinberg and E. Witten, Phys. Lett. 96B, 59 (1980).

23) J. C. Pati and Abdus Salam Phys. D10 (1974) 275 (Footnote 7); Proceedings of the EPS International Conference on High Energy Physics, held at Palermo, Italy (June 1975), Pages 154-170.

24) R. Barbieri, R. N. Mohapatra and A. Masiero, Phys. Lett. 105B (1981) 369.

25) F. Ferrara, L. Girardello and H. Niles, Phys. Lett. 125B (1983) 457.

26) J. C. Pati and Abdus Salam, (Ref. 16).

27) A. Davidson and J. C. Pati, Nucl. Phys. B175 (1980) 175.

28) J. C. Pati and Abdus Salam, Phys. Rev. D10 (1974) 275; R. N. Mohapatra and J. C. Pati, Phys. Rev. D11 (1975) 566; ibid (1975) 2558; G. Senjanovic and R. N. Mohapatra, Phys. Rev. D12 (1975) 1502; R. N. Mohapatra and G. Senjanovic, Phys. Rev. Lett. 44 (1980) 912.

29) D. Chang, R. N. Mohapatra, P. Pal and J. C. Pati, "Spontaneous Breakdown of Parity as the Origin of Isospin Breaking", Md. Phys. Publication 86-19.

30) S. Weinberg, published in Rabi Festschrift, NY Academy of Science, 1978.

31) E. J. Eichten, K. D. Lane and M. E. Peskin, Phys. Rev. Lett. 50 (1983) 811.

32) PETRA experiments reported at Kyoto Conference, August 1985. To appear in the Proceedings.

33) C. Rubbia, UA1 experiments, Proc. Kyoto Conference, August 1985 (to appear).

34) J.C. Pati (Ref. 14); O. W. Greenberg, R. N. Mohapatra and S. Nussinov, Phys. Lett 148B, (1984) 465.

35) A. Datta and J. C. Pati, ICTP preprint, IC/85/145, Phys. Lett., to be published.

36) R. Godbole, Trieste preprint (to appear).

37) For a review and relevant references, see e.g. E. Farhi and L. Susskind, Phys. Rep. (1982).

38) J. C. Pati, Proc. XXI Intern. Conf. on High Energy Physics (Paris, July 1982) C3, p. 197, Proc. 1983 Trieste Workshop (IC/83/221), pages 400-459.

39) W. Buchmuler, R.D. Peccei and T. Yanagida, Phys. Lett. 124B (1983) 67; Nucl. Phys. B227 (1983) 503; B321 (1984) 53.

40) O.W. Greenberg, R.N. Mohapatra and M. Yasue, Phys. Lett., 128B (1983) 65; Phys. Rev. Lett. 51 (1983) 1737.

41) R.N. Mohapatra, J.C. Pati and M. Yasue, Phys. Lett. 151B (1985) 251.

42) S. Dimopoulos, S. Raby and G.L. Kane, Nucl. Phys. B182 (1981) 77; N.G. Deshpande and R.J. Johnson, Phys. Rev. D27 (1983) 1193.

43) H. Terezawa, Y. Chikashige and K. Akama, Phys. Rev. D15 (1977) 480.

44) Y. Chikashige, R.N. Mohapatra and R.D. Peccei, Phys. Lett. 98B (1981) 265; G.B. Gelmini, S. Nussinov and T. Yanagida, Nucl. Phys. B219 (1983), 31.

45) J.C. Pati, unpublished.

PARTICLE PHYSICS AND COSMOLOGY

D. W. Sciama
International School for Advanced Studies, Trieste,
International Centre for Theoretical Physics, Trieste,
Department of Astrophysics, Oxford University

ABSTRACT

A review is given of the implications for particle physics of the primordial nucleosynthesis of the light elements and of the missing mass problem in astronomy and cosmology. These questions are interrelated, a central role being played by the number of neutrino types. Recent developments, observational, experimental and theoretical, in our understanding of this number are emphasised.

Introduction

It is only 2 or 3 years since the flow of ideas between cosmology and particle physics has become intense, detailed and highly specific. Nevertheless the main message of this lecture is that the union between these two subjects has already reached maturity. By this I mean that it has become not only close and symbiotic but also highly elaborate, complicated and technical, and divided into many specialisations, with experts in each not understanding what experts in others are doing and saying.

I am not an expert in any of these specialisations. What I will try to do in this lecture is to show how they are all linked to one another in a fascinating network of relationships.

I should add that I am a cosmologist and not a particle physicist, so I hope that you will be indulgent to my handling of the particle physics.

I would like to begin by listing the main processes so far discussed in which particle physics may have a decisive influence on the evolution of the universe. They are as follows:

t_{sec}	Processes	T°K	E_{GeV}
10^{-35}	Inflation	10^{27}	10^{14}
10^{-33}	Baryosynthesis	10^{26}	10^{13}
10^{-5}	Quark-hadron phase transition	10^{12}	10^{-1}
10^{2}	Light element synthesis	10^{9}	10^{-4}
10^{17}	Survival of WIMPS (weakly interacting massive particles)	3	10^{-12}

The left-hand column gives a rough indication of the time in seconds after the big bang at which the given process occurred, while the right-hand columns give the corresponding temperature of the universe in degrees absolute and in GeV energy units.

I will not have time to discuss all these processes (for details of them see the report of the ESO/CERN Symposium 1984[1]), and will concentrate on the last two, which are perhaps the least speculative and which also lead to many of the presently known contacts between astronomy, cosmology and particle physics. Before doing so I would like to list the main constraints by which cosmology in its turn influences our knowledge of particle physics. These constraints have been discussed in detail in Subir Sarkar's lecture at this workshop.

Constraints

Light element abundances

Spectrum of the 3°K background

Isotropy " " " "

Polarisation " " "

X and γ ray backgrounds

Age of the universe $\to \rho < 2\,\rho_{crit}$

Galaxy masses

" distribution

Liouville's theorem for WIMPS.

To prepare for our discussion of light element abundances and WIMP survival, we first briefly recall the main cosmological input data.

1) The Hubble Law

The recession velocity v of a galaxy is related to its distance r by the linear Hubble law

$$v = Hr \ ,$$

where the Hubble constant H is observed to be

$$H = 100 \ h \ km.sec^{-1} Mpc^{-1}$$

with

$$½ \leq h \leq 1$$

covering the main observational uncertainty.

The universe is destined to recollapse into a "big crunch" if the mean density of the universe ρ exceeds a critical density ρ_{crit} given by

$$\frac{8\pi}{3} G \rho_{crit} = H^2 \ .$$

Hence

$$\rho_{crit} \sim 2 \times 10^{-29} h^2 \ gm.cm^{-3}$$

If the critical density were in the form of baryons (which we shall soon see is most unlikely) we would have

$$n_{b,crit} \sim 10^{-5} h^2 \ cm^{-3} \ .$$

2) The 3°K background

This background has a fairly accurate thermal spectrum (a claimed deviation near the peak at ∿1 mm having not been confirmed by more recent observations). It is also highly isotropic, apart from a dipolar deviation probably due to our net motion relative to the universe as a whole. Its contribution ρ_{rad} to the density of the universe is

$$\rho_{rad} \sim 10^{-33} \ gm.cm^{-3} \ ,$$

a quantity which will enter our consideration later.

The background temperature was greater in the past, and at sufficiently early times (when $\rho_{rad} > \rho_b$), it was related to the time in the following simple manner

$$T \sim \frac{10^{10}\ {}^\circ K}{t_{sec}^{1/2}} \quad .$$

This relation holds if only one relativistic species (that is photons) contributes to the density of the universe. If other relativistic species are important (e.g. massless or low-mass neutrinos, photinos etc.) then the expansion rate of the universe would be appreciably speeded up. This point will be of decisive importance in our later discussions.

3) Light element abundances

It is believed that a major fraction of the lightest elements observed in galaxies today were formed by thermonuclear reations at a time of about 100 seconds after the big bang, when the ambient temperature in the universe was about $10^9\ {}^\circ K$. After correcting for subsequent evolutionary effects occurring in stars, one finds that the observationally required primordial mass fractions of these elements are roughly as follows:

H^1	0.77
H^2	5×10^{-5}
He^3	10^{-4}
He^4	0.23
Li^7	5×10^{-10}

The calculated abundances involve a large number of reaction networks and have to be derived numerically. The striking result of these calculations is that the derived abundances of these light elements depend very sensitively on the timescale of the expansion at a temperature $\sim 10^9\ {}^\circ K$. It is thus highly non-trivial that one can find reasonable input data which yield abundances that fit well with the required abundances given above (for a recent detailed discussion see Yang et al.[2] and for deviations from the standard model Audouse et al.[3], Applegate and Hogan[4], Centrella et al.[5]). The main input data are the baryon to photon ratio n_b/n_γ at the epoch of nucleosynthesis, and the number of neutrino types N_ν which were relativistic at that time. The ratio n_b/n_γ controls the rate of the nuclear reactions, while N_ν influences the expansion timescale via the contribution of each neutrino type to the ambient density.

One finds in this way that one can fit the observations with a choice of

$$n_{b/n_\gamma} \sim 4 \pm 2 \times 10^{-10} \; .$$

This choice has a number of consequences of importance for our discussion.

(a) Baryosynthesis must account for this value of n_{b/n_γ}. According to the current view, based on grand unified theories or their supersymmetric generalisations, this value is essentially determined by microphysical parameters such as the extent of CP violation etc. No concensus yet exists on the detailed mechanisms involved, although it is recognised that the requirement of successful baryosynthesis can impose a strong constraint on the large photon input associated with the earlier process of inflation. These problems have been recently reviewed by Kolb and Turner[6).]

(b) If the universe expands adiabatically after nucleosynthesis, so that in particular there is no significant additional input of photons, then the ratio n_b/n_γ stays constant during the expansion. Its value today would thus still be $\sim 4\text{x}10^{-10}$. Since now

$$n_\gamma \sim 400 \text{ cm}^{-3}$$

we must also have now

$$n_b \sim 1.6 \times 10^{-7} \text{ cm}^{-3} .$$

Our standard model of the hot big bang survives a significant test at this point, since this derived value of n_b today agrees well with the value obtained by smoothing out all the baryonic matter observed in stars and interstellar gas in galaxies. On the other hand, it is far less than the value $10^{-5} h^2 \text{cm}^{-3}$ required to make up the critical density. Later on this discrepancy will provide us with one of our missing mass problems.

We now consider the derived value of N_ν, the number of neutrino types which were relativistic at nucleosynthesis, that is, whose rest-masses satisfy $m_\nu < 100$ kev. The main determinant of N_ν is the helium mass fraction Y, and one finds that $\Delta Y = 0.01$ corresponds to $\Delta N_\nu = 1$. This gives a measure of the astronomical precision required to pin N_ν down precisely. There are in fact two schools of thought amongst astronomers about this question. According to one school the observed helium abundance implies that

$$2 < N_\nu < 5 \, ,$$

whereas according to the other

$$2 < N_\nu < 4 \, .$$

We shall see shortly that a great deal hangs on this difference, so it is to be hoped that the astronomical discusssion can be sharpened up in the next few years (Davidson and Kinman[7]).

We now give for comparison a brief indication of laboratory estimates of N_ν. In recent times these estimates have changed as follows:

last year	$3 < N_\nu < \sim 100{,}000$	
this year	$< \sim 20$	Z_o data
today	$< 5.4 \pm 1$	Z_o data

We see that particle physics is catching up on cosmology! However, it is important to note that the two techniques do not measure exactly the same quantity. The CERN data would count neutrinos of GeV rest-mass coupled to the Z_o in the standard way, while such neutrinos would not have been relativistic at nucleosynthesis and so would not be counted by the cosmological argument. By contrast this argument would count some low mass particle superweakly coupled to the Z_o if it were as abundant as ordinary neutrinos at nucleosynthesis, whereas such a particle would not show up in the Z_o data.

Unfortunately we still do not possess a fundamental theory which prescribes a definite value for N_ν. However constraints on N_ν have been derived in the recent theory of Candelas, Horowitz, Strominger and Witten[8] based on superstrings. This theory starts out from a ten-dimensional manifold. Six dimensions become compactified into a closed manifold, the remaining four corresponding to ordinary space-time. Supersymmetry considerations then require that the 6D manifold should be Ricci flat and have an SU(3) holonomy group. Such spaces are named after Calabi and Yau, and can be constructed explicitly by the methods of algebraic geometry although no metric is known for any of them. In this theory the number of generations is given by half the modulus of the euler characteristic of the 6D manifold. This characteristic can be computed even when the metric is not known.

In the original preprint of Candelas et al. it was conjectured, on the basis of the examples of Calabi-Yau spaces known at the time, that their Euler characteristic is always divisible by four. This would have meant that N_ν would have to be even, and so be at least four, since three neutrino types are already known. Thus if one were to adopt the more stringent cosmological constraint on N_ν, the fourth neutrino type would have to have a mass in the GeV range or greater (as we shall see later when we come to discuss the neutrino contribution to the present density of the universe). Alternatively, one could argue that perhaps the conjecture is wrong. This now seems to be the case, since I understand that yau has constructed several manifolds with an Euler number whose modulus is six. I also understand that the resulting theory suffers from some difficulties, so that more work needs to be done here. However, it does seem to be possible that by measuring the abundance of helium in the Galaxy one might be able to demonstrate the existence of six dimensional compact Ricci flat manifolds with Euler number equal to ±6!

There would be another important consequence of adopting the more stringent cosmological constraint $N_\nu < 4$. This arises because one already knows of the existence of three neutrino types, namely ν_e, ν_μ, and ν_τ. One can show from a variety of arguments that they were all relativistic and essentially stable at the time of primordial nucleosynthesis (see Sarkar's lecture), so that there would be no room for even one more particle type which was relativistic at that epoch and as abundant as neutrinos. Thus if hypothetical particles such as photinos and gravitinos actually exist and have sufficiently low mass, they must be suppressed compared with neutrinos.

In order to see what is involved here, I first remind the reader of the considerations which determine the neutrino abundance today, which is again cosmologically important. At sufficiently early times neutrino pairs were in thermal equilibrium with the radiation field as a result of weak interactions of the type

$$\nu + \bar{\nu} \leftrightarrow e^- + e^+ .$$

However these reaction rates eventually became smaller than the expansion rate of the universe, and the neutrinos then decoupled from the heat bath and became essentially non-interacting (except through gravitation). For standard neutrino interaction rates this decoupling occurs at a temperature $\sim$1 MeV.

Before decoupling we have the simple equilibrium relation

$$n_\nu \sim n_\gamma ,$$

apart from a small effect coming from the Fermi-Dirac statistics of the neutrinos, which is unimportant for us. After decoupling the neutrino system expands freely, but in essentially the same way as the photons, so we again have the simple relation

$$n_\nu \sim n_\gamma .$$

Of course we are here assuming that neutrino annihilations are negligible, which would be true if they were relativistic at decoupling, that is, if $m_\nu < 1$ MeV.

This simple argument overlooks one important point, however. Electron pairs annihilated permanently at $T \sim \frac{1}{2}$ MeV, that is, after the neutrinos decoupled. Thus the resulting annihilation products, namely, photons, would have boosted the radiation field without boosting the neutrinos. One may speak of the neutrinos as having been suppressed compared to the photons, and one finds that this suppression effect is given by

$$n_\nu \sim \frac{1}{4} n_\gamma \quad .$$

This then is the neutrino abundance which enters into the determination of N_ν from the abundances of the light elements.

4) Neutrino abundances today

Before considering the suppression of other particles relative to neutrinos, we shall explore a further consequence of the present argument, namely that $n_\nu \sim \frac{1}{4} n_\gamma$ today as well as at nucleosynthesis. Although we do not know today's date very accurately, we do know n_γ from the 3°K background. As we have seen this is given by

$$n_\gamma \sim 400 \text{ cm}^{-3}$$

Hence, today

$$n_\nu \sim 100 \text{ cm}^{-3} \quad .$$

This leads to the famous constraint

$$\sum_\nu n_\nu m_\nu \lesssim \rho_{crit}$$

or

$$\Sigma\, m_\nu \lesssim 100\ h^2 \text{ ev} \quad .$$

This constraint has many consequences. In particular if Simpson's[9] 17 kev neutrino exists, it must decay or be annihilated at sufficiently early epochs for the energy of its products to be adequently red-shifted away by the present time. Schemes satisfying this constraint have been constructed, but are rather contrived. However, several preprints are now circulating which contradict Simpson's claim, so perhaps there is no problem here, e.g. Altzitzoglou et al.[10].

Similarly for a general WIMP p if

$$m_p > 100\ h^2 \text{ ev}$$

we must have

$$n_p < n_\nu \quad .$$

This could give a second reason for the suppression of particles like photinos and gravitinos relative to neutrinos.

5) Suppression Mechanisms

Four suppresion mechanisms have been widely discussed.

(a) If the WIMP decoupled from the heat bath before μ, π ... annihilated permanently, that is, at $T \gtrsim 200$ Mev, then it would have been suppressed compared ~~to~~ neutrinos for the same reason as neutrinos are suppressed compared to photons. The calculation (Olive, Schramm and Steigman[11]) involves knowing which particle species are involved in the annihilations, and one finds suppression factors of order 10-20 if decoupling occurred somewhat before μ and π annihilated, with greater suppression factors for very much earlier decoupling.

(b) The WIMP may itself annihilate sufficiently if it were non-relativistic at decoupling. The suppression is then due to a Boltzmann factor $e^{-mp/kT}$, and one typically finds both for neutrinos and photinos that the critical density constraint leads to the requirement (Lee and Weinberg[12], Goldberg[13], Ellis et al.[14])

$$m \gtrsim 1 \text{ Gev} .$$

Of course the known types of neutrino violate this constraint, and so annihilation is not important for them.

(c) The WIMP may decay either before the present time or before the epoch of primordial nucleosynthesis. One must then ensure that the decay products do not violate any cosmological constraints. This can be a severe requirement, as Subir Sarkar discusses at this conference. (See also Audouze, Lindley and Silk[3]).

(d) The WIMP may be diluted by a significant input of photons from a delayed phase transition, of the type invoked by the inflationary hypothesis. This mechanism has been discussed particularly in relation to gravitinos (Ellis, Linde and Nanopoulos[15]).

6) Missing Mass

The missing mass problem is by now well known (Faber[16]). Less well known is the first recorded example of a missing mass problem outside the solar system. This arose from Oort's[17] analysis of the motions of stars near the sun at right angles to the galactic plane. To account for these motions Oort invoked a material density near the sun which was about twice the known density in stars and gas and which had a scale height less than about a kiloparsec. The need to have such additional matter of unknown origin has been much debated since then. A recent elaborate re-discussion by Bahcall[18] based on new observational data has given results close to Oort's original ones. It seems doubtful whether one could attribute this missing mass to neutrinos, photinos etc., since it is not clear that such non-interacting particles could relax into the required flattened distribution.

The more familiar forms of missing mass involve galactic halos and clusters of galaxies. In addition inflationary theories suggest that the density of the universe should now be very close to critical, whereas, as we have seen, the baryon density would be expected to be much less than this. A natural suggestion, therefore is that WIMPs are involved both in galaxies and in the universe as a whole. The formation of galaxies in a universe dominated by WIMPs has been much discussed but is not yet well understood. See Blumenthal et al.[19], Davis et al.[20], Melott[21], Fry and Melott[22]. The one certain result is that after decoupling the WIMPs must obey Liouville's theorem, so that their phase space density today in a galaxy cannot exceed the maximum value of this density at decoupling (Tremaine and Gunn[23]). This requirement gives the important constraint

$$m_p^4 \gtrsim h^3/Gv_o a^2 \quad ,$$

where here h is Planck's constant (not the Hubble constant), v_o is the velocity dispersion of the WIMPs, a is their core radius, and we have suppressed a dimensionaless coefficient of order unity which depends in detail on the WIMP distribution. This is a remarkable relation involving a strange mixture of quantities, namely microphysical parameters (m_p and h), a macrophysical parameter (G), and astronomical properties of individual galaxies (v_o and a).

For the Milky Way one has $v_o \sim 300$ km.sec^{-1} and a $\sim$10 kpc. One then finds

$$m_p \gtrsim 35 \text{ ev}.$$

One could thus have p = ν and also $\rho_\nu = \rho_{crit}$, if only the observed values of (some of) the neutrino masses turn out to be in the tens of ev range.

On the other hand, it has recently been claimed that dwarf galaxies lying in the outskirts of the Milky Way also possess missing mass (See Faber[16]). These galaxies have appreciably lower values of v_o and a, and the Liouville constraint would now give

$$m_p \gtrsim 350 \text{ ev} .$$

Thus the dwarf galaxies could not be dominated by the known types of neutrino, but could be dominated by a WIMP which decoupled before muons and pions annihilated, since one would then obtain precisely a suppression factor $\gtrsim$10, as required. The Oort missing matter near the sun would lead to the same Liouville constraint as the dwarf galaxies, so it would seem to be worth searching further for a relaxation mechanism which could lead to the formation of a thin disk of non-dissipative particles. Perhaps the dissipative baryons forming the conventional galaxy play a key role here.

In general the galaxy formation process is not well understood, as we have already stated. The process depends very strongly on the value of m_p, and one distinguishes between hot particles (ypically of tens of ev mass) warm particles (kev) and cold particles (GeV). In particular, whether galaxies form before or after clusters and superclusters depends on the value of m_p. It has recently been argued that neutrinos are no longer a good candidate for the missing mass because calculations of the resulting galaxy correlations are in disagreement with observation. However, this should not be taken as decisive since fundamental new ideas are still being introduced into the discussion of galaxy formation. The latest idea is the so-called biassed galaxy formation, in which only relatively large fluctuations in density lead to the formation of a galaxy (see Schaeffer and Silk[24]). This rather natural suggestion would make a major change in our understanding of galaxy formation.

We have already mentioned several candidates for the missing mass. A partial list is as follows:

Particle	Possible Mass
Axion	10^{-5} ev
Neutrino	30 ev or 1 GeV
Photino	1 kev or 1 GeV
Gravitino	?
Monopole	?
Quark nugget	2×10^{9} - 2×10^{16} gm
Shadow particle	?
Black Hole	$< 10^{7}$ $M_{\odot}$

The extent of our ignorance is illustrated by the fact that the possible masses involved range over a factor 10^{78}! No doubt future theoretical developments will help to distinguish between these possibilities, but it would be highly desirable to detect in the laboratory effects produced by the particles in the halo of our Galaxy. There are various possibilities here which have been reviewed recently by Goodman and Witten[25].

Another possibility is to detect the annihilation or decay products of the particles. Annihilation γ rays and antiporotons produced by Gev photinos have been considered by Sciama[26], and Silk and Srednicki[27] respectively, while high energy neutrinos produced by photinos trapped in the sun have been discussed by Silk, Olive and Srednicki[28].

My own favourite possibility is to detect photons emitted by decaying low-mass photinos. The reason for this preference is that, if the relevant supersymmetry parameters were favourable, the photon signal produced would be a strong one and would be easily detectable in future experiments. The idea was proposed by Cabibbo, Farrar and Maiani[29], who calculated the lifetime $\tau_{\tilde{\gamma}}$ for a photino to decay into a photon and a (lighter) goldstino or gravitino. They obtained

$$\tau_{\tilde{\gamma}} \sim 5 \times 10^{26} \left(\frac{d}{10^9 \mathrm{GeV}^2}\right)^2 \left(\frac{500\mathrm{ev}}{m_{\tilde{\gamma}}}\right)^5 \quad \mathrm{secs},$$

where d is the square of the energy at which supersymmetry is broken. If $m_{3/2} \ll m_{\tilde{\gamma}}$, the photon energy E_γ is given simply by

$$E_\gamma \sim \tfrac{1}{2} m_{\tilde{\gamma}} \ ,$$

and if photinos dominate the halo of our Galaxy their velocity dispersion $\ll$ c, so that the emitted photons would be in a sharp line which could potentially be distinguished from the continuous photon spectrum of conventional origin.

The observational possibilities have been discussed by Sciama[30),31),32)]. If we take $m_\gamma \sim 500$ ev (admittedly an unfashionable value) to satisfy the dwarf galaxy constraint, then $E_\gamma \sim 250$ ev, so that we would be dealing with a soft X-ray line in a spectral region not yet explored by X-ray telescopes of adequate energy resolution to detect a sharp line.

The value of d is undetermined in the present state of supersymmetry theory, but is constrained by the requirement that our 500 ev photinos should have interactions which ensure that they are suppressed by decoupling before muons and pions annihilate. A value $\sim 10^9 \mathrm{GeV}^2$ would be satisfactory in this respect and would lead to a lifetime $\sim 5 \times 10^{26}$ sec. Although this is much longer than the age of the universe, the resulting line would be a strong one which in the future would easily be detectable from the halo of our Galaxy, and also from extragalactic sources such as the Andromeda galaxy, M87 etc.

For example, the flux from Andromeda would be about

$$1 \text{ photon } \mathrm{cm}^{-2} \ \mathrm{sec}^{-1}.$$

By contrast, the sensitivity of proposed X-ray satellites which are planned to have good energy resolution in this region of the spectrum will be

$$10^{-5} \text{ photon cm}^{-2} \text{ sec}^{-1}.$$

We would therefore still have a measurable signal if the supersymmetry parameters were less favourable than the ones we have chosen here for purposes of illustration.

These X-ray satellites are planned for the early 1990's. Meanwhile a dedicated rocket experiment, code-named Pegasus, should be searching for the photon line in about two years from now. If the anomalous CERN events are in fact not real, the first observational evidence in favour of supersymmetry may yet come from X-ray astronomy!

Conclusions

I hope that this brief review justifies my claim that, whatever the uncertainties, the union between particle physics and cosmology has now reached maturity.

References

1) ESO/CERN Symposium on The Large Scale Structure of the Universe, Cosmology and Fundamental Physics (Ed. G. Setti and L. van Hove), 1984.
2) S. Yang, M.S. Turner, D.N. Schramm and K.A. Olive, Ap. J. 281, 493, 1984.
3) J. Audouze, D. Lindley and J. Silk Ap. J. Lett. 293, L53, 1985.
4) J.H. Applegate and C.J. Hogan, Phys. Rev. D. 31, 3037, 1985.
5) J. Centrella, R.A. Matzner, T. Rothman and J. Wilson, to be published.
6) E.W. Kolb and M.S. Turner, Ann. Rev. Nucl. Part. Sci. 33, 365, 1983.
7) K. Davidson and T. Kinman, Ap. J. Suppl. 58, No.3, 1985.
8) P. Candelas, G. Horowitz, A. Strominger and E. Witten, Nucl. Phys., to appear.
9) J.J. Simpson, Phys. Rev. Lett. 54, 1891, 1985.
10) T. Altzitzoglou, F. Calaprice, M. Dewey, M. Lowry, L. Piilononen, J. Brorson, S. Hagen and F. Loeser, Phys. Rev. Lett. 55, 799, 1985.
11) K.A. Olive, D.N. Schramm and G. Steigman, Nucl. Phys. B 180, 497, 1981.
12) B.W. Lee and S. Weinberg, Phys. Rev. Lett. 39, 165, 1977.
13) H. Goldberg, Phys. Rev. Lett. 50, 1419, 1983.
14) J. Ellis, J.S. Hagelin, D.V. Nanopoulos, K. Olive and M. Srednicki, Nucl. Phys. B. 238, 453, 1984.
15) J. Ellis, A.D. Linde and D.V. Nanopoulos, Phys. Lett. 118B, 59, 1982.
16) S. Faber in ESO/CERN Symposium 1984.
17) J.H. Oort, Bull. Astr. Inst. Nath. 6, 249, 1932.
18) J.N. Bahcall, Ap. J. 276, 169, 1984.
19) G.R. Blumenthal, S. Faber, J.R. Primack and M.J. Rees, Nat 311, 517, 1874.

20) M. Davis, G. Efstathiou, C.S. Frenk and S.D.M. White, Ap. J. 292, 317, 1985.
21) A.L. Melott, Ap. J. 289, 2, 1985.
22) J.N. Fry and A.L. Melott, Ap. J. 292, 395, 1985.
23) S. Tremaine and J.E. Gunn, Phys. Rev. Lett. 42, 407, 1979.
24) R. Schaeffer and J. Silk, Ap. J. 292, 319, 1985.
25) M.W. Goodman and E. Witten, Phys. Rev. D 31, 3059, 1985.
26) D.W. Sciama, Phys. Lett. 137B, 169, 1984.
27) J. Silk and M. Srednicki, Phys. Rev. Lett. 53, 624, 1984.
28) J. Silk, K.A. Olive and M. Srednicki, Phys. Rev. Lett. 55, 257, 1985.
29) N. Cabibbo, G.R. Farrar and L. Maiani, Phys. Lett. 105B, 155, 1981.
30) D.W. Sciama, Phys. Lett. 114B, 19, 1982.
31) D.W. Sciama, Phys. Lett. 121B, 119, 1983.
32) D.W. Sciama in Big Bang: G. Lemaitre in a Modern World (ed. A. Berger) Dordrecht: Reidel 1984.

LECTURES ON PARTICLE PHYSICS AND COSMOLOGY

Edward W. Kolb

Fermi National Accelerator Laboratory

Batavia, Illinois 60510 USA

ABSTRACT

These lectures were given in the 1985 ICTP High Energy Physics and Cosmology Workshop. The three lectures are: I) Generation of a Cosmological Baryon Asymmetry, II) Extra Dimensions and Cosmology, and III) The Saga of Cygnus X-3.

1. Production of a Baryon Asymmetry[1.1]

The overwhelming evidence is that if antimatter exists in the Universe in any appreciable amount, it must be separated from matter on a scale of clusters of galaxies, $10^{14} - 10^{15} M_\odot$.[1.2] Furthermore, this separation must be done in the early Universe when the temperature of the Universe was $T > M_N$, where M_N is the nucleon mass. If we would assume the Universe had zero net baryon number, $n_N = n_{\bar{N}}$, when $T \lesssim M_N$, the nucleons would annihilate with antinucleons. We can estimate the number of nucleons that survive annihilation by considering the time evolution of the nucleon number density, given by

$$\frac{dn_N}{dt} = \frac{dn_{\bar{N}}}{dt} = -(n_N^2 - n_N^{eq\ 2})\sigma_A|v| - 3(\dot{R}/R)n_N. \tag{1.1}$$

In Eq. (1.1) σ_A is the total nucleon-antinucleon annihilation cross section, and n_N^{eq} is the equilibrium abundance for a temperature T. The first term, $-n_N^2\sigma_A|v|$, accounts for the depletion of nucleon-antinucleon pairs by annihilation. The second term, $(n_N^{eq})^2\sigma_A|v|$, accounts for $N\bar{N}$ creation. The final term, $-3(\dot{R}/R)n_N$, accounts for the dilution of the density due to the overall expansion of the Universe. Note that if $\sigma_A|v|$ is "large", the nucleons will track the equilibrium abundance, which becomes exponentially small for $T < M_N$. If $\sigma_A|v|$ is "small", the pair creation and annihilation terms will be unimportant and the nucleon abundance will change only due to the expansion of the Universe. Large and small $\sigma_A|v|$ refer to the relative size of the $n\sigma_A|v|$ term compared to the expansion term. When $n\sigma_A|v| < (\dot{R}/R)$, the reaction rates have "frozen out." If the reaction rates have frozen out, then the ratios of the nucleon density to entropy remain constant since $\dot{n}_N = -3(\dot{R}/R)n_N$ and $\dot{s} = -3(\dot{R}/R)s$. The concept of freeze-out and subsequent conservation of number density relative to entropy is quite general.

If Eq. (1.1) is solved for nucleons, the final baryon-to-entropy ratio is less than 10^{-18}. The entropy density today is related to the photon density by*

$$s = \frac{\pi^4}{45\zeta(3)}\left\{2 + (21/4)\,(T_\nu/T_\gamma)^3\right\}n_\gamma \tag{1.2}$$

$$\approx 7n_\gamma$$

$$\approx 2800\ \mathrm{cm}^{-3}\ . \qquad (T_\gamma = 2.7\mathrm{K})\ .$$

The nucleon density today is $n_N = 1.13 \times 10^{-5}\Omega_N h^2 \mathrm{cm}^{-3}$, and n_N/s is given by†

$$\frac{n_N}{s} = 4.0 \times 10^{-9}\Omega_N h^2\ . \tag{1.3}$$

*In Eq. (1.2) we have assumed 3 families of 2-component neutrinos with temperature $T_\nu = 0.714T_\gamma$.

†The baryon density is defined as $n_B = n_N - n_{\bar{N}}$. If $n_{\bar{N}} = 0$, then $n_B = n_N$. If $n_N = n_{\bar{N}}$, then $n_B = 0$.

Of course, the exact value of n_N/s is uncertain due to the appearance of $\Omega_N h^2$, but it is certainly much greater than 10^{-18}. Therefore if $n_N = n_{\bar{N}}$, annihilations would have reduced n_N far below its observed value unless N and $\bar{N}$ are separated before T = M_N.

Any attempt to separate N and $\bar{N}$ at early times must face the horizon problem. The distance over which causal processes can act is called the particle horizon. In the standard big-bang model the horizon distance is given by

$$\begin{aligned} d_H &= 2t \\ &= (45/4\pi^3)^{1/2} g_*^{-1/2} m_{pl} T^{-2} \ . \end{aligned} \tag{1.4}$$

At temperature $T = m_N = 10^3$MeV, the horizon distance is $1.4 \times 10^5 g_*^{-1/2}$cm. The entropy in the horizon volume is $S = s(4/3)\pi d_H^3 = (5/g_*\pi^3)^{1/2}\,(m_{pl}/T) = 7 \times 10^{57} g_*^{-1}$ at $T = m_N$. Even if the nucleon to entropy ratio was equal to one, there would be only $1M_\odot$ of baryons in the horizon at $T = M_N$, and non-causal processes would have to separate N and $\bar{N}$ on supercluster scales.

If, however, $n_B = n_N - n_{\bar{N}} \gtrsim 0$ at $T \approx 1$GeV, then the Universe had an excess of nucleons over antinucleons and even perfect annihilation will leave the excess baryon. This excess is quantified as the baryon number B

$$B = \frac{n_B}{s} = \frac{n_N - n_{\bar{N}}}{s} \ . \tag{1.5}$$

If we assume that today $n_{\bar{N}} << n_N$, then $B = 4 \times 10^{-9}\Omega_B h^2$ should be a conserved number as long as baryon number is conserved and the entropy is conserved. B of 10^{-9} is a rather curious number. At temperatures greater than M_N, there were roughly equal number of photons, baryons, and antibaryons, but for every ten billion antibaryons, there were ten billion and one baryons. The single extra baryon survives annihilation to become the baryons we observe in the Universe today.

One of the most remarkable advances in the field of particle physics and cosmology is the development of a model to account for the baryon asymmetry. The necessary ingredients for the generation of a baryon asymmetry from an initially symmetric state were first pointed out by Sakharov in 1967.[1.3] The three ingredients are 1) Baryon number violation, 2) C and CP violation, and 3) Non-equilibrium conditions.

The first condition, baryon number violation, is obvious. If baryon number is an exactly conserved quantum number in all interactions, $B \approx 10^{-10}$ must simply reflect the initial conditions. The second condition, C and CP violation is necessary since baryons are odd under C and CP. The third condition, non-equilibrium conditions is somewhat more subtle. If the baryon number can change, the chemical potential is not constant, and the entropy will be maximized when the baryon chemical potential (hence baryon number) is zero. Therefore if true chemical equilibrium is obtained, the baryon number will vanish.

Grand Unified Theories (GUTs), theories that unify the strong and electroweak interactions,[1.4] have baryon number violation. Although CP violation is not fully understood at low energies, it is supposed that CP and C violation will occur in GUTs. Finally the expansion of the Universe may be responsible for non-equilibrium. GUTs and the expansion of the early Universe offer a possible mechanism for the generation of the baryon asymmetry from an initially symmetric state. I will illustrate the mechanism by a simple model introduced by Kolb and Wolfram.[1.5]

Assume a model with a real massive boson X, and a massless species $b(\bar{b})$ with baryon number 1/2 (-(1/2)). In the absence of a Bose condensate or degenerate fermions we can use Maxwell-Boltzmann statistics. We will also assume that baryon-number conserving reactions will occur rapidly compared to the timescale for changing the baryon number. We may then write

$$\begin{aligned} f_b &= \exp[-(E-\mu)/T] \\ f_{\bar{b}} &= \exp[-(E+\mu)/T] \qquad (1.6) \\ f_X &= \exp[-E/T] \end{aligned}$$

where $E^2 = p^2 + m^2$ and the number density of a particle species is $n = \int f d^3p/(2\pi)^3$.

We model CP violation by writing the amplitudes for $X \leftrightarrow bb$, $X \leftrightarrow \bar{b}\bar{b}$ as (note: CPT requires $|M(i\to j)|^2 = |M(\bar{j}\to\bar{i})|^2$)

$$|M(X\to bb)|^2 = |M(\bar{b}\bar{b}\to X)|^2 = (1+\varepsilon)M_o^2/2$$
$$|M(X\to\bar{b}\bar{b})|^2 = |M(bb\to X)|^2 = (1-\varepsilon)M_o^2/2. \tag{1.7}$$

CP is violated in X decay if $\varepsilon \neq 0$.

The X number density evolves according to

$$\dot{n}_X = -\Gamma_X(n_X - n_X^{eq}) - 3(\dot{R}/R)n_X \tag{1.8}$$

where n_X^{eq} is the equilibrium X abundance, and Γ_X is the total X-decay width ($\Gamma_X = |M_o|^2/16\pi$). The baryon number density, $n_B = n_b - n_{\bar{b}}$, evolves according to

$$\dot{n}_B = \varepsilon\Gamma_X(n_X - n_X^{eq}) - n_B n_X^{eq}\Gamma_X - n_B 2n(\sigma v) \tag{1.9}$$

where (σv) is the thermally averaged $2\leftrightarrow2$ scattering cross section without real intermediate X states. Note that the driving term in $\dot{n}_B \propto \varepsilon\Gamma_X(n_X - n_X^{eq})$ is non-vanishing only if CP is violated $(\varepsilon\neq0)$, B is violated $(\Gamma_X\neq0)$, and non-equilibrium obtains $(n_X - n_X^{eq} \neq 0)$.

The basic process for non-equilibrium is that the expansion of the Universe is too rapid for n_X to track its equilibrium value. This will occur when $T \approx m_X$ if the X interaction rate, $\propto \alpha m_X$, is less than the expansion rate, $\propto T^2 m_{pl}^{-1} = m_X^2 m_{pl}^{-1}$. This will happen only if

$$K = \alpha m_{pl}/m_X \tag{1.10}$$

is not too much larger than one. Analytic solutions give the final value of $B \propto K^{-1}$. If K is not too large m_X must be comparable to m_{pl}.

Detailed calculation of the baryon asymmetry in GUT models lead to a set of equations similar to Eqs. (1.8) and (1.9), albeit much more complicated.[1.5,1.6] The detailed studies show that an adequate baryon asymmetry can be generated for reasonable parameters (coupling, masses, CP violation) in the theories. It now seems likely that some sort of GUT interactions are responsible for generating the baryon asymmetry, hence generating the neutrons and protons that survive the early Universe to form the galaxies we observe.

2. Extra Dimensions and Cosmology

At the present time, superstrings seem to offer the best hope for a consistent, unified quantum theory of gravity.[2.1] There are five superstring theories that are known to be internally consistent. It is possible that upon further study one or more of these five theories will be rejected on the basis of consistency. It is also possible that a deep physical principle will emerge that selects a single superstring theory. There is some justification for the hope that there is but a single possible theory.

It may be that this single theory admits many possible ground states. Even if there is but one ground state that leads to the observed low-energy physical world, we must one day address the question of why that particular solution was picked out. It is likely that the answer will involve the dynamics of the expansion of the very early Universe. It is possible that in the space of all possible background field configurations, there is a restricted set of solutions that result in one or more spatial directions becoming large. After all, the real question is not why are the extra dimensions so small, but rather why are three dimensions so large.

In this talk I will consider a somewhat more mundane question of what keeps the extra dimensions static. The search for static cosmological solutions has a long and inglorious history. Soon after developing the general theory of relativity, Einstein realized that his field equations implied the Universe was dynamic, contrary to the then-current belief in an eternal, unchanging world. In order to make his equations conform to his view of the Universe, he inserted a cosmological term to obtain a static solution. Eddington, Lemaitre, and Tolman subsequently showed, however, that even this static model is unstable to expansion or collapse. The advent of modern higher dimensional theories marks a return to what one might call partially static cosmologies. In these theories, one looks for solutions with three dimensions expanding in Friedmann fashion, while D extra spatial dimensions are static and curled up into a compact manifold of unobservably small size.

All attempts to stabilize the compact dimensions have involved balancing a positive "bare" cosmological constant against the stress-energy of classical or quantum fields. However this method for stabilization results in a metastable ground state.[2.2]

For a simple example of how the stabilization works, take the ground state metric to have the form $R^1 \times Q^3 \times S^D$, where Q^3 stands for flat R^3, S^3 or the 3-hyperboloid. The metric can be written in the form $g_{MN} = \mathrm{diag}(-1, a^2(t)\,\tilde{g}_{mn}, b^2(t)\,\tilde{g}_{\mu\nu})$, where a(t) and b(t) are the scale factors for Q^3 and S^D, and $\tilde{g}_{mn}$ and $\tilde{g}_{\mu\nu}$ are metrics on the maximally symmetric unit 3-space and D-sphere. Upper case indices M,N run over all values, lower case latin indices m,n = 1,2,3 and lower case greek indices $\mu,\nu = 5,6...,4+D$. The Einstein equations are

$$R_{MN} = -8\pi\tilde{G}\left[T_{MN} - \frac{g_{MN}}{D+2}T\right] - \frac{g_{MN}\Lambda}{D+2} \qquad (2.1)$$

where T_{MN} is the stress energy of classical and/or quantum fields, T is its trace, and $\tilde{G}$ is a gravitational constant in 4+D dimensions. In the compactification schemes under consideration, the non-vanishing components of the stress tensor can be written as

$$T_{00} = \rho;\; T_{mn} = g_{mn}P_3;\; T_{\mu\nu} = g_{\mu\nu}P_D \qquad (2.2)$$

with trace $T = -\rho + 3p_3 + Dp_D$, where ρ, p_3, and p_D are polynomial functions of $b(t)$ (see below). The dynamical equations for the evolution of the scale factors are

$$3\frac{\ddot{a}}{a} + D\frac{\ddot{b}}{b} = \frac{1}{D+2}\left[\Lambda - 8\pi\bar{G}((D+1)\rho + 3p_3 + Dp_D)\right] \tag{2.3}$$

$$\frac{\ddot{a}}{a} + 2\frac{\dot{a}^2}{a^2} + D\frac{\dot{a}\dot{b}}{ab} + \frac{2k}{a^2} = \frac{1}{D+2}\left[\Lambda - 8\pi\bar{G}(-\rho + Dp_D + (1-D)p_3)\right] \tag{2.4}$$

$$\frac{\ddot{b}}{b} + (D-1)\frac{\dot{b}^2}{b^2} + 3\frac{\dot{b}\dot{a}}{ba} + \frac{D-1}{b^2} = \frac{1}{D+2}\left[\Lambda - 8\pi\bar{G}(-\rho + 3p_3 - 2p_D)\right] \tag{2.5}$$

A number of compactification schemes have been considered in the literature, and we consider two representative examples. In six-dimensional Einstein-Maxwell theory, the classical ground state of the U(1) gauge field is assumed to be a magnetic monopole configuration on the compact 2-sphere.[2.3] In this case, the terms in the energy momentum tensor are ρ, p_3, $p_D \sim 1/e^2b^4$, where e is the U(1) coupling constant. In the second model,[2.4] the energy-momentum tensor arises from one-loop quantum fluctuations in massless matter fields (due to the non-trivial topology of the spacetime), in analogy to the Casimir effect in quantum electrodynamics. In this case dimensional analysis gives $\rho = A/b^{4+D}$, $p_3 = B/b^{4+D}$, $p_D = C/b^{4+D}$, where A, B and C are model-dependent (dimensional) constants. Note that this result for T_{MN} was obtained assuming $\dot{a} = \dot{b} = 0$, $a \gg b$.

Now focus on the Casimir model. For $k = 0$, there is a static solution $a = a_o$, $b = b_o$ if $A = -B$, while conservation of energy-momentum requires $C = 4A/D$. This gives

$$b_o^{D+2} = \frac{8\pi\bar{G}A(4+D)}{D(D-1)} \tag{2.6}$$

$$\Lambda = \frac{D(D-1)(D+2)}{b_o^2(D+4)} \tag{2.7}$$

This solution is stable against small perturbations: $\delta b(t) = b(t) - b_o$ has no exponentially growing modes. Consider, however, the case of large b(t), i.e., ρ, p_3, $p_D \ll \Lambda/\bar{G}$. In the limit $b(t) \to \infty$ (and k = 0 or $a \to \infty$), the solutions become $a,b \sim e^{Ht}$ and $a,b \sim e^{-Ht}$, with $H^2 = \Lambda/(D+2)(D+3)$. Using initial conditions $b(0) \gg b_o$, $\dot{b}(0) = 0$, the approximate solution (exact in the limit $t \gg 1/H$) is $b(t) \sim b(0)\cosh Ht$; i.e., if the radius is ever large (and initially static), it subsequently increases without bound instead of relaxing to the equilibrium value b_o. This clearly represents an instability. Since it occurs whenever the compactification terms ρ, p_3, $p_D \to 0$ as $b \to \infty$, this analysis holds for the Einstein-Maxwell as well as the Casimir model.

We can take the analysis beyond the level of word calculus by regarding the radius of the extra dimensions as a scalar field in a potential in four dimensions. Start with a piece of the gravitational action, $S_k = -\int d^{4+D}x\sqrt{-g}(R_k/16\pi\bar{G})$, where R_k is that part of the Ricci scalar R containing time derivatives of b,

$$R_k = -D\left[2\frac{\ddot{b}}{b} + (D-1)\left(\frac{\dot{b}}{b}\right)^2 + 6\frac{\dot{a}\dot{b}}{ab}\right]. \tag{2.8}$$

An integration by parts and over the internal D dimensions gives $S_k = -D(D-1)m_{pl}^2\int d^4x\sqrt{-g_4}\,(b/b_o)^{D-2}(\dot{b}/b_o)^2/16\pi$, where $m_{pl}^{-2} = \bar{G}/V_o^D$ is the 4-dimensional Newton constant (V_o^D is the volume of the compact D-sphere with radius b_o), and g_4 is the determinant of the 4-dimensional part of the metric. We define the new variable $\phi(b) \equiv m_{pl}(b/b_o)^{D/2}\,((D-1)/2\pi D)^{1/2}$ which has the usual action for a homogeneous scalar field in four dimensions. With this change of variable, the equation of motion for ϕ is

$$\ddot{\phi} + 3\frac{\dot{a}}{a}\dot{\phi} + \frac{\dot{\phi}^2}{\phi} = -\frac{dV}{d\phi} \tag{2.9}$$

where, for the Casimir model,

$$V(\phi) = \frac{(D-1)\Lambda m_{Pl}^2}{8\pi(D+2)}\left\{-\left(\frac{\phi}{\phi_o}\right)^2 + \left(\frac{\phi}{\phi_o}\right)^{-8/D} + \left(\left(\frac{\phi}{\phi_o}\right)^{2(D-2)/D} - 1\right)\left(\frac{D+4}{D-2}\right)\right\}$$

where $\phi_o \equiv \phi(b_o)$ is the static solution. The integration constant has been chosen so that $V(\phi_o) = 0$. Note that at $\phi = \phi_o$, $dV/d\phi = 0$ and $d^2V/d\phi^2 > 0$, so this point is a local minimum of the potential. At large ϕ, the potential is dominated by the negative quadratic term, and it is unbounded from below.

It is natural to ask for the lifetime of the compactified state $b = b_o$ against quantum tunneling. Except for the extra term $\dot{\phi}^2/\phi$, (2.9) is the equation of motion for a homogeneous, minimally coupled scalar field in a 4-dimensional Friedmann universe; if we extend our original ansatz for the metric so that $b = b(\vec{x},t)$ (2.9) would be replaced by the equation of motion for an ordinary scalar field $\phi(\vec{x},t)$ in a potential $V(\phi)$.

In the flat space limit the tunnel action is[2.2] $S \approx 165 m_{pl}^2 \Lambda^{-1}$. In order that there is a reasonable chance for the Universe to remain in the compactified state for 15 billion years, then $\Lambda \lesssim 0.3 m_{pl}^2$, which corresponds to $b_o \gtrsim 11 \ell_{pl}$.

It is not clear if in superstring theories the compactified state will be metastable, because it is not clear what is responsible for compactification. Much further work in this direction is needed.

3. The Saga of Cygnus X-3

The reported observation in underground detectors of high-energy muons from the direction of the compact binary X-ray source CYG X-3 (2030+4047) cannot be explained by conventional physics. In this section some explanations for the effect based upon unconventional physics are reviewed.

Cygnus X-3 is believed to be a compact X-ray source in the Cygnus constellation. It is observed to be a very robust source of radiation from infrared to UHE (ultra high energy, $E > 1$ TeV). All radiation above infrared is modulated with a 4.8 hour period. This 4.8^h period is thought to be the orbital period of a binary system composed of a neutron star and a companion star of about $4\ M_\odot$.

The observed spectrum of radiation from CYG X-3 can be fit by a single power law over 13 decades in energy, from 10^3 eV to 10^{16} eV:[3.1]

$$\frac{dN_\gamma}{dE} = 3\times10^{-10}\left(\frac{E}{1\,\mathrm{TeV}}\right)^{-2.1} \mathrm{cm}^{-2}\mathrm{s}^{-1}\,\mathrm{TeV}^{-1}\ . \qquad (3.1)$$

This spectrum has roughly equal luminosity per decade of energy. Assuming a distance of 12 kpc, the total luminosity of CYG X-3 above 1 GeV is in excess of 10^{38} erg s^{-1}, making it the brightest γ-ray source in our galaxy.

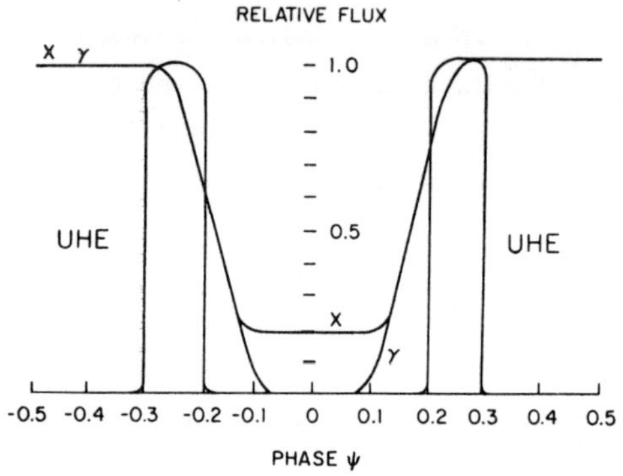

It is possible to construct an astrophysical model for the CYG X-3 system by using the phase information of the radiation. A 'theorist's rendition' of the phase information is shown in Figure 3.1. The origin of the X and γ rays is thought to be the neutron star, and the minimum of the X-ray and γ-ray radiation occurs when the neutron star is eclipsed by the main sequence star between phase -0.25 and +0.25. The absence of a zero-flux minimum for the X-rays can be understood if there is a cocoon of optical depth order unity for X-rays surrounding the system. The cocoon can back scatter X-rays during the eclipse, giving a reduced, but non-zero, X-ray flux. The cocoon will be transparent to γ-rays, giving a zero minimum γ-ray flux during eclipse. As seen in Figure 1, the UHE radiation has a 4.8^h period, but a phase structure much different than the X-rays or γ-rays. Vestrand and Eichler[3.2] and also Stecker[3.3] and Stenger[3.4] have used the phase structure to model the UHE emission. They assume the pulsar is a source of an energetic beam of primary protons. The primary protons hit the star and produce secondary π^0's (among other things), which decay to the γ's detected as the UHE flux. This mechanism will produce detectable UHE γ's if the primary beam passes through enough material to produce π^0's but not enough material to completely absorb the γ's from π^0 decay. This condition will be met for only a small fraction of the orbital period, around $\psi = \pm\ 0.25$ when the neutron star is at grazing incidence. The primary proton beam is not detected because it is dispersed by galactic magnetic fields before reaching the solar system. Models of the acceleration mechanism for the primary beam are quite complicated. They are thought to involve large $\vec{v} \times \vec{B}$ electric fields in the vicinity of the pulsar, but a completely self-consistent picture for the acceleration mechanism is very difficult to construct.[3.5] Most of our results will be independent of the details of the acceleration mechanism of the primary beam.

In order to understand the signal for new physics it is first necessary to calculate the expected flux of neutrinos from CYG X-3. The calculation reported here was done by several groups with very similar results.[3.6,3.7] The basic assumption is that the primary beam makes charged mesons, in addition to the neutral mesons, in the collision with the star. The decay of the charged mesons will result in a neutrino flux from the system.

If the UHE γ-rays originate from a source spectrum with the power law form

$$\frac{dS_\gamma}{dE_\gamma} = AE^{-n} , \tag{3.2}$$

the π^0 source spectrum should be of the form

$$\frac{dS_{\pi^\circ}}{dE_\pi} = A\ 2^{n-2}E^{-n} , \tag{3.3}$$

where the factor 2^{n-2} is from 2 photons of energy $E_{\pi^\circ}/2$. There should be $\pi^\pm$'s produced also, and the source spectrum for the $\pi^\pm$'s should be

$$\frac{dS_{\pi^+ + \pi^-}}{dE_\pi} = 2^{n-1} A E^{-n} = 2^{n-1} \frac{dS_\gamma}{dE_\gamma} \quad . \tag{3.4}$$

The energy of the neutrinos produced by $\pi^\pm$ decay will depend upon whether the $\pi^\pm$'s decay in flight or interact before decay. If the $\pi^\pm$'s decay in flight the neutrino source spectrum (and, of course, a similar antineutrino spectrum) should be related to the photon source spectrum by

$$\frac{dS_\nu}{dE_\nu} = \frac{1}{2}(1-m_\mu^2/m_\pi^2)^{n-1} \frac{dS_\gamma}{dE_\gamma} . \tag{3.5}$$

Propogation of the neutrino and photon source spectra through the star will result in the flux detected terrestrially. The absorption of neutrinos and photons depends upon the material seen by the particle traversing the star, which, in turn, depends upon the phase of the orbit. The γN cross section at high energies is nearly energy independent, so the phase structure of UHE photons should be energy independent. Unlike photons, the neutrino cross section is proportional to the energy (for E_ν < 100 TeV), and the phase diagram for the neutrino flux will depend upon E_ν. The neutrino light curve found by propagating neutrinos through a $4M_\odot$ main sequence star is shown in Figure 3.2. It should be remembered that the <u>relative</u> flux is shown in Figure 3.2 - the <u>absolute</u> flux relating different energies falls as $E^{-2.1}$. The details of the phase structure is most sensitive to the central density of the star. Observation of the neutrino light curve would offer a unique tool to probe the central density of the star.

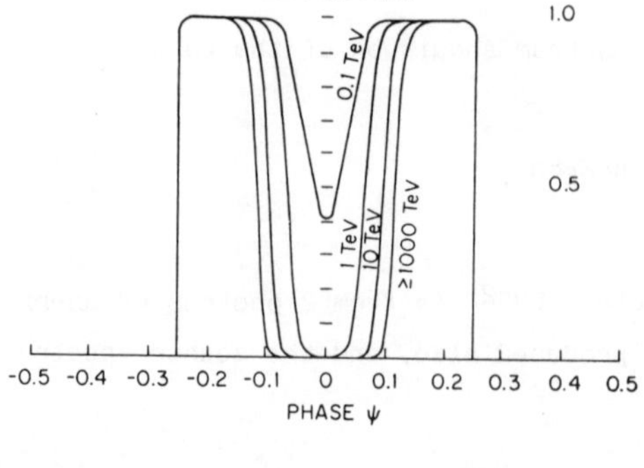

In the above calculations we have assumed the mesons decay before interacting. Since the decay length of the mesons is proportional to E/m, sufficiently energetic mesons will decay before interacting, and neutrino production will be via a beam dump mode. The decay lengths of $\pi^{\pm}$ and $K^{\pm}$ in the star are

$$(\gamma_{c\tau})_{\pi^{\pm}} = 5\times10^{6}\ (E/1\ \text{TeV})\text{cm}$$

$$(3.6)$$

$$(\gamma_{c\tau})K^{\pm} = 8\times10^{5}(E/1\ \text{TeV})\text{cm}$$

The cross section for interaction of the mesons is about 3×10^{-26} cm^2 at $E > 1$ TeV. The typical density in the envelope of the star is about 10^{-6}g cm^{-3}, so we parameterize the envelope density as $\rho = 10^{-6}\rho_{-6}$g cm^{-3}. There will be a threshold energy above which the mesons will decay before interaction, given by

$$E_T = \begin{cases} 300\ (\rho_{-6})^{-1}\ \text{TeV} & (K^{\pm}) \\ 30\ (\rho_{-6})^{-1}\ \text{TeV} & (\pi^{\pm})\ . \end{cases} \qquad (3.7)$$

The existence of this threshold should result in a feature in the neutrino spectrum at E_T, providing a unique tool to examine the density of the stellar envelope.

Using the observed UHE photon spectrum and phase width, the phase-averaged neutrino flux is expected to be

$$\frac{dN_{\nu}}{dE} = 2\times10^{-10}\left(\frac{E}{1\,\text{TeV}}\right)^{-2.1}\ \text{cm}^{-2}\text{s}^{-1} \qquad (E < E_T) \qquad (3.8)$$

The normalization of the neutrino flux is uncertain by at least an order of magnitude due to uncertainties in the photon phase width, absorption of photons, etc. However the normalization is unlikely to be off by more than two orders of magnitude, and the slope of the power law spectrum should be close to -2.1.

The neutrino flux can be detected in underground detectors either by observing a $\nu_\mu \to \mu$ conversion in the detector (which we call a contained event) or by observing a muon from a $\nu_\mu \to \mu$ conversion in the surrounding rock passing through the detector (which we call an external event). The probability that the neutrino converts in a detector of linear dimension $\ell \approx 10$ m is given by (all distances are given in terms of water equivalent)

$$\begin{aligned} P_c &= n\sigma\ell \\ &= 4\times10^{-9}(E/1\ \text{TeV}) \qquad (E \lesssim 100\ \text{TeV}) \\ &= 7\times10^{-8}\ \ell n(E/1\ \text{TeV}) \qquad (E \gtrsim 100\ \text{TeV}), \end{aligned} \tag{3.9}$$

where σ is the cross section for $\nu_\mu N \to \mu^- + X$ (for simplicity we have assumed equal cross section for ν and $\bar{\nu}$). The event rate for contained events is given by

$$\Gamma_c = A_D \int P_c \frac{dN_\nu}{dE} dE \tag{3.10}$$

$$= 6\times10^{-12}\left(\frac{E_{min}}{\text{GeV}}\right)^{-0.1} \text{sec}^{-1}$$

where E_{min} is the larger of the detector threshold and the low energy cutoff in the neutrino spectrum, and A_D is the detector area, assumed to be $4\times10^6\text{cm}^2$.

The probability that a neutrino interacts in the earth and produces a muon which passes through the detector depends upon the range of the muon. With a muon energy loss rate of

$$\frac{-dE}{dx} = 1.9\times10^{-6}\ \text{TeVcm}^{-1} + 4\times10^{-6}\ \text{cm}^{-1}\ E, \tag{3.11}$$

the range of the muon is

$$R(E) = 3\times10^5\ \ell n(1 + 2E/1\text{TeV})\text{cm}. \tag{3.12}$$

For external events, the range of the muon replaces the detector linear dimension in Eq. (3.9). If we assume the muon has half the incident neutrino energy, then the probability of an external event is

$$P_E = 10^{-6}\ (E/1\,\text{TeV})\ \ell n(1 + E/1\,\text{TeV}). \qquad (E < 100\ \text{TeV}) \qquad (3.13)$$

Notice that for E greater than a few GeV, $P_E > P_C$. The rate for external events is

$$\Gamma_E = A_D \int P_E \frac{dN}{dE}\, dE$$
$$= 3\times10^{-8}\ \text{sec}^{-1}. \qquad (3.14)$$

The fact that the external events dominate the contained events is a result of the slope of the spectrum. The total detection rate is about one per year, but as discussed above, that estimate could be off by one or possibly two orders of magnitude. For the external events, the effective size of the target is limited either by the muon range, or by the distance to the surface of the earth. It is clear that as the zenith angle of CYG X-3 increases, the muon signal due to primary neutrinos should not decrease.

Recently two experimental groups, Soudan[3.8] and NUSEX,[3.9] have reported an excess of high energy muons from the direction of CYG X-3, with a distribution of arrival times modulated with a 4.8^h period. The number of muons seen by the experiments, 84 ± 20 events in 0.96 years in Soudan and 32 events in 2.4 years in NUSEX, is much larger than the above estimates for neutrino-induced events, if one takes into account the relatively small size of the detectors. In fact, the detected muon flux is comparable to the total "photon" flux. The most striking characteristic of the signal is that the muons are not seen at large zenith angles. The zenith angle dependence of the signal strongly suggests that the muons have an atmospheric origin (or perhaps an origin in the first few hundred meters of rock). The

Soudan and NUSEX results seem to confirm previous results from the Kiel air shower experiment[3.10] of excess muons in the air showers from CYG X-3. The magnitude and zenith angle dependence of the muons rule out neutrinos as a source of the muons. If the primary particles in the air showers are photons, conventional calculations of muon production in the shower cannot account for the observed muon flux.[3.11] (Even if the air shower primaries are protons, the observed underground muon flux is too large to be accounted for by the observed air shower flux.[3.12]

The data suggest that the initiating particle must: 1) be neutral, in order to reach the solar system without being dispersed by galactic magnetic fields; 2) be light (less than a few GeV), in order to keep phase coherence with the photons over the 12 kpc distance to CYG X-3; 3) be long-lived, with a lifetime greater than weeks or months depending on γ; 4) shower in the atmosphere like a hadron, i.e., produce muons efficiently; and 5) have a flux comparable to the air shower flux of $3\times10^{-10}(E/1\mathrm{TeV})^{-2.1}$ cm^{-2} sec^{-1} TeV^{-1}. It is clear that the above profile for a particle cannot be fit by any known particle. The unknown particle postulated to fit the above profile has been given the name cygnet.[3.16] In the rest of this section I will mention some recent proposals for cygnet candidates.

<u>QUARK NUGGETS</u>[3.12,3.13] It is conceivable that there is a separate, stable phase of matter at greater than nuclear matter density. Quark nuggets would have a very small Z/A ratio, and can exist as a stable hadronic system with large A. It is not inconceivable to imagine the neutron star is, in fact, a quark star, and a source of quark nuggets. A problem with this scenario is that the conventional quark nuggets are only stable for large A, and are probably too massive to account for the phase coherence. Quark matter does lead to an enhancement over normal nuclear matter in muon production when the primary showers in the atmosphere. But the enhancement is only about a factor of two.[3.12]

<u>R-ODD PARTICLES FROM SUPERSYMMETRY</u>[3.14,3.15] In supersymmetric theories with an unbroken R-parity, the lightest

R-odd particle is stable. The photino is a good candidate for the lightest R-odd particle. It has been proposed that gluinos are produced in pp collisions along with the neutral and charged mesons responsible for the γ and ν flux. The gluinos will decay to photinos before interacting. If threshold for gluino production is low enough, which requires gluino masses less than a few GeV, the photino flux from CYG X-3 could be comparable to the photon flux at high energies. Although the photino would be a relatively light, neutral, stable particle, there are problems with this scenario. First, the photino most likely will not interact in the atmosphere, and the zenith angle dependence for muons produced by photino primaries should probably resemble that for neutrinos. Second, the mass needed to push threshold for gluino production low enough for an appreciable photino flux is very close to being ruled out by experimental data, if it has not already been ruled out. Although photinos are unlikely to be cygnets, they may prove to be detectable in future underground experiments.

H-PARTICLES[3.16] The H-particle proposed by Jaffe[3.17] is a metastable neutral strange dibaryon. The H would be a tightly bound six quark state (uuddss). The color-spin wavefunction of the H is the most symmetric, which should maximize the QCD hyperfine interaction leading to a more attractive potential than in other dibaryon systems. If the mass of the H is below $p\Lambda$ threshold, the H can only decay via double beta decay, and can have a lifetime long enough to reach the solar system from Cygnus. The H is almost unique in matching the first four profiles for a cygnet. Whether the H flux can be comparable to the total air shower flux depends upon the mechanism for cygnet production. If the incident beam from the neutron star is a proton beam, H production will be suppressed, as the $pp \to HX$ cross section is smaller than the $pp \to \pi X$ cross section because it is necessary to create two units of strangeness and an $A = 2$ system. Under these conditions, it is difficult to imagine an H flux comparable to the γ flux. If, however, the primary beam has a large strangeness fraction and consists of particles with $A > 1$, then there will not be any large suppression factors in H/Υ production, and the flux of cygnets could be a large fraction, perhaps O(1), of the total air shower flux.

There are several potential problems with the H explanation for cygnets. The mass of the H may not be below pΛ threshold (or, perhaps, not even below $\Lambda\Lambda$ threshold). The mass of the H is a question that can be settled by experiment. Even if the H flux is comparable to the total air shower flux, the secondary muon flux would be smaller than that reported by a factor of 2-10. This is a problem for any explanation of the observation, not just for the H scenario. One possible reason for the discrepancy could be an intrinsic variability in the source. The astrophysical environment of CYG X-3 is much more complicated than the simple picture presented here. If anything, it is surprising that the system is as stable as observed. The flux may change between the measurements of the air shower flux and the detection of the underground neutrinos. Finally, there is no explanation for the angular spread seen in the data. The NUSEX signal is seen in a $10° \times 10°$ window in celestial coordinates, much larger than the 0.5° expected angular resolution. Again, this is a problem for any explanation of the signal.

This work was supported in part by the DOE and NASA.

References

1.1) For a review of big-bang baryogenesis, see e.g. E. W. Kolb and M. S. Turner, Ann. Rev. Nucl. Part. Sci. 33, 645 (1983).

1.2) G. Steigman, Ann. Rev. Astr. Astro. 14, 339 (1976).

1.3) A. D. Sakharov, Zh. Eksp. Teor. Fiz. Pis'ma 5, ' 32 (1967).

1.4) See G. G. Ross, Grand Unified Theories (Benjamin Cummings, Menlo Park, 1985).

1.5) E. W. Kolb and S. Wolfram, Nucl. Phys. B172, 224 (1980).

1.6) J. N. Fry, K. A. Olive, and M. S. Turner, Phys. Rev. D 22, 2953 (1980).

2.1) For a review of superstrings, see e.g. M. Green, Nature 314, 409 (1985).

2.2) Most of this section is the result of collaboration with J. Frieman, reported in J. Frieman and E. W. Kolb, Fermilab Pub-85/68-A.

2.3) Z. Horvath, L. Palla, E. Cremmer, and J. Scherk, Nucl. Phys. B127, 57 (1977); S. Randjbar-Daemi, A. Salam, and J. Strathdee, Nucl. Phys. B124, 491 (1983). Similar models include compactification with an SU(2) Yang-Mills instanton on S^4 (S. Randjbar-Daemi, A. Salam and J. Strathdee, Phys. Lett. 132B, 56 (1983)) and an antisymmetric 3rd rank tensor field in 11-dimensional supergravity (but without a cosmological constant) (P. G. O. Freund and M. A. Rubin, Phys. Lett. 97B, 233 (1980).) The six-dimensional model has been described in a cosmological context by M. Gleiser, S. Rajpoot, and J. G. Taylor, Phys. Rev. D 30, 756 (1984) and by Y. Okada, Phys. Lett. 150B, 103 (1985).

2.4) T. Applequist and A. Chodos, Phys. Rev. Lett. 50, 141 (1983); Phys. Rev. D 28, 772 (1983); P. Candelas and S. Weinberg, Nucl. Phys. B237, 397 (1983); C. Ordonez and M. A. Rubin, Univ. of Texas preprint (1984). These models have been discussed in a cosmological setting by I. G. Moss, Phys. Lett. 140B, 29 (1984); D. Bailin, A. Love, and C. E. Vayonakis, Phys. Lett. 142B, 344 (1984); M. Yoshimura, Phys. Rev. D 30, 344 (1984); KEK preprint KEK-TH-89 (1984).

3.1) A review of the CYG X-3 observations can be found in N. Porter, Nature 305, 179 (1983).

3.2) W. T. Vestrand and D. Eichler, Ap. J. 261, 251 (1982).

3.3) F. W. Stecker, Ap. J. 228, 919 (1979).

3.4) V. J. Stenger, Ap. J. 284 (1984).

3.5) See, e.g. M. A. Ruderman and P. G. Sutherland, Ap. J. 196, 51 (1975).

3.6) E. W. Kolb, M. S. Turner, and T. P. Walker, Phys. Rev. D15, to appear (1985).

3.7) M. V. Barnhill, T. K. Gaisser, T. Stanev, and F. Halzen, preprint MAD/PH/243; V. S. Berezinsky, C. Castagnole, and P. Galeotti, preprint (1985).

3.8) M. L. Marshak, etal., Phys. Rev. Lett. 54, 2079 (1985).

3.9) G. Battistoni, etal., Phys. Lett. 155B, 465 (1985).

3.10) M. Samorski and W. Stamm, Ap. J. 268, L17 (1983).

3.11) T. Stanev and Ch. Vankov, Phys. Lett., to appear (1985).

3.12) M. V. Barnhill, T. K. Gaisser, T. Stanev, and F. Halzen, preprint MAD/PH/259 (1985).

3.13) G. L. Shaw, G. Benford, and D. J. Silverman, preprint 85-14 (1985).

3.14) V. J. Stenger, preprint HDC-3-85 (1985).

3.15) R. W. Robinett, Phys. Rev. Lett. 55, 469 (1985).

3.16) G. Baym, E. W. Kolb, L. McLerran, T. P. Walker, and R. L. Jaffe, preprint 85/98-A (1985).

3.17) R. Jaffe, Phys. Rev. Lett. 38, 195 (1977).

PARTICLE PHYSICS AND THE STANDARD COSMOLOGY

Subir Sarkar

Department of Astrophysics, University of Oxford, OX1 3RQ, U.K.
and
HEP Theory Group, Rutherford Appleton Laboratory, OX11 0QX U.K.

ABSTRACT

We review the constraints imposed by the standard cosmological model on physics beyond the standard $SU(3)_C \times SU(2)_L \times U(1)_Y$ model, with particular attention to supersymmetry and supergravity.

1. INTRODUCTION

The standard 'Big Bang' cosmological model rests on three major empirical pieces of evidence.[1)] The Hubble expansion of galaxies is directly observed upto a redshift of $z \sim 4$, when the universe was about a tenth of its present age, $t \sim 10^9$ yr ($T \sim 10^{-3}$ eV). The 2.7^0K microwave background radiation must have last scattered off matter at $t \sim 5 \times 10^5$ yr ($T \sim 0.3$ eV), while its blackbody spectrum must have been established earlier than $t \sim 10^5$ sec ($T \sim 5$ KeV). Finally, the cosmic abundance of ^{4}He and possibly also of the trace elements, D, ^{3}He and ^{7}Li are well accounted for in terms of nuclear reactions which must have occurred at $t \sim 10^2$ sec ($T \sim 0.1$ MeV), following the breakdown of thermal equilibrium between neutrons and protons at $t \sim 1$ sec ($T \sim 1$ MeV). The standard SU(3) x SU(2) x U(1) model enables the expansion history to be traced further back, past the QCD deconfinement/chiral phase transitions at $t \sim 10^{-5}$ sec ($T \sim 300$ MeV), upto the phase transition at $t \sim 10^{-10}$ sec ($T \sim 100$ GeV) associated with spontaneous electroweak symmetry breaking.[2)] Grand Unification models suggest there was another phase transition at $t \sim 10^{-35}$ sec ($T \sim 10^{14}$ GeV) when the symmetry between the strong and electroweak force was spontaneously broken. The observed ratio of baryons to photons today may have resulted from the generation of a small excess of matter over antimatter by out-of-equilibrium B-, C- and CP- violating processes at that epoch.[3)] Moreover if the universe had been dominated by the energy density of a metastable vacuum during such spontaneous symmetry breaking, then the resulting exponential expansion ('inflation') would have naturally led to the remarkable homogeneity, isotropy and (near) critical density of the universe today.[4] Alternatively, inflation may have been associated with the compactification of extra spatial dimensions in Kaluza-Klein theories at temperatures approaching the Planck scale, $T \sim 10^{19}$ GeV ($t \sim 10^{-43}$ sec).[5] Quantum corrections to gravity become important at such epochs and it remains for a fully consistent theory of quantum gravity, perhaps based on superstrings[6], to explain how the universe may have emerged from the pre-Planckian era. The requirement that such new physics beyond the standard model should not lead to violation of the standard cosmology imposes many interesting constraints on the physical parameters of such theories, often those inaccessible to laboratory experiment.

2. THE STANDARD COSMOLOGY

The large-scale homogeneity and isotropy of the universe observed locally, if also true globally, imply that space-time is maximally symmetric with the Robertson-Walker metric

$$ds^2 = -dt^2 + R^2(t)\left[dr^2/(1-kr^2) + r^2\,(d\Theta^2 + \sin^2\Theta\, d\Phi^2)\right], \qquad (1)$$

where $k = 0, \pm 1$ is the 3-space curvature signature. The scale-factor $R(t)$ which relates physical to co-ordinate distance evolves according to the Einstein equation,

$$H^2 \equiv (\dot{R}/R)^2 = 8\pi\rho/3M_p^2 - k/R^2 + \Lambda/3 \qquad (2)$$

where the energy density $\rho = \rho_M + \rho_R$ includes both (non-relativistic) 'matter' and (relativistic) 'radiation', $M_p \equiv G_N^{-\frac{1}{2}} \simeq 1.22 \times 10^{19}$ GeV, and $\Lambda = 8\pi\rho_{vac}/M_p{}^2$ is a possible cosmological constant reflecting a non-zero vacuum energy density, ρ_{vac}. Assuming $\Lambda = 0$, the present age relates to the present Hubble parameter as

$$t_o = f(\Omega)\;/\;H_o \quad ; \quad \Omega = \rho/\rho_c \quad , \qquad (3)$$

where the critical density, $\rho_c = 3H^2M_p^2/8\pi \simeq 1.88 \times 10^{-29}\ h^2$ gm cm^{-3} ($\sim 10.5\ h^2$ KeV cm^{-3}) and $h \equiv H_o/10^2$ Km sec^{-1} Mpc^{-1} = 9.78×10^9 yr. H_o. In the standard model the universe has been 'matter-dominated' ($\Omega \simeq \Omega_M$) for most of its lifetime, with $f(\Omega_M)$ given in ref.1). There are models in which the universe has largely been 'radiation-dominated' ($\Omega \simeq \Omega_R$), in which case $f(\Omega_R) = (1 + \Omega^{\frac{1}{2}})^{-1}$. (*) Radioactive isotope dating

*) The present density of visible radiation is negligible, $\rho(2.7\,^0K) \simeq 4.5 \times 10^{-34}$ gm cm^{-3}, but there may be 'invisible' radiation in goldstone bosons, massless neutrinos etc. created by decays of massive particles.

requires the age of the Galaxy to exceed 10^{10} yr, [7] while the age of the oldest stars (in globular clusters) is inferred to be $(16 \pm 3.5) \times 10^9$ yr. [8] Observations of the present expansion rate suggest $0.4 \lesssim h \lesssim 1$. [9] Thus if $t_o > 10^{10}$ yr, $h > 0.4$, then

$$\Omega_M h^2 < 1.02 \quad , \qquad \Omega_R h^2 < 0.333 \quad , \tag{4a}$$

for a MD and RD universe respectively. For a higher minimum age, $t_o > 1.6 \times 10^{10}$ yr ,

$$\Omega_M h^2 < 0.18 \quad , \qquad \Omega_R h^2 < 0.045 \quad . \tag{4b}$$

However if $\Lambda \neq 0$, these bounds are considerably degraded; then requiring $H_o t_o > 0.49$, [10]

$$\Omega_M < 9.4 \quad , \qquad \Omega_R < 6.8 \tag{5}$$

Direct observations of that component of matter clustered similarly to visible galaxies suggest [11]

$$\Omega_M \simeq 0.1 - 0.5 \tag{6}$$

but this does not include matter which is differently distributed. Note that 'inflationary' models require (to high accuracy) that, [4]

$$\Omega + \Lambda / 3H^2 = 1 \tag{7}$$

3. PARTICLE SURVIVAL

In the standard Friedmann-Robertson-Walker model (with $\Lambda=0$), the early universe ($t \lesssim 4 \times 10^{10}$ sec $[\Omega_M h^2]^{-2}$) was 'radiation-dominated' by relativistic particles with the energy density,

$$\rho = (g_*/2)\ \rho_\gamma = g_*\ \pi^2 T^4/30 \tag{8}$$

so that,

$$H \equiv (\dot{R}/R) = -\ (\dot{T}/T) \simeq 1.66\ g_*^{\frac{1}{2}}\ T^2/M_p \quad . \tag{9}$$

The effective number of degrees of freedom sums over all relativistic bosonic (B) and fermionic (F) helicity states,

$$g_* = \sum_B g_B\ (T_B/T)^4 + 7/8 \sum_F g_F\ (T_F/T)^4 \quad . \tag{10}$$

The temperature $T_{B,F}$ may differ from the photon temperature T since a particle x will stay in thermal equilibrium ($T_x = T$) only as long as its interaction rate, $\Gamma_{int} = n\ \langle\sigma v\rangle$, exceeds the expansion rate H. It will 'decouple' when $\Gamma(T_d) \simeq H\ (T_d)$, with the equilibrium abundance,

$$n_x^{eq}(T_x = T_d) = (g_x/2)\ n_\gamma(T=T_d) \cdot \begin{cases} 1 & \text{.. Boson} \\ 3/4 & \text{.. Fermion} \end{cases},$$

$$n_\gamma = 2\zeta(3)\ T^3/\pi^2 \quad , \tag{11}$$

if it is still relativistic ($m_x < T_d$). Subsequently as T drops below various mass thresholds, the corresponding particles annihilate heating the photons (and other interacting particles), but not the 'decoupled' x particles, so that T_x/T decreases from unity. By conservation of entropy one can determine the increase in the number of photons, $N_\gamma = R^3 n_\gamma$, in a comoving volume,

$$T_x/T = \left[N_\gamma(T_d)/N_\gamma(T)\right]^{1/3} = \left[g'(T)/g'(T_d)\right]^{1/3} , \qquad (12)$$

where g' is the number of interacting degrees of freedom. Thus the present abundance of x is given in relation to the present photon number density as,

$$n_x(T_x) / n_\gamma(T_o) = \left[n_x^{eq}/n_\gamma\right]_{T_d} \cdot \left[N_\gamma(T_d) / N_\gamma(T_o)\right]. \qquad (13)$$

Numerical values of g_* and $(T_x/T_o)^{-3} = N_\gamma(T_o)/N_\gamma(T_d)$ are given below for the standard SU(3) x SU(2) x U(1) model, following ref. 12).

TABLE 1

T_d or T_f	Particle Content	$g_*(T)$	$[N_\gamma(T_o)/N_\gamma(T_d)]$
$T < m_e$	γ, 3 ν_L'S	3.36	1
$m_e < T < m_\mu$	add $e^\pm$	43/4	2.75
$m_\mu < T < m_\pi$	add $\mu^\pm$	57/4	3.65
$m_\pi < T < T_c$ *)	add π's	69/4	4.41
$T_c < T < m_s$	γ, $3\nu_L$'s, $e^\pm$, $\mu^\pm$, $\overset{(-)}{u}$, $\overset{(-)}{d}$, g	205/4	13.1
$m_s < T < m_c$	add $\overset{(-)}{s}$	247/4	15.8
$m_c < T < m_\tau$	add $\overset{(-)}{c}$	289/4	18.5
$m_\tau < T < m_b$	add $\tau^\pm$	303/4	19.4
$m_b < T < m_t$	add $\overset{(-)}{b}$	345/4	22.1
$m_t < T < m_W$	add $\overset{(-)}{t}$	387/4	24.8
$T > m_W$, m_Z	add $W^\pm$, Z	423/4	27.1
	H^0(?)	427/4(?)	

*) T_c is the temperature above which hadrons can presumably be described in terms of quark and gluon degrees of freedom.

If x is massive it may become non-relativistic while still in thermal equilibrium, with its number density now depleted by a Boltzmann factor

$$n_x^{eq} = (g_x/2\pi^2) \int p^2 dp / \left[\exp\,(p^2 + m^2)^{1/2}/T \pm 1\right] \qquad (14)$$

$$\simeq g_x\,(m_x T/2\pi)^{3/2} \exp\,(-m_x/T) , \quad \text{for } T < m_x$$

Since x must relax to its equilibrium density through annihilations alone, it will go out of chemical equilbrium when its annihilation rate

$\Gamma_{ann} = n_x \langle\sigma v\rangle_{x\bar{x}}$, falls behind the expansion rate, H. The governing Boltzmann equation is [12, 13],

$$dn_x/dt = -(3\,\dot{R}/R)n_x - \langle\sigma v\rangle_{x\bar{x}}\,[n_x^2 - (n_x^{eq})^2]. \qquad (15)$$

A good approximation to exact solution of eq.(15) is to determine the 'freeze-out' temperature by setting $n_x^{eq}(T_f)\,\langle\sigma v\rangle_{x\bar{x}} = H\,(T_f)$; this yields [13]

$$T_f \simeq m_x/(40.7 + \ln\{g_x[\langle\sigma v\rangle\, m_x]_{GeV^{-1}}/g_*^{\frac{1}{2}}\}). \qquad (16)$$

The surviving abundance, in relation to photons, is

$$\frac{n_x^{eq}\,(T_x = T_f)}{n_\gamma\,(T = T_f)} \simeq \frac{2.3\times10^{-17} g_*^{\frac{1}{2}}}{[\langle\sigma v\rangle m_x]_{GeV^{-1}}} \cdot [1 + (\ln\{g_x\langle\sigma v\rangle\, m_x/g_*^{\frac{1}{2}}\})/40.7] \;. \qquad (17)$$

The abundance today can be determined by allowing for subsequent photon heating as in eq. (13), using Table 1.

4. CONSTRAINTS FROM THE ENERGY DENSITY

The calculated surviving density of stable ($\tau \gg 10^{17}$ sec) particles, or the decay products of unstable particles, can be directly searched for if they are observable and/or compared with the maximum observationally permitted energy density, $\Omega_M h^2 \lesssim 1$, corresponding to $\rho_M \lesssim \rho_c$ (h=1) $\simeq 10.5$ KeV cm^{-3} (eq. 4a).

Strongly Interacting Particles:

With $\langle\sigma v\rangle_{N\bar{N}} \sim 300$ GeV^{-2}, nucleons 'freeze-out' at $T_f \sim 20$ MeV with an abundance,[14]

$$n_N/n_\gamma = n_{\bar{N}}/n_\gamma \simeq 3.5\times10^{-19} \;, \qquad (18a)$$

whereas observations of luminous matter in galaxies imply[15] $\Omega_N \gtrsim 0.01$, i.e.

$$\eta \equiv n_N/n_\gamma \simeq 2.81\times10^{-8}\,\Omega_N h^2\,\Theta^{-3} \gtrsim 4\times10^{-11} \;, \qquad (18b)$$

assuming $h \gtrsim 0.4$ and $\Theta \equiv (T_o/2.7^0K) \gtrsim 1$. This enormous discrepancy and the observational lack of comparable amounts of antimatter[16], imply the existence of baryon number violating processes, presumably at the GUT epoch, which created an initial baryon asymmetry, $(n_B - n_{\bar{B}})/(n_B + n_{\bar{B}}) \approx 10^{-8}$, so that at low temperatures $(n_B - n_{\bar{B}})/\,n_\gamma \approx 10^{-10}$ [3]. Note that if any other particle was created with a similar asymmetry, $\varepsilon = (n_x - n_{\bar{x}})/n_\gamma \sim 10^{-10}$, then there is an upper limit to its mass,

$$m_x \lesssim \rho_M/\varepsilon n_\gamma \simeq 250 \text{ GeV} \qquad (19)$$

If there exist new heavy stable quarks(Q), then for $m_Q > 10$ GeV, assuming $\langle\sigma v\rangle \sim \pi\alpha_s^2/m_Q^2$, 'freeze-out' occurs before confinement in hadrons at $T_c \sim 0.2 - 0.4$ GeV, with $n_Q/n\gamma \approx 10^{-16}\,(m_Q/\text{GeV})$.[17]

Subsequently as T drops below T_c, the hadrons (H) formed annihilate again with $\langle\sigma v\rangle_{H\bar{H}} \sim \langle\sigma v\rangle_{N\bar{N}}$, yielding $n_H/n_N \sim 2 \times 10^{-10}$ for the relic abundance today.[18] However searches for anomalously heavy isotopes of hydrogen [19], carbon[20], oxygen[21] and sodium[22] have yielded severe upper limits on the abundance per nucleon of $< 10^{-26} - 10^{-29}$ ($m \sim 12$-1200 GeV), $< 2 \times 10^{-15}$ ($m < 10^5$ GeV), $< 10^{-16} - 10^{-18}$ ($m < 54$ GeV) and $< 5 \times 10^{-12}$ ($m > 10^2$ GeV) respectively. Thus it is unlikely that such heavy quarks can be stable. In particular, gluinos or squarks are not favoured as the lightest supersymmetric particle (LSP) since hadrons containing these would then be stable (if R-parity is unbroken).[23] Note that the mass density constraint alone would require,

$$m_Q \lesssim 10^{12} \text{ GeV} \tag{20}$$

Experimental claims[24] of unconfined fractional charges at a level $\gtrsim 10^{-20}$/nucleon have inspired suggestions that such a relic abundance of quarks may arise if QCD is broken[25] or in unconventional GUTs[26].

If quarks (and leptons) are composite at a scale Λ_p then there may be a stable particle with $m \sim \Lambda_p$ and $\langle\sigma v\rangle \sim 1/\Lambda_p^2$, so that the mass density constraint would imply[27]

$$\Lambda_p \lesssim 300 \text{ TeV} \tag{21}$$

Uncoloured Particles

For stable particles with $\langle\sigma v\rangle \sim \pi\alpha^2/m^2$, the relic density is $\sim 10^{-12}$ (m/GeV) cm^{-3} for $m > 10$ GeV [17]. The mass density constraint then requires [28]

$$m \lesssim 3 \text{ TeV} \tag{22}$$

If these were charged leptons and hence coupled to matter through electromagnetic interactions, then further annihilations may have occured when bound structures (stars, planets..) formed. However such annihilations could not have reduced the abundance below $\sim 10^{-8}$/nucleon, without generating an excessive γ-ray background[28].(*)

Such an abundance is incompatible with the upper limit on anomalously

---

(*) If there was an early (Pop III) generation of very massive stars, then annihilations within them may have reduced the abundance to $\sim 10^{-19}$/nucleon without attendant generation of γ-rays; thus the experimentally observed[24] fractional charges may well be leptons[29].

heavy hydrogen in sea water, $< 10^{-26}-10^{-29}$/nucleon, for $m \sim 12-1200$ GeV [19]. Thus such heavy leptons cannot be stable and similarly winos, charged higgsinos or sleptons cannot be the LSP[23].

Weakly Interacting Particles

Neutrinos: Massless, left-handed neutrinos with neutral current interactions, $\langle\sigma v\rangle \sim G_F^2 T^2$ for $T \ll M_Z$, 'decouple' at $T_d \sim 3.5$ MeV (except ν_e which has additional charged current interactions and decouples slightly later at $T_d \sim 2$ MeV). Thus even massive neutrinos would be relativistic at decoupling if $m_\nu \lesssim 1$ MeV, with a present abundance, $n_\nu/n_\gamma = 3\, g_\nu/22$ (eq. 13); for Majorana masses $g_\nu = 2$ while for Dirac masses $g_\nu = 4$. However in the standard model, right-handed neutrinos are isosinglets with no gauge interactions and left-right transitions are suppressed by a factor $\sim (m_\nu/T)^2$, so g_ν is ~ 2 even for Dirac neutrinos when $m_\nu \ll T_d$.[30] Then the mass density constraint implies,

$$\Sigma\ (g_\nu/2)\ m_\nu \ < 97 \text{ eV } \left[\Omega_M h^2\ \theta^{-3}\right] \quad , \tag{23}$$

for the well known[31] limit on the sum of the masses of stable, light ($m_\nu \ll 1$ MeV), left-handed neutrino types. Right-handed neutrinos which interact with effective strength $G < G_F$ 'decouple' earlier leading to a smaller present abundance, so their mass limit is higher than that for left-handed neutrinos by upto a factor of ~ 10 (see eq.13 & Table 1)[32].

Very massive ($m_\nu \gg 1$ MeV) Dirac ($g_\nu = 4$) neutrinos with $\langle\sigma v\rangle \sim G_F^2\, m_\nu^2\, N_A/2\pi$ 'freeze-out' at $T_f \sim m_\nu/20$ yielding a relic abundance $\rho_\nu \sim 43$ KeV cm^{-3} $(m_\nu/\text{GeV})^{-1.85}$. $(N_A/14)^{-0.95}$. $(g_*/10.75)^{0.48}$. θ^3, for $T_f < m_\mu$, where N_A is the number of annihilation channels into lighter fermions [33]. This implies,

$$m_\nu^D \ > 1.8 \text{ GeV } (\Omega_M h^2 \theta^{-3})^{-1}. \tag{24a}$$

If such neutrinos are of Majorana type ($g_\nu = 2$) then their annihilation cross-section is suppressed due to identical particle effects resulting in a relic density which is higher by a factor $\sim 3-10$.[34] Thus the lower bound is altered to

$$m_\nu^M > 5 \text{ GeV} \quad , \quad \text{for } \Omega_M h^2 \ \sim\ 1 \tag{24b}$$

Note that for $m_\nu > M_Z$, propogator mass effects cause the annihilation cross-section to begin decreasing (assuming no new interactions, e.g. with the Higgs field) leading to an absolute upper limit on the mass of ~ 3 TeV.[13] A smaller bound obtains if the neutrinos have a chemical potential of the same order as nucleons (eq. 19)[35].

Thus no stable neutrino may have a mass between ~ 100 eV and ~ 2 GeV unless it has additional interactions which enable it to annihilate faster than is allowed by the standard model. For example in the majoron model where global lepton number conservation is spontaneously violated by a Higgs triplet acquiring a vacuum expectation value, v, [36] rapid annihilations into the associated goldstone bosons (majorons) can reduce ρ_ν below observational bounds for any m_ν.[37] However majoron emission from stellar objects would cause unacceptably fast evolution unless $v < 0.8$ MeV,[38] and this is also the upper bound on m_ν in this model.

Unstable neutrinos which decay into charged particles (for $m_\nu \gtrsim 1$ MeV) or photons are severely constrained by various observations to be discussed later, but decays to light neutrinos or goldstone bosons can be constrained only by dynamical arguments. Requiring that the red-shifted energy density of the decay products (assumed massless) does not exceed the maximum allowed for a radiation dominated universe today (eq. 4a) implies an upper limit on the lifetime[39, 44]

$$\tau < 1.5 \times 10^7 \text{ yr } (m_\nu/\text{KeV})^{-2} (g_\nu/2)^{-2} (\Omega_R h^2/0.333)^2 \Theta^{-6}, \quad \text{for } m_\nu \lesssim 1 \text{ MeV}. \tag{25}$$

This limit is more stringent if the decay products have masses and eventually matter-dominate the universe. In fact, density fluctuations could not have grown by gravitational clumping to give the observed large-scale structure if the universe has been mostly radiation-dominated. Thus a more stringent constraint is that the universe should have become matter-dominated at a redshift $z_m > 10^3$ so that fluctuations have grown by at least a factor 10^3;(*) this yields[40]

$$\tau < 60 \text{ yr } (m_\nu/\text{KeV})^{-2} (g_\nu/2)^{-2} (\Omega_M h^2)^{3/2} \Theta^{-9/2} (z_m/10^3)^{-3/2}, \quad \text{for } m_\nu \lesssim 1 \text{ MeV}. \tag{26}$$

The decay $\nu \to 3\nu'$ through flavour-changing neutral currents[41] or Higgs scalars [42] in extensions of the standard model cannot satisfy these bounds without unnatural 'fine tuning' to avoid conflict with

---

(*) Density fluctuations grow proportionally as R(t), i.e. $(1+z)^{-1}$, in the MD era, and are going non-linear ($\Delta\rho/\rho > 1$) today on a comoving scale $\sim 5h^{-1}$ Mpc. However their amplitude at recombination ($z \sim 10^3$) is restricted by the microwave background anisotropy to be $\Delta\rho/\rho < 10^{-3}$.

laboratory experiment [43, 44] and to suppress radiative decays [45]. Decays into majorons are prohibitively slow [46] unless different lepton numbers are assigned to different generations [47] in a somewhat contrived variant of the 'singlet' majoron model [48]. If global family symmetry is spontaneously violated at a scale F thus generating a massless 'familon' (f), the decay $\nu \to \nu' + f$ occurs with the lifetime $\tau \simeq 3.3 \times 10^{15}$ sec $(m_\nu/\mathrm{KeV})^{-3}$ $(F/10^{10}\ \mathrm{GeV})^2$. [49] Experimental limits on $\mu \to ef$ [50] and $K \to \pi f$ [51] require $F > 6.5 \times 10^9$ GeV and $F > 2.7 \times 10^{10}$ GeV respectively, so consistency with eq.(20) is possible but consistency with eq. (25) is not, for $m_\nu \lesssim 1$ MeV. The recent experimental claim of a 17 KeV neutrino emitted in tritium β-decay [52] has inspired much discussion of its cosmological implications [53]; however other experiments have refuted this claim. [54] Finally, it has been suggested that unclustered, relativistic particles from the recent non-radiative decay of a massive neutrino may dominate the universe today thus reconciling 'inflationary' models ($\Omega = 1$) with observations which suggest that clustered matter only contributes $\Omega_M \approx 0.2$ [55]. However such models imply excessive fine-scale anisotropy in the microwave background radiation [56] as well as excessive peculiar (non-Hubble) velocities within the local supercluster of galaxies [57].

<u>Scalar Neutrinos</u>: Sneutrinos can annihilate rapidly through zino exchange if the Majorana component of the zino mass exceeds a few GeV. [58] Then the relic mass density, which is independent of the sneutrino mass, can lie below observational bounds, so the sneutrino could well be the LSP, from cosmological considerations alone.

<u>Photinos & Higgsinos</u>: For these majorana fermions (χ) the non-relativistic annihilation cross-section is p-wave suppressed, thus involving an additional momentum dependence, $\langle \sigma v \rangle \sim \hat{a} + \hat{b}T/m_\chi$, where $\hat{a}$, $\hat{b}$ involve m_χ, final state fermion masses and propagator sfermion masses. [23, 59] The analogue of the neutrino mass bound (eq. 24) for a stable photino or Higgsino mass eigenstate is then (for $\Omega_M h^2 \lesssim 1$) [23]

$$m_{\tilde{\gamma}} > 0.5\ ,\quad 1.8,\quad 5\quad \mathrm{GeV} \tag{27}$$

accordingly as, $m_{\tilde{\ell},\tilde{q}} = 20,\ 40,\ 100$ GeV,

$$\begin{aligned} m_{\tilde{H}^0} &\gtrsim m_b \quad, \quad \text{for } v_1 \neq v_2 \\ &\gtrsim m_t \quad, \quad \text{for } v_1 = v_2 \end{aligned} \tag{28}$$

where v_1, v_2 are the doublet Higgs vacuum expectation values.

Alternatively, these particles could be relativistic at

decoupling with annihilation cross-section similar to that of neutrinos, so that for $\Omega_M h^2 \lesssim 1$ [23)]

$$m_{\tilde{H}0} \;,\; m_{\tilde{\gamma}} \lesssim 100 \text{ eV} \quad . \qquad (29)$$

However such a light photino or Higgsino is phenomenologically disfavoured[23, 59)].

Gravitinos: Helicity ± 3/2 gravitinos interact only gravitationally and hence decouple very early at temperatures approaching M_P. Subsequent entropy generation then dilutes their abundance just as for neutrinos, but by a much larger factor, so the mass bound on a stable gravitino is related to that on a stable neutrino (eq. 23) as,

$$m_{3/2} \lesssim 100 \text{ eV} \left[g' (T_d \sim M_P) \;/\; g' (T_d \sim 1 \text{ MeV}) \right] . \qquad (30)$$

The dilution factor is at least ~ 10 (see Table 1), which suggests $m_{3/2} \lesssim 1$ KeV [60,61)], and it could in principle be ~ 100-1000, which would relax the bound to $m_{3/2} \lesssim 100$ KeV [62)]. However, if the universe underwent a period of inflation after the Planck era, then the primordial gravitino abundance is reduced to almost zero.[4)] Subsequently the dissipation of the scalar field driving the inflation reheats the universe to a temperature T_R. This regenerates through 2-body scattering processes, a gravitino abundance [63)]

$$n_{3/2}/n_\gamma \simeq 1.38 \times 10^{-11} \,(T_{3/2}/T_\gamma)^3 \; T_{R9} \,(1 - 0.018 \ln T_{R9}), \qquad (31)$$

$$\text{for } m_{3/2} \ll T_R \equiv T_{R9} \cdot 10^9 \text{ GeV},$$

where $(T_{3/2}/T_\gamma)^3 = 172/10065$ for $T < m_e$, in the minimal supersymmetric standard model. Then, in order that such gravitinos do not exceed the mass density today, we require [64)]

$$T_R < 1.3 \times 10^{14} \text{ GeV } (m_{3/2}/\text{GeV})^{-1}. \qquad (32)$$

In most phenomenological supergravity models, the natural scale for the gravitino mass is $0\,(M_W)$ (so as not to resurrect the 'hierarchy' problem) and such gravitinos would be unstable against decay into lighter supersymmetric particles, $3/2 \to \tilde{\gamma} + \gamma, \; \tilde{g} + g \ldots$. Constraints on such decays are discussed later.

In a class of models where the gravitino (and the photino) may be light ($\lesssim 1$ MeV) [62)], the supersymmetry breaking scale f is bounded from below by requiring that the emission of these particles by stars does not speed up stellar evolution excessively [65)],

$$f \gtrsim 50 - 100 \text{ GeV}. \qquad (33)$$

Axions: The absence of CP violation in strong interactions is presumably due to breaking of a U(1) Peccei-Quinn symmetry at $T \sim f_a$, which generates a massless axion [66)]. When T drops below ~ 1 GeV, the axion acquires a mass, $m_a \sim m_\pi f_\pi / f_a$, through QCD instanton effects and coherent oscillations are set up in the axion field due to its initial misalignment with the potential minimum. The surviving energy density in such oscillations would be excessive unless $f_a \lesssim 10^{12}$ GeV,[67] corresponding to,

$$m_a \gtrsim 10^{-5} \text{ eV} \tag{34a}$$

(This bound would be relaxed if there are additional sources of entropy generation at $T \lesssim 1$ GeV [68] or if the initial misalignment is small [69]. Moreover, requiring that the energy loss from stellar objects due to axion production (by Compton and Primakoff processes) not be excessive implies $f_a \gtrsim 10^7 - 10^9$ GeV [38, 70], i.e.

$$m_a \lesssim 10^{-2} - 1 \text{ eV} , \tag{34b}$$

as long as $m_a \lesssim 0.2$ MeV [70].

An additional cosmological constraint arises from requiring that the present horizon should not contain many different strong CP-conserving vacua since the domain walls separating these would speed up the expansion catastrophically; the implications for axion models are reviewed in ref. 66).

Monopoles: Grand Unified symmetry breaking at $T \sim M_{GUT}$ generates monopoles of mass $m_M \simeq M_{GUT}/\alpha$, with an abundance [71]

$$n_M/n_\gamma \gtrsim 5 \times 10^{-10} , \tag{35a}$$

whereas requiring that they do not contribute excessively to the mass density today implies [71]

$$n_M/n_\gamma \lesssim 5 \times 10^{-24} \, (m_M/10^{16} \text{ GeV})^{-1} . \tag{35b}$$

This embarassing disparity is best resolved by the proposal that the universe underwent a period of inflation which diluted the initial monopole abundance to a negligible level [4]. The subsequent reheating to a temperature T_R regenerates an abundance [72]

$$n_M/n_\gamma \simeq 3 \times 10^3 \, (m_M/T_R)^3 \exp(-2 \, m_M/T_R), \tag{35c}$$

which satisfies the mass density constraint for $m_M/T_R \gtrsim 35$, a natural enough expectation.

Stringent astrophysical constraints on the monopole flux today follow by requiring that they should not destroy the galactic magnetic, field or heat up condensed objects (white dwarfs, neutron stars etc.) excessively by catalysing nucleon decay; these are reviewed in ref. 73). However, with a sufficiently large flux, monopoles may in fact, maintain the galactic magnetic field rather than destroy it [74] and the nucleon catalysis rate may be much lower than has been assumed[75] so the validity of these bounds remain questionable.

5. THE COSMIC BACKGROUND RADIATION

Observations of the microwave background radiation provide two classes of constraints on the early universe. Firstly, deviations of the spectral occupation number from that of a perfect blackbody ($\eta_\gamma = \lfloor \exp \nu/T - 1 \rfloor^{-1}$) are a sensitive probe of any non-thermal energy release (such as from particle decays) from the thermalisation era ($t \sim 10^5$ sec, $T \sim 5$ KeV) onwards [76]. Secondly, measurements of possible fluctuations from true isotropy ($\lfloor T(\Theta) - T(\Theta') \rfloor / T(\Theta) \sim \Delta T/T$) are a probe of corresponding fluctuations in the gravitational potential at the recombination epoch ($t \sim 5 \times 10^5$ yr, $T \sim 0.3$ eV) when photons 'decoupled' from matter. [77]

The type of possible spectral distortion depends on the epoch of energy release in photons or charged particles (of magnitude $\Delta\rho_\gamma$) to the thermal background photons (ρ_γ). Thermalisation of photons is achieved through Compton scattering against thermal electrons and ions (of density ρ_N), and attendant radiative processes such as bremsstrahlung and double Compton scattering which have efficiencies proportional to ρ_N^2 and $\rho_N\rho_\gamma$ respectively. Any energy release, however large, before a redshift $z_T \simeq 3.9 \times 10^6 (\Omega_N h^2)^{-1/3}$, is completely thermalised by double-Compton scattering, and no spectral distortions are produced [78] (*). Energy release occuring after this time but before $z_1 \simeq 1.9 \times 10^4 (\Omega_N h^2)^{-\frac{1}{2}}$, is incompletely thermalised, resulting in a Bose-Einstein spectrum, ($\eta_\gamma = \lfloor \exp (\nu/T + \mu) - 1 \rfloor^{-1}$), with a chemical potential $\mu \simeq 1.4\, \Delta\rho_\gamma/\rho_\gamma$, for $\mu \ll 1$. [76] Any subsequent energy release up until the recombination era, $z_{rec} \simeq 1100$ will not be thermalised; it will heat the thermal electrons creating a Compton-distorted spectrum parametrized by a quantity $u \simeq \Delta\rho_\gamma/4\rho_\gamma$, for $u \ll 1$ [76]. After recombination, and before the universe is reionised again (at $z \sim 10$) similar distortions will result from particle decays only if the energy release can be sufficiently degraded to create a non-thermal population of electrons.

---

*) A different limit ($z_T \sim 10^8$) was obtained in refs. 76, 79), assuming bremsstrahlung to be the dominant photon creation process. However, double-Compton scattering dominates if $\Omega_N h^2 \lesssim 1$.

Recent spectral measurements [80] do not support the earlier suggestion of an excess at the Wien peak [81] and confirm a Planck spectrum with $T = 2.72 \pm 0.02\ {}^0K$.[82] The 1σ upper limits on possible distortions are [82] $\mu < 4 \times 10^{-4} - 4 \times 10^{-3}$; $u < 4 \times 10^{-3} - 7 \times 10^{-3}$, accordingly as $\Omega_N h^2$ varies between 0.25 and 2.5×10^{-3}. Assuming $\Omega_N h^2 = 0.025$ (as suggested by primordial nucleosynthesis), the corresponding limits on energy release are, [82]

$$\begin{aligned} \Delta\rho_\gamma/\rho_\gamma &< 7.1 \times 10^{-4} \quad , \quad \text{for } 10^5 \lesssim z \lesssim 10^7 \\ &< 2.8 \times 10^{-2} \quad , \quad \text{for } 10^3 \lesssim z \lesssim 10^5 \\ &< 1.6 \times 10^{-2} \quad , \quad \text{for } 10 \lesssim z \lesssim 10^3 \end{aligned} \tag{36}$$

Thus if a massive particle, mass m_x, lifetime τ, converts a fraction of its mass density into photons (or charged particles which rapidly transfer their energy to photons), then

$$n_x/n_\gamma < 2.2 \times 10^{-6}\ f^{-1}\ (\tau/\text{sec})^{-\frac{1}{2}}\ (m_x/\text{GeV})^{-1}, \quad \text{for } 10^5 \lesssim \tau \lesssim 2 \times 10^9 \text{ sec}, \tag{37a}$$

$$n_x/n_\gamma < 8.7 \times 10^{-5}\ f^{-1}\ (\tau/\text{sec})^{-\frac{1}{2}}\ (m_x/\text{GeV})^{-1}, \quad \text{for } 2 \times 10^9 \lesssim \tau \lesssim 2 \times 10^{13} \text{ sec}. \tag{37b}$$

Hence, given the dependence of τ on m_x, the relic abundance of any particle can be constrained. For example, for massive gravitinos ($m_{3/2} \equiv m_{100} \cdot 10^2$GeV) of lifetime [63] $\tau\ (3/2 \to \tilde{\gamma} + \gamma) \simeq 4 \times 10^8\ m_{100}^{-3}$ sec, the relic abundance and hence the reheating temperature (see eq. 31) is constrained to be, [64, 83]

$$\begin{aligned} T_R &< 9.8 \times 10^9\ m_{100}^{\frac{1}{2}} \text{ GeV} \quad , \quad \text{for } 0.6 \lesssim m_{100} \lesssim 16 \\ &< 4.2 \times 10^{10}\ m_{100}^{\frac{1}{2}} \text{ GeV} \quad , \quad \text{for } 0.03 \lesssim m_{100} \lesssim 0.6 \end{aligned} \tag{38}$$

Conversely, in the case when n_x/n_γ is known as a function of m_x, the lifetime τ can be constrained. For example, for massive neutrinos decaying into charged particles or photons, the perturbation $\Delta\rho_\gamma/\rho_\gamma$ exceeds unity for $m_\nu \lesssim 100$ MeV, so that such decays must either have occured very early, [78]

$$\tau_\nu \lesssim 10^4 - 10^5 \text{ sec}, \tag{39}$$

or should not have occured yet (i.e. $\tau_\nu \gg t_o \sim 3 \times 10^{17}$ sec). The latter possibility is only relevant for neutrinos which are light enough to be consistent with the mass density constraint today, i.e. $m_\nu \lesssim 100$ eV. Their radiative decays are in any case expected to have a long lifetime, $\tau\ (\nu \to \nu' + \gamma) > 2.4 \times 10^{34}$ sec $(m_\nu/\text{eV})^{-5}$, because of GIM suppression in the standard model, but shorter lifetimes may be possible in various extended models. [84] Such decays would generate UV photons and direct observational limits on the UV background [85], as well as the requirement that neutral hydrogen clouds near our Galaxy should not have been ionised away [86], imply that

$$\begin{aligned} \tau &\gtrsim 10^{23} - 10^{25} \text{ sec} , \quad m_\nu \sim 4 - 100 \text{ eV} \\ &\gtrsim 10^{18} - 10^{20} \text{ sec} , \quad m_\nu \sim 0.2 - 4 \text{ eV} \end{aligned} \qquad (40)$$

Constraints on heavier superweakly interacting particles with suppressed abundances consistent with the mass density constraint have been studied in ref. 87). In addition to the constraints from the microwave background (eq. 36), the absence of distinctive spectral features in the infrared, optical and UV backgrounds rules out radiative decays occuring between recombination ($t \sim 10^6$ yr) and the present ($t \sim 10^{10}$ yr); however the decay photons may be hidden in interstellar dust emission if $\tau \approx 10^6$ yr $(m/\text{eV})^{-3/2}$ [87]

The observational upper limits on fluctuations in the microwave background, $\Delta T/T \lesssim 10^{-5} - 10^{-3}$ on angular scales ranging from an arcmin upward, [88] imply that the universe was extremely smooth at recombination. This contrasts with the great deal of structure displayed today by galaxies, on scales upto clusters ($\sim 5\ h^{-1}$ Mpc) and superclusters ($\sim 30\ h^{-1}$ Mpc).[89] Quantitative estimates of the expected fluctuations and clustering properties enable constraints to be placed on the properties of the dominant constituent of the universe. An universe dominated by nucleons is ruled out [90] as is one dominated by massive ($\lesssim 100$ eV) stable neutrinos [91, 92]. In models [55] where an unstable particle dominates the universe during the growth of structure and subsequently decays to provide unclustered matter, the decays are constrained to occur before a redshift $z_d \lesssim 4 - 10$. [56] This rules out most such models. Universes dominated by 'cold' particles (i.e. with negligible velocity dispersion, e.g. axions, ~ few GeV neutrinos or LSP, quark nuggets...) are also constrained by the small-scale anisotropy. [92, 93]

If the initial spectrum of fluctuations arises at the inflationary epoch, their amplitude at horizon crossing is restricted to be $< 3.4 \times 10^{-6}$ [94] from the observed dipole anisotropy of the microwave background. This implies that the mass of the scalar field driving inflation must be $\lesssim (10^{-6} - 10^{-5})\ M_P$. [95] Alternatively, if the fluctuations are generated by a self-similarly evolving network of cosmic strings (1-D topological defects possibly created during spontaneous symmetry breaking [96]), then the predicted anisotropies are below the observational limits for a string tension $\mu \sim 2 \times 10^{-6} M_p^2$, which is suggested by independent considerations about clustering [97].

As mentioned earlier, radiation backgrounds other than the microwave background can also provide a variety of constraints. Long lived relics which decay into muons would contribute to the cosmic ray induced muon background recorded in proton decay detectors [98]; this was used to infer the bound $\tau \gtrsim 10^{31}$ yr (n_x/n_N) (m_x/m_N).[28] These detectors also record neutrino interactions and may be able to detect high energy neutrinos from the annihilations of photinos [99] or from nucleon decay catalysed by monopoles[100], which have accreted within the Sun. Photinos annihilating in the galactic halo have also been proposed as a source of the low energy antiprotons observed in cosmic rays [101], though, given the astrophysical uncertainties, it is more prudent to consider the antiproton flux as setting an upper limit on such annihilations. Heavy unstable neutrinos ($m_\nu \sim 1$-10 MeV) produced in the Sun, which decay into positrons, are constrained to decay sufficiently slowly so as not to exceed the positron flux measured at Earth, $\tau \gtrsim 10^9$ sec $(m_\nu/\text{MeV})^{-4}$. [102] The observational limits on solar X- and γ- ray fluxes restrict the radiative decays of neutrinos produced in the Sun, $\tau \gtrsim 7 \times 10^9$ sec $(m_\nu/\text{eV})^{-1}$. [103] The radiative decays of neutrinos produced in other astrophysical sources such as white dwarfs and supernovae are also restricted by observational limits on the X- and γ- ray backgrounds, $\tau \gtrsim 10^{12}$ sec $(m_\nu/\text{eV})^{-1}$.[103] Supernova energetics rule out neutrino radiative lifetimes between $\sim 10^{-3}$ sec and $\sim 10^3$ sec for $m_\nu \lesssim 10$ MeV . [104] The annihilations of heavy neutrinos in the Galactic halo may contribute to the isotropic γ-ray background, while a spectral feature at ~ 1 MeV in the background spectrum has been ascribed to the radiative decays (after recombination) of ~ 2-5 GeV gravitinos [105]. Again it is best to interpret the γ-ray flux as setting a limit on the annihilations or decays of such particles, rather than as evidence for their presence.

6. PRIMORDIAL NUCLEOSYNTHESIS

The weak interaction rate keeping neutrons and protons in thermal equilibrium, $\Gamma_{weak} \propto \tau_n^{-1} \sim G_F^2 T^5$, falls behind the expansion rate, $H \propto g_*^{\frac{1}{2}} T^2$, at $T_F \sim 0.7$ MeV ($\propto g_*^{1/6} \tau_n^{1/3}$) and the n/p ratio 'freezes out' at $(n/p)_{T=T_F} = \exp -[(m_n - m_p) / T_F] \sim 1/6$. [106]. Nuclear reactions begin when the universe cools to $T \sim 0.1$ MeV so that the number of photons (per nucleon) in the Wien tail of the Planck spectrum which are capable of photodissociating the lightest bound nucleus deuterium ($n_\gamma [E \geqslant Q_D]/n_N \sim \eta^{-1} \exp[-Q_D/T]$) falls below unity. The two-body strong interaction rates which govern the synthesis of n and p into D and ^{3}He, and subsequently into ^{4}He, themselves 'freeze-out' at $T \sim 0.01$ MeV. By this time almost all neutrons have been synthesized into ^{4}He which therefore has a primordial mass fraction $Y_p(^4\mathrm{He}) \simeq 2\,(n/p) / (1+n/p) \sim 0.25$ (since n/p has by now become $\sim 1/7$ through β-decay). Thus Y_p depends strongly on the expansion rate ($\propto g_*^{\frac{1}{2}}$) and the weak interaction rate ($\propto \tau_n^{-1}$) but only weakly on the nucleon density ($n_N = \eta\, n_\gamma$) [106],

$$Y_p(^4\mathrm{He}) = 0.23 + 0.011 \ln(\eta/10^{-10}) + 0.013\,(N_\nu - 3) + 0.014\,(\tau_{\frac{1}{2}} - 10.6), \quad \text{for } 1.5 \lesssim (\eta/10^{-10}) \lesssim 10 , \tag{41}$$

where $\tau_{\frac{1}{2}}$ is the neutron half-life in minutes and N_ν is the number of two-component, light ($m_\nu \ll 1$ MeV) neutrino species, parametrising $g_*(T \sim 1\ \text{MeV}) = (43/4)\,[1 + (7/43)\,(N_\nu - 3)]$. The "left over" abundances of D & ^{3}He are a strong decreasing function of the nucleon density, $(D + {}^3\mathrm{He}) / H \simeq 5 \times 10^{-4}\, \eta^{-1.4}$ for $\eta \sim (0.2 - 10) \times 10^{-10}$, while an increase in the expansion rate $t \rightarrow t' = t / \xi$, $\xi \geqslant 1$, is equivalent to a decrease in the nucleon density, i.e. $(D + {}^3\mathrm{He}) / H = f(\eta/\xi)$ [106]. A very small abundance of lithium, $^7\mathrm{Li}/H \sim 10^{-10} - 10^{-9}$, is also produced and is a double-valued function of η, having a shallow minimum at $\eta \sim 3 \times 10^{-10}$. However, the absence of stable elements with A = 5 and 8, as well as the increasing nuclear coulomb barriers as the universe cools, prevent the synthesis of other heavy elements. According to ref. 107), the weak rate is unknown to ~ 2% ($\tau_{\frac{1}{2}} = 10.6 \pm 0.2$ min) while uncertainties in the nuclear reaction cross-sections make the calculated abundances uncertain by $\lesssim 1\%$ for ^{4}He, $\lesssim 10\%$ for D, ^{3}He and upto a factor ~ 2-3 for ^{7}Li, after allowing [108] for radiative and coulomb corrections, plasma effects etc.

Observationally, η may vary in the wide range between $\sim 4 \times 10^{-11}$ and $\sim 2.8 \times 10^{-8}$ corresponding to the bounds $\Omega_N > 0.01$ and $\Omega_N h^2 < 1$ respectively, assuming $2.7 < T_o < 2.8$ ^{0}K. and $h > 0.4$. The primordial ^{4}He abundance by mass, inferred from observations of unevolved ('metal poor') galaxies and HII regions, is claimed to be [109]

$$Y_p\ (^4\text{He}) = 0.245 \pm 0.003 \quad , \tag{42}$$

corresponding to a '3σ' upper limit of 0.254. Assuming that primordial D is burnt ('astrated') in stars to ^{3}He, a fraction g of which survives stellar processing, the primordial abundance (by number) of D + ^{3}He is related to the abundance at a subsequent time t as [107]

$$\lfloor (D+^3\text{He})/H \rfloor_p < \lfloor (D+^3\text{He})/H \rfloor_t + (^1/g - 1) \lfloor ^3\text{He}/H \rfloor_t . \tag{43}$$

Measurements [110] of ^{3}He/^{4}He in meteorites and lunar rocks are interpreted to imply $\lfloor (D + {}^3\text{He}) / H \rfloor_\odot \simeq (3.6 \pm 0.6) \times 10^{-5}$, $\lfloor {}^3\text{He} / H \rfloor_\odot \simeq (1.4 \pm 0.4) \times 10^{-5}$, for the presolar abundances. Using the '1σ' upper limits for these abundances, and the theoretical constraint $g > 0.25$ suggested by stellar evolution models, then yields [107]

$$\lfloor (D+^3\text{He})/H \rfloor_p < 10^{-4} \tag{44}$$

Observations of deuterium in the interstellar medium [111] provide a lower limit to the primordial abundance (assuming that there has been no net production of D subsequently)

$$\lfloor D/H \rfloor_p > 10^{-5} \tag{45}$$

Finally the rather uniform abundance of lithium seen in old (Pop II) stars suggests a primordial value [112]

$$\lfloor ^7\text{Li}/H \rfloor_p \simeq (0.8 - 1.8) \times 10^{-10} \tag{46}$$

The bounds on the D, ^{3}He abundances (egs. 44 & 45) constrain η to be in the range $\sim 3 \times 10^{-10} - 10^{-9}$, consistent with the range $\sim 1.5 \times 10^{-10} - 5 \times 10^{-10}$ suggested by the ^{7}Li abundance (eq. 46). The predicted abundance of ^{4}He for the standard model ($N_\nu = 3$) is $Y_p = 0.233 - 0.255$ for $\eta = (1.5 - 10) \times 10^{-9}$, in agreement with the observed value (eq. 42). This impressive concordance [107] is persuasive evidence in support of the standard model of primordial nucleosynthesis and suggests that nucleons alone cannot make up the critical density, since $\eta < 10^{-9}$ implies

$$\Omega_N < 0.25, \tag{47}$$

for $h > 0.4$, $T_o < 2.8\ ^0$K. The expansion rate at freeze-out or equivalently N_ν can be similarly constrained using the relation [106]

$$Y_p\ (^4\text{He}) = 0.261 - 0.018 \log \lfloor 10^5 (D + {}^3\text{He})/H \rfloor + 0.014\ (N_\nu - 3) + 0.014\ (\tau_{\frac{1}{2}} - 10.6) \ . \tag{48}$$

Requiring $\tau_{\frac{1}{2}} > 10.4$ min, $Y_p < 0.254$, $(D + {}^3\text{He})/H < 10^{-4}$ constrains the number of light ($m_\nu \lesssim 0.1$ MeV), stable ($\tau_\nu >> 10^2$ sec) two-component ($g_\nu = 2$) neutrinos, [107)]

$$(N_\nu - 3) < 1 \qquad (49)$$

(In fact systematic uncertainties in the observationally inferred elemental abundances may permit upto 3 additional neutrino types [113)]) Accepting this constraint, it follows that no particle which is relativistic at nucleosynthesis (e.g. ~ 1 KeV gravitinos or photinos [61, 62)], $\lesssim$ 100 eV right-handed neutrinos [114)]) may contribute more than the energy density of one neutrino type,

$$\Sigma\, g_B\, (T_F/T_\nu)^4 + (7/8)\, \Sigma\, g_F\, (T_F/T_\nu)^4 < 7/4 \qquad (50)$$

Thus new light particles should interact sufficiently weakly and decouple early enough to adequately reduce T/T_ν; for example three ν_R's which decouple before $T \sim T_c$ are allowed, as are upto twenty gravitini which decouple before $T \sim M_W$. [115)]

The presence during nucleosynthesis of an additional non-relativistic particle increases the synthesized abundances, though not in the same way as an additional particle, because the temperature dependence of its energy density is different ($\propto m\, T^3$ instead of T^4). [116)] For example, a neutrino mass $m_\nu \sim 0.1 - 10$ MeV would increase Y_p more than a 'massless' neutrino ($m_\nu \lesssim 0.1$ MeV), though, since, $\rho_\nu \propto m_\nu^{-3}$, there is no effect on the synthesized abundances if the mass is high enough, $m_\nu \gtrsim 25$ MeV. Thus if there are three light neutrinos, (*) a fourth heavy neutrino must have $m_\nu > 20$ MeV in order to be consistent with the abundances which can be tolerated observationally [116)]. This implies the constraint [64)]

$$n_X/n_\gamma < 1.6 \times 10^{-4}\, (m_X/\mathrm{GeV})^{-1} \qquad (51)$$

on any non-relativistic particle (massive gravitinos, monopoles..) present at nucleosynthesis. (Note that this constraint is often incorrectly interpreted; for example in ref. 71) the energy density of magnetic monopoles during nucleosynthesis is required to be less than that of a <u>relativistic</u> 2-component neutrino, which implies $n_M/n_\gamma \lesssim 5 \times 10^{-3}\, (m_M/\mathrm{GeV})^{-1}$. However, the effect of a non-relativistic particle is quite different from that of a relativistic particle, hence the true constraint eq. 51) differs considerably).

---

*) The experimental limits [117)], m('ν_e') < 46 eV, m('ν_μ') < 250 KeV, m('ν_τ') < 70 MeV, would appear to allow the possibility of a heavy ν_τ, but this is (almost) ruled out [78)] as discussed later.

If such a massive particle decays subsequently and creates new photons then η would be decreased from its value at nucleosynthesis.[116] Requiring that η should not be reduced below its observational limit today implies that the particle should decay early enough to release sufficiently few thermalised photons,[64]

$$\begin{aligned} n_x/n_\gamma &< 3.3 \times 10^{-3}\,(m_x/\mathrm{GeV})\,(\tau/\mathrm{sec})^{-1/3}\,f^{-1/2}, \\ &\quad \text{for } 10^2 \lesssim \tau/\mathrm{sec} \lesssim 3.8 \times 10^5\,f^{-3/2}\,; \\ &< 2.8 \times 10^{-2}\,(m_x/\mathrm{GeV})^{-1}\,(\tau/\mathrm{sec})^{-1/2}\,f^{-3/4}, \\ &\quad \text{for } \tau \gtrsim 3.8 \times 10^5\,f^{-3/2}\ \mathrm{sec}. \end{aligned} \tag{52}$$

This constraint imposes the limit $T_R \lesssim 10^{14}$ GeV pertaining to massive gravitino ($m_{3/2} \sim M_W$) generation and requires for neutrinos of mass $\sim$ 1-10 MeV,[78, 116]

$$\tau_\nu \lesssim 200 - 2000\ \mathrm{sec}. \tag{53}$$

More stringent constraints follow by requiring that the stable final state decay products (γ, $e^\pm$, $\overset{(-)}{p}$..) do not directly alter the synthesized abundances. For example, energetic photons (generated through $3/2 \to \tilde{\gamma} + \gamma$; $\nu \to e^- e^+ \nu'$, $e^\pm\gamma \to e^\pm\gamma'$) may cause photofission,[118, 78]

$$\begin{aligned} &\gamma + D \to p + n \quad (Q_D \sim 2.2\ \mathrm{MeV})\ ; \\ &\gamma + {}^4He \to D,\ {}^3He.. + X \quad (Q_{He} \sim 20\ \mathrm{MeV})\ , \end{aligned} \tag{54}$$

while slow antiprotons (generated through $3/2 \to \tilde{g} + g$, $\tilde{g} \to q\bar{q}\,\tilde{\gamma}$) would annihilate on heavy nuclei,[119]

$$\bar{p} + {}^4He \to D,\ {}^3He, \ldots + X \tag{55}$$

The photofission rate can be computed by following through the history of the primary γ's or $e^\pm$'s as they generate electromagnetic cascades in the radiation-dominated thermal plasma.[64, 120] When the energy density of the decaying particle is large (as for heavy neutrinos) the decays should occur early enough to avoid excessive photofission of D[121]; for $m_\nu \sim 5 - 100$ MeV,

$$\tau_\nu \lesssim (2 \times 10^3 - 10^4)\ \mathrm{sec}\ , \tag{56}$$

Conversely, when decays occur late (as for gravitinos), the energy density should be sufficiently small to avoid excessive photoproduction of D + ^{3}He by photofission of ^{4}He[64],

$$n_x/n_\gamma < 3.1 \times 10^{-16}\,(m_x/\mathrm{GeV})^{-1}\,f^{-1} \tag{57}$$

In the case of gravitino decay, this implies a rather low reheating temperature after inflation[64]

$$T_R < 2.5 \times 10^8\,(m_{3/2}\,/\,100\ \mathrm{GeV})^{-1} \tag{58}$$

Furthermore, if a fraction r antiprotons are produced per decay, then to avoid excessive production of D + ^3He through ^4He annihilation requires [64, 122)

$$n_X / n_\gamma < 2 \times 10^{-12}\ r^{-1} \tag{59}$$

This implies, [64)

$$T_R < 1.8 \times 10^{10}\ \text{Gev} \tag{60}$$

if the decay channel $3/2 \rightarrow \tilde{g} + g$ exists for gravitinos.

7. DISCUSSION

The cosmological bounds discussed above can be usefully combined with laboratory bounds to test the viability of proposed phenomenological models.

For example, for massive neutrinos, experimental upper limits on the charged current mixing angles, $\nu_\ell = \Sigma\, U_{\ell i}\nu_i$, $\ell \equiv e, \mu, \tau ..,$ $i = 1,2,3...$, imply lower limits on the lifetime, $\tau\,(\nu_i \rightarrow \ell\ell'\nu_{i'})$ $\simeq \tau\,(\mu \rightarrow e\,\nu\,\bar{\nu}).\ |U_{\ell i}|^{-2}.\ (m_\mu/m_{\nu i})^5.$ [123)]
Such limits on $|U_{ei}|^2$ from a search for secondary peaks in π_{e2} decay [124)] are inconsistent with the lifetime limit $\tau\,(\nu_i \rightarrow e^- e^+ \nu_e)$ $\lesssim 10^2 - 10^4$ sec, for $1 \lesssim m_\nu \lesssim 10$ MeV, derived from the entropy generation and microwave background constraints (eqs. 39 & 53). [78)]
Hence no neutrino, stable or unstable, can have such a mass, which is of interest in left-right symmetric $SU(2)_L \times SU(2)_R \times U(1)_{B-L}$ models[125)].
For the tau neutrino in particular, a beam dump experiment ($pN \rightarrow F + X$, $F \rightarrow \tau\nu_\tau$, $\nu_\tau \rightarrow e^- e^+ \nu_e$) suggests $\tau \gtrsim 200 - 1000$ sec for $m_{\nu 3} \sim 30 - 200$ MeV [126)], while the D photofission argument was used to argue that $\tau \lesssim 100$ sec [78, 127)]. Thus it appeared that ν_3 must be too light to have charged current decays, i.e. $m_{\nu 3} < 1$ MeV. However, the photofission bound is weakened to $\tau \lesssim 2 \times 10^3 - 10^4$ sec after correctly accounting for photon energy degradation by pair production on energetic thermal photons in the Wien Tail, [121)] so that a narrow mass range, ~ 30-70 MeV, would appear to be allowed for an unstable tau neutrino, with $\tau \sim 200 - 2000$ sec. [128)] However new limits [129)] on $|U_{ei}|^2$ will rule out most of this region, so that such a heavy ν_3 seems rather unlikely. Thus a mass hierachy of the form $m_\nu \sim m_\ell^{\,2}/M$ would be acceptable only if neutrino masses are effectively generated at a rather large mass scale, $M \gtrsim 10^8 - 10^{10}$ GeV. In some models, the existence of Goldstone bosons (majorons, familons), into which neutrinos

can decay, considerably weakens the applicability of the above limits, but these models are severely constrained by the lifetime limit from galaxy formation (eq. 26).

The bound of O (10^8 GeV) on the reheating temperature after inflation (eq. 58) is an embarassment for conventional supersymmetric cosmology in which the cosmological baryon asymmetry is generated by the out-of-equilibrium decays of particles with B-violating interactions. Since these particles mediate proton decay, they must be rather heavy ($m \gtrsim 10^{16}$ GeV [dim-5 operators] or $m \gtrsim 10^{11}$ GeV [dim-6 operators]) to satisfy the present experimental bound, $\tau_p > 10^{31}$ yr. [130] Since the baryon asymmetry can only be produced during or after reheating, this implies that a thermal Boltzmann distribution for such heavy particles would yield far too few of them. Perhaps the baryon asymmetry is created non-thermally, for example directly through the decay of the scalar field driving the inflation. [131] Alternatively, the gravitino must be sufficiently heavy ($\gtrsim 10^4$ GeV) in order to decay before nucleosynthesis [132] or be the LSP and therefore stable. [64] (In the latter case the unstable particle bounds presumably apply to the photino; however these are easy to satisfy since the electroweak interactions of the photinos enable them to annihilate efficiently [23] leaving behind an acceptably small density. [64]) Either of these possibilities can be realised in supergravity models with non-minimal kinetic terms for the chiral and/or vector superfields. [133] Interestingly enough, such 'no-scale' models emerge naturally in the compactification to four space-time dimensions of $N = 1$, $d = 10$ supergravity which is the zero slope limit of the superstring. [134]

The maximum baryon-to-entropy ratio which can be generated in GUT baryosynthesis is, $n_B/s \sim 10^{-6}$. [3] Thus any subsequent generation of entropy is constrained by the requirement that n_B/s should not be reduced below $\sim 10^{-10}$, as required by primordial nucleosynthesis and present observations. This imposes lifetime limits on heavy unstable particles such as new superheavy fermions, N, which appear in many GUTs (SO(10), E(6) etc.); if these decays involve heavy bosons X, then $M_N \gtrsim 5 \times 10^{-3} M_X (M_X/M_p)^{1/3}$. [135] Unacceptably large entropy production is also associated with supersymmetry breaking in a 'hidden' gauge singlet sector (the 'Polyoni problem'); Suppressing this imposes restrictions on the coupling of the hidden to the 'visible' gauge non-singlet sector.[136] Finally the Weinberg-Salam phase transition leads to entropy production and similarly constraining this implies the bound $m_H > 9$ GeV on the Higgs mass.[137]

ACKNOWLEDGEMENTS

I would like to thank John Ellis, Amanda Sarkar and Dennis Sciama for many dicussions and for much encouragement.

REFERENCES

1) S. Weinberg, 'Gravitation & Cosmology' |Wiley, 1972|;
Ya. B. Zeldovich & I.D. Novikov, 'The Structure & Evolution of the Universe' |Univ. of Chicago Press, 1983|.

2) S. Weinberg, Physica Scripta 21 (1980) 773;
A.D. Linde, Rep. Prog. Phys. 42 (1979) 389.

3) E.W. Kolb & M.S. Turner, Ann. Rev. Nucl. Part. Sci. 33 (1983) 645.

4) R.H. Brandenberger, Rev. Mod. Phys. 57 (1985) 1;
A.D. Linde, Rep. Prog. Phys. 47 (1984) 925, Comments. Astrophys. 10 (1985) 229.

5) C. Wetterich, Nucl. Phys. B252 (1985) 309;
E.W. Kolb, Proc. APS meeting (Santa Fe, 1984) in press.

6) M.B. Green & J. Schwarz, Phys. Lett. 149B (1984) 117.

7) F.K. Thielemann, J. Metzinger & H.V. Klapdor, Astr. Astrophys. 123 (1983) 162;
E.M.D. Symbalisty & D.N. Schramm, Rep. Prog. Phys. 44 (1981) 293.

8) I. Iben & A. Renzini, Phys. Rep. 105 C (1984) 329.

9) N. Bartel et al., Nature 318 (1985) 25;
P.W. Hodge, Ann. Rev. Astr. Astrophys. 19 (1981) 357.

10) R.L. Zimmerman & R.W. Hellings, Astrophys. J. 241(1980) 475.

11) M. Davis & P.J.E. Peebles, Ann. Rev. Astr. Astrophys. 21 (1983) 109;
P.J.E. Peebles, Astrophys. J. 284 (1984) 439.

12) G. Steigman, Ann. Rev. Nucl. Part. Sci. 29 (1979) 313.

13) A.D. Dolgov & Ya. B. Zeldovich, Rev. Mod. Phys. 53 (1981) 1.

14) Ya. B. Zeldovich, Adv. Astr. Astrophys. 3(1965) 241;
H.Y. Chiu, Phys. Rev. Lett. 17 (1966) 712.

15) J.E. Felten, Comments Astrophys. 11 (1985) 53;
S.M. Faber & J.S. Gallagher, Ann. Rev. Astr. Astrophys. 17(1979) 339.

16) G. Steigman, Ann. Rev. Astr. Astrophys. 14 (1976) 339.

17) S. Wolfram, Phys. Lett. 82B (1979) 65.

18) C.B. Dover, T.K. Gaisser & G. Steigman, Phys. Rev. Lett. 42 (1979) 1117;
R.V. Wagoner, I. Schmitt & P.M. Zerwas, Phys. Rev. D27 (1983) 1696.

19) P.F. Smith et al., Nucl. Phys. B206 (1982) 333.

20) A. Turkevich, K. Wielgoz & T.E. Economou, Phys. Rev. D30 (1984) 1876.

21) R. Middleton, R.W. Zurmuble, J. Klein & R.V. Kollarits, Phys. Rev. Lett. 43 (1979) 429.

22) W.J. Dick, G.W. Greenlees & S.L. Kaufman, Phys. Rev. Lett. 53 1984) 431.

23) J. Ellis, J.S. Hagelin, D.V. Nanopoulos, K. Olive & M. Srednicki, Nucl. Phys. B238 (1984) 453;
H. Goldberg, Phys. Rev. Lett. 50(1983) 1419.

24) G.S. La Rue, J.D. Phillips & W.M. Fairbank, Phys. Rev. Lett. 46 (1981) 967.

25) E.W. Kolb & M.S. Turner, Phys. Lett. 115B (1982) 99;
E.W. Kolb, G. Steigman & M.S. Turner, Phys. Rev. Lett. 47 (1981) 1357;
R.V. Wagoner, & G. Steigman, Phys. Rev. D20 (1979) 825.

26) S.M. Barr, D.B. Reiss & A. Zee, Phys. Rev. Lett. 50 (1983) 317;
H. Goldberg, T. Kephart & R. Vaugh, Phys. Rev. Lett. 47 (1981)1429.

27) I. Bars, M.J. Bowick & K. Freese, Phys. Lett. 138B (1984) 159.

28) J. Ellis, T.K. Gaisser & G. Steigman, Nucl. Phys. B177 (1981) 427.

29) H. Goldberg, Phys. Rev. Lett. 48 (1982) 1518.

30) S.L. Shapiro, S.A. Teukolsky & I. Wasserman, Phys. Rev. Lett. 45 (1980) 669.

31) S. Gershtein & Ya. B. Zeldovich, JETP Lett. 4 (1966) 120;
R. Cowsik & J. McClelland, Phys. Rev. Lett. 29 (1972) 669;
A.S. Szalay & G. Marx, Astr. Astrophys. 49 (1976) 437;
J. Bernstein & G. Feinberg, Phys. Lett. 101B (1981) 39.

32) K.A. Olive & M.S. Turner, Phys. Rev. D25 (1982) 213.

33) B.W. Lee & S. Weinberg, Phys. Rev. Lett., 39 (1977) 165;
D.A. Dicus, E.W. Kolb & V.L. Teplitz, Phys. Rev. Lett. 39 (1977) 168;
M.I. Vysotskii, A.D. Dolgov & Ya. B. Zeldovich, JETP Lett. 26 (1977) 188.

34) L.M. Krauss, Phys. Lett. 128B (1983) 37.

35) P. Hut & K. Olive, Phys. Lett. 87B (1979) 144.

36) G. Gelmini & M. Roncadelli, Phys. Lett. 99B (1981) 411.

37) H. Georgi, S.L. Glashow & S. Nussinov, Nucl. Phys. B193 (1981) 297;
S.L. Glashow & A. Manohar, Phys. Rev. Lett. 54 (1985) 2306;
E.W. Kolb & M.S. Turner, Phys. Lett. 159B (1985) 102.

38) M. Fukugita, S. Watamura & M. Yoshimura, Phys. Rev. D26 (1982) 1840.

39) D.A. Dicus, E.W. Kolb & V.L. Teplitz, Astrophys. J. 221 (1978) 327.

40) G. Steigman & M.S. Turner, Nucl. Phys. B253 (1985) 375.

41) Y. Hosotani, Nucl. Phys. B191 (1981) 411.

42) M. Roncadelli & G. Senjanovic, Phys. Lett. 107 B (1981) 59.

43) A.A. Natale, Phys. Lett. 141B (1984) 323.

44) P.B. Pal, Nucl. Phys. B227 (1983) 237.

45) M. Gronau & R. Yahalom, Phys. Rev. D30 (1984) 2422;
B.H.J. McKellar & S. Pakvasa, Phys. Lett. 122B (1983) 33.

46) J. Schechter & J.W.F Valle, Phys. Rev. D25 (1982) 774.

47) G.B. Gelmini & J.W.F. Valle, Phys. Lett. 142B (1984) 181.

48) Y. Chikashige, R.N. Mohapatra & R.D. Peccei, Phys. Rev. Lett. 45 (1980) 1926.

49) F. Wilczek, Phys. Rev. Lett. 49 (1982) 1549;
G. Gelmini, S. Nussinov & T. Yanagida, Nucl. Phys. B219 (1983) 31.

50) H.M. Steiner et al., Proc. XXII HEP conf. (Leipzig, 1984) p. 208.

51) Y. Asano et al., Phys. Lett. 107B (1981) 159.

52) J.J. Simpson, Phys. Rev. Lett. 54 (1985) 1891.

53) M. Dugan, G. Gelmini, H. Georgi & L. Hall, Phys. Rev. Lett. 54 (1985) 2302;
J.W.F. Valle, Phys. Lett. 158B (1985) 49;
A. Nelson, Phys. Lett. 159B (1985) 389;
B. Grinstein, J. Preskill & M.B. Wise, Phys. Lett. 159B (1985) 517;
B. Holdom, Phys. Lett. 160B (1985) 303.

54) T. Altzizoglu et al., Phys. Rev. Lett. 55 (1985) 799;
T. Ohi et al., Phys. Lett. 160B (1985) 322;
A. Apalikov et al., ITEP-114, Moscow (1985).

55) M.S. Turner, G. Steigman & L.M. Krauss, Phys. Rev. Lett. 52 (1984) 2090;
M. Fukugita & T. Yanagida, Phys. Lett. 144B (1984) 386;
G. Gelmini, D.N. Schramm & J.W.F. Valle, Phys. Lett. 146B (1984) 311.

56) N. Vittorio & J. Silk, Phys. Rev. Lett. 54 (1985) 2269;
M. Turner, Phys. Rev. Lett. 55(1985) 549.

57) G. Efstathiou, Mon. Not. R. astr. Soc. 213 (1985) 29p.

58) L.E. Ibanez, Phys. Lett. 137B (1984) 160;
J.S. Hagelin, G.L. Kane & S. Raby, Nucl. Phys. B241 (1984) 638.

59) J. Ellis, J.S. Hagelin & D.V. Nanopoulos, Phys. Lett. 159B (1985)26;
J. Ellis, Proc. Lepton-Photon Symp. (Kyoto, 1985) in press.

60) H. Pagels & J.R. Primack, Phys. Rev. Lett. 48 (1982) 223.

61) D.W. Sciama, Phys. Lett. 118B (1982) 327, 137B (1984) 169.

62) P. Fayet, Proc. First ESO-CERN Symp. (Geneva, 1983) p. 35.

63) J. Ellis, J.E. Kim & D.V. Nanopoulos, Phys. Lett. 145B (1984) 181.

64) J. Ellis, D.V. Nanopoulos & S. Sarkar, Nucl. Phys. B259 (1985) 175.

65) J. Ellis & K. Olive, Nucl. Phys. B223 (1983) 252;
M. Fukugita & N. Sakai, Phys. Lett. 114B (1982) 23.

66) P. Sikivie, Proc. Fourth Workshop on Grand Unif. (Philadelphia, 1983) p. 214, Proc. Fifth Workshop on Grand Unif. (Providence, 1984) p. 108.

67) J. Preskill, M.B. Wise & F. Wilczek, Phys. Lett. 120B (1983) 127;
L.F. Abbott & P. Sikivie, Phys. Lett. 120B (1983) 133;
M. Dine & W. Fischler, Phys. Lett. 120B (1983) 137.

68) P.J. Steinhardt & M.S. Turner, Phys. Lett. 129B (1983) 51;
K. Yamamoto, Phys. Lett. 161B (1985) 289.

69) S.Y.Pi, Phys. Rev. Lett. 52 (1984) 1725.

70) D.A. Dicus, E.W. Kolb, V.L. Teplitz & R.V. Wagoner, Phys. Rev. D18 (1978) 1829, D22 (1980) 839;
N. Iwamoto, Phys. Rev. Lett. 53 (1984) 1198.

71) J. Preskill, Phys. Rev. Lett. 43 (1979) 1365, Ann. Rev. Nucl. Part. Sci. 34 (1984) 461.

72) M.S. Turner, Phys. Lett. 115B (1982) 95;
G. Lazarides, Q. Shafi & W.P. Trower, Phys. Rev. Lett. 49 (1982) 1756.

73) M.S. Turner, Proc. Monopole '83 (Ann Arbor, 1983) p. 127, Proc. First Aspen Winter Conf. (Aspen, 1985) in press;
E.W. Kolb, Ann. N.Y. Acad. Sci., 422 (1984) 33;
M. Fukugita, Proc. Workshop on Grand Unif. & Cosmology (KEK, 1984) p. 38.

74) J. Arons & R.D. Blandford, Phys. Rev. Lett. 50 (1983) 544;
R.T. Farouki, S.L. Shapiro & I. Wasserman, Astrophys. J. 284 (1984) 282;
D. Chernoff, S.L. Shapiro & I. Wasserman, Cornell preprint CRSR 835 (1985).

75) A.S. Goldhaber, in 'Magnetic Monopoles' [Plenum, 1983] p. 1, Proc. Fourth Workshop on Grand Unif. (Philadelphia, 1983) p. 115.

76) A.F. Illarianov & R.A. Sunyaev, Sov. Astr. 18 (1975) 691;
R.A. Sunyaev & Ya. B. Zeldovich, Ann. Rev. Astr. Astrophys. 18 (1980) 537.

77) J. Silk, GIFT School on Relativistic Astrophysics & Cosmology (Catalonia, 1983) p. 249, Proc. Inner Space-Outer Space (Fermilab, 1984) in press;
C.J. Hogan, N. Kaiser & M.J. Rees, Phil. Trans. R. Soc. Lond. A 307 (1982) 97.

78) S. Sarkar & A.M. Cooper, Phys. Lett. 148 B (1984) 347.

79) J.E. Gunn, B.W. Lee, I. Lerche, D.N. Schramm & G. Steigman, Astrophys. J. 223 (1978) 1015;

80) G.F. Smoot et al., Astrophys. J. 291 (1985) L 23;
J.B. Peterson, P.L. Richards & T. Timusk, Phys. Rev. Lett. 55 (1985) 332;
D.M. Meyer & M. Jura, Astrophys. J. 297 (1985) 119.

81) D.P. Woody & P.L. Richards, Astrophys. J. 248 (1981) 18.

82) C. Witebsky, Ph.D. Thesis, Univ. of California (Berkeley, 1985);
G. De. Amici, G.F. Smoot, S.D. Freidman & C. Witebsky, LBL preprint 19323 (1985).

83) R. Juszkiewicz, J. Silk & A. Stebbins, Phys. Lett. 158B (1985) 463;
P. Salati, Phys. Lett. 163B (1985) 236.

84) P.B. Pal & L. Wolfenstein, Phys. Rev. D 25 (1982) 766;
J.F. Nieves, Phys. Rev. D 28 (1983) 1664;
A. de Rujula & S.L. Glashow, Phys. Rev. Lett. 45 (1980) 942.

85) F.W. Stecker & R.W. Brown, Astrophys. J. 257 (1982) 1;
R. Kimble, S. Bowyer & P. Jacobsen, Phys. Rev. Lett. 46 (1981) 80;
H.L. Shipman & R. Cowsik, Astrophys. J. 247 (1981) L 111.

86) A. Melott & D.W. Sciama, Phys. Rev. Lett. 46 (1981) 1369.

87) J. Silk & A. Stebbins, Astrophys. J. 269 (1983) 1.

88) D.H. Wilkinson, Proc. Inner Space-Outer Space (Fermilab, 1984) in press.

89) J.R. Primack, Proc. Intern. School of Physics 'Enrico Fermi' (Varenna, 1984) in press;
G. Efstathiou & J. Silk, Fund. Cosmic Phys. 9 (1983) 1.

90) M.L. Wilson & J. Silk, Astrophys. J. 243 (1984) 14;
M.L. Wilson, Astrophys. J. 273 (1983) 2.

91) S.D.M. White, C. Frenk & M. Davis, Astrophys. J. 274 (1983) L1;
N. Kaiser, Astrophys. J. 273 (1983) L17.

92) J.R. Bond & G. Efstathiou, Astrophys. J. 285 (1984) L45;
N. Vittorio & J. Silk, Astrophys. J. 285 (1984) L39, 293 (1985)L1.

93) M. Davis, G. Efstathiou, C. Frenk & S.D.M. White, Astrophys. J. Suppl. (1985) in press.

94) L.F. Abbott & M.B. Wise, Astrophys. J. 282 (1984) L 47.

95) S.W. Hawking, Phys. Lett. 150 B (1985) 339;
V.A. Rubakov, M.V. Sazhin & A.V. Veryaskin, Phys. Lett. 115 B (1982) 189.

96) T.W.B. Kibble, Nucl. Phys. B252 (1985) 277;
A. Vilenkin, Phys. Rep 121 C (1985) 263.

97) R.H. Brandenberger & N. Turok, NSF-ITP-85-88 (1985);
N. Turok, Phys. Rev. Lett. 55 (1985) 1801.

98) P.H. Frampton & S.L. Glashow, Phys. Rev. Lett. 44 (1980) 1481.

99) J. Silk, K. Olive & M. Srednicki, Phys. Rev. Lett. 55 (1985) 257.

100) J. Arafune & M. Fukugita, Phys. Lett. 133B (1983) 380.

101) J. Silk & M. Srednicki, Phys. Rev. Lett. 53 (1984) 624.

102) D. Toussaint & F. Wilczek, Nature 289 (1981) 777.

103) G.G. Raffelt, Phys. Rev. D31 (1985) 3002, MPI-PAE/PTh 9/85 (1985);
R. Cowsik, Phys. Rev. Lett. 39 (1977) 784, Phys. Rev. D19 (1979) 2219.

104) S.W. Falk & D.N. Schramm, Phys. Lett. 79B (1978) 511.

105) K. Olive & J. Silk, Phys. Rev. Lett. 55 (1985) 2362.

106) A.M. Boesgaard & G. Steigman, Ann. Rev. Astr. Astrophys. 23 (1985) 319;
D.N. Schramm & R.V. Wagoner, Ann. Rev. Nucl. Sci. 27 (1977) 37.

107) J. Yang, M.S. Turner, G. Steigman, D.N. Schramm & K. Olive, Astrophys. J. 281 (1984) 493.

108) D.A. Dicus, E.W. Kolb, A.M. Gleeson, E.C.G. Sudarshan, V.L. Teplitz & M.S. Turner, Phys. Rev. D26 (1982) 2694.

109) D. Kunth & W.L.W. Sargent, Astrophys. J. 273 (1983) 81.

110) J. Geiss & H. Reeves, Astron. Astrophys. 18 (1972) 126;
D.C. Black, Geochim. Cosmochim. Acta 36 (1972) 347;
E. Anders, D. Heymann & E. Mazor, Geochim. Cosmochim. Acta 34 (1970) 127.

111) C. Laurent, Proc. ESO Workshop on Primordial Helium (Munich, 1983) p. 335.

112) M. Spite, J.P. Maillard & F. Spite, Astron. Astrophys. 141 (1984) 56.

113) J. Ellis, K. Enqvist, D.V. Nanopoulos & S. Sarkar, CERN preprint TH. 4303 (1985).

114) J.A. Bagger, S. Dimopoulos, E. Masso & M.H. Reno, Nucl. Phys. B258 (1985) 565.

115) K. Olive, D.N. Schramm & G. Steigman, Nucl. Phys. B180 (1981) 497.

116) E.W. Kolb & R.J. Scherrer, Phys. Rev. D25 (1982) 1481;
D.A. Dicus, E.W. Kolb, V.L. Teplitz & R.V. Wagoner, Phys. Rev. D17 (1978) 1529.

117) E. Bergqvist, Proc. Lepton Photon Symp. (Kyoto, 1985) in press.

118) D. Lindley, Mon. Not. R. Astr. Soc. 188 (1979) 15p.

119) Ya. B. Zeldovich, A.A. Starobinskii, M. Yu. Khlopov & V.M. Chechetkin, Sov. Astr. Lett. 3 (1978) 110;
I.V. Falomkin et al., Nuovo Cimento 79 A (1984) 19;
F. Balestra et al., CERN preprint EP/84-108 (1984).

120) D. Lindley, Mon. Not. R. Astr. Soc. 193 (1980) 593;
F.A. Aharonian, V.G. Kirillov - Ugryumov & V.V. Vardanian, ERFI preprint 83 - 676 (1983).

121) D. Lindley, Astrophys. J. 294 (1985) 1.

122) M. Yu. Khlopov & A.D. Linde, Phys. Lett. 138 B (1984) 265.

123) K. Sato & M. Kobayashi, Prog. Theor. Phys. 58 (1977) 1775;
E.W. Kolb & T. Goldman, Phys. Rev. Lett. 43 (1979) 897.

124) D.A. Bryman et al., Phys. Rev. Lett. 50 (1983) 1546.

125) R.N. Mohapatra, Proc. 3rd LAMPF II Workshop (Los Alamos, 1983).

126) F. Bergsma et al., Phys. Lett. 128 B (1983) 361.

127) L.M. Krauss, Phys. Rev. Lett. 53 (1984) 1976.

128) S. Sarkar, Proc. Intern. HEP Conf. (Bari, 1985) in press.

129) J. Deutsch, Proc. Intern. HEP Conf. (Bari, 1985) in press.

130) B.A. Campbell, J.E. Ellis & D.V. Nanopoulos, Phys. Lett. 141B (1985) 229;
P. Nath. A.H. Chamseddine & R. Arnowitt, Phys. Rev. D32 (1985) 2348.

131) P.J. Steinhardt & M.S. Turner, Phys. Rev. D29 (1984) 2162;
G.D. Coughlan, G.G. Ross, R. Holman, P. Ramond, M. Ruiz-Altaba & J.W.F. Valle, Phys. Lett. 158B (1985) 401, 160 B (1985) 249.

132) S. Weinberg, Phys. Rev. Lett. 48 (1982) 1303.

133) J. Ellis, K. Enqvist, D.V. Nanopoulos & M. Tamavakis, Phys. Lett. 156B (1985) 381;
J. Ellis, C. Kounnas & D.V. Nanopoulos, Phys. Lett. 143B (1984) 410

134) E. Witten, Phys. Lett. 155B (1985) 151;
E. Cohen, J. Ellis, C. Gomez & D.V. Nanopoulos, Phys. Lett. 160B (1985) 62;
J. Ellis, Proc. Sixth Workshop on Grand Unif. (Minneapolis, 1985) in press.

135) J.A. Harvey, E.W. Kolb, D.B. Reiss & S. Wolfram, Nucl. Phys. B177 (1981) 456.

136) G.D. Coughlan, W. Fischler, E.W. Kolb, S. Raby & G.G. Ross, Phys. Lett. 131B (1983) 59.

137) A.H. Guth & E.J. Weinberg, Phys. Rev. Lett. 45 (1980) 1131;
E. Witten, Nucl. Phys. B177 (1981) 477.

DARK MATTER AND GALAXY FORMATION*

R. Valdarnini

International School for Advanced Studies,

Strada Costiera,11, 34014 Trieste,

ITALY.

ABSTRACT

A short review is given on the current status of galaxy formation theory with dark matter made of massive collisionless particles. These particles might be massive neutrinos or other, more massive, weakly interacting quanta which decouple earlier. Collisionless particles which decouple when nonrelativistic are termed cold dark matter. The massive neutrinos hypothesis is almost completely ruled out by a number of difficulties. Numerical simulations for the evolution of large scale structure show that the predictions of a cold dark matter model, with $\Omega = 1$, are consistent with the present galaxy clustering if galaxies form at the high peaks of the primeval dark matter density field (biased galaxy formation). The large scale peculiar velocity fields of the model are in excess of that recently observed, but observational uncertainties do not allow us to reach firm conclusions.

An improvement in observations for the large scale velocity fields and clustering features at early epochs is likely to set severe constraints for different dark matter models.

1. Introduction

1.1 Experimental evidence for dark matter

In recent years there has been a growth of experimental evidence

* Summer Workshop on High Energy Physics and Cosmology, 10 June-19 July, 1985, Trieste, Italy.

that a large fraction of the total matter content in the universe is in the form of a dark matter (e.g. Faber and Gallagher 1979). Until now dark matter has been detected only indirectly through its gravitational interaction with visible systems. The ratio of dark to luminous matter on a given mass scale can be expressed in terms of the M/M_{lum} ratio, where M_{lum} is the total baryonic mass for the considered system. The M/M_{lum} ratio is found to be roughly constant over a wide range of mass scales (Faber 1982), $M/M_{lum} \simeq 10-15$, from spiral galaxies up to cluster scales ($\simeq$1 Mpc).

To estimate the mean matter density in the universe the cosmological density parameter $\Omega = \rho/\rho_c$ is defined, where ρ is the present mean matter density and $\rho_c = 3H_o^2/8\pi G$ is the critical density necessary to close the universe. There are two main ways to estimate Ω: the first makes use of the cosmic virial theorem (Davis et al. 1978, Peebles 1979a), the other estimates Ω from the local anisotropies in the Hubble flow due to the gravitational collapse of the Local Group towards the Local Super Cluster (LSC). In the first method the assumption of virial equilibrium on the considered scale yields a relation between the relative peculiar galaxy pair velocities and Ω; in determining this relation the two-point correlation function $\xi(r)$ (see below) is used.

Typically one finds (Davis et al. 1978, Peebles 1979a) $\Omega \simeq 0.2-0.3$ on length scales $r \simeq 1$ Mpc. In the second approach the motion of the Local Group is supposed to be determined from the LSC matter over density. Observations of Aaronson et al. (1980), Hart and Davies (1982), who did not find appreciable anisotropies in the Hubble flow at redshift $cz \simeq 3000-5000$ Km sec^{-1}, support this view. From the Local Group motion it is found that $\Omega \simeq 0.4 \pm 0.1$ (Davis et al. 1980, Davis and Peebles 1983a) on length scales $r \simeq 10$ Mpc.

In the methods quoted here to estimate Ω the assumption is implicit that light is a tracer of the mass distribution. If there is a dark matter component which is more weakly clustered than the visible matter we will not detect it and Ω would be underestimated. Then these measurements for Ω are not a difficulty for the inflationary scenario (Guth 1981), which suggests $\Omega = 1$.

We have not yet discussed of what dark matter could be made. Using the standard big-bang model it is possible to set upper limits on the baryonic contribution to Ω through the observed Helium and Deuterium abundances; Yang et al. (1984) found that $0.01 < \Omega_b < 0.14$. From this upper limit on Ω_b it appears unlikely that baryons can account for the total amount of dark matter inferred from large-scale dynamical estimates for Ω. Further there are various arguments which suggest (Peebles 1979b) that the dark haloes of spiral galaxies are not baryonic.

In what follows we shall assume as a fundamental hypothesis that dark matter is of a non-baryonic nature. Within this framework even primordial black holes are a possibility (Carr 1977), but only massive neutrinos or other weakly interacting massive collisionless quanta will be considered.

1.2 General framework

In the gravitational instability theory galaxies form through the growth of primeval matter density fluctuations $\delta(\vec{x},t) \equiv \frac{\delta\rho}{\rho} = \rho(\vec{x},t)/\bar{\rho}-1$ (see Jones 1976 for a review). Primeval matter inhomogeneities could be generated through vacuum fluctuations in the De Sitter phase of an inflationary universe (Brandenberger 1985 and references quoted therein). However we shall discuss here only the general features for the linear ($\delta \ll 1$) and non-linear ($\delta \gg 1$) evolution of $\delta\rho/\rho$.

A common assumption is that at early epochs $\delta(\vec{x},t)$ was described by a random Gaussian process with zero mean $\langle\delta(\vec{x},t)\rangle = 0$ and variance $\sigma^2 = \langle\delta^2(\vec{x},t)\rangle$; here brackets denote an average over the ensemble (cf. Peebles 1980 sect. 18, hereafter LSS).

To study the time evolution of the density field it is more convenient to introduce the Fourier transform

$$\delta_{\vec{k}} = \int \delta(\vec{x},t)\, e^{i\vec{k}\cdot\vec{x}}\, d^3x \qquad (1.1)$$

where it is assumed that initially the $\delta_{\vec{k}}$ have random phases and power spectrum

$$|\delta_{\bar{k}}|^2 \propto K^n \quad , \tag{1.2}$$

n is called the spectral index. Eq. (1.2) is a general statement that there were no characteristic scales in the original fluctuation spectrum. An estimate of the density fluctuation on the mass scale $M \propto \lambda^3$ is

$$\frac{\delta\rho}{\rho} \simeq \left[\int_0^{\lambda^{-1}} K^2 |\delta_k|^2 dK \right]^{1/2} \propto M^{-\frac{1}{2}-\frac{n}{6}} \quad . \tag{1.3}$$

A safe assumption is n > -3, that is $\delta\rho/\rho$ on small scales are larger than that on large scales. The time evolution of $\frac{\delta\rho}{\rho}$ in the linear phase, with dark matter made of massive neutrinos or other massive collisionless particles, is followed solving numerically the Einstein-Boltzmann equations for perturbations in the metric as well as density perturbations in radiation, baryonic matter and massive particles.

The integration is performed starting from high redshifts, $z \approx 10^8-10^9$ when the perturbations are well in the linear regime, down to final redshifts $z \approx 10^3-10^2$, before the onset of non-linear phases.

The small angular scale temperature anisotropies $\Delta T/T$ are one of the main observational parameters of the linear theory. These small scale temperature anisotropies are left by density fluctuations on the cosmic background radiation during the matter-radiation decoupling epoch. The present upper limit is $\Delta T/T < 3.10^{-5}$ at angular scales of 4.5' (Uson and Wilkinson 1984). In sect. 3 we give a comparison of the computed $\Delta T/T$ for different dark matter models with the present upper limit.

When the matter distribution approaches non-linearity then N-body methods are best suited to follow its evolution. In these methods a set of N collisionless particles is left free to evolve in time under the action of its gravitational forces. Initial conditions for the matter distribution are given according to the calculations of the linear theory. The potential ϕ of the system is calculated either by direct summation or solving the Poisson equation over a mesh (particle-mesh codes, PM). The latter technique has been improved by Hockney and Eastwood (1981) with a direct summation for the interparticle forces at short distances (particle-particle/particle-mesh codes, P^3M). In dealing with N-body simulations of this kind one must be sure that the final results are not masked by numerical effects such as the finite number of particles, finite length of the mesh or relaxation effects. For a detailed discussion of these sorts of problems, and on P^3M codes, see Efstathiou et al. (1985).

In a cosmological N-body simulation the final (or at earlier times depending on the physical scaling) particle distribution should represent the observed large-scale clustering pattern of matter. Then one of the main output parameters of the integration is the two-point correlation function $\xi(r)$ of the particle distribution. $\xi(r)$ is defined as the joint probability dP of finding galaxies in the volumes δV_1 and δV_2

$$dP = \bar{n}^2 \left[1 + \xi(r)\right] \delta V_1 \, \delta V_2 \, , \tag{1.4}$$

where $\bar{n}$ is the mean number density of galaxies. Observationally it is found that (Davis and Peebles 1983b)

$$\xi(r) = (r/r_o)^{-\gamma} \, , \; \gamma = 1.77 \pm 0.4, \; r_o = (5.4 \pm 0.3) h^{-1} \, \mathrm{Mpc} \, , \tag{1.5}$$

where h is the Hubble constant in units of 100 Km $\mathrm{sec}^{-1}\mathrm{Mpc}^{-1}$ and r_o is called the clustering length. The hypothesis that dark matter is made of massive collisionless particles has been tested through various large-scale N-body simulations. The main results are discussed in sect. 2.3 and 3.2.

2. Massive neutrinos

2.1 Cosmological consequences

The experimental detection of a non-zero rest mass for the electronic neutrino (Lyubimov et al. 1980), of the order of tens of eV, has led many authors to consider the hypothesis that dark matter is made of massive neutrinos. Let us first examine the cosmological consequences of a non-zero rest mass for neutrinos. We assume that neutrinos are relativistic when decoupled and their phase-space distribution is that of Fermi-Dirac with zero chemical potential. After neutrino decoupling both the neutrino temperature and momentum fall as (1+z), then we can apply the relativistic Fermi-Dirac distribution to calculate the present neutrino number density. The neutrino temperature T_ν today is related to that of radiation by $T_\nu = (4/11)^{1/3} T_r$, because after they decouple neutrinos do not share in the heating caused by e^+e^- pair annihilation (Weinberg 1972). The neutrino contribution to Ω is

$$\Omega_\nu = \frac{\rho_\nu}{\rho_c} = 10^{-2} h^{-2} \Sigma_i m_{\nu_i} \quad , \tag{2.1}$$

where i = e,μ,τ , $m_{\nu i}$ are in eV units and the photon temperature today is 2.7°K.

From the present upper limit to Ω_b (Ω_b < 0.14, Yang et al. 1984) neutrinos give the largest contribution to Ω for $\Sigma_i m_{\nu i}$ > 2.4 eV (Schramm and Steigman 1981). An upper limit to the sum of the neutrino masses can be worked out from the present observational constraints on Ω, h and the age of the Universe t_o. In the standard big-bang model with a zero cosmological constant $\Omega = 2q_o$, where q_o is the deceleration parameter. With the present estimates for q_o (Tamman et al. 1979) an upper limit to Ω is $\Omega < 2$. Measurements of the Hubble parameter yield $0.5 < h < 1$ (Sandage and Tamman 1984, De Vaucoulers 1984), while the estimated globular cluster ages set $t_o > 13.10^9$ yr (Tamman et al. 1979, Schramm 1982). From the above constraints and the relation between t_o and Ω (Weinberg 1972) it is possible to conclude that (Zel'dovich and Syunyeev 1980)

$$\Sigma_i m_{\nu_i} \leq 30 \text{ eV} . \tag{2.2}$$

If the sum of the neutrino masses would exceed this limit massive neutrinos could still be a solution to the dark matter problem with a non-zero cosmological constant, which is the hypothesis suggested by Zel'dovich and Syunyaev (1980). This hypothesis will not be considered here. From (2.1) we see that massive neutrinos with rest mass of the order of $m_\nu \simeq 10$ eV yield a density parameter Ω of order unity, in the same range as the observed one for dark matter. Let us consider the impact of the massive neutrinos hypothesis on the galaxy formation theory.

2.2 Galaxy formation

At present the existence of isothermal density perturbations at early epochs is not favoured by grand unified theories, GUT. In these theories the specific entropy is a constant which depends only on micro-physical parameters and any pre-existing isothermal perturbation would have been erased at the baryosynthesis epoch ($T \simeq 10^{28}$ °K, Weinberg 1980). Then we shall assume that primeval neutrino density perturbations were adiabatic. The evolution of neutrino density perturbations in the linear phases has been widely studied in the literature (Doroshkevich et al. 1980, Bond et al. 1980, Wasserman 1981, Peebles 1982, Bond and Szalay 1983). A neutrino Jeans mass $M_{J\nu}$ can be calculated if one replaces in the Jeans mass the sound velocity c_s^2 with $\langle v_\nu^2 \rangle/3$, where $\langle v_\nu^2 \rangle^{1/2}$ is the neutrino velocity dispersion (this is a very crude argument, in reality one must resort to a numerical solution of the relativistic collision-less Boltzmann equation for the neutrino distribution, e.g. Peebles 1982).

Neutrinos become nonrelativistic at a redshift $z_{\nu_{NR}} \simeq 4.10^4 \, m_\nu/30$ eV, of the same order as the redshift $z_{eq} \simeq 4.10^4 \Omega h^2$ when the universe becomes matter dominated. At $z \gg z_{\nu_{NR}}$ $\langle v_\nu^2 \rangle^{1/2} \simeq c/\sqrt{3}$ and the value of $M_{J\nu}$ is close to that of the horizon mass $M_H \simeq \rho(ct)^3 \propto (1+z)^{-3}$. For $z \ll z_{\nu_{NR}}$ $\langle v_\nu^2 \rangle^{1/2} \simeq$ $\simeq p_\nu/m_\nu \propto (1+z)$ and $M_{J\nu} \propto (1+z)^{3/2}$. Then the neutrino Jeans mass reaches its maximum value $M_{\nu c} \simeq 10^{15} \, (m_\nu/30 \text{ eV})^{-2} \, M_\odot$ at $z \simeq z_{\nu_{NR}}$ (cfr. Fig. 1 of Bond et al. 1980). Neutrino density perturbations with $M < M_{\nu c}$ are erased by Landau damping as they cross the horizon. Thus the only neutrino density perturbations which survive ν- derelativistization are those with $M > M_{\nu c}$. This mass scale corresponds to a neutrino coherence length $\lambda_{\nu c} \simeq 40 (m_\nu/30\text{eV})^{-1}$ Mpc. For all mass scales M inside the horizon at $z \ll z_{\nu_{NR}}$ the rms neutrino density perturbations on mass scale M will be

$$\begin{cases} \delta_\gamma \propto M^{-\frac{1}{2}-\frac{n}{6}} & M \geq M_{\gamma_c} \\ \delta_\gamma \simeq 0 & M < M_{\gamma_c} \end{cases} \tag{2.3}$$

The mass scale $M_{\nu c}$ corresponds to a cluster-super cluster scale and in the massive neutrino picture it is the first scale to undergo collapse and to reach non-linearity.

After recombination at $z < z_r \simeq 10^3$ baryons will be caught inside neutrino perturbation potential wells and they will collapse together with neutrinos. At redshift z_{nl} of order unity the collapse of the perturbation is one-dimensional, as in the Zel'dovich (1970) pancake theory. The pancake collapse will produce a strong compression of baryons in the midplane, a shock wave will form which will propagate outwards. The infalling baryons will be shock heated and will dissipate their energy though radiative processes. One-dimensional numerical simulations for the baryon+neutrino pancake collapse with $\Omega_\nu = 0.9$, $\Omega_b = 0.1$ (Shapiro et al. 1984) have shown that at $z \simeq 1$ only 10% of the shock-heated baryons inside the pancake will cool enough to form galaxies, the remaining baryons will be in the form of hot intergalactic gas ($T \simeq 10^6$ °K) with $\Omega_{IGM} \simeq 0.1$. In this scenario galaxies are formed after the collapse of the neutrino pancake on mass scale $M_{\nu c}$, through the cooling and fragmentation of the baryonic pancake. The M/M_{lum} ratio is expected to increase with the considered scale, since more neutrinos will be caught inside deeper potential wells.

A constraint on the neutrino mass from the scale of the collapsed system can be worked out from phase space considerations (Tremaine and Gunn 1979, hereafter TG). Because of the Liouville theorem the final neutrino phase space density cannot exceed its maximum value when neutrinos decouple. From this constraint TG found that

$$m_\nu \gtrsim 10^2 \left(\frac{100\ \mathrm{Km\,sec^{-1}}}{\sigma}\right)^{1/4} \left(\frac{1\ \mathrm{kpc}}{r_c}\right)^{1/2} eV, \tag{2.4}$$

here σ is the velocity dispersion and r_c the core radius of the collapsed system. From the cosmological upper limit to m_ν eq. (2.4) becomes just an equality for a galactic system. In a one-dimensional numerical simulation Mellot (1983) has shown that about 12% of neutrinos inside the pancake have a low velocity dispersion ($\simeq 10^2$ Km sec^{-1}), and are able to collapse over galactic scales. This is the general picture for the formation of galaxies in a neutrino dominated universe. In the next section we shall quote various problems which arise in this picture.

2.3 Difficulties

The neutrino scenario suffers serious problems both on small scales (below the galactic ones) and on large cluster scales. The recent experimental evidence for dark matter over dwarf galaxy scales (Faber and Lin 1983, Aaronson 1983) from eq. (2.4) implies a neutrino mass $m_\nu > 500$ eV. This does not agree with the upper limit to m_ν from eq. (2.2). These measurements for dwarf galaxies have not been confirmed (Cohen 1983), but a positive detection of dark matter would definitely rule out neutrinos as sole dark matter constituents. On large scales a first difficulty is that in the neutrino model the M/M_{lum} ratio should increase with the considered scale, which conflicts with the observed trend (Faber 1982): $M/M_{lum} \simeq 10\text{–}15$ stays constant from galaxies up to cluster scales.

Large scale N-body simulations for the nonlinear evolution of neutrino density perturbations have been made by a number of authors (Centrella and Mellot 1983, Klypin and Shandarin 1983, Frenk et al. 1983, White et al. 1983). One of the main results, for initial spectra with n=0, is that it is not possible to have consistency between the large value of the neutrino clustering length $\lambda_{\nu c}$ and the present clustering length of galaxies r_o. This difficulty can be avoided if galaxies form at very recent epochs ($z<1$). In a neutrino model with n=1 another difficulty (Kaiser 1983) is that the present large scale ($\simeq$ 50 Mpc) matter velocities are too large with respect to the observed values ($\simeq$ 100 Km $\sec^{-1}$). The latter two difficulties are excluded with very steep initial spectra ($n \simeq 4$). However a n=1 spectrum is theoretically favoured since it implies that density perturbations of different wavelengths will have the same amplitude as they cross the horizon, i.e. the primeval spectrum has no preferrred scales. (Zel'dovich 1972). A n=1 spectrum is also predicted in the inflationary universe (see Brandeberger 1985). These difficulties do not finally exclude neutrino models for dark matter, but they put serious doubts on the idea that dark matter is made of massive neutrinos.

3. Other particles

3.1 Galaxy formation

Dark matter (DM) could also be made of neutral weakly interacting particles, more massive than neutrinos. We call these particles X. They might be photinos (Cabibbo et al. 1981), gravitinos (Pagels and Primack 1982), right-handed neutrinos (Olive and Turner 1982) or axions (Preskill et al. 1983). X particles interact more weakly than neutrinos and decouple earlier. If X particles decouple when they are relativistic an upper limit to m_X is, assuming that there is no entropy generation after X decoupling,

$$m_X \leq \Omega_X h^2 (g_{*d}/100)(g_X/1.5)^{-1}\ \mathrm{KeV}\ . \tag{3.1}$$

Here g_{*d} is the total effective number of degrees of freedom at X decoupling and g_X is the number of X helicity states. The simplest grand unified theory, SU(5), gives $g_{*d} < 100$, thus m_X cannot be much larger than some KeV. In this case free-streaming processes set the critical damping scale $M_{Xc} \simeq 10^{12}(m_X/\mathrm{KeV})^{-2} M_\odot$ in the X fluctuation spectrum. Because of the different free-streaming scales massive neutrinos are termed hot DM and X particles with $m_X \simeq 1$ KeV warm DM. If X particles decouple when they are nonrelativistic their velocity dispersion is so low that free-streaming scales are not of cosmological interest. The latter are termed cold DM. At present cold DM candidates are photinos, gravitinos (with masses $m_X > 1$ GeV) or axions; the only X particles which could be relativistic when decoupled are right-handed neutrinos.

Let us consider the linear evolution of warm DM fluctuations (Bond et al. 1982, Blumenthal et al. 1982, Valdarnini 1985). The growth of X density perturbations with $M_{Xc} < M < M_{eq} \equiv M_H(z = z_{eq}) \simeq 10^{16} M_\odot$ is suppressed as they cross the horizon down to $z = z_{eq}$ since at $z > z_{eq}$ the universe is still radiation dominated (Meszaros 1974). For X density perturbations in the mass range $M_{Xc} < M < M_{eq}$ this corresponds to a flattening by a factor $M^{2/3}$ in the final spectrum. Then at $z \ll z_{eq}$ the final rms X density perturbations on mass scale M will be

$$\begin{cases} \delta_X \propto M^{-\frac{1}{2}-\frac{n}{6}} & M > M_{eq} \\ \delta_X \propto M^{-(n-1)/6} & M_{X_c} \leq M \leq M_{eq} \\ \delta_X \simeq 0 & M < M_{X_c} \end{cases} \tag{3.2}$$

In the cold DM case damping scales are unimportant and at $z < z_{eq}$ $\delta_X \propto M^{-(n-1)/6}$ for all mass scales $M < M_{eq}$. In this case a minimum mass scale is set by the post-recombination baryon Jeans mass $M_{Jb} \simeq 10^6 M_\odot$. On mass scales $M > M_{Jb}$ baryonic pressure will prevent δ_b to reach δ_X after recombination (Bond and Szalay 1983, Peebles 1984, Blumenthal et al. 1984). Baryonic structures inside DM perturbations will be erased when larger scales virialize unless they cool on time scales shorter than dynamical times (White and Rees 1978). As a result of this constraint, for a n=1 cold DM spectrum, only protogalaxies with total mass $10^8 M_\odot < M < 10^{12} M_\odot$ should have the baryonic material separated from DM (Blumenthal et al. 1984). This mass range roughly corresponds to that observed for galaxies. If $m_X \simeq 1$ KeV non-linearity is achieved first on galactic scales and large scale structures will form later through non-dissipative gravitational clustering of smaller scales.

The same picture holds in cold DM models, where the smallest scale is given by the baryon Jeans mass. In the warm and cold DM scenarios galaxies cluster together with their dark haloes. The M/M_{lum} ratio is expected to be roughly constant from galaxies up to cluster scales, in better agreement with observations with respect to the predictions of a neutrino scenario. It must be stressed that in these DM models (warm and cold) light and mass are tightly correlated.

3.2 Comparison with observations

The experimental evidence for DM in dwarf galaxies ($M_{DG} \simeq 10^7 M_\odot$-$10^8 M_\odot$) could be a problem for warm DM models. On these scales warm DM particles satisfy the TG constraint but X density fluctuations with $M \simeq M_{DG}$ have been previously erased by Landau damping. Then it is unclear how dwarf galaxies have gained their dark haloes.

The small angular scale temperature anisotropies in the cosmic background radiation have been calculated for different DM models (Bonometto et al. 1984, hot and warm DM; Bond and Efstathiou 1984, hot and cold DM; Vittorio and Silk 1984, cold DM). The DM fluctuation spectra have been normalized to the present clustering requiring the rms mass fluctuation $\delta M/M$ (LSS, sect.26) to be one at a scale $R \simeq 8h^{-1}$Mpc (Peebles 1984). For $\Omega = 1$ and $\Omega_b = 0.03$, in all the considered DM models, the computed $\Delta T/T$ are below the present small scale upper limit $\Delta T/T < 3.10^{-5}$ at 4.5' (Uson and Wilkinson 1984). Low-density universes ($\Omega \simeq 0.2$-0.4) are excluded since the small scale anisotropy $\Delta T/T$ would exceed the observational limit. Reheating of the intergalactic medium after recombination is unlikely to affect these conclusions (e.g. Bond and Efstathious 1984).

N-body simulations have been performed by Davis et al (1985), with a P^3M code, to follow the evolution of large scale structure in cold DM dominated universes. Linear initial conditions are given by a cold DM fluctuation spectrum with n=1. In these simulations the slope of the two-point correlation function $\xi(r)$ agrees with that observed ($\gamma = 1.8$) at a time identified with the present epoch. At this time, for $\Omega = 1$ ($\Omega = 0.2$), we have $\xi=1$ at $r = 5h^{-1}$ Mpc if $h = 0.25$ ($h=1.1$). The value $h \simeq 0.25$ is outside the allowed observational range, the correlation function of the simulations agrees in slope and amplitude with that observed only if $\Omega \simeq 0.2$ and $h \simeq 1$. However cold DM models with $\Omega \simeq 0.2$ and $h \simeq 1$ led to small scale temperature fluctuations which violate the present upper limit. As a further difficulty in this model the age of the universe is $t_o \simeq 8.10^9$ yr, below the lower limit set by globular cluster ages.

These difficulties are avoided if galaxies form only at the high peaks of the primeval DM density fields (Bardeen 1984, Kaiser 1984). Physical mechanisms which can lead to a threshold for galaxy formation have been discussed by Bardeen(1984). This biasing hypothesis implies that present galaxies are mainly in high density regions and more strongly clustered than DM. Davis et al. (1985) found that the predictions of a $\Omega=1$ cold DM simulation, with biased galaxy formation, are consistent with the present galaxy clustering. In the simulation they identify all peaks of the initial density field with amplitude in excess of $\nu = 2.5$ times the rms density fluctuation, then the particle with the nearest position to each peak is identified as a galaxy. In this DM model the small scale fluctuations $\Delta T/T$ are below the present small scale upper limit. The difficulties for the biasing hypothesis arise on large scales ($\simeq$ 50 Mpc). In a Ω=1 cold DM model with biased galaxy formation the large scale peculiar velocity fields (LSVF) are not in agreement with observations. Recent data on LSVF are given by Sandage and Tamman (1985) and De Vaucoulers and Peters (1985). The latter data are in conflict with the predictions of a cold DM scenario if the dark and luminous matter distribution is not tightly correlated (Vittorio and Silk 1985). However the present observational uncertainities in the LSVF data do not allow us to firmly rule out the biasing galaxy formation hypothesis.

An alternative possibility could be that DM is made of two species of massive collisionless particles (Bonometto and Valdarnini 1985), with comparable energy densities today. The model predicts an evolution of the clustering length r_o on redshift which might be observationally detectable. Thus no definite conclusions can yet be drawn: an improvement in observations for the LSVF, and better analysis for deep galaxy clustering, are needed in order to check different DM models.

REFERENCES

Aaronson, M., Mould, J., Huchra, J. 1980, Ap. J. 237, 655.
Aaronson, M., 1983, Ap. J. Lett. 266, L11.
Beardeen, J., 1984, in Inner space/Outer space (Chicago: Univ. of Chicago Press).
Blumenthal, G.R., Pagels, H., Primack, J.R., 1982, Nature, 299, 37.
Blumenthal, G.R., Faber, S.M., Primack, J.R., Rees, M.J., 1984, Nature, 311, 517.
Bond, J.R., Efstathiou , G., Silk, J., 1980, Phys. Rev. Lett. 45, 1980.
Bond, J.R., Szalay, A.S., Turner, M.S., 1982, Phys. Rev. Lett. 48, 1636.
Bond, J.R., Szalay, A.S., 1983, Ap. J., 274, 443.
Bond, J.R., Efstathiou , G., 1984, Astrophys. J. Lett. 285, L45.
Bonometto, S.A. Lucchin, F., Valdarnini, R., 1984, Astr. and Astrophys., 140, L27.
Bonometto, S.A., Valdarnini, R., 1985, ISAS preprint 52/85/A, to be published in Ap. J. Lett.

Brandenberger, R.H., 1985, Rev. Mod. Phys., 57, 1.
Cabibbo, N., Farrar, G.R., Maiani, L., 1981, Phys. Lett. 105B, 155.
Carr, B.J., 1977, Astr. and Astrophys. 56, 377.
Centrella, J., Mellott, A.L. 1983, Nature, 305, 196.
Cohen, J.G., 1983, Astrophys. J. Lett. 270, L41.
Davis, H., Geller, M.J., Huchra, J., 1978, Ap. J. 221, 1.
Davis, H., Tonry, J., Huchra, J., Latham, D.W., 1980, Ap. J. Lett. 238, L113.
Davis, H., Peebles, P.J.E., 1983a, Ann. Rev. Astr. and Astrophys. 21, 109.
Davis, H., Peebles, P.J.E., 1983b, Ap. J. 267, 465.
Davis, H., Efstathiou , G., Frenk, C.S., White, S.D.M., 1985, Ap. J. 292, 371.
De Vaucoulers, G., 1984 in Cluster and Groups of Galaxies (eds. Mardirossian,F., Giuricin, F., Mezzetti, M.), Dordrecht, Holland.
De Vaucolulers, G., Peters, W.L., 1985, Ap. J. 287, 1.
Doroshkevich, A.G., Zel'dovich, Ya.B., Syunyaev, R.A. Khlopov, Y.M., 1980, Sov. Astr. Lett. 6, 252.
Efstathiou , G., Davis, H, Frenk, C.S., White, S.D.M. 1985, Ap. J. Suppl. 57, 241.
Faber, S.M., Gallagher, J.S. 1979, Ann. Rev. Astr. Astrophys., 17, 135
Faber, S.M., 1982, in Astrophysical Cosmology (eds. Bruck, H., Coyne, G., Longair, M.S.), Pontifical Scientific Academy, Vatican, 1982.
Faber, S.M., Lin, D.N.C. 1983, Ap. J. Lett. 266, L17.
Frenk, C.S. White, S.D.M., Davis, M. 1983, Ap. J. 271, 417.
Guth, A., 1981, Phys. Rev. D, 23, 347.
Hart, L. Davies, R., 1982, Nature, 297, 191.
Hockney, R.W., Eastwood, J.W., 1981, Computer simulations using particles, McGraw Hill.
Jones, B.J.T., 1976, Rev. Mod. Phys., 48, 107.
Kaiser, N., 1983, Ap. J. Lett. 273, L17.
Kaiser, N., 1984, in Inner space/Outer space (Chicago: Univ. of Chicago Press).
Klypin, A.A., Shandarin, S.F., 1983, MNRAS, 204, 891.
Lyubimov, V.A., Novikov, E.G., Nozik, V.Z., Tretyakov,E.F., Kozik, V.F., 1980, Phys. Lett., 94B, 266.
Mellott, A.L., 1983, Ap. J. 264, 59.
Meszaros, P., 1974, Astr. and Astrophys., 37, 225.
Olive, K.A., Turner, M.S., 1982, Phys. Rev. D., 25, 213.
Pagels, H.R., Primack, J;R., 1982, Phys. Rev. Lett., 48, 223.
Peebles, P.J.E., 1979a, Astron. J., 84, 730.
Peebles, P.J.E., 1979b, in Physical Cosmology (eds., Balian, R., Andouze, J., Schramm, D.N.), North Holland, Amsterdam.
Peebles, P.J.E., 1980, The Large Scale Structure of the Universe, Princeton University Press.
Peebles, P.J.E., 1982, Ap. J., 258, 415.
Peebles, P.J.E., 1984, Ap. J. 277, 470.
Preskill, J., Wise, M.S., Wilczeck, F., 1983, Phys. Lett., 120B, 127.
Sandage,A., Tamman, G.A., 1984, in First ESO-CERN Symp.: Large-Scale Structure of the Universe (1984).
Sandage, A., Tamman, G.A., 1985, Ap. J., 294, 81.

Schramm, D.N., Steigman, G., 1981, Ap. J. 243, 1.
Schramm, D.N., 1982, Phil. Trans. R. Soc. London, A307, 43
Shapiro, P.R., Struck-Marcell, C., Mellott, A.L., 1983, Ap. J. 275, 413.
Tamman, G.A., Sandage, A., Yahil, A. 1979, in Physical Cosmology (eds. Balian, R., Andouze, J., Schramm, D.N.), North Holland, Amsterdam.
Tremaine, S., Gunn, J.E., 1979, Phys. Rev. Lett. 42, 407.
Uson, J.M., Wilkinson, D.T., 1984, Ap. J. Lett. 277, L1.
Valdarnini, R., 1985, Astr. and Astrophys. 143, 94
Vittorio, N., Silk, J., 1984, Ap. J. Lett. 285, L39.
Vittorio, N., Silk, J., 1985, Ap. J. 293, L1.
Wasserman, I., 1981, Ap. J., 248, 1.
Weinberg, S., 1972, Gravitation and Cosmology, Wiley.
Weinberg, S., 1980, Phys. Scripta, 21, 773.
White, S.D.M., Rees, M.J. 1978, MNRAS, 183 , 341.
White, S.D.M., Frenk, C.S., Davis, M., 1983, Ap. J. Lett. 274, L1.
Yang, J., Turner, M.S., Steigman, G., Schramm, D.N., Olive, K.A., 1984, Ap. J., 281, 493.
Zel'dovich, Ya.B., 1970, Astr. and Astrophys. 5, 84.
Zel'dovich, Ya.B., 1972, MNRAS, 160, 1p.
Zel'dovich, Ya.B., Syunyaev, R.A., 1980, Sov. Astr. Lett. 6, 249.

Supersymmetry, Monojets, and Dark Matter

G.L. Kane*

Randall Laboratory of Physics
University of Michigan,
Ann Arbor, MI 48109-1120, USA.

Abstract

These lectures cover a range of subjects, related primarily by the goal of finding clues to physics beyond the Standard Model. First the ways in which evidence for supersymmetry might appear are derived and examined, and the complications of real experiments are considered. Then the three candidates photino, sneutrino, and higgsino for the lightest supersymmetric partner, which could consistute the dark matter, are considered. Ways to detect dark matter by detecting annihilation products or energy transferred in collisions are examined in detail, and cross sections are given.

*Research supported in part by the U.S. Department of Energy.

Now that the Standard Model of particle physics is so well established, we need clues to what will help us understand why the Standard Model takes the form it does, why various regularities are present, and what is the origin of mass. One clue is the baryon asymmetry of the universe, which leads in the direction of grand unified theories. Another clue is the hierarchy problem, the apparent presence of physics on two or more widely separated scales (100 GeV and about 10^{15} GeV). The beautiful mathematical theory of supersymmetry has been invoked as a physical way to maintain a hierarchy. If supersymmetry plays such a role, we would expect to find experimental evidence for it on the scale of 100 GeV.

In these lectures we will examine how to look for evidence for supersymmetry, and the current status of the search.(1) If superpartners did exist, the lightest one (LSP) could be stable. It would have been in equilibrium with other matter in the early universe, and there would be enough of it around to account for the dark matter which is probably being observed by astronomers. We will also discuss how to detect some forms of dark matter directly.

THE SPECTRUM OF SUPERSYMMETRY

The minimal spectrum of new particles expected if supersymmetry were a symmetry of nature realized on the weak scale is that each of the quarks, leptons, and gauge and Higgs bosons should have a partner whose properties are identical to the particle except that it differs in spin by half a unit. A useful notation and nomenclature is indicated in Table 1. The names indicate both the origin of the particle and, for the partners of gauge bosons and Higgs bosons, how the partner interacts; because of weak interaction eigenstates mixing to form mass eigenstates, the interaction properties can be subtle, but they are important to determine observability. Left-(L) and right-handed(R) fermions are separated since they have different weak interactions.

Table 1

The Minimal Particle Spectrum in a Supersymmetric Theory.

Normal Particles	Superpartners	Names	
g	$\tilde{g}$	gluino	
u_L	$\tilde{u}_L$	up-squark	
u_R	$\tilde{u}_R$	.	
.	.	.	
.	.	.	
e_L	$\tilde{e}_L$	selectron	
.	.	.	
.	.	.	
$W^{\pm}$	$\tilde{W}^{\pm}$	wino whiggsino	charginos
$H^{\pm}_{1,2}$	$\tilde{h}^{\pm}_{1,2}$	higgsino	
γ	$\tilde{\gamma}$	photino	neutralinos
Z	$\tilde{Z}$	zino	
$H^{o}_{1,2}$	$\tilde{h}^{o}_{1,2}$	higgsino	

If supersymmetry were an unbroken symmetry, the partners would have the same mass, color, charge, etc. as the particles, and would have easily been detected. Since they have not, we consider them to be heavier because of symmetry breaking. If supersymmetry is relevant to explaining the weak scale, presumably it makes sense to expect the partners to have mass $\sim M_W$. If partners are not discovered on the weak scale eventually the limits on their masses will be pushed up to well beyond the weak scale and they will become irrelevant to understanding physics on the weak scale.

Because the way in which the supersymmetry is broken is not understood, the masses of the superpartners are unknown. It is necessary to proceed by assuming a set of masses (in practice that means masses for mainly $\tilde{g}$, $\tilde{q}$, $\tilde{w}$, $\tilde{\ell}^{\pm}$, $\tilde{\nu}$, $\tilde{\gamma}$), calculating predictions, and testing them.

Fortunately, the couplings of the superpartners are known, because they are mainly gauge couplings. In fact, they are the actual measured gauge couplings, so no uncertainty arises at all. The rule that emerges from the theory is that every vertex in the standard theory has added to it vertices obtained by replacing particles by their partners in pairs, keeping the interaction strength the same. Thus the $\bar{q}qg$ coupling gives $\tilde{q}\tilde{g}q$ and $\tilde{q}\tilde{q}g$ couplings all of strength g_3 ($g_3^2/4\pi = \alpha_s$), the $We\nu$ coupling gives $\tilde{w}\tilde{e}\nu$, $\tilde{w}e\tilde{\nu}$, and $W\tilde{e}\tilde{\nu}$ couplings all of strength g_2, etc. With this rule one can draw all the Feynman graphs that are relevant to whatever process is being considered, and estimate any cross section or decay.

Since supersymmetric partners occur in pairs, they can either be pair-produced, or a single one can be produced via $\tilde{q}$ or $\tilde{g}$ in a proton (the other supersymmetric particle goes down the beam pipe with other soft partners). We will consider both mechanisms; it turns out that if gluinos are lighter than about 25GeV, the gluino content of a proton is significant.

Once a superpartner is produced, it will decay into normal particles plus a lighter superpartner; the lightest superpartner will be stable (one can consider modified cases where the lightest superpartner is unstable, and carry through the analysis in a way analagous to the following, but little changes and we will not pursue this alternative here). The lightest superpartner can be a partner of a gauge boson or Higgs boson, or it can be a sneutrino. For our purposes, it will not matter much what is the lightest superpartner (until we consider dark matter), as its essential property is that it normally escapes collider detectors. That is because to interact it must excite a superpartner in the detector, and the partners of quarks and leptons are heavy, so the interaction cross section is at most of order $\alpha^2/\tilde{m}^2$ with $\tilde{m} \gtrsim$ 20GeV, giving too small a cross section to see.

Thus the basic signature of the production of supersymmetric partners is missing momentum, accompanied by jets or charged leptons in characteristic patterns. If a squark is produced it will decay into $q\tilde{g}$ and $q\tilde{\gamma}$ if both are kinematically allowed, in a ratio (ignoring phase space)

$$\frac{\Gamma(\tilde{q}\to q\tilde{\gamma})}{\Gamma(\tilde{q}\to q\tilde{g})} = \frac{3e_q^2\,\alpha}{4\alpha_s}$$

For $\tilde{u}$ this is $\alpha/3\alpha_s \simeq 0.02$. If $\tilde{m}_q \gg \tilde{m}_W, M_q$ then

$$\frac{\Gamma(\tilde{q} \to q\tilde{w})}{\Gamma(\tilde{q} \to q\tilde{g})} \simeq \frac{1}{2}\,\frac{3\alpha_2}{8\alpha_s} \simeq 0.05 \ ,$$

where the 1/2 appears because only $\tilde{q}_L$ has a coupling to $\tilde{w}$.

Similarly, if $\tilde{\ell}^\pm$ appear, they will decay $\tilde{\ell}^\pm \to \ell^\pm\tilde{\gamma}$

Gluinos will mainly decay via a virtual $\tilde{q}$, $\tilde{g} \to \bar{q}q\tilde{\gamma}$, with a branching ratio (that depends on other masses) of perhaps 1% to $g\tilde{\gamma}$.

If sleptons or sneutrinos are lighter than winos or zinos the latter will have useful signatures,

$$\tilde{w} \to \tilde{\ell}^\pm\nu \to \ell^\pm \not{p}_T$$

or

$$\tilde{w} \to \ell^\pm\tilde{\nu},$$

(where $\not{p}_T$ = missing momentum),

or

$$\tilde{z} \to \tilde{\ell}^+ \ell^- \to \ell^+ \ell^- \not{p}_T .$$

If sleptons and sneutrinos and also $\tilde{g}$ are heavier than winos and zinos, the latter still may have a useful signature,

$$w \tilde{\to} q\bar{q}\tilde{\gamma} \tilde{\to} jj\not{p}_T ,$$

$$\tilde{z} \to \bar{q}q\tilde{\gamma} \to jj\not{p}_T .$$

If only g are lighter than $\tilde{w}$, $\tilde{z}$ the latter will decay into $jjj\tilde{\gamma}$, becoming essentially unobservable.

THE STANDARD ASSUMPTION - THE PHOTINO AS LSP

The question of the LSP in supersymmetric theories is a model dependent question. In principle one must know the neutralino (neutral gaugino and higgsino) mass matrix (for a review see ref. 1). By diagnalizing, one obtains the LSP which may be some linear combination of photino, zino and higgsino. We shall start with the assumption that the LSP is the photino. This is relevant for the signatures of scalar-quarks and gluinos. For definiteness, let us suppose for the moment that that $M_{\tilde{g}} > M_{\tilde{q}}$. If the $\tilde{\gamma}$ is the LSP, then once scalar-quarks and gluinos are produced, they will decay via $\tilde{q} \to q\tilde{\gamma}$ and $\tilde{g} \to \bar{q}\tilde{q} \to q\bar{q}\tilde{\gamma}$. The photino escapes escapes and is interpreted as missing energy. There are two alternative possibilities. First, the LSP is not a pure photino, but it is a mixture of photino and other neutralino states. The only changes which occur are minor -some decay rates are changed due to the appearance of mixing angle factors. As long as the mixing angles are not unusually small, all the results we obtain in this paper are basically unchanged. The second possibility that the photino is not the LSP. This may or may not dramatically change our results depending on how the photino decays. For example, if the scalar-neutrino is the LSP, then $\tilde{\gamma} \to \nu + \tilde{\nu}$. But, both the ν and $\tilde{\nu}$ will not be observed, so the phenomenology will be identical to the case where the $\tilde{\gamma}$ is the LSP. One the other hand, if the LSP is a Higgsino $\tilde{H}$, the phenomenology can be vastly different. Then the $\tilde{\gamma}$ would decay dominantly via $\tilde{\gamma} \to \gamma + \tilde{H}$ thereby softening considerably the missing transverse energy of the events.

PRODUCTION OF SUPERPARTENERS

There are two kinds of ways to produce superpartners at a hadron collider.

A. Starting with quarks and gluons in the proton, we have

$$\sigma(p\bar{p} \to \tilde{a}\tilde{b}X) = \int dx_1 dx_2 F_c(x_1) F_{\bar{d}}(x_2) \hat{\sigma}(c\bar{d} \to \tilde{a}\tilde{b})$$

where F_c is the structure function for constituent c in a proton, F_d the structure function for constituent d in a proton, and $\hat{\sigma}$ the constituent cross section. Fig.1 shows a number of constituent processes to produce superpartners. Which ones dominate depends on masses, Kinematic regions, cuts, etc. It is important to include all subprocesses in the calculations. The results given below are based on calculation of all cross sections to produce $\tilde{q}\tilde{q}$, $\tilde{g}\tilde{g}$, $\tilde{g}\tilde{q}$, $\tilde{w}\tilde{g}$, $\tilde{g}\tilde{\gamma}$, etc., using the EHLQ structure functions for calculations,(2) with $\Lambda = 0.29$ GeV. Then events are generated by a Monte Carlo integration technique and final-state particles are decayed, using a procedure given in ref 3.

Ultimately an important test of whether supersymmetric partners are being produced will be the determination of the spins of the new particles. While it is clearly too early to be doing that, the spins do affect the distributions and rates so their effects are being included. At the present level of calculation no polarization effects are expected; transverse polarizations are absent since we are assuming parity conservation. Supersymmetric processes in general do not conserve parity, since masses of $\tilde{q}_L$ and $\tilde{q}_R$ are not in general equal. By including such mass differences longitudinal polarizations would be generated, but at the present time such effects are highly model dependent and we will neglect them.

B. If colored superpartners were light they would occur in hadrons as well as normal colored quarks and gluons. Since squarks are known to be heavier than about 20 GeV, they will not appear in significant quantities in a proton. The probability of a gluon splitting into a gluino pair is 6 times larger than the probability of it splitting into a quark pair of the same mass, so even if $M_{\tilde{g}}$ is 20-25 GeV there is a significant gluino content to the proton.

EXPERIMENTAL CUTS

In a real experimental environment it is necessary to apply cuts to the data in order to exclude conventional sources of anomalous events. The most important cut is on the minimum $\not{p}_T$ required to count the event as a monojet event. Data has been presented in various ways, with fixed cuts such as $\not{p}_T > 40$ GeV, and with a "4σ" cut. We will discuss how to implement these in the theory in the following section; it is subtle to make the "4σ" cut in a way where theory is being correctly compared to experiment. It is a difficult procedure and not all analyses yet agree, though convergence is occurring.

In addition, an anomalous event must have one jet with $p_T^j > 25$ GeV. Since $\not{p}_T$ and p_T^j are both large, clearly that biases all other activity in the event to be softer. A jet is defined as all activity in a bin

$\Delta\eta^2 + \Delta\phi^2 \leq 1$, where η is rapidity and ϕ is azimuthal angle.

Some geometrical cuts are applied, as described in the UA1 paper, and we have included those in our analysis.

THEORETICAL CUTS

In discussing the data it is useful to consider three different cuts on $\not{p}_T$, (a) $\not{p}_T$>40 GeV, (b) $\not{p}_T$>32 GeV, and (c) $\not{p}_T > 2.8\sqrt{E_T}$ (i.e. the 4σ cut). The first two are straightforward. The recently reported data is most complete for the $\not{p}_T$ > 40 GeV cut. The $\not{p}_T$ > 32 GeV results are presented as a comparison and as a prediction of how the number of monojet events changes as the cut is varied.

For the cuts on jet p_T we use the same values as the experimental analysis, p_T^j > 12 GeV for others. This is a subtle point, as it is not clear that a 25 GeV parton corresponds to a 25 GeV jet in the detector. Future analyses can improve on this.

The most subtle aspect of the analysis is defining the 4σ cut. The experimental missing energy fluctuations are characterized by a $\sigma = 0.7\sqrt{E_T}$, i.e. the amount of energy missing in an event goes as $\sqrt{E_T}$. To be safe in assigning an event to be new physics, it is required that $\not{p}_T$ be greater than 4σ. The problem is how to relate what happens in a theory event to E_T. So-called "minimum bias" collider events (those with no large p_T activity) have a distribution in E_T with a mean of $\langle E_T\rangle \simeq$ 20-25 GeV. Further, normal QCD jet-jet events are observed to have the property that when the two jets are removed, the remaining event has $\langle E_T\rangle \simeq$ 40-50 GeV.

Now suppose the monojets are due to the production of supersymmetric partners. Then presumably the total E_T is given by

$$E_T^{Tot} = E_T^o + E_T^{q,g}$$

where E_T^o is related to what is observed in QCD jet-jet events after the two jets are removed, and $E_T^{q,g}$ is the total energy associated with all of the partners that arise from decay of supersymmetric particles, whether they pass the jet cuts or not. We take E_T^o to have a mean of 40 GeV (using the lower side of the range because there is one less jet to radiate), and we use a distribution about E_T^o to make the simulation as realistic as possible(5).

This procedure gives a larger $E_T = E_T^{Tot}$ with which to calculate σ, so fewer events pass the cuts, and we do not find too many monojets arising from light gluinos that are pair produced. In this connection, it should be noted that as the gluino mass decreases it becomes less probable that a given $\tilde{g}\tilde{g}$ event passes the cuts, while the cross section for making the $\tilde{g}\tilde{g}$ pair grows. Since the probability of passing the cuts gets quite small ($\lesssim 10^{-3}$ when $\tilde{M}g \simeq$ 5 GeV) it is necessary to generate a large number of Monte Carlo events to have reliable results.

RESULTS FOR MONOJETS AND MISSING-ENERGY EVENTS

Our results(5) are the consequence of the full analysis described in previous sections. All supersymmetric processes for the production of scalar quarks and gluinos were calculated along with all possible decay channels, and in the end they were summed.

It is necessary to calculate rates for all combinations of $M_{\tilde{g}}$ and $M_{\tilde{q}}$. Detailed modeling of the new 1984 UA1(6) cuts and triggers is incorporated together with simulation of some detector resolutions and efficiencies. Among the cuts are those which eliminate "back-to-back" events. Fragmentation and gluon bremsstrahlung were accounted for. The integrations were done by Monte Carlo techniques, but there were a variety of analytic checks for every process.

THE 1984 DATA

Let us first summarize the newly reported 1984 data from the UA1 Collaboration.[6] They found 23 monojets with at least 15 GeV of missing transverse energy in their 1984 run. They identified 9 as having the characteristics of the $W \rightarrow \tau\nu(\tau \rightarrow \nu + \text{hadrons})$ source. Of the remaining 14 events, UA1 estimates that 6-8 events are due to background sources. We will take 13 events as the 90% confidence limit for the number of events which could represent new physics. Since their integrated luminosity in 1984 was about 270 nb^{-1}, the 90% confidence upper limit for monojet production due to new physics is then 4.8 events/100 nb^{-1}. This is actually quite conservative since (as discussed below) our distributions show that most of these events are unlikely to be from supersymmetric sources so that the real limit might be as low as perhaps 2 events/100nb^{-1}.

Theoretical calculations of monojet rates are subject to the fine distinction between monojets and dijets. It is better to use the combined monojet + dijet + multijet rate. The UA1 Collaboration reports that after the back-to-back cuts, 2 dijet and no multijet events remain from the 1984 run where they estimate backgrounds at 2 events. The total number of missing-energy events (after τ subtraction) is then 16 with backgrounds at 8-10 events. Suppose we again take 13 events to be the 90% confidence level limit giving the limit for the rate for missing-energy events with $E_T^{miss} > 15$ GeV to be 4.8 events/100 nb^{-1}, while the dijet rate's limit is 2 events/100 nb^{-1}. One can make a higher cut on E_T^{miss} in order to eliminate most backgrounds. Of the reported missing-energy events in the 1984 run, only 6 would pass an $E_T^{miss} > 40$ GeV cut. If we assume a background of 2 events, this leads to 8 events at the 90% confidence level or 3 events/100 nb^{-1}.

LIMITS ON SCALAR-QUARK AND GLUINO MASSES

Results(5) are summarized in the contour plots, Fig. 2. They should only be compared with the 1984 UA1 data. Let us momentarily ignore the region at very low gluino masses where rates are low due to fragmentation effects. From the monojet rate for $E_T^{miss} > 15$ GeV and the data described above, one can set the limits(5) $M_{\tilde{q}} > 50$-60 GeV depending on $M_{\tilde{g}}$ and $M_{\tilde{g}} > 45$-55 GeV depending on $M_{\tilde{q}}$. That is, if the squark and gluino masses were below these numbers, a detectable signal would have been seen.

From all missing energy events and the above data, we can set(5) $M_{\tilde{q}} > 60$-70 GeV depending on $M_{\tilde{g}}$ and $M_{\tilde{g}} > 50$-70 GeV depending on $M_{\tilde{q}}$. This significant improvement in the limits occurs because, for these large masses, supersymmetry predicts that dijet production should dominate over monojet production even with the back-to-back cuts. So these results combining all missing-energy events are both more limiting and more reliable (since they need make no distinction among numbers of jets).

If one wishes instead to assume that we can accurately separate monojets and dijets, then the above results suggest that it will be useful to examine the dijet rate separately. This is, in effect, done by subtracting Fig. 2a from Fig 2b. Using the above data we then find

$$M_{\tilde{q}} > \begin{cases} 65\ \text{GeV} & M_{\tilde{g}} \approx 150\ \text{GeV} \\ 75\ \text{GeV} & M_{\tilde{g}} \approx 80\ \text{GeV} \end{cases}$$

$$M_{\tilde{g}} > \begin{cases} 60\ \text{GeV} & M_{\tilde{q}} \approx 100\ \text{GeV} \\ 70\ \text{GeV} & M_{\tilde{q}} \approx 80\ \text{GeV} \end{cases} \qquad (7.1)$$

The limits quoted above would change by $\Delta M \approx 5$ GeV if the predictions were off by 50%.

There are a number of uncertainties which enter into the calculation, both from theoretical sources and experimental sources. Examples of the theoretical uncertainties are those associated with perturbative QCD. There are uncertainties introduced due to our lack of knowledge of the transverse-energy distribution of the remainder, i.e., that part of an event not included in the observed jets. Other sources of uncertainty such as smearing and modeling of UA1 cuts and triggers were also discussed above. In general, we expect the uncertainty is less than a factor of two. This implies an uncertaintly in the mass limits of roughly 5 GeV. However, in the particular case where the gluino is light (say $M_{\tilde{g}} \lesssim 20$ GeV), much more care must be given to the estimation of uncertainties.

ON THE QUESTION OF LIGHT GLUINOS

We have chosen to interpret the data in terms of limits. Before addressing the question of whether some of the observed events may actually be due to supersymmetry, it will be useful to return to the subject of the very light gluino. The predicted event rates drop of as $M_{\tilde{g}}$ becomes very small. Very light gluinos lose much of their energy due to fragmentation and gluon bremsstrahlung, and therefore they lead to very little missing energy. As a result very few pass the E_T^{miss} cuts.

The calculations for very low mass gluinos are subject to much larger uncertainties due to fragmentation and to the surviving events being on the tails of the E_T^{miss} and E_T^{jet} distributions. We would predict for $M_q \approx 100$ GeV 26 events/100 nb^{-1} for a 5 GeV gluino and 13 events/100nb^{-1} for a 3 GeV gluino (these are all monojets; dijet production is negligible). While these numbers are much larger than the 4.8 events limit, one cannot neglect the uncertainties intrinsic to theoretical calculations for light gluinos. These uncertainties could be as much as a factor of 4 or 5 if added linearly. In spite of this large uncertainty, the predicted event rate is large enough to conclude that $M_{\tilde{g}} = 5$ GeV is ruled out and that $M_{\tilde{g}} = 3$ GeV is very marginal. If the photino mass is nonzero the photino would carry off even more energy, and our results would be strenghtened.

An additional input on this subject comes from the beam-dump experiments. A recent BEBC experiment(7) gets the limits $M_{\tilde{g}} > 3$-4 GeV at the 90% confidence level (depending on the value of $M_{\tilde{q}}$). Therefore, what some authors have referred to as a "window" allowing light gluinos, is at best a "peephole," and mostly likely is ruled out.

This conclusion could not be reached in earlier papers analyzing the 1983 data for two primary reasons. The new missing-energy trigger in the 1984 run is extremely important for $\tilde{g}\tilde{g}$ production when $M_g \lesssim 10$ GeV. It raises predictions in this case by an order-of-magnitude, whle experimentally this trigger does not dramatically change the observed rate. The calculation of the $\tilde{g}\tilde{g}g$ process by Herzog and Kunszt(8) was not available for the earlier analyses, and it increases the predictions for 5 GeV gluinos by a factor of 3. While other recent refinements bring down the rate a little, the end result is that because of the higher rates predicted, it is now possible (or almost possible) to rule out light gluinos.

SUPERSYMMETRY-LOST OR FOUND?

Could some of the observed monojet events be due to the production of gluinos or scalar quarks? The UA1 Collaboration(6) cannot rule out the possibility that 6-8 of the monojets come from new physics. There are, however, two factors which argue against the monojets coming from gluino or scalar quark production. Both are consequences of the fact that the appropriate event rate (2-3 monojets/100 nb^{-1}) only occurs for large $M_{\tilde{q}}$ or $M_{\tilde{g}}$ ($\approx$ 60 GeV). For such masses we would predict[5] 4-6 dijets/100nb^{-1}, and these certainly have not been observed. Furthermore, at these masses one would expect significant numbers of monojets with $E_T^{miss} > 45$ GeV, and only one was observed in the 1984 run.

The two observed dijet events (surviving the back-to-back cuts) in the 1984 run have $E_T^{miss} \geqslant 55$ GeV. Although there is a roughly equal background expected, these backgrounds are unlikely to have so much E_T^{miss}. A 70-90 GeV scalar quark could give dijets with such characteristics and with this rate, and would produce very few monojets. Clearly, however, such speculation must await considerably more statistics.

IMPLICATIONS OF MINIMAL LOW-ENERGY SUPERGRAVITY MODELS

Throughout this discussion we have treated the gluino and scalar-quark masses as independent parameters. Furthermore, we took the photino to be the LSP and assumed that its mass could be neglected. If we are willing to adopt a particular approach to low-energy supergravity model building, we can constraing certain parts of the $M_{\tilde{q}}$ - $M_{\tilde{g}}$ plane.

For illustration purposes, let us consider a class of models[9] which have been often referred to as being "minimal" supergravity models. These models are "minimal" in two respects. First, they consist of the minimal number of particles: the Standard Model particles with two Higgs doublet and their supersymmetric partners. Second, these models depend on a minimal set of parameters: the gravitino mass($m_{3/2}$), the gluino mass $M_{\tilde{g}}$, μ (a supersymmetric Higgsino mass), A (a parameter related to the super-Higgs mechanism), and v_2/v_1 (the ratio of vacuum expectation values of the two Higgs fields). These models are obtained in two steps. First, starting with an $SU(3) \times SU(2) \times U(1)$ (or grand unified) gauge theory coupled to supergravity, an effective renormalizable field theory at the Planck mass (M_P) is obtained in the limit of $M_P \to \infty$. Second, renormalization group equations are used in order that we may obtain the effective theory at the Planck scale does not break the $SU(2) \times U(1)$ gauge symmetry. Thus, in order that the $SU(2) \times U(1)$ electroweak group be broken in the standard way in the low-energy effective theory (such that $SU(3)_{color} \times U(1)_{EM}$ remain conserved), it must happen that the parameters of the theory, which evolve via the renormalization group, satisfy an approprite set of conditions. These conditions are quite restrictive and tend to reduce much of the freedom in the choice of parameters of the model. For example, if we assume that $m_t \approx 40$ GeV (as may be indicated by recent results of UA1), then it follows that $v_1 \approx v_2$ and $B\mu m_{3/2} \sim O(m_{3/2}^2)$, where $B \equiv A - 1$ at the Planck scale and is a number of order unity. The renormalization group analysis of such models has been performed, and the following results have been obtained:

$$M_{\tilde{\gamma}} \approx \frac{1}{6} M_{\tilde{g}}$$

$$M^2_{\tilde{q}_L} = m^2_{3/2} + C_{\tilde{q}_L} M^2_{\tilde{g}} + m_Z^2(T_{3q} - e_q \sin^2\theta_w)\cos 2\beta$$

$$M^2_{\tilde{q}_R} = m^2_{3/2} + C_{\tilde{q}_R} M^2_{\tilde{g}} + m_Z^2 e_q \sin^2\theta_w \cos 2\beta$$

$$M^2_{\tilde{\ell}_L} = m^2_{3/2} + C_{\tilde{\ell}_L} M^2_{\tilde{g}} + m_Z^2(T_{3\ell} - e_\ell \sin^2\theta_w)\cos 2\beta$$

$$M^2_{\tilde{\ell}_R} = m^2_{3/2} + C_{\tilde{\ell}_R} M^2_{\tilde{g}} + m^2_Z e_\ell \sin^2\theta_W \cos 2\beta$$

where $\tan\beta \equiv v_2/v_1$ and the constants are given by: $C_{\tilde{q}_L} = 0.85$, $C_{\tilde{\ell}_L} = 0.08$, and $C_{\tilde{\ell}_R} = 0.02$. T_{3i} and e_i ($i = q,\ell$) are the weak isospin and electric charge (in units of e) of the quarks and leptons respectively. Strictly speaking, these equations must be modified somewhat for the third generation; we refer the reader to ref. 9 for details. In models where the gluino mass is not too large, the LSP is approximately a pure photino with the mass shown above. We noted above that in models where m_t is not too heavy, it follows that $v_1 \approx v_2$. This in turn implies that $\cos 2\beta \approx 0$ and therefore the scalar-quark and scalar-lepton masses depend on two parameters. It follows that

$$M^2_{\tilde{\ell}_R} \approx M^2_{\tilde{q}} - 0.8\, M^2_{\tilde{g}} \quad .$$

This is an interesting constraint, in that it implies that the gluino cannot be much heavier than the scalar quark.

Another constraint can be obtained by considering the cosmological implications of a light photino. In particular, since photinos are the LSP (and thus stable), their annihilation rate must be sufficiently efficient to reduce their abundance in the early universe to a cosmologically acceptable level. (Another cosmologically acceptable solution -to have the photino nearly massless like the neutrino -is unacceptable since it would imply a nearly massless gluino, which is almost certainly ruled out.) A calculation of Ellis et al.[10] shows that $M_{\tilde{\gamma}} \gtrsim 0.5$ GeV if $M_{\tilde{q}} \gtrsim 20$ GeV and $M_{\tilde{\gamma}} \gtrsim 5$ GeV if $M_{\tilde{q}} \gtrsim 100$ GeV. (The efficiency of photino annihilation decreases as the scalar-quark mass increases.) This leads to a lower limit on the gluino mass as a function of the scalar-quark mass.

The two constraints discussed above substantially limit the region of the $M_{\tilde{q}}$ - $M_{\tilde{g}}$ plane which is consistent with the minimal low-energy supergravity model above.

AN ALTERNATIVE: THE LIGHT HGGSINO

It would be misleading to finish the discussion of low-energy supergravity models without indicating the possibility of other scenarios. We have emphasized earlier that many conclusions (and the strict mass limits) depend on the assumption that the photino is the LSP. It is of interest to consider whether it is possible to construct models where this assumption is not valid and what the implications are for another candidate for the LSP.

Suppose the LSP is a light Higgsino. In the minimal model discussed above, the Higgsino mass turns out to be $M_{\tilde{H}} \sim \mu \sim O(m_{3/2})$. Hence, unless the gluino mass is large enough (implying a large value for $M_{\tilde{\gamma}}$), the Higgsino will not be the LSP. However, as shown in ref. 11 one can easily generalize the minimal model in such a way that the parameter μ is not constrainted to be $O(m_{3/2})$. In such a model, the Higgsino will be the LSP as long as $\mu < M_{\tilde{\gamma}}$.

Let us summarize some of the phenomenological implications of this alternative scenario. First, the photino will tend to be the second lightest supersymmetric particle, and hence only two decay channels are available: $\tilde{\gamma} \to f\bar{f}\tilde{H}$ or $\tilde{\gamma} \to \gamma\tilde{H}$. The latter decay occurs via a one-loop Feynman diagram. The three-body tree-level decay of the photino occurs via the exchange of a virtual $\tilde{f}$. Because the $\tilde{H}f\tilde{f}$ vertex is proportional to the mass of fermion (m_f), this decay rate is negligible and the decay $\tilde{\gamma} \to \gamma\tilde{H}$ is the dominant one. Calculations of the two-body decay yield approximately:

$$\tau_{\tilde{\gamma}} \approx 10^{-13} \text{ sec. } \left(\frac{1 \text{ GeV}}{M_{\tilde{\gamma}}}\right)^3$$

which indicates that the photino decay is prompt unless $M_{\tilde{\gamma}}$ is sufficiently light. Note that because the photino is now unstable, the cosmological limits for the photino obtained by Ellis et al. no longer apply. Instead, one now finds cosmological limits on the Higgsino mass: either $M_{\tilde{H}} \lesssim 100$ eV or $M_{\tilde{H}} \gtrsim m_b$. Since the Higgsino and gluino masses are logically independent, there is no phenomenological reason which rules out a massless Higgsino.

Scalar-quark and gluino decays are, however, unchanged. In principle, one could have $\tilde{g} \to q\bar{q}\tilde{H}$ and $\tilde{q} \to q\tilde{H}$. However, as stated above, the $\tilde{H}\tilde{f}f$ vertex is proportional to m_f and hence these decay rates can be neglected as compared with the standard ones involving the photino. Two cases can be envisioned. If the photino is long lived then it will escape the collider detectors, and there is no change in any of the results obtained. However, if the photino decays promptly, $\tilde{\gamma} \to \gamma\tilde{H}$, then the phenomenology changes drastically. First, the limits on supersymmetric masses obtained above are significantly weakened. We may obtain an estimate of the new limits by noting that Dawson[12] has investigated the implication of missing energy events for supersymmetric models which violate R-parity. In these models, the photino is the LSP but is unstable and decays via $\tilde{\gamma} \to \gamma\nu$. Thus, the signature is identical to the case we are considering here, so we may use her results. Dawson finds (see Fig. 11 of ref. 12) that the number of events which pass the UA1 cuts and triggers is roughly a factor of five less than for a stable photino, approximately independent of $M_{\tilde{q}}$ and $M_{\tilde{g}}$. Then the allowed regions are

$$M_{\tilde{g}} \lesssim 5 \text{ GeV} \quad \text{or} \quad M_{\tilde{g}} \gtrsim 40 \text{ GeV}$$

$$M_{\tilde{q}} \gtrsim 45\text{-}60 \text{ GeV}$$

The better limit is obtained as we take the gluino mass approaching the scalar-quark mass. Note that the "light-gluino window" has returned. That is , we no longer feel able to rule out gluino masses of order $3 \lesssim M_{\tilde{g}} \lesssim 5$ GeV since the number of predicted events passing the UA1 cuts has been significnatly reduced.

Suppose we accept the possibility that a few of the monojets could be due to supersymmetry where the Higgsino is the LSP. As argued above, when scalar quarks and gluinos are produced, they decay into photinos which subsequently decay into Higgsinos: $\tilde{\gamma} \to \gamma\tilde{H}$. Thus, these events should contain photons! Can this be ruled out? At present, the answer seems to be negative. The photon could not be easily distinguished from a π^0 in the UA1 detector, so these events would just exhibit extra observed neutral energy. A signature which could confirm or exclude such a model is the presence of events where one photino gives a hard photon plus a soft $\tilde{H}$, and the other photino gives a soft photon and a hard $\tilde{H}$, so the full event has large E_T^{miss}, an isolated hard photon and one or more jets.

Note, however, that some limits do exist because $e^+e^- \to \tilde{\gamma}\tilde{\gamma}$, $\tilde{\gamma} \to \gamma\tilde{H}$ can take place yielding $e^+e^- \to \gamma\gamma$ + missing energy. Such a process has been searched for at PETRA; no events of this type above background have been seen.(28) This implies that the cross-section for $e^+e^- \to \tilde{\gamma}\tilde{\gamma}$ cannot be too large. Since this process occurs via exchange of a scalar-electron, the absence of this process (assuming an unstable photino which decays radiatively) puts a limit on the scalar-electrom:

$M_{\tilde{e}} \gtrsim 100$ GeV. This implies that $M_{\tilde{q}} \gtrsim 100$ GeV. Thus, in the case where the Higgsino is the LSP, supersymmetry cannot be the explanation for monojets unless the gluino is very light, $M_{\tilde{g}} \lesssim 5$ GeV.

If the sneutrino is the LSP, photinos again decay, $\tilde{\gamma} \to \nu\tilde{\nu}$ at one loop. Since ν and $\tilde{\nu}$ escape collider detectors, in this case there is no change in the conclusions of the analysis where photinos are the LSP.

Now that we have examined the present situation in the experimental search for supersymmetry, and seen how the various possibilities for LSP can affect the different experiments, we turn to considering the dark matter question. There we will be more general and include other possibilities as well as the LSP.

DARK MATTER

The LSP would have been in equilibrium with other particles in the early universe. Its interactions with normal particles are usually mediated either by Z^0's and W's, or by other heavy particles, so they are often quite small. Consequently, a large number of such particles would have survived from the early universe, as non-luminous matter which constitutes a significant fraction of the mass of the universe.

It should be emphasized that such dark matter would generally be hypothesized to exist on the basis of extensions of particle physics beyond the Standard Model. We approach the subject from the particle physics point of view, studying how such objects interact, what effects they might have, and how they might be detected. Even if there had been no hint of dark matter from astronomy experiments, the subject would have emerged from the present direction of particle physics; it is even more exciting that indications(13) of the existence of some form(s) of dark matter come from astronomy and cosmology.

The kinds of effects one can hope for depend significantly on numbers. Consequently, generic analyses or cross sections given by dimensional arguments are not precise enough, especially since cancellations often occur (e.g. in the photino-nucleon cross section). Further, as we have seen above, strong new lower limits exist on masses of supersymmetric partners that participate in some processes, so some cross sections are smaller than they might have been.

In this section we give detailed results from ref 14 for two types of particles,

(i) heavy neutrinos coupled as normal sequential ones, and

(ii) possibilities in a supersymmetric theory, where the lightest supersymmetric particle (LSP) is taken to be stable. The supersymmetric candidates for LSP are photino, scalar-neutrino (sneutrino), and higgsino, the appropriate partner of the Higgs boson. Some references which consider supersymmetric partners as dark matter include refs 14-22.

We briefly consider effects such dark matter might have when concentrated in the galaxy or the sun, and possible experimental approaches to directly detect dark matter with GeV masses. We do not consider dark matter in the form of axions, or any of the relevant information from the field of galaxy formation.

ANNIHILATION CROSS SECTIONS

The basic annihilation mechanisms are of course characteristic of the theory. Heavy neutrinos that are assumed to be in $SU(2)_L$ doublets can annihilate through a Z^0 (Fig. 3a), or via a (presumably Cabibbo suppressed) $W^{\pm}$ exchange (Fig. 3b) to a lighter charged lepton. We consider only the case where the latter mechanism is numerically negligible, as would be expected in most models.

The LSP annihilation depends strongly on what the LSP is. For photinos the sfermion exchange dominates, with f = any quark or lepton, as in Fig. 4. For higgsinos there are two contributions, Fig. 5a and Fig. 5b. Fig. 5a in general dominates since $\tilde{h}$ carries electroweak quantum number $T_3 \neq 0$; the $\tilde{h}\tilde{h}Z^\circ$ coupling is proportional to $1-v_2/v_1$, where v_1 and v_2 are the two vacuum expectation values required in a supersymmetric theory, so in models where $v_2 \approx v_1$ this mechanism can be quite small. Fig. 5b involves $\tilde{h}\tilde{f}f$ couplings with factors $m_f/2m_W$, so the rates are suppressed by $(m_f/2m_W)^4$ and are usually quite small; we neglect contributions of Fig. 5b. Sneutrinos do have normal electroweak quantum numbers and can annihilate through a Z°, but the dominant mechanism is the exchange of a Majorana zino as in Fig.6; note the annihilation is[17] $\tilde{\nu}\tilde{\nu} \to \nu\nu$, not $\tilde{\nu}\bar{\tilde{\nu}} \to \nu\bar{\nu}$. This is because the Majorana nature of $\tilde{Z}$ gives a factor of $1/m_{\tilde{Z}}$ in the amplitude (rather than $1/m_{\tilde{Z}}^2$), so the cross section only has one G_F rather than the normal G_F^2. Lepton number is not violated since $\tilde{\nu}$ and ν both carry the same lepton number, number, presumably a different one for each flavor. Fermion number does change, which is all right since the Majorana nature of Z allows that to happen, but that is not observable if equal numbers of $\tilde{\nu}$ and $\bar{\tilde{\nu}}$ exist since $\bar{\tilde{\nu}}\bar{\tilde{\nu}} \to \bar{\nu}\bar{\nu}$ occurs equally often.

To encompass all the above cases, and to present some techniques generally, we denote a general stable dark matter particle by χ. Whenever necessary, a $\bar{\chi}$ will be understood if needed.

An additional annihilation effect is present for the masses of interest, in the GeV region.[14] Whenever $2m_\chi = m_x$, where x is a vector meson resonance (x = ρ, ω, ρ , ψ, ϒ), the rate for $\chi\chi \to Z \to \bar{f}f$ is enhanced, e.g. for Fig. 3a, Fig. 5a, as shown in Fig. 7. Then an analysis following Lee and Weinberg[24] gives a density of dark matter candidates which is shown for heavy neutrinos in Fig. 8 and for higgsinos in Fig. 9. We see that the usual result is obtained that masses above some limit are allowed, but isolated masses below that limit are also allowed. For example, heavy neutrinos of mass 1.55 GeV are allowed, as are photinos of mass 275 MeV. The vector meson resonances decay dominantly into pions and kaons.

A similar effect occurs[14] whenever the annihilation is into final state quarks, e.g. in Fig. 3b, or Fig. 4, as shown in Fig. 10,. Then the quarks can resonate. The rate gets suppressed by a factor of the wave function of the x resonance at the origin, but enhanced by M_x/Γ_x, and gives a large enough annihilation cross section to reduce the χ density below the critical density, so again certain isolated masses are allowed. The results are shown in Fig. 11 for photinos.

For heavy neutrinos, note[14] that $m_{\nu H} > 3.3$ GeV is required from the continuum curve rather than the "2 GeV" usually quoted for the Lee-Weinberg limit. That is because we use the neutral current annihilation that a sequential ν_H would naively be expected to have, rather than the generic charged current cross section of ref. 23.

The annihilation cross sections are useful in (at least) two situations. (1) They allow a calculation of the density of any given dark matter candidate as we have seen. (2) As discussed in refs. 18, 19, the annihilation products include photons, antiprotons, positrons, and neutrinos that may be detectable under appropriate conditions. A further point that should be emphaiszed is that a given dark matter candidate leads to characteristic numbers of the various detectable particles. Most obviously, for example, sneutrinos give essentially only monoenergetic neutrinos in their annihilation; if the $\overline{p}$'s of ref. 24 are cosmological, then sneutrinos cannot be the dark matter. For other candidates the $\overline{p}$ fluxes differ, and the ratios of positron to $\overline{p}$ fluxes differ. So if any signal is detected, it will be possible to decide experimentally what is being observed.

Scattering Cross Sections

A similar situation holds for energy transfer by scattering. For every dark matter candidate, the cross section on quarks and leptons is calculable. It may depend on basic (not yet measured or calculated) parameters of the theory, such as the ratio of the two vacuum expectation values in a supersymmetric theory. For practical consequences it is necessary to calculate the cross sections on a proton and a neutron. That is somewhat subtle(14) since cancellations occur, and the naive quark model does not work well here.

Goodman and Witten(20) have discussed the basic ideas about using energy transferred in a scattering process to detect the presence of dark matter. We add here various considerations, including an additional cross section, for higgsinos, and a more realistic calculation of all dark matter cross sections on protons and neutrons (which, unfortunately, suppresses the cross sections by a factor which can be as much as five).

Any dark matter candidate with a mass M_{DM} on the GeV scale can transfer an energy of the order of $(2M_{DM}v)^2/2M_N$ in a collision with a nucleus N (which can be a proton or a neutron or a heavier nucleus) at relative velocity v. In the simplest view one can imagine we are moving with a velocity $v \simeq$ 230Km/sec with respect to the galaxy, while the dark matter, of mass M_{DM}, is at rest. In practice there will be a spectrum of velocities, and directional considerations, which have not yet been discussed. If the above energy transfer can be of order 100 eV, in principle detectors should exist to observe the collisions.

Next it is necessary to calculate the cross sections to determine whether the probability of a collision is large enough. The collisions on a quark or lepton (fermion, f) are as shown in Fig. 12; they can either proceed via a neutral current interaction, exchanging a Z°, or exchange of a new object such as a scalar fermion in supersymmetry. More subtle is the way cross sections on quarks are combined to give those on protons or neutrons (which can then be combined further to give the cross section on any nucleus). There is one subtlety.[25] Using quark model arguments does not work well for the axial vector isosinglet. For example, the matrix element $\langle p|\overline{u}\vec{\gamma}\gamma_5 u + \overline{d}\vec{\gamma}\gamma_5 d|p\rangle$ is is given by $2\langle p|\vec{s}|p\rangle$ in the naive quark model. If the D and F SU(3) coefficients are explicitly written, the coefficient of $2\langle p|\vec{s}|p\rangle$ is 3F-D. The naive quark model sets 3F-D = 1, while experimentally it is about 0.45, which enters squared in rates. Keeping track of this effect, we arrive at the cross sections in Table II, which we believe can be used as a reliable basis for considering experiments. The resulting cross sections are small, but may be large enough to permit workable detectors to be constructed.

TABLE II

Scattering cross sections for dark matter candidates on protons (p) and neutrons (n) are given. The table gives the coefficient λ in $\sigma = \lambda G_F^2 M_{DM}^2 M_P^2/1.45\pi(M_{DM} + M_p)^2 \simeq 1.16\,\lambda 10^{-38} M_{DM}^2/(M_{DM} + M_p)^2$ cm. The numbers assume all heavy superpartner masses are given by M_W. Larger cross sections will not be likely for supersymmetric partners since lower limits of about 70 GeV already exist for squarks. The quantity Δ in the higgsino entry depends on the presently unknown ratio of vacuum expectation values in a supersymmetric theory, $\Delta = (1-v_1^2/v_2^2)/(1+v_1^2/v_2^2) = \cos 2\beta$; Δ can range from a rather small number of order 0.01-0.02, up to about 0.5.

	p	n
ν_H	1.7	1.7
$\tilde{\gamma}$	0.7	0.05
$\tilde{\nu}$	0.01	0.7
$\tilde{h}$	3.4Δ	3.4Δ

The event rate is given by R = flux x σ x K, K = number of nuclei in the the target. The flux is $\rho v/M_{DM}$. The cross section σ is that on a nucleus, which presumably scales as the nucleon number A. Then the resulting rate is

$$R \simeq 6\left(\frac{2\ \text{GeV}}{M_{DM}}\right)\left(\frac{\rho}{10^{-24}\,\text{gm/cm}^3}\right)\left(\frac{v}{230\text{km/sec}}\right)\left(\frac{\sigma}{A\times10^{-38}\text{cm}^2}\right)\ \text{events/kg/day.}$$

This may be a large enough rate to be encouraging.

One very important thing to note from Table II is how different the ratio of cross section on neutrons and cross section on protons can be, and how the ratio is characteristic of different dark matter candidates. As a result, it is probable that if a signal is ever found for dark matter it will be possible to determine whether it can be any of the proposed candidates, and if not what kind of particle is required.

CURRENT PARTICLE PHYSICS AND THE LSP

As dicusssed in the introduction, if nature were supersymmetric on the scale of electroweak interactions, the lightest supersymmetric particle (LSP) may be stable and would then necessarily provide some dark matter. As shown in Figs. 8, 9, 11, for large ranges of masses the LSP contribution to dark matter gives $\Omega_{LSP} \sim 1$.

If evidence for supersymmetry is found in particle physics experiments, presumably the mass of the LSP will be directly measured, and Ω_{LSP} can be calculated. When particle theories are better understood perhaps the masses of superpartners (or the equivalent particles in a different theory) can be calculated and therefore Ω_{LSP} can be predicted.

In fact, it is important to note that particle theories are likely to naturally contain more than one form of dark matter. For example, a supersymmetric grand unified theory will probably contain a LSP with mass of a few GeV, it will probably contain (3) massive neutrinos with masses in the eV or KeV range, and Goldstone bosons (axions) that get some mass from non-perturbative effects and are needed because of the strong CP problem. While the masses and the fractions of Ω_{cr} in each kind of dark matter might seem arbitrary, presumably they are determined by the theory and the relative amounts will be in principle calculable. Most likely the fraction of Ω that is baryonic is equally calculable and related to the rest, as it should be determined by the same underlying theory.

Of course, we are not quite at that stage yet. Good models exist where $\tilde{\gamma}$, $\tilde{h}$, or $\tilde{\nu}$ is the LSP. There are some restrictions from data. As we saw above, if $\tilde{\gamma}$ or $\tilde{\nu}$ = LSP, the UA1 data implies $m_{\tilde{q}} \gtrsim 70$ GeV and $m_{\tilde{g}} \gtrsim 65$ GeV. The ASP data[26] implies a lower limit on $m_{\tilde{e}}$ of about 50 GeV for $m_{\tilde{\gamma}} = 0$, decreasing to no limit as $m_{\tilde{\gamma}}$ increases to 12 GeV. If

$\tilde{h}$ = LSP, photinos decay to $\gamma\tilde{h}$, so the above limits do not hold; data on $e^+e^- \to \gamma\gamma$ + missing momentum imply(27) $m_{\tilde{e}} \gtrsim 100$ GeV, and the UA1 data allows a signal to be present with $m_{\tilde{g}} \simeq 5$ GeV, $m_{\tilde{q}} \simeq 110$ GeV, or requires $m_{\tilde{q}} > 50$ GeV, $m_{\tilde{g}} \gtrsim 45$ GeV.

These numbers already have some consequences for dark matter candidates. If, for example, the $\bar{p}/p$ ratio of ref. 24 turns out to be cosmological in origin, and therefore presumably due to dark matter, then

(i) obviously $\tilde{\nu} \neq$ LSP since $\tilde{\nu}$ mainly annihilates to ν;

(ii) it cannot be arranged that $\tilde{\gamma}$ = LSP and that $\Omega_{\tilde{\gamma}} \simeq 1$ since having $m_{\tilde{q}} \gtrsim 70$ GeV reduces the $\tilde{\gamma}$ annihilation cross section enough that $m_{\tilde{\gamma}} \gtrsim m_b$, and the $\bar{p}$ flux from $\tilde{\gamma}$ annihilation is not sufficient to give the 10^{-4} ratio.

Whatever the dark matter, it can be seen from Table II that the cross sections for WIMP's concentrated in the sun to transfer energy energy away from the core and lower its temperature are about 10^{-38}cm^2, which are about two orders of magnitude less than those remarked on as optimal in ref. 28. Whether sufficiently large concentrations can be achieved to overcome these small cross sections is not clear to us at present.

Acknowledgement

I am grateful for my collaborations with H. Haber, R.M. Barnett, and Iraj Kani. I appreciate very much discussions with them and with G. Altarelli, R. Arnowitt, V. Barger, A. de Rujula, J. Ellis, S. Ellis, J. Gunion, F. Paige, F. Yndurian, K. Eggert, A. Honma, A. Kernan, E. Radermacher, J. Rohlf, D. Hegyi, and G. Tarlé. I am grateful to A. Salam, J. Pati, and D. Sciama for discussions and hospitality at the school.

REFERENCES

1. For a recent review similar to the approach here, and exclusive references, see H.E. Haber and G.L. Kane, Phys. Rep. 117 (1985) 75.

2. E. Eichten, I. Hinchliffe, K. Lane, and C. Quigg, Rev. Mod. Phys. 56 (1984) 579.

3. R.M. Barnett and H.E. Haber, Proceedings of the 1984 Division of Particles and Fields Study on the Uhlijaha of the SSC, Snowmass, Co, ed. R. Donaldson and J. Morfin.

4. G. Arnison et al., Phys. Lett. 139B (1984) 115.

5. R.M. Barnett, H.E. Haber, and G.L. Kane, LBL-20102, August 1985.

6. J. Rohlf, invited talk at the 1985 Division of Particles and Fields Conference, Eugene, Oregon, August 1985; C. Rubbia, invited talk at the 1985 Lepton-Photon Conference, Kyoto, Japan, August 1985.

7. A.M. Cooper et al., CERN/EP85-97.

8. F. Herzog and Z. Kunszt, Phys. Lett. 157B (1985) 430.

9. For a review of such models H.P. Niles, Phys. Rep. 110C (1984) 1; P. Nath, R. Arnowitt, and A.H. Chamseddine, "Applied N=1 Supergravity," ICTP Series in Theoretical Physics, Vol. I, (World Scientific, Singapore, 1985); C. Kounnas, A.B. Lahanas, D.V. Nanopoulos and M. Quiras, Phys. Lett 132B (1983) 95; Nucl. Phys. B236 (1984) 438.

10. J. Ellis, J.S. Hagelin, D.V. Nanopoulos, K. Olive and M. Srednicki, Nucl. Phys B238 (1984) 453.

11. H.E. Haber, G.L. Kane and M. Quiros, University of Michigan preprints U TH-85-8, UM TH 85-12 (1985).

12. S. Dawson, Berkeley preprint LBL-19460 (1985).

13. For recent reviews see Joel R. Primack, "Dark Matter, Galaxies, and Large Scale Structure in the Universe", Lectures at the International School of Physics "Enrico Fermi", Varenna Italy, June 1984 SLAC-Pub 3387; Keith A. Olive and David N. Schramm, "The Cosmology/Particle Physics Interface", Fermilab-Pub-85/113-A, August 1985.

14. G.L. Kane and Iraj Kani, Michigan preprint UM TH 85-20.

15. D. Sciamia, Phys. Lett. 114B (1982) 19; Phys Lett 137B (1984) 169.

16. H. Goldberg, Phys. Rev. Lett. 50 (1983) 1419; John Ellis, J.S. Hagelin, D.V. Nanopoulos, K. Olive, and M. Srednicki, Nucl. Phys. B238 453.

17. J.S. Hagelin, G.L. Kane, and S. Raby, Nuc. Phys. B241 (1984) 638.

18. J. Silk and M. Srednicki, Phys. Rev. Lett. 53 (1984) 624.

19. John S. Hagelin and G.L. Kane, to be pubished in Nuc. Phys.

20. Mark W. Goodman and Edward Witten, Phys. REv. D31 (1985) 3059.

21. Katherine Freeze, "Can Scalar Neutrinos or Massive Dirac Neutrinos be the Missing Mass", Harvard-Smithsonian Center for Astrophysics preprint, June 1985.

22. F.W. Stecker, S. Rudaz, and T.F. Walsh, "Galactic Antiprotons from Phohnos", Univ. of Minn. preprint UMN-TH-520/85.

23. B.W. Lee and S. Weinberg, Phys Rev. Lett 39 (1977) 165. See also P. Hut, Phys. Lett. 69B (1977) 85, and Jeremy Bernstein, Lowell S. Brown, and Gerald Feinberg, "The Cosmological Heavy Neutrino Problem Revisited", CU-TP-316.

24. A. Buffington et al., AP. J. 248 (1981) 1179.

25. C. Cahn and G.L. Kane, Phys. Lett. 71B (1977) 348.

26. R. Wilson, Talk at the DPF annual meeting, Eugene, Oregon, August 1985; D. Burke, private communication.

27. E. Ros, Talk at the 13th International Winter Meeting on Fundamental Physics, Cuenca, Spain, April 1985; Sau-lan Wu, Invited talk at the DPF annual meeting, Eugene Oregon, August 1985.

28. David N. Spergal and William H. Press, Ap. J. 294 (1985) 663.

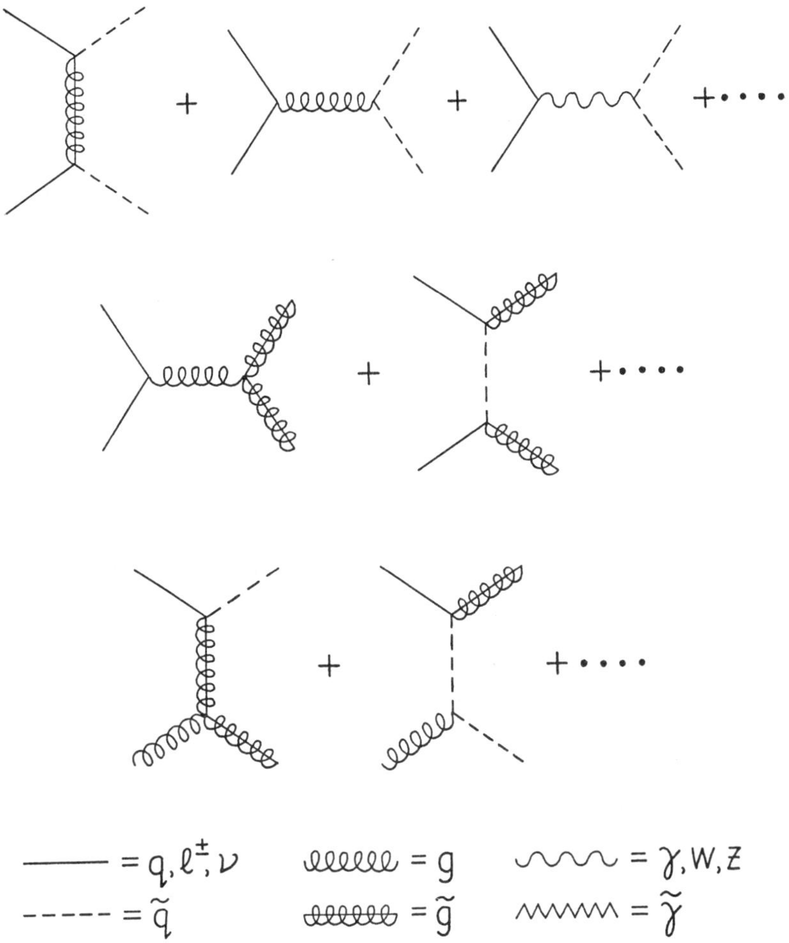

Fig. 1 Ways to produce superpartners are shown.

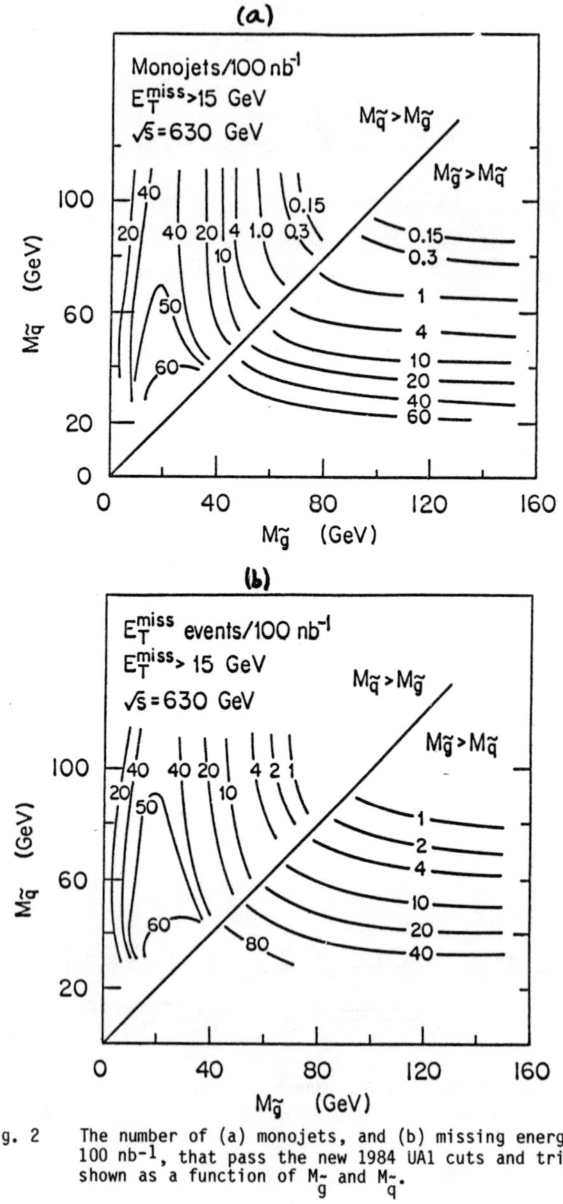

Fig. 2 The number of (a) monojets, and (b) missing energy events, per 100 nb^{-1}, that pass the new 1984 UA1 cuts and triggers are shown as a function of $M_{\tilde{g}}$ and $M_{\tilde{q}}$.

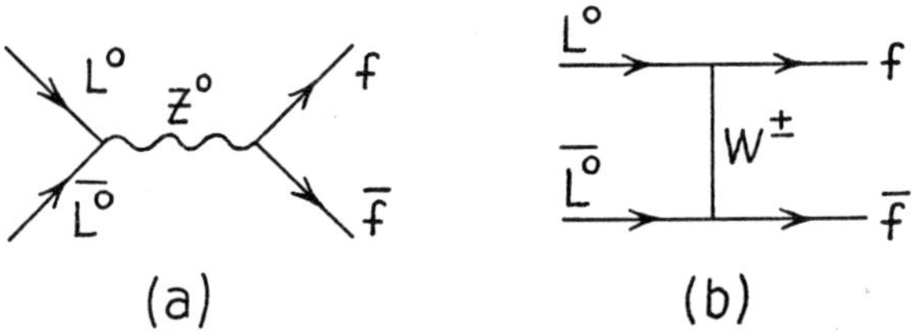

Fig. 3 Massive neutrinos can annihilate via these processes. Diagram b is probably suppressed by mixing angle factors.

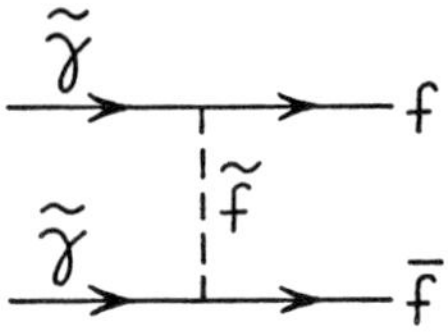

Fig. 4 Photinos annihilate mainly by this process.

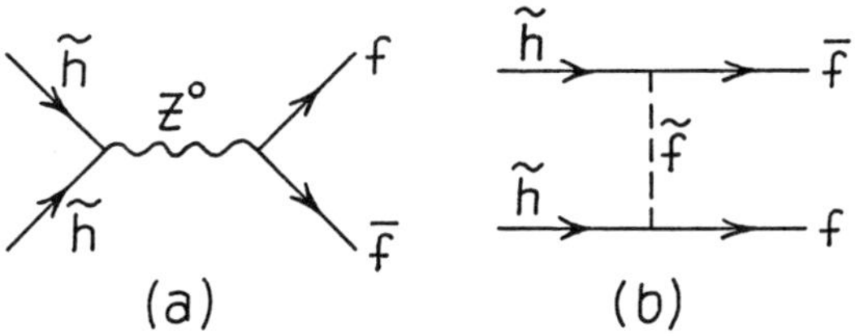

Fig. 5 Higgsino's annihilate by these processes. (a) is proportional to $(1-v_2/v_1)^2$ and (b) is proportional to fermion masses squared.

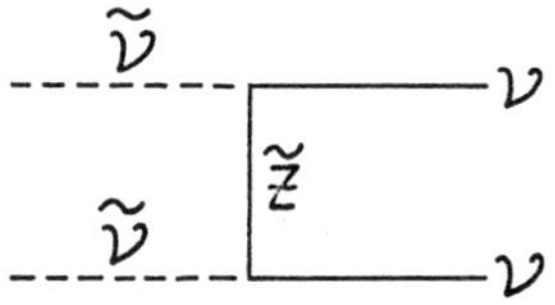

Fig. 6 Sneutrinos annihilate mainly via Majorana zino exchange if it is not suppressed, giving a rate independent of $m_{\tilde{\nu}}$.

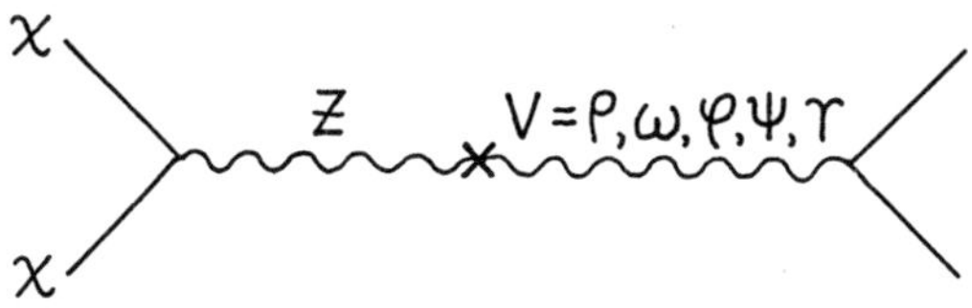

Fig. 7 When $2m_\chi \approx m_x$, any dark matter particle χ can have enhanced annihilation into a vector meson x.

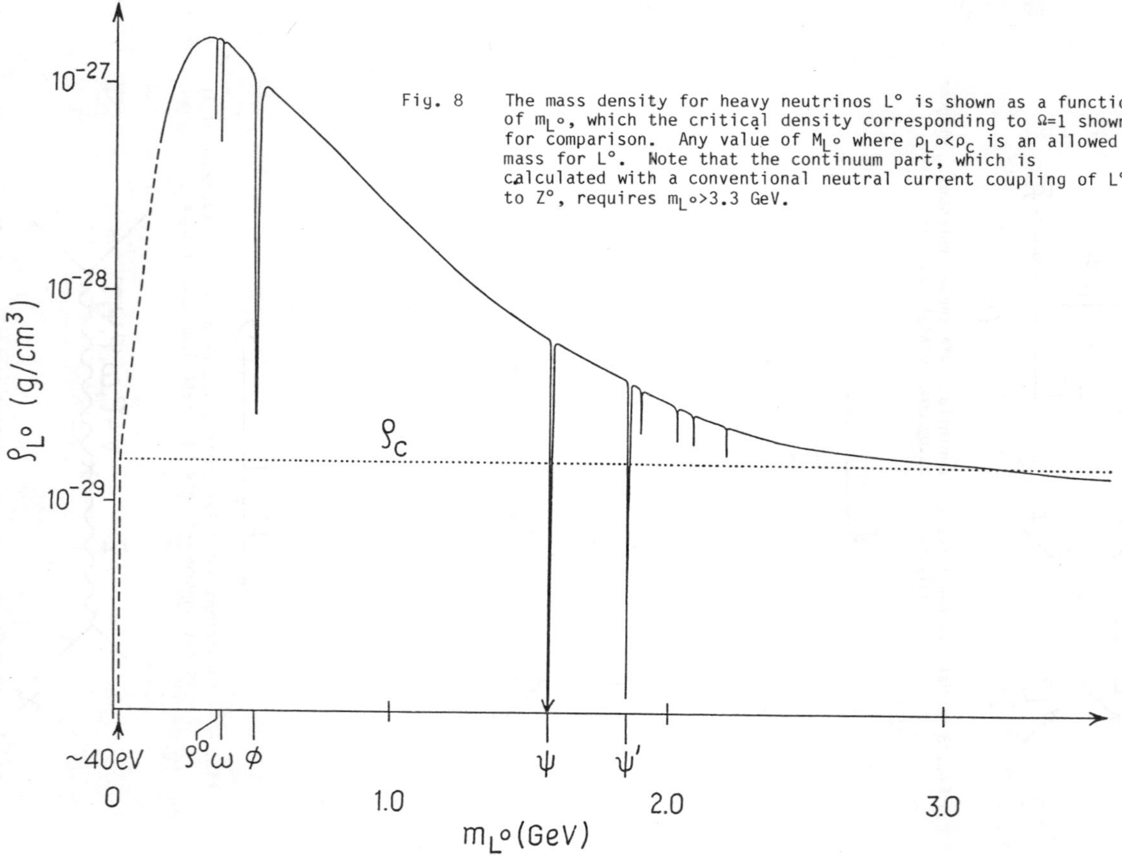

Fig. 8 The mass density for heavy neutrinos L° is shown as a function of m_{L°, which the critical density corresponding to $\Omega=1$ shown for comparison. Any value of M_{L° where $\rho_{L^\circ}<\rho_c$ is an allowed mass for L°. Note that the continuum part, which is calculated with a conventional neutral current coupling of L° to Z°, requires $m_{L^\circ}>3.3$ GeV.

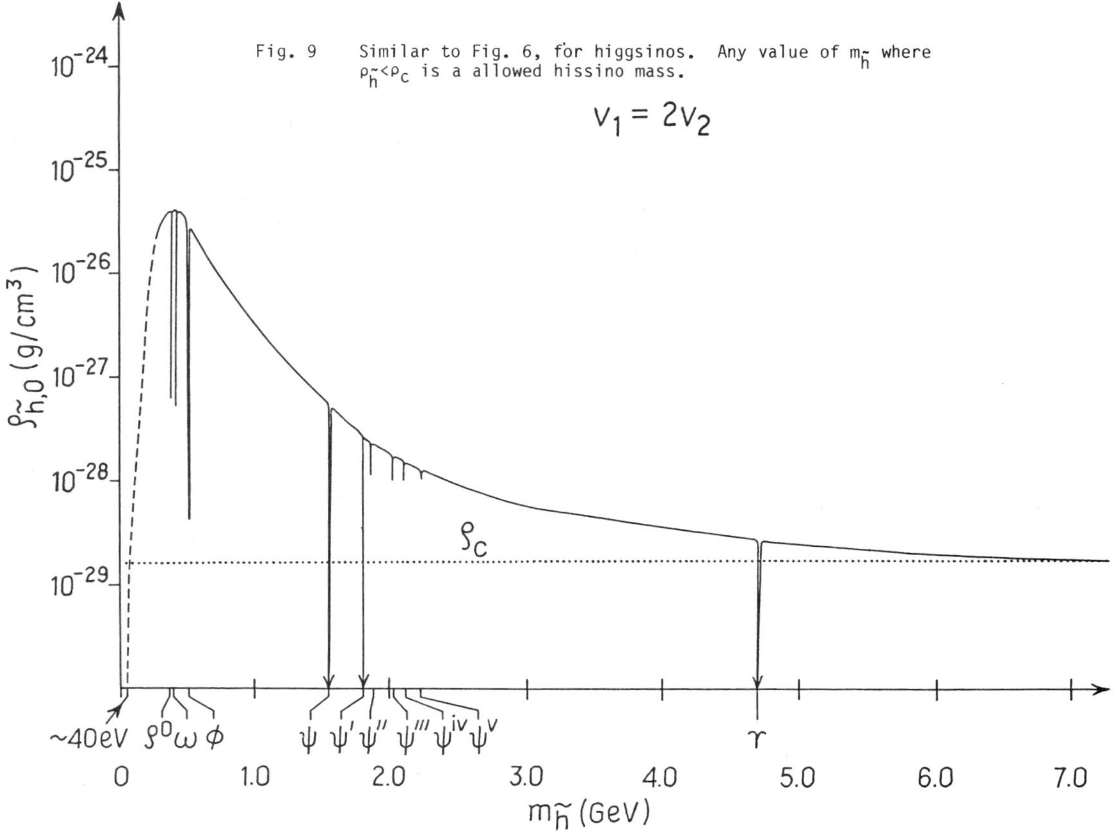

Fig. 9 Similar to Fig. 6, for higgsinos. Any value of $m_{\tilde{h}}$ where $\rho_{\tilde{h}} < \rho_c$ is a allowed hissino mass.

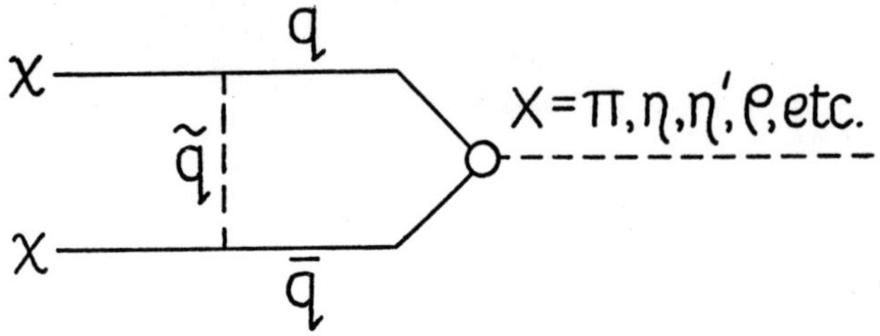

Fig. 10 When annihilation is through sfermion exchange the strongest coupling is to pseudoscalar resonances, which for some states enhance the annihilation and allow masses below the continuum.

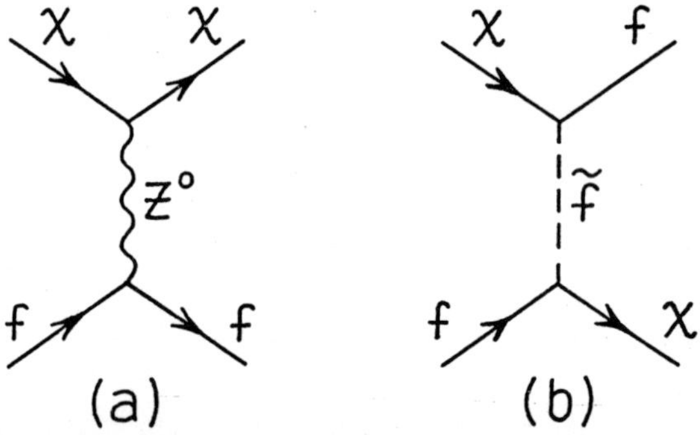

Fig. 12 These are the dominant mechanisms for scattering of dark matter candidates on quarks and leptons.

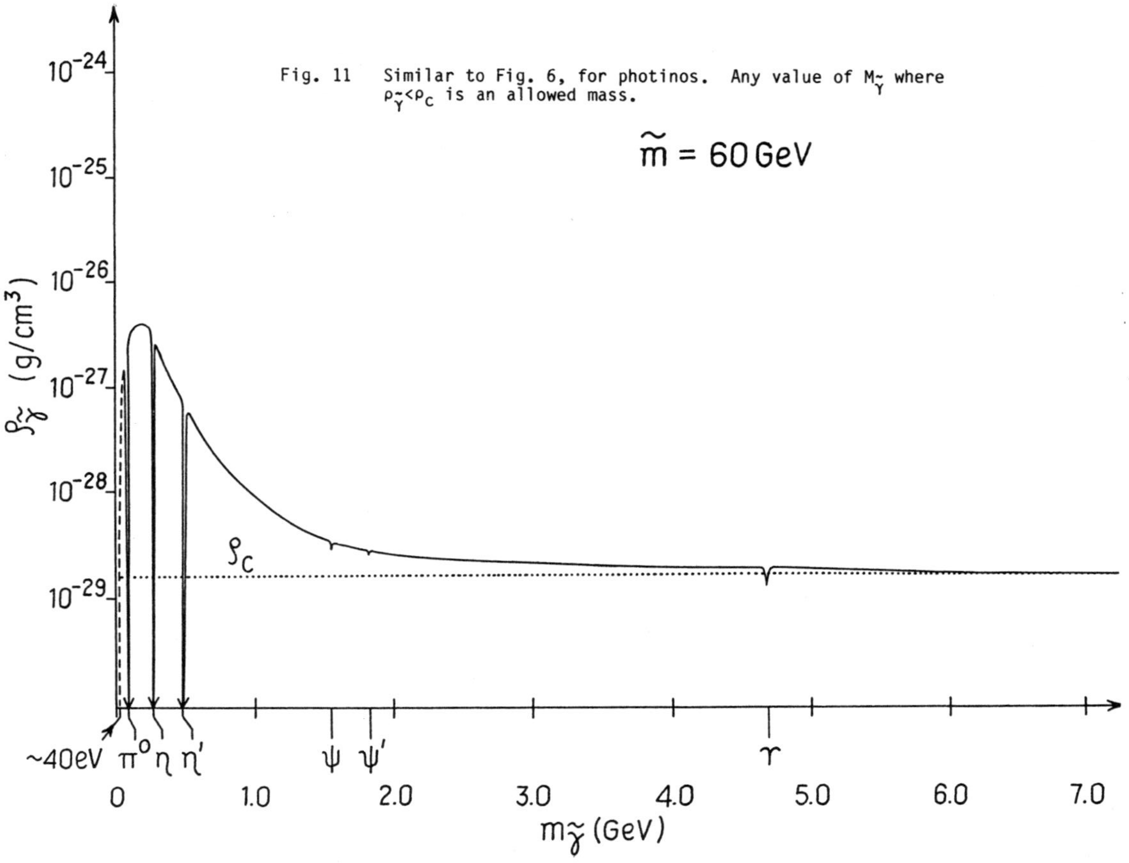

Fig. 11 Similar to Fig. 6, for photinos. Any value of $M_{\tilde{\gamma}}$ where $\rho_{\tilde{\gamma}} < \rho_c$ is an allowed mass.

THE OBSERVATION AND PHENOMENOLOGY OF GLUEBALLS*

S.J. Lindenbaum

Brookhaven National Laboratory
Dept. of Physics, Bldg. 510A
Upton, New York 11973 U.S.A.

and

City College of New York
Dept. of Physics
138th St. and Convent Ave.
New York, New York 10031

INTRODUCTION

The standard model explains the electro-weak interaction by $SU(2)_L \times U(1)$. The critical features of the electro-weak interaction at presently attainable energies have been successfully quantitatively described by this standard electro-weak model.[1)]

The hadronic interactions which have been traditionally theoretically intractable, are explained in the standard model by QCD.[2)] The heart of QCD is the locally gauge invariant non-abelian group $SU(3)_{color}$ which has eight colored massless gauge bosons, which self-interact. Their interaction exhibits asymptotic freedom and is described by a running coupling constant. The experimental values of Λ are of the order of 100 MeV. Thus the asymptotic freedom is quite "precious" even at moderate q^2 and energies. Color confinement and the running coupling constant make the existence of multi-gluon resonances or glueballs[3)] inescapable. This has been quantitatively demonstrated by Lattice Gauge calculations.[4)] In fact in a pure Yang Mills Theory[5)] of $SU(3)_{color}$, glueballs would be the only hadrons in the world. The addition to the theory of quarks interacting via gluons to yield QCD should in no way remove the glueballs.

Thus in the soft QCD sector where the quark-gluon and gluon-gluon coupling constants become strong, the fact that the Particle Data Group tables contain hundreds of quark-built meson and baryon states and there is no glueball section is a serious missing link in QCD. Thus, in spite of the fact that perturbative QCD has had many successes which include quantitative ones to ~ 10-20%. The glueball missing link must be found if the theory is to survive.

---

* This research was supported by the U.S. Department of Energy under Contract Nos. DE-AC02-76CH00016 (BNL) and DE-AC02-83ER40107 (CCNY).

Evidence has been presented for glueball candidates.[6-9,16,17] However, in only one case the BNL/CCNY $I^G J^{PC} = 0^+2^{++}$ states, $g_T(2050)$, $g_{T'}(2300)$ and $g_{T''}(2350)$, the glueball resonance hypothesis (i.e. these states are produced by 1-3 glueballs) has been shown to well explain the very striking and unusual characteristics of the data naturally within the context of QCD; whereas the published alternative explanations have been shown to be both incorrect and ruled out by the data.[10-12] Thus we have very probably discovered glueballs and found the missing link in QCD.

In the workshop lecture on which this paper is based, the experimental evidence and the relevant phenomenology of glueballs was reviewed.

SEARCHING FOR GLUEBALLS

I. <u>USE AN OZI[13,6-8] SUPPRESSED CHANNEL WITH VARIABLE MASS SUCH AS THE REACTION $\pi^- p \rightarrow \phi\phi n$.</u> The breakdown of the OZI suppression signals a glueball The OZI suppression is a filter which allows resonating gluons or glueballs to pass, while strongly rejecting conventional quark-built hadronic states.† Therefore as has been previously concluded, the BNL/CCNY[7-8,10-12,14] $g_T(2050)$, $g_{T'}(2300)$ and $g_{T''}(2350)$ are produced by glueballs.

II. <u>LOOK IN A CHANNEL ENRICHED IN GLUONS SUCH AS THE RADIATIVE DECAY OF THE J/ψ AND SEARCH FOR NEW PHENOMENA</u> such as the iota(1440),[9a] the $\theta(1640)$,[9b] and the $\zeta(2220)$.[15]

III. Pattern recognition of a decuplet - a $q\bar{q}$ nonet + glueball $\rightarrow$ decuplet with characteristic splitting and mixing. The $g_S(1240)$[16] and the G(1590)[17]†† are examples.

IV. Double Pomeron exchange.[7]

V. $\phi\phi$ inclusive.[18]

Method III, pattern recognition of a decuplet has glueball candidates[16-17] which are relatively weak and inconclusive.

---

† Provided $\phi\phi$ system $J \geq 1$ so that vacuum mixing is neglectable, otherwhise this vacuum mixing could possibly lead to violations of Zweig suppression, since it can lead to large departures from ideal mixing.

Method IV, double Pomeron exchange has no stand-alone glueball candidates but some indications for them.[7)]

Method V, the $\phi\phi$ inclusive experiment of Booth et al.,[18)] shows consistency with the BNL/CCNY $J^{PC} = 2^{++}$ resonances in moments activity, and fitting the shape of the correlated $\phi\phi$ spectrum, requires two resonances consistent with the BNL/CCNY states.

It is generally agreed that the most prominent glueball candidates are the BNL/CCNY g_T, $g_{T'}$, and $g_{T''}$ found in the (OZI forbidden) reaction $\pi^- p \to \phi\phi n$, and the SLAC J/ψ radiative decay candidates (i.e. the iota and the θ).

In the case of the iota and the θ, plausible alternative explanations[19)] other than the glueball resonance hypothesis have been proposed and published and have not been refuted. Thus these candidates are inconclusive.

However in the case of the BNL/CCNY g_T, $g_{T'}$ and $g_{T''}$, We have been able to refute[10-12)] published alternatives. I will show that these alternatives are either incorrect, or do not fit the data, or both.

Thus in my opinion the glueball resonance explanation is the only viable one (which fits the data) of those proposed and published after several years of effort by various authors. Hence I conclude the g_T, $g_{T'}$ and $g_{T''}$ are produced (to a very high probability) by glueball(s) and will spend the major portion of this paper discussing them.

THE OZI RULE[23)]

In the u,d,s quark system it has been well established experimentally and via phenomenological analyses, that $q\bar{q}$ meson nonets exist for $J^{PC} = 0^{-+}$, 1^{--}, 2^{++} and 3^{--}. Except for the 0^{-+} nonet, all those with $J \geq 1$ are nearly ideally mixed.[24)] Ideal mixing is characterized by the requirement that the singlet state be composed of an $s\bar{s}$ pair exclusively and that the singlet of the octet states contains no strange quarks. An ideally mixed nonet is conveniently representable by Zweig's Quark Line Diagrams.

---

†† The G(1590) and the $S^{*\prime}(1750)$ may be the same particle.

The disconnected Zweig diagrams for decay and the production of the ϕ and f' are shown in Figs. 1 and 2. The connected Zweig diagrams in QCD (Fig. 3) are characterized by a continuous flow of color carried by the quark lines, which allows a series of single gluon exchanges which involve strong collective soft glue effects to create or annihilate $q\bar{q}$ pairs relatively easily, and thus gives us the relatively highly probable Zweig-allowed decay and production processes. However, when the diagram became disconnected (Figs. 1 and 2) the $s\bar{s}$ pair in the ϕ has to be annihilated or created by at least three hard gluons to conserve all quantum numbers including color and by at least two hard gluons in the case of the f'. Asymptotic freedom strongly decouples hard glue from quarks. This has been observed to occur at relatively moderate gluon energies such as those involved in the three-gluon decay of the ϕ and thus is often referred to as "precocious" asymptotic freedom. The resultant relatively weak coupling constants of the hard gluons naturally explains the observed OZI suppression factors ~ 100 for both ϕ and f' decay and production, and the even larger OZI suppression in the decay of the J/ψ and Υ.

In meson states built from $q\bar{q}$ pairs, departures from ideal mixing can only be expected to occur when flavor changing diagrams which convert $s\bar{s}$ quark pairs into $u\bar{u}$ or $d\bar{d}$ quark pairs or vice versa have the connecting gluons relatively strongly coupled.

The relatively heavy $s\bar{s}$ pairs and precocious asymptotic freedom can explain why this does not happen in $J \geq 1$ nonets, assuming there is no flavor mixing mechanism. However if there is a flavor mixing mechanism as for example vacuum effects (i.e. instantons, etc.) which as Novikov et al.[25)] have pointed out are expected to be important for the $J^{PC} = 0^{-+}$ nonet, we can get a badly mixed nonet. Novikov *et al.*[25)] estimate vacuum effects are important for $J = 0$ but are unimportant for $J \geq 1$. This is certainly consistent with the experimental results which show all established nonets with $J \geq 1$ are nearly ideally mixed whereas the well-established $J^{PC} = 0^{-+}$ nonet is far from ideally mixed.

In QCD there is only one other basic mixing mechanism, namely the presence of glueballs with the same quantum numbers near enough to the nonet singlet masses and with the appropriate width to effectively mix with the singlets. This glueball mixing mechanism could destroy ideal

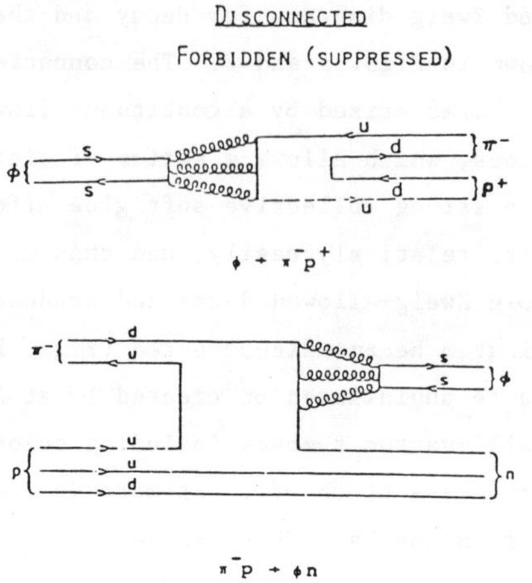

Figure 1
Zweig disconnected diagrams (suppressed reaction) for the u,d,s, quark system. The helixes represent gluons bridging the disconnection for the decay and production of the ϕ.

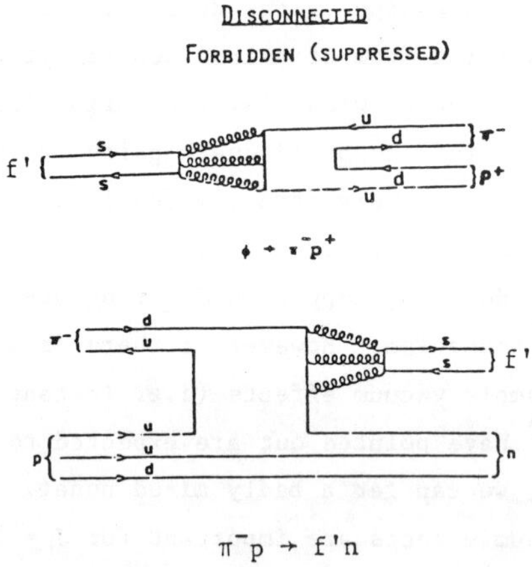

Figure 2
Zweig disconnected diagram (suppressed reaction) for the u, d, s quark system for the decay and production of the f'.

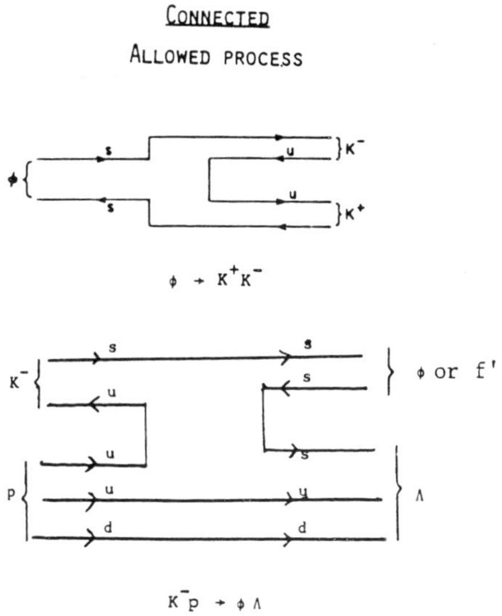

Figure 3

Zweig connected (allowed reaction) diagrams for the u,d,s quark system.

mixing and lead to a badly mixed $J \geq 1$ nonet. Furthermore a glueball resonance is expected to be a collection of strongly coupled gluons since the gluons can easily split. As they move apart and become softer, their couplings increase (i.e. infrared slavery). Thus glueballs like other hadronic resonance are expected to be relatively strongly coupled.

It is a well-known experimental fact that in all Zweig disconnected diagrams in the ϕ, f', J/ψ and Υ systems the OZI rule appears universal as illustrated in Figs. 1 through 5. Furthermore the OZI suppression in the J/ψ and Υ systems is much larger than that for the ϕ and f'. With u,d,s,c quarks we expect sixteen-plets and with u, d, s, c, b quarks twenty-five plets. If $J \geq 1$ and there are no glueballs with the same quantum numbers near enough to the singlets, with the right width to cause appreciable mixing, these higher multiplets should be nearly ideally mixed and the u,d,s nonets contained within them will also be nearly ideally mixed.

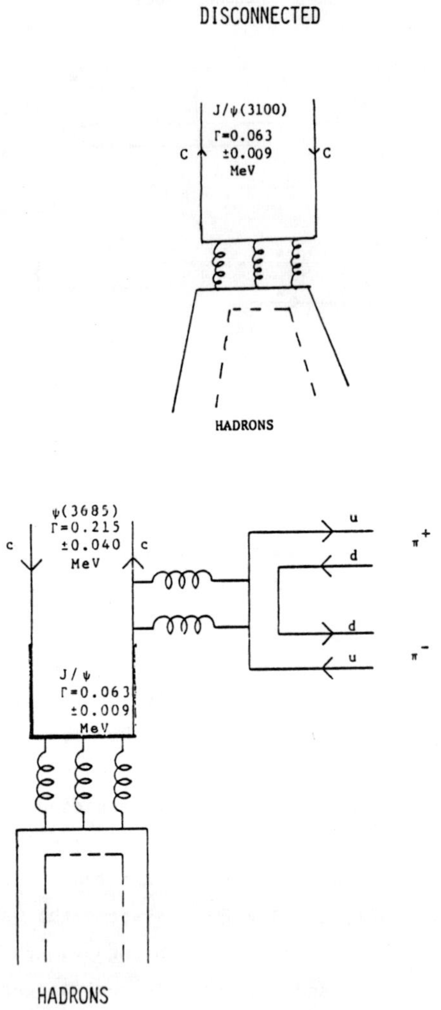

Figure 4
Zweig disconnected diagrams in the J/ψ and excited ψ states.

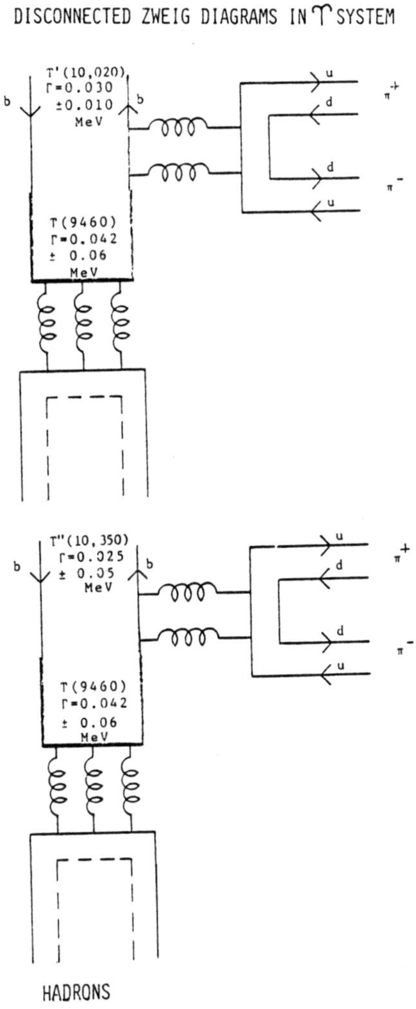

Figure 5
Zweig disconnected diagrams in the ϒ system.

As previously discussed,[14,8,26] the OZI rule appears on paper to be defeatable by two-step processes, each of which are OZI allowed. For example:

In decay,

1) $\phi \to K^+K^- \to \rho\pi$ or 3π

2) $f' \to K^+K^- \to \pi\pi$

3) $f' \to \eta\eta \to \pi^+\pi^-$

In production,

4) $\pi^-p \to K^+K^-n \to \phi n$

5) $\pi^-p \to K^+K^-n \to f'n$

6) $\pi^-p \to \eta\eta n \to f'n$

7) $\pi^-p \to \phi K^+K^-n \to \phi\phi n^\dagger$

Reactions 1-6 are experimentally determined to be OZI suppressed clearly demonstrating that such two-step processes do not break the OZI suppression. One should note that in reactions 2, 3, 5 and 6 the intermediate step is considerably above threshold whereas in reactions 1 and 4 are just above threshold. In order to properly consider such two-step processes, the QCD dynamics, the overall quantum numbers of the system, and cancellations between all possible intermediate steps should be taken into account. The reason they probably are suppressed is partially due to the fact that hadronization in the first step takes place at the outer regions of the confinement region where the first $q\bar{q}$ pair is far apart (and probably moving away from each other) and the coupling is strong. Then for the two-step process to occur (except in reaction 3), a $q\bar{q}$ pair of quarks have to return to short distances where annihilation takes place and then annihilate. This is expected to be discriminated against dynamically and in fact the K^+ and K^- or other particles formed in the intermediate step may already be color singlets (i.e. have their own bags) thus further inhibiting the subsequent needed annihilation of the $q\bar{q}$ pair for the two step process to break the OZI suppression.

In the case of reaction 3, each η can in principle be created via its $s\bar{s}$ component and then by vacuum mixing transform to $u\bar{u}$ and $d\bar{d}$

---

† The OZI suppression is expected if the $\phi\phi$ system has $J \geq 1$. For $J = 0$, there is the possibility of vacuum mixing.

components and then by a quark rearrangement interaction π^+ and π^-. This would partially break the OZI suppression with a process analogous to that proposed by Donoghue[22] to break the OZI suppression in $\pi^- p \to \phi\phi n$. The fact that the coupling of the f' to $\pi^+\pi^-$ has experimentally been observed to be small in both decay and production demonstrations that such processes are unlikely enough so that they do not materially affect the OZI suppression. In the context of QCD the only way you can break the OZI suppression in the (u,d,s) $q\bar{q}$ quark states is by changing the near ideal mixing observed in the nonets for $J \geq 1$ by a strong enough flavor changing mixing mechanism which converts $u\bar{u}$ or $d\bar{d} \rightleftarrows s\bar{s}$. Vacuum effects[25] can do this for J = 0 (e.g., the $J^P = 0^-$ nonet). However, they are not expected to and do not appear to affect $J \geq 1$ nonets, and thus would not be expected to appreciably affect the $J^{PC} = 2^{++}$ $\phi\phi$ states observed in the BNL/CCNY experiment. The only other known basic flavor-changing mechanism is a glueball.

Figures 6, 7, and 8 are the Zweig Quark Line Diagrams for the three reactions studied by the BNL/CCNY group. Figure 9 shows a scatter plot of K^+K^- masses from the BNL/CCNY experiment which used the BNL MPS II. We see the general ≈ uniform background from the reaction a) $\pi^- p \to K^+K^-K^+K^- n$ which is OZI allowed, and the two ϕ bands representing b) $\pi^- p \to \phi K^+K^- n$ which is also OZI allowed. Where the two bands cross we have the Zweig forbidden reaction $\pi^- p \to \phi\phi n$. Although there are two ϕ mesons instead of one, one would expect this reaction to be more or less as forbidden as $\pi^- p \to \phi n$ provided the $\phi\phi$ system does not have the quantum number J = 0 so that the vacuum can mix flavors.

The black spot where the two ϕ bands cross shows an obviously more or less complete breakdown of the Zweig suppression. This has been quantitatively shown to be so in these reactions[26] and also by comparing K^- induced ϕ and $\phi\phi$ production.[27]

The black spot when corrected for double counting and resolution is ≈ 1,000 times the density of reaction (a) and ≈ 50 times the density of reaction (b). If one projects out the ϕ bands, even with rather wide cuts ± 14 MeV, there is a huge $\phi\phi$ signal which is ≈ 10 times greater than the background from reaction (b). The recoil neutron signal is also very clean, ≈ 97% neutron (see Figs. 10a and 10b).

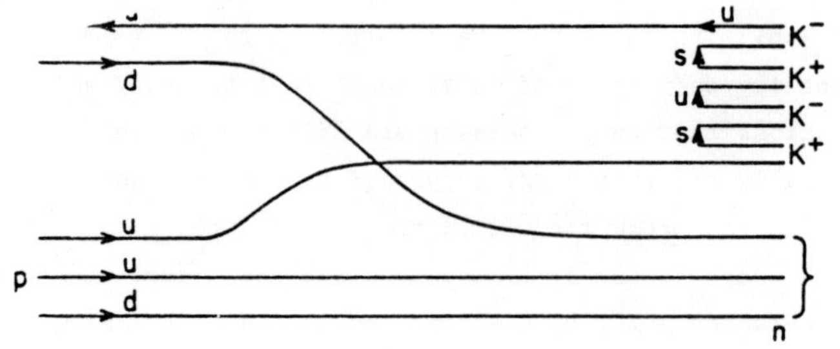

Figure 6
The Zweig quark line diagram for the reaction $\pi^- p \to K^+K^-K^+K^-n$, which is connected and OZI allowed.

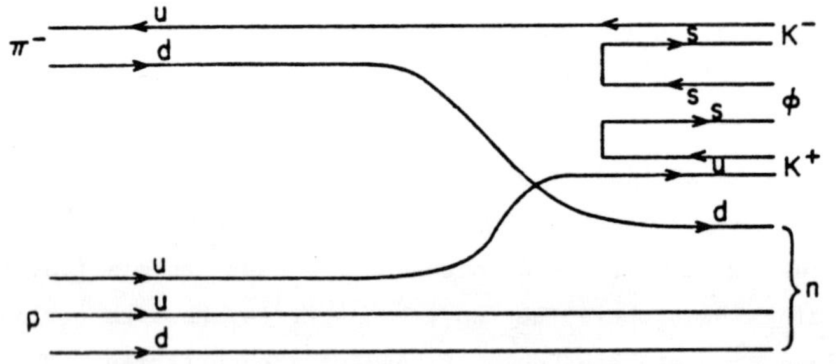

Figure 7
The Zweig quark line diagram for the reaction $\pi^- p \to \phi K^+K^-n$, which is connected and OZI allowed.

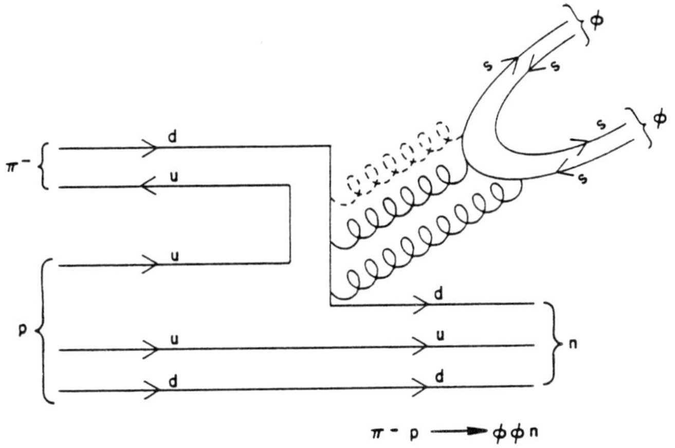

Figure 8a
The Zweig quark line diagram for the reaction $\pi^- p \to \phi\phi n$ which is disconnected (i.e. a double hairpin diagram) and is OZI forbidden. Two or three gluons are shown connecting the disconnected parts of the diagram depending upon the quantum numbers of the $\phi\phi$ system. For the g_T's, $J^{PC} = 2^{++}$, and only two gluons are required. From the data analysis they come from the annihilation of the incident π^- and a π^+ exchanged between the lower and the upper parts of the diagram.

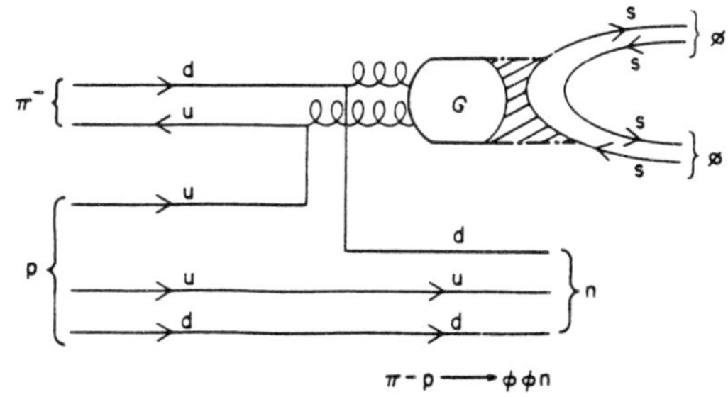

Figure 8b
The $J^{PC} = 2^{++}$ glueball intermediate state in $\pi^- p \to \phi\phi n$. The dash-dot lines with crosshatch lines region indicates that we don't know details of the glueball hadronization into $\phi\phi$.

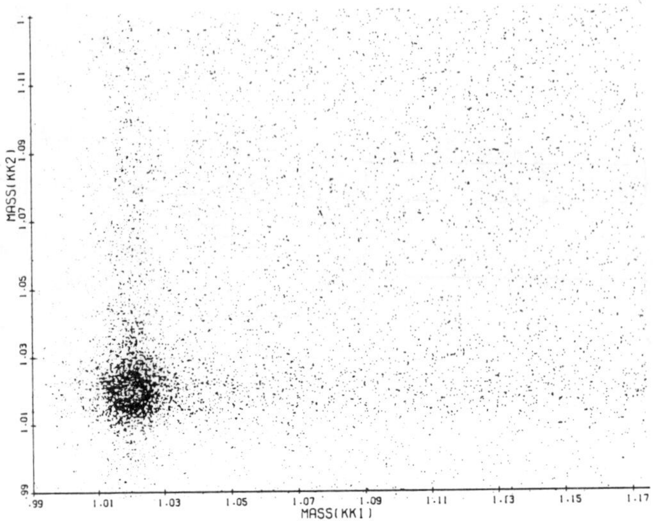

Figure 9:
Scatter plot of K^+K^- effective mass for each pair of K^+K^- masses. Clear bands of $\phi(1020)$ are seen with enormous enhancement (black spot) where they overlap (i.e. $\phi\phi$) showing essentially complete breakdown of OZI suppression.

Figure 10c shows the acceptance corrected $\phi\phi$ mass spectrum in the ten mass bins which were used for the partial wave analysis.

All waves with J = 0 - 4, L = 0 - 3, S = 0 - 2, P = ± and η (exchange naturality) = ± were allowed in the partial wave analysis. Thus 52 waves were considered. The incident π^- lab momentum vector and the lab momentum vectors of the four kaons completely specified an event. The Gottfried-Jackson frame angles β(polar) and γ(azimuthal) are shown in Fig. 11a. These and the polar angles (θ_1,θ_2) of the K^+ decay in the ϕ rest systems relative to the ϕ direction and the azimuthal angles α_1 and α_2 of the K^+ decay direction in the ϕ_1,ϕ_2 rest systems (see Fig. 11b) were used to specify an event.

In the PWA, the standard LBL/SLAC isobar model program was used but modified so that the spectator particle was replaced by the second ϕ. Bose statistics was satisfied by the requirements L + S must be even for the $\phi\phi$ system. Because of the narrowness of the ϕ (less than the experimental resolution $\Gamma_{\phi\phi} \approx 8$ MeV), the partial wave analysis is

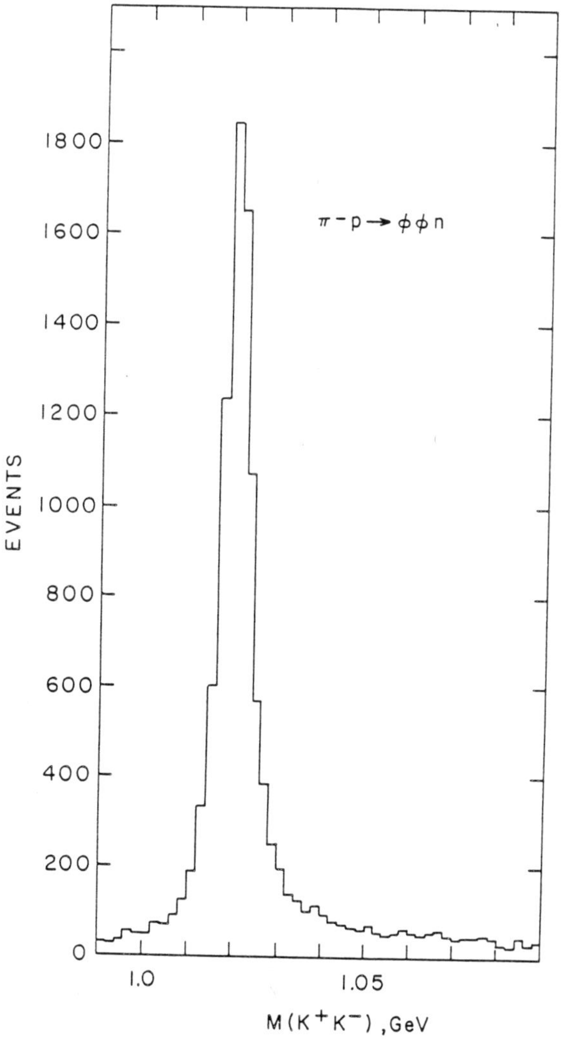

Figure 10a
The effective mass of each K^+K^- pair for which the other pair was in the ϕ mass band.

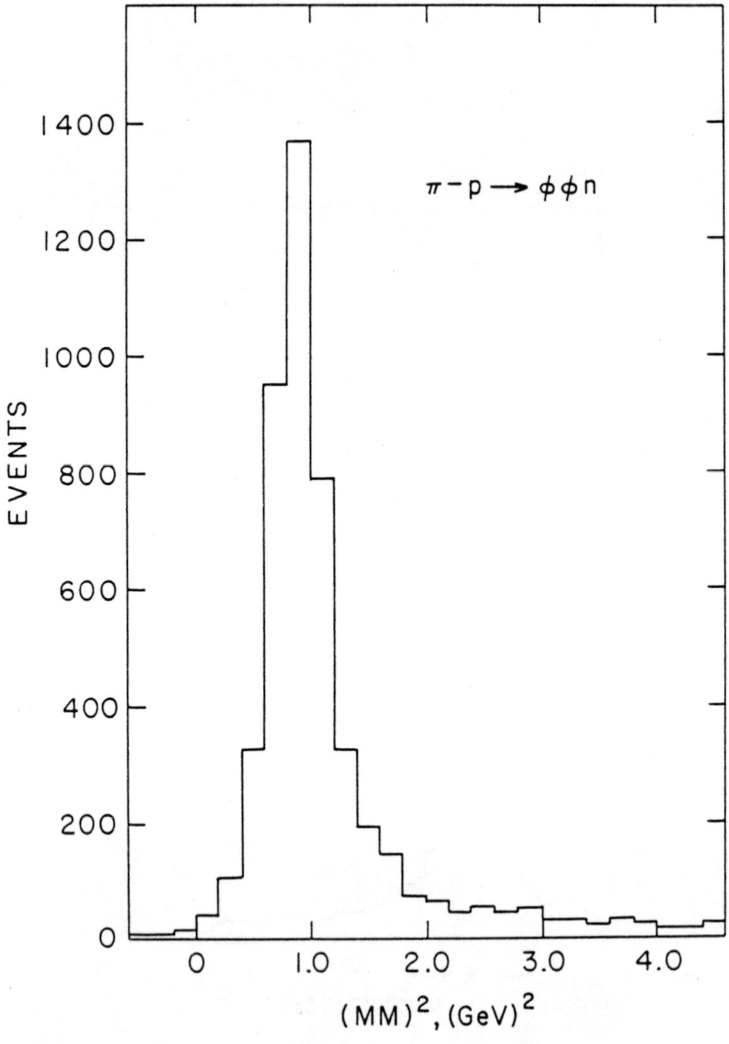

Figure 10b
The missing mass squared for the neutral system recoiling from the $\phi\phi$.

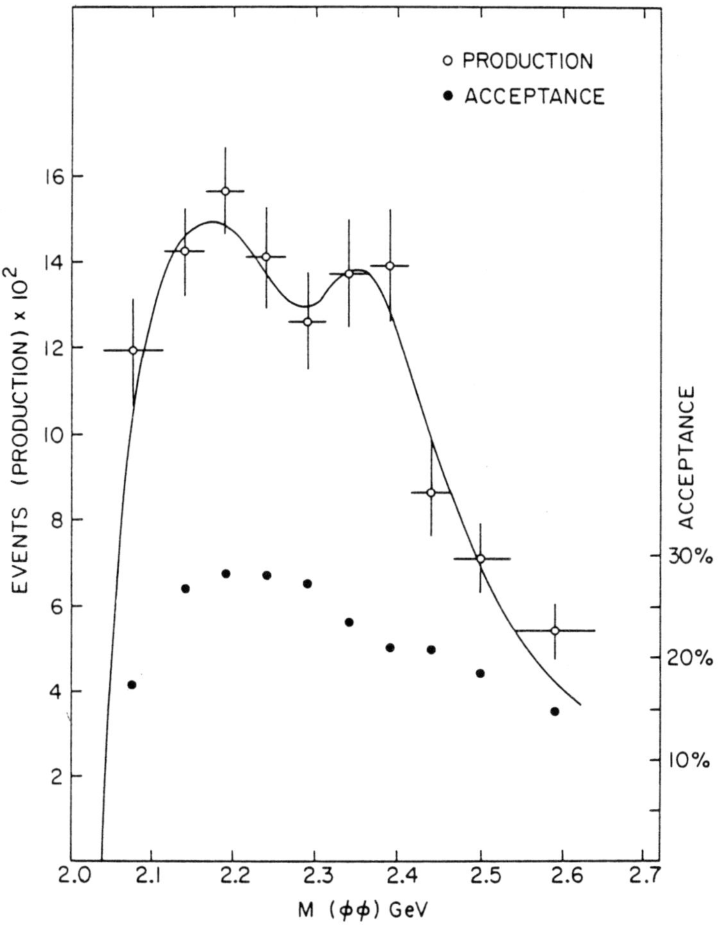

Figure 10c

The ϕϕ mass spectrum corrected for acceptance. The solid line is the fit to the data with the three resonant states to be described later. The points at the bottom of the diagram are the acceptance for each mass bin to be read with the scale at the right.

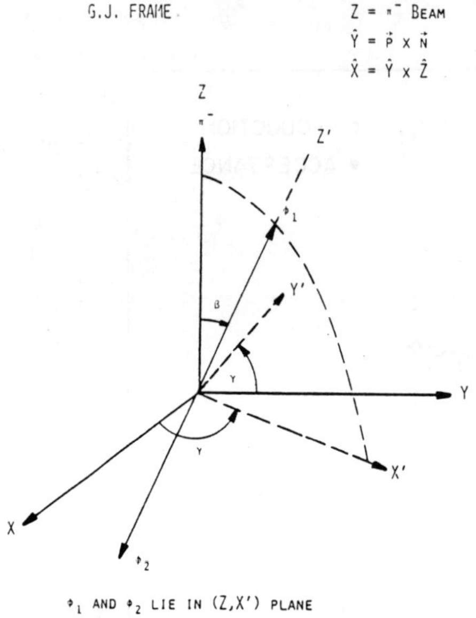

Figure 11a: The Gottfried-Jackson frame with polar angle β and azimuthal angle γ.

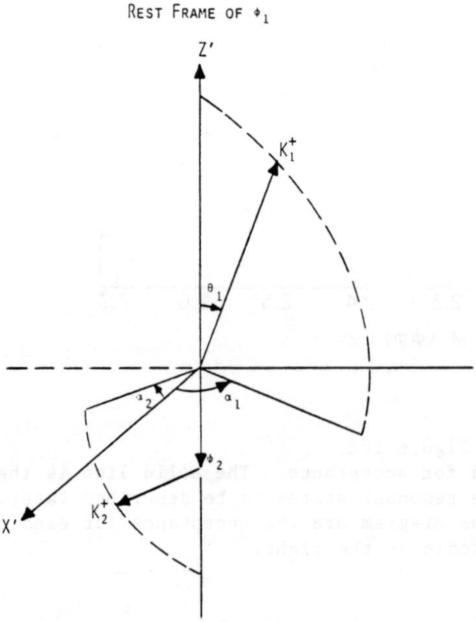

Figure 11b: The ϕ_1 rest frame with the polar angle θ_1 of the decay $K_1{}^+$ (relative to ϕ direction) and the azimuthal angle α_1 of the decay $K_1{}^+$.

model independent and only depends on the angular characteristics of the partial waves. Due to the spins of the two narrow decaying ϕ's, this partial wave analysis is much more powerful and selective than any other previously employed.† Thus, although we have allowed all waves up to L = 3 and J = 4 (i.e. 52 waves), we find a unique solution (Figs. 12a and 12b) which consists of only three $J^{PC} = 2^{++}$ waves, an S-wave with S = 2, a D-wave with S = 2, and a D-wave with S = 0, which is a good fit. All three waves have $J_Z = 0$ in the Gottfried-Jackson frame and the exchange naturality = (-1), the characteristics of pion exchange. The observed $(d\sigma/dt')_{\phi\phi} = e^{(9.4 \pm 0.7)t'}$ for $t' < 0.3$, the low t'-region which contains most of the data, is shown in Fig. 13. The only charged particle exchange which will give this is pion exchange. The best two-wave fit is ≈ 30 σ away. Our selected three-wave fit is ≈ 15 σ better than the next best three-wave fit.

The few percent background was estimated to be almost entirely composed of the reaction $\pi^- p \rightarrow \phi K^+ K^- n$. The wide cuts that were used allowed the background to be ≈ 13% of the $\phi\phi n$ events. This ensured that no biases in ϕ selection were introduced and also allowed the possibility of finding a coherent wave in this background to serve as a reference for our phase motion. However, the partial wave analysis of this background in the region where the K^+K^- mass was slightly larger than that of the ϕ revealed that ≈ 65% of it was structureless and incoherent. Only approximately 7% of the background was 2^{++} which had an ≈ 0 amplitude in the threshold region and peaked at ≈ 2.4 GeV (see Fig. 14). Thus the $\phi K^+ K^-$ background reaction was totally different than the 2^{++} observed in the $\phi\phi$ system. There was ≈ 28% of 1^{--} background which are expected quantum numbers for a $\phi K^+ K^-$ system where all particles are in an S-wave with respect to each other and the production is via π-exchange. Thus in the $\phi K^+ K^-$ background we are dominated by the structureless incoherent background one would expect to be the result of the addition of the many possible partial waves. The 2^{++} wave intensity is a small fraction of the total.

The partial wave amplitudes and phase behavior of the $\phi\phi$ system (shown in Figs. 12a and 12b) clearly suggest that these three waves are produced by resonances. A two-pole K-matrix fit which allows all

---

† See Refs. 8 and 14 for a discussion of selectivity.

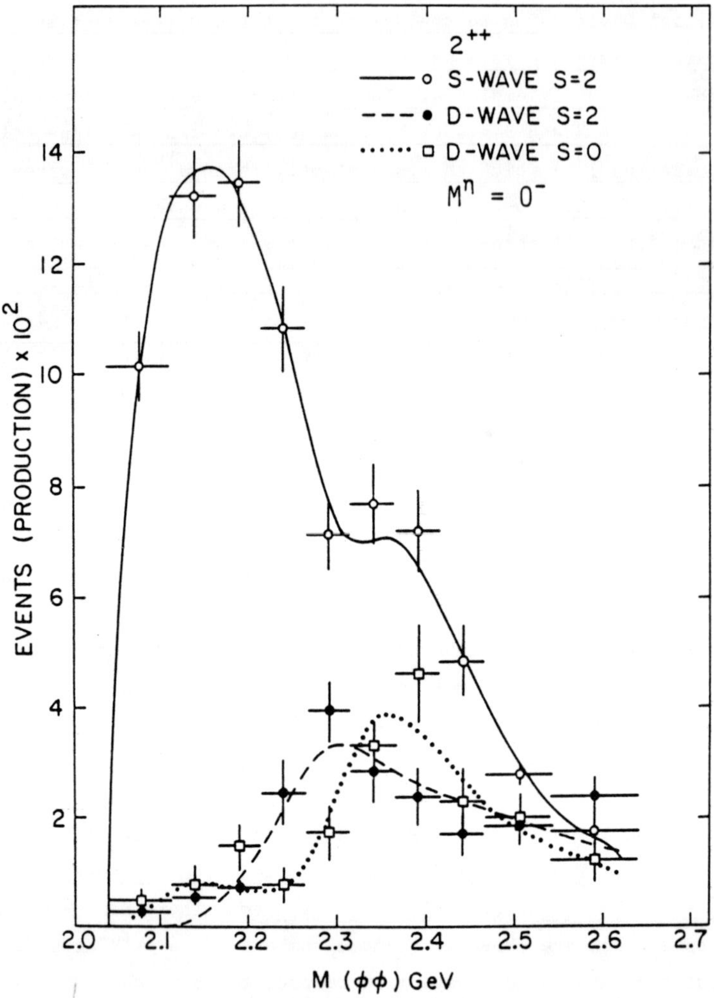

Figure 12a: The three $J^{PC} = 2^{++}$ partial waves at production in 50 MeV mass bins (except at ends). $J_Z = 0$ in the Gottfried-Jackson frame and the exchange naturality is (-) corresponding to pion exchange for all three waves. The smooth curves are derived from a three-pole K-matrix fit.

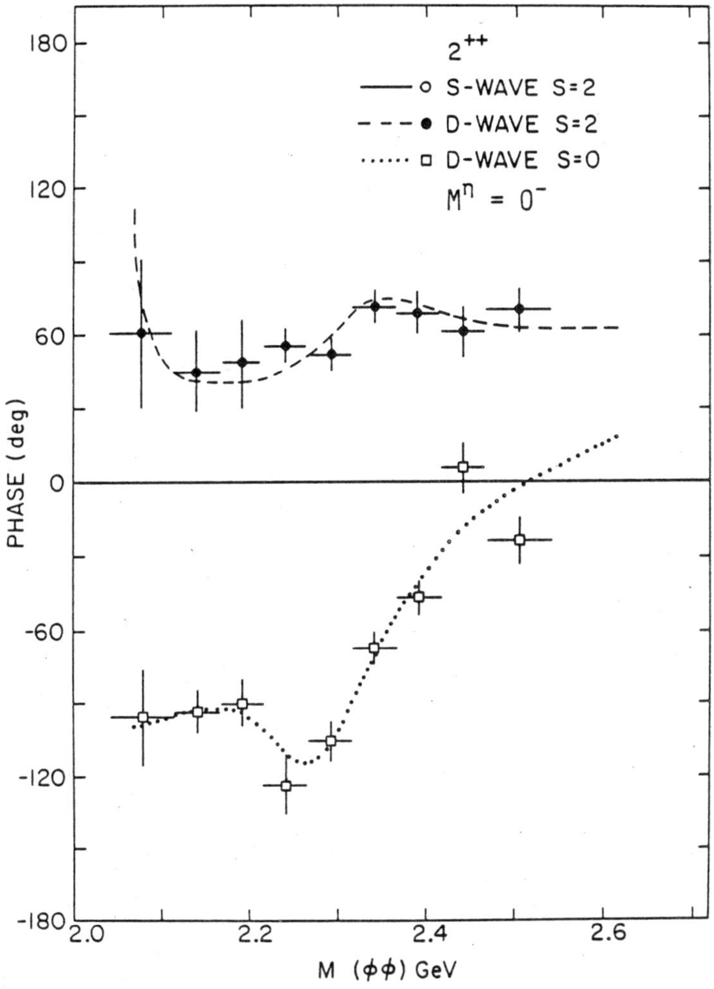

Figure 12b: D-S phase difference from the partial wave analysis vs. φφ mass. The smooth curves are derived from a three-pole K-matrix fit.

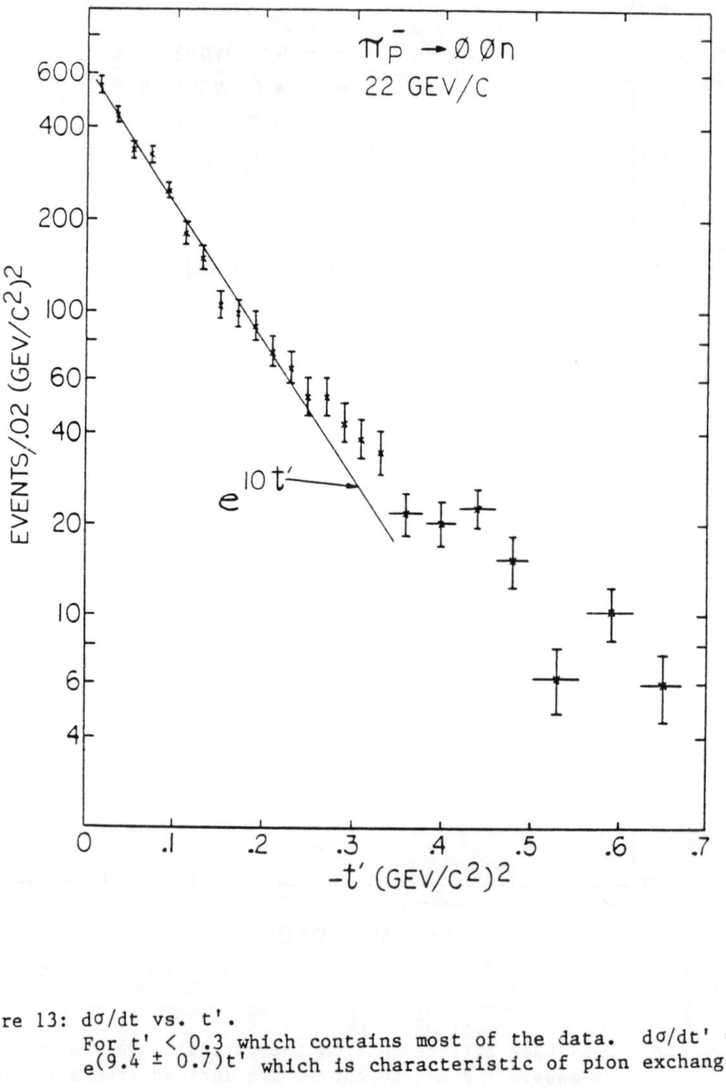

Figure 13: dσ/dt vs. t'.
For $t' < 0.3$ which contains most of the data. $d\sigma/dt' = e^{(9.4 \pm 0.7)t'}$ which is characteristic of pion exchange.

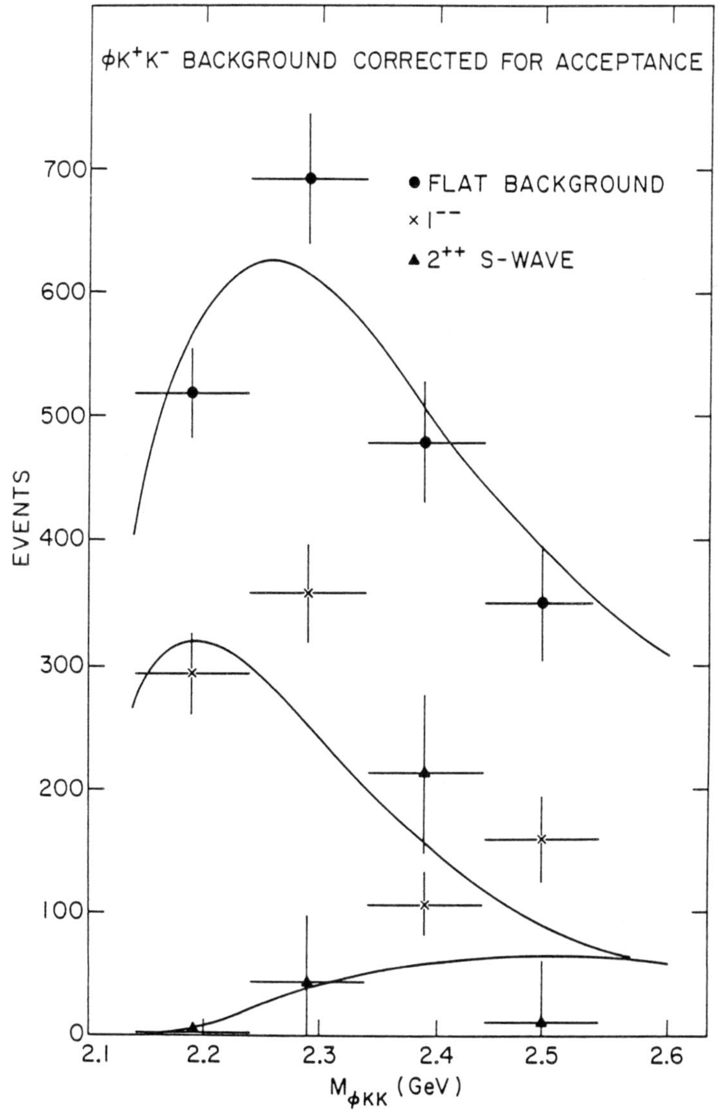

Figure 14: The partial wave intensities in the background reaction $\pi^- p \to \phi K^+ K^- n$.

three observed waves to mix in each pole was ≈ 15σ away from a good fit.

A three-pole K-matrix fit gave a good fit. The Argand diagram for this fit is shown in Fig. 15. A four-pole K-matrix fit did not lead to any further improvement. The three-pole K-matrix fit was used to fit the data which was contained in 90 angular variable bins for each of the ten mass bins used, thus yielding a total of 900 data bins. The fit to these 900 data bins was $\lesssim$ a one σ fit. The resonance parameters of this fit are given in Table I.

Due to the small background and the fact that the background is mostly incoherent, the S-wave which dominates the $g_T(2050)$ must be used as a phase reference. The phase difference of this and the other two D-waves precisely match the 3-pole K-matrix fit and thus clearly demonstrate that all three states have the pole behavior which is the best and only critical definition of a resonance.

Attributing the production of these states to 1-3 primary $J^{PC} = 2^{++}$ glueballs has explained all their features in a clear-cut and simple manner.[7,8,14)]

It should be noted that the mixing of waves is substantial in these three $J^{PC} = 2^{++}$ states and the exact wave content and parameters of each resonance or K-matrix pole is therefore somewhat sensitive to details and somewhat uncertain. However from the glueball physics point of view we are at present mostly interested in the quantum numbers and general characteristics of the parameters of the resonant states and not very concerned about their exact values and wave contents.

GLUEBALL MASSES

The constituent (i.e. gluon has effective mass) gluon models would predict three low lying $J^{PC} = 2^{++}$ glueballs. The mass estimates from the lattice gauge groups cover the range ≈ 1.7 to 2.5 GeV for $J^{PC} = 2^{++}$ glueballs.[4,28-29)] with which we are clearly consistent.

T.D. Lee has analytically calculated J = 2 glueballs in the strong coupling limit[30)] and obtains three glueball states which correspond to our three states. His strong coupling calculation gives the mass differences between these three states in terms of two

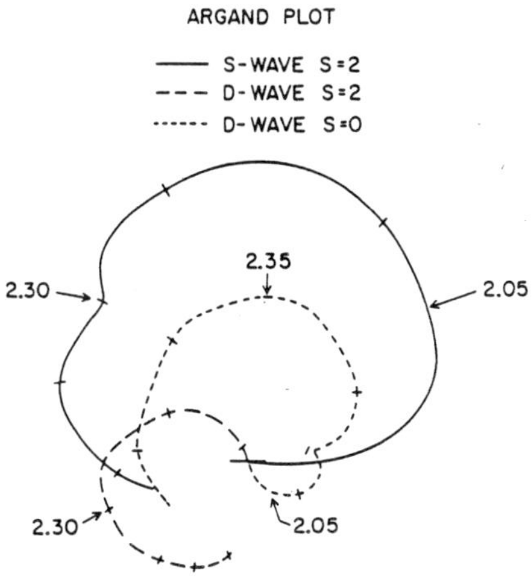

Argand plot from K-matrix fit.

parameters, one being essentially the effective strength of the coupling and the other a mass scale parameter. In order to try to adjust his strong coupling calculation to the real world of intermediate coupling we took the mass of the 0^{++} glueball as ≈ 1 GeV from the Lattice Gauge calculations, and fit our three masses with the other parameter and found a reasonable fit.

Many years ago I employed a similar procedure in the case of the Pauli-Dancoff strong coupling calculations of the nucleon isobars. In that case when I put in the known f^2 and a reasonable value for the cut-off, the strong coupling calculation results gave reasonable agreement with the experimental observations on nucleon isobars.

GLUEBALL WIDTH

In hadrons, the hadronization process consists of creation of one or more $\bar{q}q$ pairs. This must occur near the outer region of confinement involving strongly interacting soft glue, including collective interactions, if we are to have resonances decay with typical hadronic widths ($\Gamma_{hadronic}$ ~ 100 to several hundred MeV). It appears that hadronic widths are more or less independent of the number of particles in the dominant decay mode.

Table 1

Resonance parameters of the three-pole K-matrix fit

$$M_1 = 2.050^{+.090}_{-.050} \quad \Gamma_1 = .200^{+.160}_{-.050} \approx 50\%^{+10\%}_{-10\%} \text{ data}$$

S-wave, S = 2 ≈ $98\%^{+02\%}_{-70\%}$ coupling sign (+) defined

D-wave, S = 2 ≈ 0% + 50% coupling sign (−)

D-wave, S = 0 ≈ $2\%^{+25\%}_{-02\%}$ coupling sign (−)

$$M_2 = 2.300^{+.020}_{-.100} \quad \Gamma_2 = .200^{+.060}_{-.050} \approx 20\%^{+20\%}_{-10\%} \text{ data}$$

S-wave, S = 2 ≈ $30\%^{+20\%}_{-20\%}$ coupling sign (+)

D-wave, S = 2 ≈ $50\%^{+20\%}_{-10\%}$ coupling sign (+)

D-wave, S = 0 ≈ $20\%^{+20\%}_{-20\%}$ coupling sign (+)

$$M_3 = 2.350^{+.020}_{-.030} \quad \Gamma_3 = .270^{+.090}_{-.130} \quad 30\%^{+10\%}_{-20\%} \text{ data}$$

S-wave, S = 2 ≈ $40\%^{+10\%}_{-20\%}$ coupling sign (+)

D-wave, S = 2 ≈ $05\%^{+15\%}_{-05\%}$ coupling sign (−)

D-wave, S = 0 ≈ $55\%^{+20\%}_{-15\%}$ coupling sign (+)

For example the $\rho(770) \to \pi\pi$ requires production of one quark pair. The width of the $\rho(770)$ is $\Gamma_\rho = 154 \pm 5$ MeV. The $\rho'(1600) \to 4\pi$ requires the production of three quark pairs.† Yet $\Gamma_{\rho'} \approx 300 \pm 100$ MeV. Hence even though production of two-three additional quark pairs is required the $\Gamma_{hadronic}$ actually increases. This example clearly shows that hadronization easily occurs via collective soft glue effects and this is the basis of typical hadronic widths.

A glueball is a resonating multi-gluon system. The glue-glue coupling is stronger than the quark-glue coupling and thus it would be

---

† Even assuming $\rho'(1600) \to \rho\pi\pi$ requires initial production of two additional $q\bar{q}$ pairs.

expected, via gluon splittings before the final hadronization, to have a similar hadronization process to a $q\bar{q}$ hadron. Therefore a glueball would be expected to have typical hadronic widths. This is clearly to be expected for ordinary (non-exotic) J^{PC} states. In the case of exotic J^{PC} states, this arguement may not hold since no one yet knows what suppresses the unobserved exotic sector. Therefore Meshkov's oddballs[19)] may be narrow.

The observed characteristics of the reaction $\pi^- p \to \phi\phi n$ are very unusual and striking in the following respects:

1. The expected OZI suppression is completely broken in a very unusual manner, since virtually all of the cross section is composed of three resonant $\phi\phi$ states, the $g_T(2050)$, the $g_{T'}(2300)$ and the $g_{T''}(2350)$, all with the same quantum numbers $I^G J^{PC} = 0^+ 2^{++}$. In contrast, hadronic reactions in other channels show no such selective quantum numbers or resonance phenomena. For example the highest statistics[31)] experiment on $\pi^- p \to K^o_s K^o_s n$, which also exhibits π-exchange characteristics, has a slowly varying non-resonant 2^{++} amplitude behavior with essentially no phase motion. The 0^{++} and 4^{++} amplitudes are also both populated in contrast to $\pi^- p \to \phi\phi n$ where only 2^{++} is populated. A second three times higher statistics experiment[42] gives similar results.

2. The OZI allowed background reaction $\pi^- p \to \phi K^+ K^- n$ which is unexpectedly only a few percent of the OZI forbidden $\pi^- p \to \phi\phi n$ has been partial wave analyzed in the $K^+ K^-$ mass region just slightly heavier than the ϕ mass and ≈ 65% of the reaction consists of a structureless, incoherent, flat in all angular distributions, background which is expected for the superposition of many waves. Approximately 28% of the reaction has $J^{PC} = 1^{--}$ which are expected quantum numbers for a $\phi K^+ K^-$ system where all particles are in a relative s-wave. Thus this is a kinematic effect. Only ≈ 7% of the cross section is a 2^{++} S-wave* with an amplitude near zero at the threshold and peaking at ≈ 2.4 GeV, and exhibiting no evidence of resonant behavior. Thus it is entirely different than the $\phi\phi$ 2^{++} amplitudes. This reaction has about the same threshold and similar kinematics to the $\pi^- p \to \phi\phi n$

---

* The $K^+ K^-$ system and the ϕ are in a relative S-wave. The $K^+ K^-$ pair are in a relative P-wave.

reaction, thus threshold effects would be quite similar in the two reactions and it is clear that the striking characteristics of the $\phi\phi$ data cannot be attributed to such a naive mechanism.

The above characteristics of the data can be very well explained naturally within the context of QCD, if one assumes 1-3 primary glueballs with $J^{PC} = 2^{++}$ produce these states.[7,8,10)]

One or two primary $J^{PC} = 2^{++}$ glueballs could break the OZI suppression and mix with nearby $q\bar{q}$ states with the same quantum numbers and similar masses. However the simplest explanation is that we have a triplet of $J^{PC} = 2^{++}$ glueballs. These would be in the right mass range predicted from Lattice Gauge calculations[4,28-29)] and would fit the prediction of three distinct masses made by T.D. Lee.[30)] As is earlier described, we can adapt his calculation to fit the splitting.

CONCLUSION ON $\pi^- p \to \phi\phi n$

To prove that these states are produced by glueballs, as in all proofs one requires the appropriate input axioms.

<u>If one assumes as input axioms</u>:

1. QCD is correct;

2. The OZI rule is universal for weakly coupled glue in Zweig disconnected diagrams where the disconnection is due to the introduction of new flavors of quarks. All we need under Axiom 2 is the more restrictive statement that the OZI rule is operative in the reaction $\pi^- p \to \phi\phi n$ in the absence of a resonating glue system (i.e. a glueball) for $J \geq 1$** for the $\phi\phi$ system (i.e. disconnected system).

<u>Then the g_T, $g_{T'}$, and $g_{T''}$ states we observe must represent the discovery of 1-3 glueballs</u>.

Note that axiom (2) allows only resonating glue (i.e.glueballs) to break the Zweig suppression. However, one or two

---

** To avoid the possibility of vacuum mixing mechanisms affecting the OZI rule. It should also be noted that when a vacuum mixing effect is present such as in the 0^- nonet η and η' you cannot draw a unique Zweig diagram due to the mixed nature of the $q\bar{q}$ states. Therefore you do not actually have Zweig disconnected diagrams thus the statement in the first sentence under 2 is technically correct but may lead to

primary glueball(s) could break down the OZI suppression and possibly mix with two quark or other possible states.

These axioms strikingly agree with the data in the ϕ, f', J/ψ and Υ systems, and merely represent modern QCD practice. Thus the glueball resonance hypothesis naturally explains the data within the context of QCD.

ALTERNATIVE EXPLANATIONS AND CRITICISMS

Let us now examine alternative explanations and criticisms of the glueball resonance explanation which have been published. Recent differences[32] regarding the degree of OZI forbiddeness of the reaction $\pi^- p \rightarrow \phi\phi n$ observed by BNL/CCNY have been resolved[33] and it was concluded that these resonances would be OZI forbidden if they were of the $q\bar{q}$ type and therefore they constituted strong evidence for glueball(s).

Gomm[20] has argued that two gluon exchanges are not intrinsically OZI suppressed in the case of tensor mesons ($J^{PC} = 2^{++}$) since the hadron wave function is antisymmetric in space. Thus the quark and anti-quark are not likely to annihilate into gluons. This arguement is self inconsistent. When an $s\bar{s}$ pair annihilates into 2 gluons, precocious asymptotic freedom will lead to an OZI suppression. If the antisymmetric wave function inhibits the $s\bar{s}$ annihilation, the observed overall suppression would merely be increased. Thus the observed breakdown of the OZI suppression in the $J^{PC} = 2^{++}$ g_T, $g_{T'}$, and $g_{T''}$ states would be even more difficult to explain. The production ($\pi^- p \rightarrow f'n$) and decay of the $J^{PC} = 2^{++}$ f' (a two-gluon exchange) clearly exhibits an OZI-like suppression[34] of the same order as in the equivalent ϕ interactions. This clearly demonstrates that both the production and decay of $J^{PC} = 2^{++}$ mesons are as inhibited as those of $J^{PC} = 1^{--}$ mesons.

As one measure of the breakdown of the OZI suppression we have used the ratio:

a) $$\frac{\sigma(\pi^- p \rightarrow \phi\phi n)}{\sigma(K^- p \rightarrow \phi\phi\Lambda)} = \frac{1}{3}\ ^{\dagger}$$

since the numerator is OZI forbidden while the denominator is OZI allowed. Reference 20 states that this ratio is comparable to that of similar reactions, e.g.

b) $\sigma(\psi \to \phi\pi\pi)/\sigma(\psi \to \omega\pi\pi) = 0.21 \pm 0.10$

and this is used to imply two-gluon exchanges are not suppressed. In Ref. 20 it is apparently not realized that both of the reactions in b) proceed via three-gluon exchange, and are both OZI forbidden and thus have no relation to the previous ratio a) where the numerator is OZI forbidden and the denominator is OZI allowed. Since gluons are flavor blind, it is not unexpected that the $\phi\pi\pi$ and $\omega\pi\pi$ final states in radiative J/ψ decay (reaction b) should have the same order of magnitude cross section which is what is observed.

The fact that the final state quark lines in the numerator involves a disconnected Zweig diagram while that in the denominator does not, is not simply interpretable since the $c\bar{c}$ annihilation creates three hard gluons and these three hard gluons which are ≈ flavor blind then produce the final quark states. The final state hadronization process of course always involves numerous soft gluons which can take care of color conservation among the new particles produced in the final state. This paper contains many other examples of confusion and erroneous statements. Reference 20 estimates the relative importance of quark-antiquark annihilation into three versus two gluons by comparing the annihilation diagrams (Fig. 1 of Ref. 20) which split m_ω and m_ρ and m_π and m_η, and concludes there is in general no suppression of disconnected diagrams if only two gluons need to be exchanged. This comparison in Ref. 20 is used as a general measure of two-gluon versus three-gluon exchanges, but is not realistic and very unreliable since the 0^- nonet is badly mixed by vacuum effects (instantons, etc.) whereas $J \geq 1$ nonets appear (as expected) to be relatively uninfluenced by vacuum effects[25]) and are observed to be ≈ ideally mixed.

Ref. 20 then tries to explain the BNL/CCNY observations via kinematical effects which lead to mass peaks. There is no explanation in that kinematical approach for the breakdown of the OZI suppression, the selection of one set of quantum numbers for the three partial wave resonances which exhibit the classic phase as well as amplitude behavior generated by poles.

---

† The measured cross section is the sum of $\phi\phi\Lambda$ and $\phi\phi\Sigma^0$. We have divided by a factor of 2 to obtain $\phi\phi\Lambda$.

The treatment naively focuses on giving possible qualitative and incomplete arguements for the kinematical generation of two peaks (one an S-wave and one a D-wave) in the $\phi\phi$ mass spectrum and ignores the fact that we have detailed partial wave amplitude and phase behavior for three peaks (one S- and two D-waves), and three resonances which is not explained. The treatment is incorrect in many other respects. For example, in generating the mass values of the S and D mass peaks, Ref. 20 assumes that the two ϕ masses are each produced by two hard gluons which are approximately colinear and argues that the production process should be similar to that where "hard gluons decay into lepton pairs with $\langle P_\perp \rangle \approx 0.6$ GeV and that higher transverse momenta are strongly suppressed." This is then used to estimate the values of the mass peaks, this type of process would not explain the forward peripheral nature of the $\phi\phi$ system observed by BNL/CCNY where $(d\sigma/dt')_{\phi\phi} \approx e^{(9.4 \pm 0.7)t'}$, and in fact would lead to much higher t'-values for the $\phi\phi$ system than observed. The Figure caption of Fig. 3 (Ref. 20) and the entire Fig. 4 are very misleading. The quark lines in Fig. 4 (showing $K^-p \rightarrow \phi\phi\Lambda$) are correct but showing two hard gluons each producing a ϕ would only contribute a very small part of the cross section. The quark line diagram of Fig. 4 is a classic Zweig connected diagram which would proceed by the single gluon and collective soft glue exchanges at a rate which would overwhelm the two hard gluon process Gomm illustrates.

Karl et al.[21] attempts to explain the $\pi^-p \rightarrow \phi\phi n$ data by a semi-classical time sequential pair creation model. One should note that a proper quantum mechanical treatment could easily wash out the mass peaks obtained with this method (although a lump at threshold might remain).

The unexpected selection of only $J^P = 2^+$ in the $\phi\phi$ system is attributed in Ref. 21 to "At these energies the annihilation is dominated by $u\bar{u}$ pairs with $J_Z = \pm 1$ which can only have angular momentum 2^+ or larger." As we can see from Fig. 8 (and the PWA), in the exclusive process we are observing $\pi^-p \rightarrow \phi\phi n$, we are annihilating a π^- and a π^+ (i.e. π exchange) thus there is no inherent net J_Z in the annihilation system which is ($u\bar{d} = \pi^+$ annihilating $\bar{u}d = \pi^-$). This is borne out clearly by our partial wave analysis which selects three waves all with $J^{PC} = 2^{++}$, $J_Z = 0$ (in the Gottfried-Jackson frame) and naturality (-) [see Fig. 12a]. These are the characteristics of π-exchange. Furthermore the peripheral nature of our reaction $d\sigma/dt'(\pi^- p \to \phi\phi n) = e^{(9.4 \pm 0.7)t'}$ for $t' < 0.3$ is clearly indicative of π-exchange. This information was clearly stated in our paper[8] that Ref. 21 quoted as a basis for their work. Hence the Ref. 21 mechanism for selecting $J^P = 2^+$ is clearly ruled out by the characteristics and analysis of the experiment. It should be noted that $J_z = \pm 1$ is rejected by 27σ in our partial wave analysis clearly indicating that the proposal of Ref. 21 is ruled out.

The mechanism of Ref. 21 could not explain the breakdown of the OZI suppression and the clear-cut resonant phase and amplitude behavior of our data. If such effects[21)] were to occur at all they would occur at the level of the OZI suppression, not the much higher cross section level corresponding to the breakdown of the OZI suppression. Furthermore, crude qualitative treatments which do not attempt to explain our detailed quantitative data and PWA are not a satisfactory explanation.

Donoghue[22)] in his Yukon Conference paper and in his summary talk at the Maryland Conference, and also his Bari talk proposed to explain our data (at least the first one or two partial waves) by kinematical hard gluon production which falls off rapidly combined with threshold effects. Among the obvious deficiencies of these arguements are 1) There is no mechanism for selecting $J^{PC} = 2^{++}$ only. For example, an S-wave $\phi\phi$ system* could have $J^{PC} = 0^{++}$ and 2^{++}, and a D-wave $\phi\phi$ system could have $J^{PC} = 0^{++}, 2^{++}, 4^{++}$. Ref. 22 does not explain the complete lack of those other than $J^P = 2^+$ waves in our data. Furthermore ad hoc assumptions and free-hand drawings as Donoghue made[11)] are not a satisfactory explanation of our quantitative detailed partial wave analysis and fitting of the data.

2) Donoghue grants that the third PWA peak phase motion is clear enough so that it should be considered a resonance, however he questions the resonance status of the first and possibly the second. Although we do use the S-wave as a phase reference, the phase differences between it and the second partial wave (D-wave with S = 2)

---

* All J^{P+} quantum numbers are possible for the $\phi\phi$ system. We have made the further restriction of π-exchange.

are precisely what one needs to explain the data. Secondly, if you accept the first D-wave dominated pole as a resonance (as Donoghue did),[11,22b)] the S-wave must have resonance phase behavior also to reproduce the observed D-S behavior.

The precise phase differences given in Fig. 12b would be very difficult (if at all possible) to generate by a non-resonant mechanism.

3) Donoghue, as stated in his Maryland Conference summary talk,[22)] concluded that the observed break of the OZI suppression must occur due to unitarity by the two-step real process 1) $\pi\pi \to \eta\eta \to \phi\phi$. He states both of these steps can proceed by a simple quark interchange. Thus $\pi^- p \to \eta\eta n \to \phi\phi n$ would not be OZI suppressed and proceed quite strongly. He does not take into account QCD dynamics, cancellations, and the overall quantum numbers of the system in his incorrectly applied unitarity arguement. If we consider this two-step process, the $u\bar{u}$ and $d\bar{d}$ quarks in each η have to be in a 0^{-+} system for the vacuum mixing which is responsible for the transition $u\bar{u}$ (or $d\bar{d}$) to $s\bar{s}$ to occur. Then in order for the two η mesons in the intermediate step to become two ϕ mesons, each η must then change to a 1^{--} system. This requires complex dynamical interactions which we expect would be strongly suppressed and could not account for the more-or-less complete breakdown of the OZI suppression, and its selectivity. Even if we adopt his naive view of the two-step process, there is no reason the $\phi\phi$ system would select only $J^{PC} = 2^{++}$. $J^{PC} = 0^{++}$ and 4^{++} would also occur and they do not. Furthermore this two-step process could proceed by Zweig allowed single and multiple soft glue exchanges and thus his mechanism for creating dynamical peaks would evaporate.

There are other real two-step processes, such as 1) $\pi^- p \to \eta\eta n \to f'n$, 2) $f' \to \eta\eta \to \pi\pi$, 3) $\pi^- p \to K^+K^- n \to \phi n$, and 4) $\pi^- p \to K^+K^- n \to f'n$, which do not break the OZI suppression. All of these two-step processes can occur as real processes, Reactions 1, 2 and 4 being considerably above threshold. The reason (except in case 2) they probably are suppressed is at least partially due to the fact that hadronization in the first step takes place at the outer regions of the confinement region where the first $q\bar{q}$ pair is far apart and the coupling is strong. Then for the two-step process to occur, a $q\bar{q}$ pair of quarks have to return to short distances where annihilation takes place and then annihilate. This is probably discriminated against dynamically and in fact the K^+ and K^- or other particles formed in the intermediate step may already be color singlets (i.e. have their own bags) thus further inhibiting the subsequent needed annihilation of the $q\bar{q}$ pair for the two step process to break the OZI suppression. In the case of Reaction 3, each η can in principle be created via its $s\bar{s}$ component and then by vacuum mixing transform to $u\bar{u}$ and $d\bar{d}$ components and by quark rearrangement interactions π^+ and π^-. The fact that the coupling of the f' to $\pi^+\pi^-$ has experimentally been observed to be small in both production and decay demonstrates that such processes are unlikely enough so that they do not materially affect the OZI suppression.

In the context of QCD the only way you can appreciably break the OZI suppression in the (u,d,s) $q\bar{q}$ quark states is by changing the near ideal mixing observed in the nonets for $J \geq 1$ by a flavor changing mixing mechanism which converts $u\bar{u}$ or $d\bar{d}$ to $s\bar{s}$. Vacuum effects[25)] can do this for J = 0 (the $J^P = 0^-$ nonet). However, they are not expected to and do not appear to affect $J \geq 1$ nonets. The only other known basic flavor changing mechanism is a glueball. Clearly Donoghue's proposed mechanism would not do this for a $J^{PC} = 2^{++}$ $\phi\phi$ state.

In any event there is no experimental evidence for two-step OZI allowed processes significantly breaking OZI suppression. However, Donoghue, following his own philosophy, as expressed in his summary talk[22b)] and transparencies,[11)] naively invents a model for OZI and other suppression which is totally in disagreement with many known experimental facts, and in conflict with QCD. In this model, quark exchange dominates, and each $q\bar{q}$ pair produced or destroyed is simply counted and gives a factor $\varepsilon \ll 1$. He then uses this model (in many cases calculated incorrectly) to explain various ratios. One obvious example why the model is unrealistic is that it is well established that the $\rho'(1600)$, for which the predominant decay mode is 4π (which requires the creation of three additional $q\bar{q}$ pairs) would have a suppression of ε^2 compared to the $\rho(770)$. Even if one takes account of the fact that the decay mode is probably $\rho\pi\pi$ then the ρ' would still be suppressed by a factor $\varepsilon \ll 1$ compared to the ρ. Thus its decay width would be much smaller than the ρ decay width, whereas in fact the ρ' is $\approx$ twice as wide as the ρ.

The ratio

$$B\frac{(\rho'(1600) \rightarrow 2\pi)}{(\rho'(1600) \rightarrow 4\pi)} = \frac{23 \pm 7}{60 \pm 7} \approx 0.38$$

According to Donoghue's model, this ratio should be $1/\varepsilon^2$ (possibly $1/\varepsilon$ for $\rho\pi\pi$) where $\varepsilon \ll 1$ and thus should be extremely large.

The fact that most hadronic widths are of the same order more-or-less independent of the number of quark pairs in the dominant decay mode shows the naivety of this model. Making jets would certainly be very difficult with Donoghue's model.

What Donoghue seems to have overlooked is that creation of additional $q\bar{q}$ pairs (of the u,d type) by hadronization where there is no disconnected Zweig diagram, seems experimentally to cost you a factor near 1† whereas creation or annihilation of a new type of $q\bar{q}$ pair in a disconnected Zweig diagram costs you a big factor (i.e. $\varepsilon \ll 1$). This is explainable by general characteristics of QCD because in the hadronization corresponding to a connected Zweig diagram, collective soft glue effects can easily create additional $q\bar{q}$ pairs, whereas precocious asymptotic freedom in QCD gives hard gluons a small coupling constant and leads naturally to the OZI rule.

Let us for the sake of arguement grant Donoghue his desired breakdown of the OZI suppression. If that were to occur, he would be dealing with an OZI allowed process which can proceed via a series of soft single gluon exchanges and soft multi-gluon collective effects. Thus his hard gluon mechanism (combined with a threshold effect) for generating the mass peaks does not occur. Secondly, there would be no mechanism for selecting $J^{PC} = 2^{++}$ only.

Furthermore, we have analyzed the OZI allowed background process $\pi^- p \rightarrow \phi K^+K^- n$ in the K^+K^- mass region just above the ϕ mass and found that $\approx$ 65% is incoherent (i.e., flat structureless background) $\approx$ 28% is a 1^{--} wave and only $\approx$ 7% is a broad $J^{PC} = 2^{++}$ wave which is absent at threshold and has a broad peak at about 2.4 GeV. Since the K^+K^- pair have almost the ϕ mass, this process should exhibit

---

† There is some phenomenological penalty for creating the first $s\bar{s}$ pair even in a Zweig connected diagram.

threshold effects similar to any which occur in $\phi\phi$ if the OZI suppression is broken. Due to the vast differences in the ϕK^+K^-n and the $\phi\phi n$ data threshold enhancement agruements to explain $\phi\phi n$ are not plausible.

In summary, threshold enhancement effects would not select one J^P, would not break the OZI suppression and would not give the characteristics exhibited by our data.

To summarize, the alternative explanations to our conclusion that the g_T, $g_{T'}$ and $g_{T''}$ are resonances have been treated qualitatively, crudely, incompletely and incorrectly. They have not been seriously compared by the authors to the BNL/CCNY data which is quantitative, detailed and has considerable statistics. They have not explained the breakdown of the OZI suppression accompanied by the selection of virtually only three $J^{PC} = 2^{++}$ partial wave amplitudes which exhibit pole behavior. As we have pointed out, these explanations have serious errors and deficiencies and definitely do not explain the critical features of the BNL/CCNY observations as Refs. 8 and 10 do.

Thus the g_T, $g_{T'}$ and $g_{T''}$ are three resonances with $I^GJ^{PC} = 0^+2^{++}$ which break down the OZI suppression and contain practically all of the reaction cross section in their mass region. These facts can be well explained by the conclusion that they are produced by glueball(s),[8,10] whereas the alternative explanations[20-22] discussed and similar ones have been shown[11-12] to be incorrect and do not fit the experimental facts.

GLUEBALL CANDIDATES FROM THE J/ψ RADIATIVE DECAY

The radiative decay of the J/ψ is thought to occur in leading order via the usual three gluons emitted in the annihilation of the $c\bar{c}$ pair where one is replaced by a photon. Thus it has been argued[35] that the two-gluon system could recoil from the photon and preferentially form a glueball. The first and most discussed glueball candidate of this type is the iota (1440)[9a] since it is a $J^{PC} = 0^{-+}$ state seen in place of the hadronic E(1420) with $J^{PC} = 1^{++}$. The status of the iota (1440) with $J^{PC} = 0^{-+}$, $M \approx 1440^{+20}_{-15}$ and $\Gamma \approx 55^{+28}_{-30}$ as of July 1982 was reviewed in the Paris Conference.[36] Some concern was expressed that the ITHEP[25] calculations on instanton effects would move a 0^{-+} glueball up to 2.0-2.5 GeV mass region. The alternative that the iota (1440) is a radial excitation etc. rather than a glueball has also been discussed.[19] The question of whether the iota is really different from the E seen in earlier hadronic production experiments[37] has also been raised.

The recent E/iota results in hadronic interactions also appear to be experiment dependent. The Dionisi et al.[38] and Armstrong et al.[39] results on the centrally produced D and E from π^+p and pp interactions at 85 GeV/c find the conventional $J^{PC} = 1^{++}$ "hadronic E".

However a recent experiment [40] in the reaction $\pi^- p \to K^+ K_s^0 \pi^- n$ at 8 GeV/c found, from a Dalitz plot analysis, that although the E region has a large 1^{++} (S-wave mostly $K^*\bar{K}$) its intensity continues to rise through the E region and thus does not exhibit behavior typical of a resonant state.

This data show an E(1420) resonance in the $J^{PC} = 0^{-+}$ wave amplitude coupling mostly to $\delta\pi$ and some $K^*\bar{K}$. They concluded this is not inconsistent with the analysis of Baillon et al.[37] Thus the status of the iota is still unclear.

Another glueball candidate of this type is the θ(1700).[9b] $J^{PC} = 2^{++}$ was favored originally with a 95% C.L. but (I understand that recently this has been improved). The resonance parameters were M ≈ 1700 ± 50. $\Gamma \approx 160 \pm 50$. New data in the radiative J/ψ decay were recently reported by the Mark III collaboration.[15] They observed the iota in the $K^+K^-\pi^0$ and also $K_s^0 K_s^0 \pi^0$ mode. The Breit-Wigner fit parameters determined were M = 1.46 ± 0.01 GeV and $\Gamma = 0.097 \pm 0.025$ GeV. In the case of the θ, the Breit-Wigner parameters were determined as M = 1.719 ± 0.006 GeV. $\Gamma = 0.117 \pm 0.023$ GeV. The iota and θ situation did not appear to change substantially from the earlier review.[36] The essentially new development was the evidence for a new narrow structure [ξ(2200)].[15] Reference 41 does not find evidence for the ξ(2200).

A recent review of the glueball candidates by Fishbane and Meshkov[19b] concluded that the iota and θ were probably not glueballs but they considered alternative explanations in which they could be.

Let us consider why the BNL/CCNY $\phi\phi$ states have not been seen in the radiative decay of the J/ψ. The new MK III results observe $J/\psi \to \gamma\phi\phi$.[15] Their detection efficiency for $\phi\phi$ is very low in the mass region of the $g_T(2050)$, $g_{T'}(2300)$ and $g_{T''}(2350)$. Thus they find only

~ 10 events in this mass region. However if one corrects their $\phi\phi$ mass spectrum for the detection efficiency it is not inconsistent with the shape of the mass spectrum seen by BNL/CCNY. However one should note we are comparing ≈ 4,000 observed events to ~ 10. It appears that the MK III can only observe strong signal, narrow, high mass $\phi\phi$ states such as the decay of the η_c, and thus is not likely to be able to observe the BNL/CCNY states.

The DM2 group[41)] has reported at the Bari Conference ~ 50 $\gamma\phi\phi$ events in the mass region of the BNL/CCNY experiment. At present due to the limited statistics they are unable to say whether this signal is related to the resonant structures (i.e., the g_T, $g_{T'}$, $g_{T''}$) observed by BNL/CCNY.

WHY HAVE THE g_T's NOT BEEN SEEN IN OTHER CHANNELS?

One can also raise the question why some other decay mode of the g_T, $g_{T'}$ and $g_{T''}$ have not been seen in other hadronic production experiments or in particular in the radiative decay of the J/ψ since this is considered to be a gluon enriched channel.

First it should be noted that in a related experiment $\pi^- Be \rightarrow \phi\phi$ inclusive[18] the data are found to be consistent with the $g_{T'}$ and $g_{T''}$ and needs two Breit Wigner resonances to explain the results. This channel would only be expected to be partially Zweig suppressed since the Zweig suppression would not apply if a $K\bar{K}_{\Sigma}$or $K(^{\Lambda})$ pair were created.

All other hadronic production experiments involve OZI-allowed channels therefore one would expect the g_T's to be submerged in the many other hadronic states one could expect. Thus their detection would likely require very large statistics and even then it might be quite difficult to separate these from the other hadronic states.

Figure 16 shows the results of the analysis of a 23 GeV/c $\pi^- p \rightarrow K_s^o K_s^o n$ experiment.[31)] In the mass region of the g_T's, the $J^{PC} = 2^{++}$ amplitude behavior is smooth and structureless and shows no phase motion. Furthermore, the 0^{++} and 4^{++} states are also populated unlike the selection of only $J^{PC} = 2^{++}$ in $\pi^- p \rightarrow \phi\phi n$. This experiment has had its statistics raised by a factor of ~ 3 recently[42)] and the results are the same. This is what I would expect when the effects of

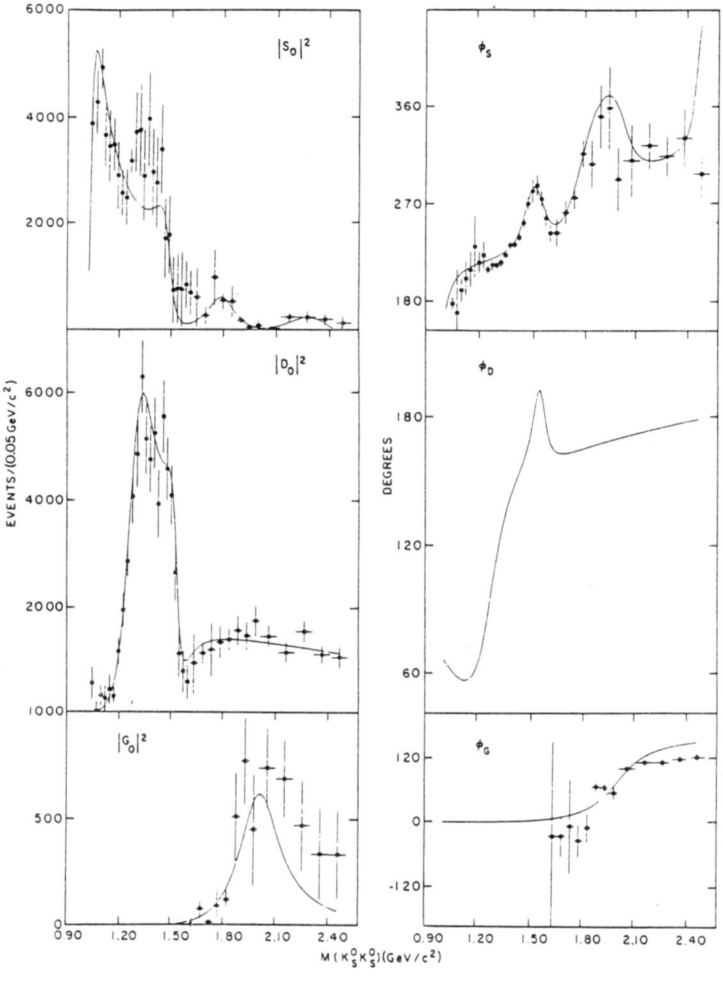

Figure 16: The square of the moduli of the S_0, D_0, and G_0 amplitudes together with their absolute phases from the best fit as functions of $K_s^0K_s^0$ effective mass for $t' < 0.1\ (GeV/c)^2$. The solid curves are the results of our preferred mass-dependent fit in the same t' interval (Ref. 31).

the OZI suppression filter action are eliminated as they are in this reaction.

As to why the g_T's have not yet been seen in the radiative J/ψ decay I would suggest the following:

We argue the Zweig suppression in our channel (with a pure glue intermediate state) should filter out other hadronic states and give a highly enriched sample of glueballs. What we found in the data is certainly consistent with this, namely we find three new states with the same quantum numbers and nothing else accompanied by very little background.

In the J/ψ radiative decay ≈ 90% of the observed states are known conventional ones and thus it is at most an inefficient filter for glueballs. If it really were almost completely glueball dominated, then we could have the reaction $J/\psi \to \gamma$ [glueball(s)] dominating. Since I would expect glueballs to be relatively strongly coupled (note perturbation theory arguements do not apply to resonant states of glue or anything else) then the width of the J/ψ would be expected to be broadened. The reason is that if the glueball is strongly coupled, the only suppressant left is the weak coupling of the γ. Thus widths of as much as a few MeV instead of .06 MeV might be expected for the J/ψ. Hence the fact that the J/ψ is so narrow implies to me that its radiative decay is not dominated by glueballs, which is consistent with the experimental observations. Furthermore perturbation theory gives the experimental ratio of the radiative decay to the total decay width. If glueballs were strongly coupled in the radiative decay channel I would expect the percentage of radiative decay to be higher than that predicted by perturbation theory where the two gluons are weakly coupled. Sinha[44] concludes that 2^{++} glueballs are weakly coupled to the radiative J/ψ decay.

When one considers the very limited statistics gathered in the J/ψ radiative decay channels combined with the inefficient glueball filter nature of this channel, I am not surprised that the g_T's have not yet been seen in it.

In regard to the g_T's, it is also worth noting that Chanowitz and Sharpe[43)] have concluded that strange quarks may well be favored in glueball decay and in particular in the $\phi\phi$ S-wave.*

Furthermore I would like to point out that except for color, the quantum numbers of a gluon and a ϕ are the same. Thus one can imagine that gluons would like to go into ϕ mesons just like photons like to go into vector mesons (i.e. similar to VDM). Of course the color must be changed to a singlet but such color rearrangements might perhaps easily be accomplished by soft gluon exchanges in the final hadronization. Thus this may also be another reason why the g_T's if they are glueballs are only seen in the $\phi\phi$ decay mode. In this regard if sufficient statistics are gathered in $J/\psi \to \gamma\phi\phi$ some evidence for the g_T, $g_{T'}$, and $g_{T''}$ states may be seen.

CONCLUSIONS ON THE STATUS OF GLUEBALL STATES

One can prove the g_T, $g_{T'}$, and $g_{T''}$ are glueballs with the appropriate input axioms. Then as we concluded previously the $g_T(2050)$, $g_{T'}(2300)$ and $g_{T''}(2350)$ are produced by 1-3 primary $J^{PC} = 2^{++}$ glueballs, if you assume as input axioms:

1. QCD is correct.

2. The OZI rule is universal for weakly coupled glue in disconnected Zweig diagrams where the disconnection is due to the creation or annihilation of new flavor(s) of quark(s), and $J \geq 1$ for the disconnected system (to avoid possible vacuum mixing effects).

We have previously stated that the BNL/CCNY $g_T(2050)$, $g_{T'}(2300)$ and $g_{T''}(2350)$ are naturally explained within the context of QCD by concluding they are produced by 1-3 primary glueballs. One or two broad primary glueballs could in principle break down the OZI suppression and mix with one or two quark states which accidentally have the same quantum numbers and nearly the same mass. However the simplest explanation of the rather unusual characteristics of our data is that we have found a triplet of $J^{PC} = 2^{++}$ glueball states.

---

* They also have meikton states breaking the OZI suppression and possibly being associated with our states as well as glueballs. However this arguement depends on bag calculations and the dynamical mechanisms are not clear.

Alternatives to the Glueball resonance explanation have been discussed earlier and found to be incorrect or do not fit the data or both.

The iota(1440) and the $\theta(1700)$ observed in J/ψ radiative decay are glueball candidates, the pros and cons of which have been discussed briefly here and more extensively in the references cited. Other glueball candidates[16-17] are relatively weak ones, and recent glueball searches[7] have not yet led to definite candidates.

REFERENCES

1. A. Salam, Elementary Particle Theory, Nobel Symposium, Ed. N. Svartholm (Wiley Interscience 1968); S. Glashow, J. Ioliopulos, L. Maianai, Phys. Rev. D 2, 1285 (1970); S. Weinberg, Phys. Rev. Lett. 19, 1264 (1967).

2. H. Fritzch and M. Gell-Mann, XVI Int. Conf. on High Energy Physics, Chicago-Batavia, 1972, Vol. 2, pp. 135; H. Fritzch, M. Gell-Mann and H. Leutwyler, Phys. Lett. 47B, 365 (1973); S. Weinberg, Phys. Rev. Lett. 31, 49 (1973); S. Weinberg, Phys. Rev. D 8, 4482 (1973); D.J. Gross and F. Wilczek, ibid, 3633 (1973).

3. a) Fritzch and Minkowski, Nuovo Cimento 30A, 393 (1975).
b) R.P. Freund and Y. Nambu, Phys. Rev. Lett. 34, 1645 (1975).
c) R. Jaffee and K. Johnson, Phys. Lett. 60B, 201 (1976).
d) Kogut, Sinclair and Susskind, Nucl. Phys. B114, 199 (1975).
e) D. Robson, Nucl. Phys. B130, 328 (1977). f) J. Bjorken, SLAC Pub. 2372.

4. a) C. Michael and I.. Teasdale, Glueballs From Asymmetric Lattices, Liverpool Univ. Preprint LTH-127, March 1985.
b) Ph. de Forcrand, G. Schierholz, H. Schneider and M. Teper, Phys. Lett. 152B, 107 (1985).
c) S.W. Otto and P. Stolorz, Phys. Lett. 151B, 428 (1985).
d) G. Schierholz and M. Teper, Phys. Lett. 136B, 64 (1984).
e) Berud Berg, The Spectrum in Lattice Gauge Theories, DESY Preprint 84-012, February 1984.

5. C.N. Yang and R.L. Mills, Phys. Rev. 96, 191 (1954).

6. S.J. Lindenbaum, C. Chan, A. Etkin, K.J. Foley, M.A. Kramer, R.S. Longacre, W.A. Love, T.W. Morris, E.D. Platner, V.A. Polychronakos, A.C. Saulys, Y. Teramoto, C.D. Wheeler. A New Higher Statistics Study of $\pi^- p \to \phi\phi n$ and Evidence for Glueballs. Proc. 21st Intern. Conf. on High Energy Physics, Paris, France, 26-31 July 1982, Journal de Physique 43, P. Petiau and M. Porneuf, Editors (Les Editions de Physique, Les Ulis, France), pp. C3-87 - C3-88; A. Etkin, K.J. Foley, R.S. Longacre, W.A. Love, T.W. Morris, E.D. Platner, V.A. Polychronakos, A.C. Saulys, C.D. Wheeler, C.S. Chan, M.A. Kramer, Y. Teramoto, S.J. Lindenbaum. The Reaction $\pi^- p \to \phi\phi n$ and Evidence for Glueballs. Phys. Rev. Lett. 49, 1620-1623 (1982).

7. S.J. Lindenbaum. Hadronic Production of Glueballs. Proc. 1983 Intern. Europhysics Conf. on High Energy Physics, Brighton, U.K., July 20-27, 1983, J. Guy and C. Costain, Editors (Rutherford Appleton Laboratory), p. 351-360.

8. S.J. Lindenbaum. Production of Glueballs. Comments on Nuclear and Particle Physics 13, #6, 285-311 (1984).

9. a) Edwards et al., Phys. Rev. Lett. 49, 259 (1982);
b) Edwards et al., Phys. Rev. Lett. 48, 458 (1982).

10. Lindenbaum, S.J. The Glueballs of QCD and Beyond. Invited Lecture. Proc. 22nd Course of the International School of Subnuclear Physics on "Quarks, Leptons and their Constituents," Erice, Trapani-Sicily, Italy, 5-15 August 1984, (to be published).

11. Lindenbaum, S.J. and Longacre, R.S. The Glueball Resonance and Alternative Explanations of the Reaction $\pi^- p \rightarrow \phi\phi n$. Hadron Spectroscopy - 1985 (International Conference, Univ. of Maryland), S. Oneda, Editor, AIP Conf. Proc. 132, p. 51-66 (American Institute of Physics, New York, 1985).

12. Lindenbaum, S.J. and Longacre, R.S. The Glueball Resonance and Alternative Explanations of the Reaction $\pi^- p \rightarrow \phi\phi n$. Phys. Lett. B (in press).

13. A. Etkin, K.J. Foley, J.H. Goldman, W.A. Love, T.W. Morris, S. Ozaki, E.D. Platner, A.C. Saulys, C.D. Wheeler, E.H. Willen, S.J. Lindenbaum, M.A. Kramer, U. Mallik, Phys. Rev. Lett. 40, 422-425 (1978); Phys. Rev. Lett. 41, 784-787 (1978).

14. S.J. Lindenbaum, Status of the Glueballs. Invited Lecture. Proc. of the 21st Course of the International School of Subnuclear Physics on "How Far we are From the Electronuclear Interactions and the Other Gauge Forces", Erice, Trapani-Sicily, 3-14 August, 1983 (to be published).

15. a) K. Einsweiler. Proc. 1983 Intern. Europhysics Conf. on High Energy Physics, Brighton, U.K., July 20-27, 1983, J. Guy and C. Costain, Editors (Rutherford Appleton Laboratory), p. 348-350;
b) D. Hitlin, Radiative Decays and Glueball Searches. Proc. of the 1983 Int. Symposium on Lepton and Photon Interactions at High Energies, Cornell University, August 4-9, 1983, David G. Cassel and David L. Kreinick, Editors, pp. 746-778.
c) C. Heusch, Proc. 22nd Course of the International School of Subnuclear Physics on "Quarks, Leptons and their Constituents," Erice, Trapani-Sicily, Italy, 5-15 August 1984, (to be published).

16. A. Etkin, K.J. Foley, R.S. Longacre, W.A. Love, T.W. Morris, S. Ozaki, E.D. Platner, V.A. Polychronakos, A.C. Saulys, Y. Teramoto, C.D. Wheeler, E.H. Willen, K.W. Lai, S.J. Lindenbaum, M.A. Kramer, U. Mallik, W.A. Mann, R. Merenyi, J. Marraffino, C.E. Roos, M.S. Webster, Phys. Rev. D 25, 2446 (1982).

17. Binon et al., Il Nuovo Cimento 78A, 313 (1983).

18. a) Booth et al., Angular Correlations in the $\phi\phi$ System. Proc. of the XXII Intern. Conf. on High Energy Physics, Leipzig, July 1984.
b) Booth et al., A High Statistics Study of the $\phi\phi$ Mass Spectrum. Proc. of the XXII Intern. Conf. on High Energy Physics, Leipzig, July 1984. See paper by Chung, et al.

18. c) Booth et al., A High Statistics Study of the $\phi\phi$ Mass Spectrum, to be published in Zeitschrift der Physik.

19. a) S. Meshkov, Proc. of the Seventh Intern. Conf. on Experimental Meson Spectroscopy, April 14-16, 1983, Brookhaven National Laboratory, S.J. Lindenbaum, Editor, AIP Conf. Proc. No. 113, p. 125-156.
b) P.M. Fishbane and S. Meshkov, Comments on Nuclear and Particle Physics 13, 325 (1984).

20. H. Gomm, Phys. Rev. D30, 1120 (1984).

21. G. Karl, W. Roberts and N. Zagury, Phys. Lett. 149B (1984) 403; G. Karl, Hadron Spectroscopy - 1985 (International Conference, Univ. of Maryland, S. Oneda, Editor, AIP Conf. Proc. 132, p. 73 (American Institute of Physics, New York, 1985).

22. a) J. Donoghue, The Status of Unusual Meson Candidates, Proc. of the Yukon Advanced study Institute: The Quark Structure of Matter, August 11-27, 1984, (to be published); UMHEP-209.
b) J. Donoghue, Theory Summary, Hadron Spectroscopy - 1985 (International Conference, Univ. of Maryland, S. Oneda, Editor, AIP Conf. Proc. 132, p. 460 (American Institute of Physics, New York, 1985).

23. a) S. Okubo, Phys. Lett. 5, 165 (1963); Phys. Rev. D16, 2336 (1977).
b) G. Zweig, CERN REPORTS TH401 and 412 (1964).
c) J. Iizuba, Prog. Theor. Physics, Suppl. 37-38, 21 (1966); J. Iizuba, K. Okuda and O. Shito, Prog. Theor. Phys. 35, 1061 (1966).

24. Particle Data Group, Review of Particle Properties, Review of Modern Physics 56, No. 2, Part II (April 1984).

25. V.A. Novikov et al., Nucl. Phys. B191, 301 (1981).

26. a) S.J. Lindenbaum, Hadronic Physics of $q\bar{q}$ Light Quark Mesons, Quark Molecules and Glueballs. Proc. of the XVIII Course of the International School of Subnuclear Physics, July 31-August 11, 1980, Erice, Trapani, Italy, Subnuclear Series, Vol. 18, "High Energy Limit", Ed. A. Zichichi, pp. 509-562; b) S.J. Lindenbaum, Il Nuovo Cimento, 65A, 222-238 (1981).

27. S.J. Lindenbaum. The Discovery of Glueballs. Surveys in High Energy Physics, Vol. 4, 69-126, John M. Charap, Editor (Harvard Academic Publishers, London, 1983).

28. a) B. Berg and A. Billoire, Nucl. Phys. B221, 109 (1983); B226, 405 (1983).
b) M. Teper, Proc. 1983 Intern. Europhysics Conf. on High Energy Physics, Brighton, U.K., July 20-27, 1983, J. Guy and C. Costain, Editors (Rutherford Appleton Laboratory); Preprint LAPP-TH-91 (1983).

29. See Ref. 19 for a fuller discussion of this topic.

30. T.D. Lee. Time as a dynamical variable CU-TP-266; Talk at Shelter Island II Conf., June 2, 1983; also, Proc. of the 21st Course of the International School of Subnuclear Physics on "How Far we are From the Electronuclear Interactions and the Other Gauge Forces", Erice, Trapani-Sicily, 3-14 August 1983, to be published.

31. Etkin, A., Foley, K.J., Longacre, R.S., Love, W.A., Morris, T.W., Ozaki, S., Platner, E.D., Polychronakos, V.A., Saulys, A.C., Teramoto, Y., Wheeler, C.D., Willen, E.H., Lai, K.W., Lindenbaum, S.J., Kramer, M.A., Mallik, U., Mann, W.A., Merenyi, R., Marraffino, J., Roos, C.E., Webster, M.S. Amplitude Analysis of the $K_s^oK_s^o$ System Produced in the Reaction $\pi^-p \rightarrow K_s^oK_s^on$ at 23 GeV/c. Phys. Rev. D 25, 1786-1802 (1982).

32. a) H.J. Lipkin, Phys. Lett. 124B (1983) 509; Nucl. Phys. B224, 147 (1984).
b) S.J. Lindenbaum, Phys. Lett. 131B (1983) 221; also Ref. 11.

33. S.J. Lindenbaum and H.J. Lipkin, Phys. Lett. 149B (1984) 407.

34. A.J. Pawlicki et al., Phys. Rev. D15, 3196 (1977); Particle Data Group, Reviews of Modern Physics 56, No. 2, Part II, April 1984 which lists $f' \rightarrow 2\pi$ as possibly seen, thus the suppression is clearly large.

35. M. Chanowitz, Phys. Rev. Lett. 46, 981 (1981).

36. E. Bloom. Proc. 21st Intern. Conf. on High Energy Physics, Paris, France, 26-31 July 1982, Journal de Physique 43, P. Petiau and M. Porneuf, Editors (Les Editions de Physique, Les Ulis, France), pp. C3-407.

37. P. Baillon. Resonance in the $KK\pi$ System Below 1.6 GeV/c^2. Experimental Meson Spectroscopy -- 1983 (Seventh International Conference, Brookhaven), S.J. Lindenbaum, Editor, AIP Conf. Proc. No. 113, p. 78-106 (American Institute of Physics, 1984).

38. Dionisi et al., Nucl. Phys. B169, 1 (1980).

39. T.A. Armstrong $(\pi^+/p)p \rightarrow \pi^+/p\ K\bar{K}\pi p$ at 85 GeV/c, CERN/EP 84-88; also submitted to XXII Intern. Intern. Conf. on High Energy Physics, Leipzig, July 17-25, 1984.

40. S.U. Chung et al., Phys. Rev. Lett. 55, 779 (1985).

41. B. Jean Marie, DM2 Results on Hadronic and Radiative J/ψ Decays, Proc. of the Intern. Europhysics Conf. on High Energy Phsyics, Bari, Italy, 18-24 July 1985 (to be published); Preprint, LAL 85-27, July 1985. This paper contains other results relevant to the iota, θ.

42. Brandeis/BNL/CCNY/Duke/Notre Dame Collaboration, private communication.

43. M.A. Beg. Dynamical Symmetry Breaking and Hypercolor. Proc. XXth Intern. Conf. on High Energy Physics, July 1980, Madison, Wisconsin, pp. 489-492 (AIP Conf. Proceedings No. 68, Part I).

44. R. Sinha, Isoscalar Meson Mixing and Glueballs, Preprint (1985).